"十三五"国家重点出版物出版规划项目
现代机械工程系列精品教材
普通高等教育"十一五"国家级规划教材

ENGINEERING DRAWING

(English-Chinese Bilingual Edition)

Third Edition

工 程 制 图

(英汉对照)

第 3 版

Editors in Chief: Lin Hu, Rong Cheng
Associate Editors in Chief: Qin Fu, Wei Sun
Participating Editors: Fenghong Wang, Xiaobo Peng, Mingli Zhang, Xue Fan
Auditors in Chief: Jinchang Chen (Chinese), Weiyin Ma (English)

主　编　胡　琳　程　蓉
副主编　付　芩　孙　炜
参　编　王枫红　彭小波　张明莉　范　雪
主　审　陈锦昌（中文）　马维银（英文）

本教材获深圳大学教材出版资助，为 UOOC 联盟指定参考用书

机械工业出版社

This book is a textbook of China's "Thirteenth Five-Year-Plan" for National Key Publications Publishing——modern high quality textbooks of machinery engineering series, as well as one of China's National "Eleventh Five-Year-Plan" textbooks for ordinary higher education. This book adopts the format of Chinese-English bilingual comparison, aiming to fill the gap in the field of bilingual textbooks of Engineering Drawing and create favorable conditions for bilingual teaching.

This book has 10 chapters and introduces the knowledge of Engineering Drawing systematically, including basic knowledge of drawing, basic knowledge of orthogonal projection, solids and their intersections, composite solids, axonometric projections, general principles of representation, expression for commonly used machinery components, detail drawings, assembly drawings, and other drawings etc. Some digital resources are attached by means of QR codes in some difficult chapters, which includes detailed answers for the exercises and animation demonstration of 3D models, for the convenience of teachers to explain problems and tutor students.

Digital resources attached to the book includes PowerPoint files, 3D models, problem solving instructions and referable exercise answers. Teachers who choose this book as the textbook can download them free of charge from the website of China Machine Press Education (www.cmpedu.com).

This book can be used as a textbook on Engineering Drawing for students majoring in mechanical engineering and similar majors (72~108 class hours) and students majoring in science, engineering, economy and management (36~64 class hours), as well as students of some engineering majors in higher vocational schools, higher professional colleges and adult higher education. This book is also a good choice as a reference book for engineering designers, mechanical and manufacturing engineers.

本书是"十三五"国家重点出版物出版规划项目——现代机械工程系列精品教材、普通高等教育"十一五"国家级规划教材。全书采用英汉对照的编排形式,填补了国内工程制图双语教材的空白,为双语教学创造了良好的条件。

全书共分10章,系统地介绍了制图基本知识、正投影基础知识、立体及其交线、组合体、轴测图、图样画法、常用机件的表达、零件图、装配图、其他工程图等内容。教材中的疑难章节配有相应的二维码资源,包括详细题解和相应的立体模型动画演示等,可方便教师讲解习题或辅导学生。

本书配有电子课件、三维模型、解题指导和习题参考答案等,选用本书作为教材的教师可以从机械工业出版社教育服务网(www.cmpedu.com)免费注册下载。

本书可作为机械类和近机械类各专业的"工程制图"课程(72~108学时)的教材,也可作为高等院校理、工、经、管各专业"工程制图"课程(36~64学时)的教材和高职高专成人高等教育等相关工科专业的教材,还可作为从事工程设计与制造的有关技术人员和企业管理人员的自学参考书。

图书在版编目(CIP)数据

工程制图:英汉对照/胡琳,程蓉主编.—3版.—北京:机械工业出版社,2018.8
(2024.6重印)

"十三五"国家重点出版物出版规划项目 现代机械工程系列精品教材 普通高等教育"十一五"国家级规划教材

ISBN 978-7-111-62046-4

Ⅰ.①工⋯ Ⅱ.①胡⋯ ②程⋯ Ⅲ.①工程制图-高等学校-教材-英、汉 Ⅳ.①TB23

中国版本图书馆 CIP 数据核字(2019)第 032050 号

机械工业出版社(北京市百万庄大街22号 邮政编码100037)
策划编辑:刘小慧 责任编辑:刘小慧 徐鲁融 武 晋 余 皞
责任校对:陈 越 封面设计:张 静
责任印制:常天培
北京铭成印刷有限公司印刷
2024年6月第3版第11次印刷
210mm×285mm・21.5印张・686千字
标准书号:ISBN 978-7-111-62046-4
定价:59.80元

电话服务 网络服务
客服电话:010-88361066 机 工 官 网:www.cmpbook.com
　　　　　010-88379833 机 工 官 博:weibo.com/cmp1952
　　　　　010-68326294 金 书 网:www.golden-book.com
封底无防伪标均为盗版 机工教育服务网:www.cmpedu.com

Foreword to the Third Edition

Since the first edition of the book was published in 2005, it has become a piece of unique academic work in the field with its innovative contents, wide practical use, and featured bilingual education. It received great support and was widely accepted by readers. The book was awarded the First Prize of Excellent Textbook of Shenzhen University. In 2008, it was selected as one of China's National "Eleventh Five-Year-Plan" Textbooks for Regular Higher Education. As a supporting teaching material, the bilingual electronic teaching plan published together with the book has also been very popular among readers and awarded several national and provincial prizes of Excellent Multimedia CAI Software for Higher Education.

This book has been revised in its third edition according to the newly formulated and improved national standards, as well as feedbacks and suggestions from readers. The basic structure of this book remains unchanged except the removal of Chapter 11 on "Computer Graphics". The contents of the book have been revised or extended following the new national standards. Two dimensional bar code resources, such as detailed answers of exercises and animations of 3D models have been added along with difficult contents in this book, which can help teachers explain exercises and answer questions from students.

Primary participants who have contributed to the editing and revision of the third edition are Rong Cheng and Xue Fan.

This book is appointed as the preferred reference book by UOOC alliance. Support from Shenzhen University through the Teaching Material Development Grant is gratefully acknowledged. Comments, criticisms and submissions from readers are always welcome.

Authors

第3版前言

本书是一本内容新颖、实用性强、富有特色的双语教材，自2005年首次出版以来，深受广大读者的欢迎和厚爱，并荣获深圳大学优秀教材一等奖，2008年被评为普通高等教育"十一五"国家级规划教材。与本书配套的双语电子教案也极受欢迎，曾多次获得全国及广东省高等教育优秀多媒体教学软件奖。

为了全面贯彻国家颁布的最新标准，同时考虑到读者的反馈意见和建议，本书在第2版的基础上进行了修订。修订时删除了第2版中"第11章 计算机绘图"，其余维持原章节不变；按照现行国家标准更新修改或补充了相关的内容，并且在疑难章节部分配备了相应的二维码资源，包括详细题解和相应的立体模型动画演示等，可方便教师讲解习题或辅导学生。

参加本书再版编写和修订的人员有程蓉、范雪。

本书为UOOC联盟指定参考书。本书在编写和修订过程中得到深圳大学教材建设基金资助，在此表示诚挚的谢意，并衷心希望广大读者继续对本书提出宝贵意见。

编　者

Foreword to the Second Edition

Since the first edition of the book was published in 2005, it has become a piece of unique academic work in the field with its innovative contents, wide practical use, and featured bilingual education. It received great support and was widely accepted by readers. The book was awarded the First Prize of Excellent Textbook of Shenzhen University. In 2008, it was selected as one of China's National "Eleventh Five-Year-Plan" Textbooks for Regular Higher Education. As a supporting teaching material, the bilingual electronic teaching plan published together with the book has also been very popular among readers and awarded several national and provincial prizes of Excellent Multimedia CAI Software for Higher Education.

This book has been revised in its second edition according to the newly formulated and improved national standards, as well as feedbacks and suggestions from readers. The basic structure of the book remains unchanged except that a new chapter on "Other Drawings" was added as the Chapter 10. The contents of the book have been revised or extended following the new national standards. Further typical examples used by home and abroad universities on part drawings and assembly drawings and their representation methods have been illustrated in the book. The Chapter 11 on "Computer Graphics", which was the 10th chapter in the first edition, has been rewritten, in which AutoCAD2008 is used as the demo software.

For the sake of comparative reading, the second edition adopts a two-column format, one column being the English contents while the other one being the corresponding Chinese version. And all tables and figures in the book are labeled bilingually in English and Chinese.

Primary participants who have contributed to the editing and revision of the second edition are Lin Hu, Rong Cheng, Xiaobo Peng, Wei Sun, Qin Fu, Fenghong Wang, Mingli Zhang, with Lin Hu being responsible for the general editing and final assembly of the entire book.

Support from Shenzhen University through the Teaching Material Development Grant is gratefully acknowledged. Comments, criticisms and submissions from readers are always welcome.

Authors

第 2 版前言

本书是一本内容新颖、实用性强、富有特色的双语教材，自 2005 年首次出版以来，深受广大读者的欢迎和厚爱，并荣获深圳大学优秀教材一等奖，2008 年被评为普通高等教育"十一五"国家级规划教材。与本书配套的双语电子教案亦极受欢迎，曾多次获得全国及广东省高等教育优秀多媒体教学软件奖。

为了全面贯彻国家颁布的最新标准，同时考虑到一些院校和读者的反馈意见和建议，本书增加了"第 10 章 其他工程图"，按新国家标准更新修改或补充了相关的内容，并适当增加了国内外高校使用的几种其他类型的零件图和装配图典型图例及其表达方法。原"第 10 章 计算机绘图"顺延为"第 11 章 计算机绘图"，并采用 AutoCAD 2008 版本的绘图软件重新进行了编写。

本书在排版形式上做了较大改进，采用在同一页里中英文对应编排的形式；所有图、表均采用中英文合在一起的编排形式，以方便学生对照阅读。

参加本书再版编写和修订的人员有胡琳、程蓉、彭小波、孙炜、付芩、王枫红、张明莉。胡琳任本书主编。

本书在编写和修订过程中得到深圳大学教材建设基金资助，在此表示诚挚的谢意，并衷心希望广大读者继续对本书提出宝贵意见。

编　者

Foreword to the First Edition

The higher education in China is now upgrading into a new revolution phase, in which the bilingual education is emphasized in aspects of teaching foundation courses, professional foundation courses and professional courses in order to improve the internationalization of the higher education and the cultivation of advanced innovative specialists. Therefore, the bilingual education is a key of the higher education to parallel with the world, a challenge of new era, a major trend in the revolution of higher education. It is a strategic choice which must be made by china's higher education in 21st century. This book has contributed a solid foundation for bilingual education and met the requirements of current internationalizing phase in the revolution of the higher education.

According to the "Basic Requirements in Teaching Descriptive Geometry and Engineering Drawing Course" developed by the Engineering Drawing Advisory Committee for Higher Engineering Education, and based on the features of the reform of teaching Engineering Drawing courses in recent years, the book is well compiled and organized, focusing on training twenty-first era's senior engineering application-oriented personnel. This book concentrates on modern manufacturing technologies, and as a guideline, emphasizes comprehensible learning style and aims at the enhancement of practical abilities. The book endeavors to establish refined, high and comprehensive qualities. This book can be used as a textbook on Engineering Drawing for students majoring in mechanical engineering or subjects closely related to that (72 to 108 class hours), university or college students of different specialties such as science, engineering, economy and management (36 to 64 class hours), and students of engineering-related subjects in vocational schools, advanced technical schools, broadcast colleges, vocational colleges, correspondence colleges and part-time colleges. It is also a very good reference book for engineering designers, mechanical design and manufacturing engineers.

The book is written based on the characteristics of Engineering Drawing as a foundation course, inheriting the essence of traditional teaching contents, integrating extensive experiences of many years teaching from the authors, and focusing on the education requirements of the new era in order to improve the comprehensive and creative abilities of students. Meanwhile, it strives to use the limited teaching resources to motivate students' learning interests without adding extra loads to students as well as lecturers. Therefore, the book contributes to the evolution of education in Engineering Graphics, which is being reformed from a "knowledge-skill focused" education style to a more comprehensive "knowledge-skill-methodology-capability-ability focused" education style. This book owns the following features:

1. Each page of the bilingual contents is represented in a double-column format, with English in one column while Chinese of the same contents in the other.

2. Practicability is implemented in terms of applicability and advancement. It simplifies the contents on descriptive geometry, but emphasizes on topics relevant to practical engineering applications, such as the basic projection theory, representation of shapes and bodies, and the ability in reading engineering drawings. Plenty of typical examples and corresponding analysis and solutions are given for the purpose of training students in aspects of methodologies, capabilities, and the skill of problem solving.

3. The Training of freehand drawing skills is emphasized. Plenty of well-designed exercises are provided to ensure sufficient training in terms of freehand drawing and basic drawing skills, and moreover, to make sense of the relations among freehand drawing, instrument drawing and computer-aided drawing.

4. To enhance students' ability of reading the national standards, and train them to be conditioned to work with up-to-date national standards, this book adopts the newly published national standards, the *Technical Drawings* and the *Mechanical Drawings*, which can be referred in the text and the appendices.

5. The supporting CAI multimedia electronic teaching plan edited by Lin Hu and Rong Cheng is published simultaneously, which covers the complete content of the book. The multimedia teaching plan takes advantage of several CAI softwares and adopts animation technology for shape illustrations, which is vivid, intuitional, heuristic and apt to inspire the learning interests of students. It is the MS PowerPoint format and opens to be modified, allowing teachers to add, delete and re-organize its contents, or even recompose it according to their own styles to satisfy various kinds of individualized teaching requirements.

6. Another supporting material of the book, EXERCISE WORKBOOK OF ENGINEERING DRAWING edited by Lin Hu and Rong Cheng, is released with an exercise CD (for students). The CD contains animations of 3D solid models in the exercises, which can help students to review and master the corresponding contents after school. The exercises in the workbook cover a wide range in categories and vary from the simple to the difficult while with an appropriate average difficulty, owning characteristics of typicality as well as universality. Besides, another CD (for teachers) containing solutions and standard answers of the exercise CD is also provided. Teachers using the textbook can ask the publisher for it.

This book is primarily edited by Prof. Lin Hu of Shenzhen University as the Editor in Chief. Rong Cheng (Shenzhen University), Qin Fu (Jianghan University) and Wei Sun (South China University of Technology) participate as the Associate Editors in Chief. The other editors include Fenghong Wang (South China University of Technology), Xue Fan and Xiaobo Peng (Shenzhen University) and Mingli Zhang (North China Institute of Aerospace Engineering). The book is primarily reviewed by Prof. Jinchang Chen of South China University of Technology, the Vice Chairman of the Engineering Graphics Education Steering Committee of the Ministry of Education (China), the Director of the GuangDong Engineering Graphics Society as Chinese reviser, and Prof. Weiyin Ma of City University of Hong Kong as English reviser, both of whom have provided thorough reviews and revisions to this book, and contributed many invaluable suggestions. We gratefully acknowledge them for their significant contributions.

This book contains various references from home and abroad publications and teaching materials, whose authors are highly acknowledged. For those who have provided their helps and kindness for this book, we announce our sincerely acknowledgement to all of them.

This book has proudly acquired special financial support from the Teaching Material Development Grant of Shenzhen University. We therefore gratefully announce our special acknowledgement.

Due to various limitations, this book may contain mistakes and all criticisms and corrections from all experts and readers are welcome.

Authors

第1版前言

当前，我国的高等教育进入了新一轮的改革阶段：大力开展基础课、专业基础课和专业课的双语教学，加速推进我国高等教育的国际化和培养高素质创造性应用型人才。因此，双语教学是我国高等教育与国际接轨，迎接新世纪挑战和教育改革发展的必然趋势，也是中国高等教育在21世纪必须做出的战略选择。本书的编写为双语教学打下了一个坚实的基础，及时地满足了高等教育与国际接轨这一教育改革形势发展的需要。

本书是依照高等学校工科制图课程教学指导委员会制订的"画法几何及工程制图课程教学基本要求"，结合近年来工程图学课程教学改革的特点编写而成的，立足培养面向21世纪的高级工程应用型人才。它面向现代制造技术，并紧紧围绕以"学"为中心、以"素质提高"为目的的指导思想，力求简明扼要、质量上乘、覆盖面广。本书可作为机械类及近机械类各专业的"工程制图"课程（72~108 学时）教材，也可作为高等院校理、工、经、管各专业的"工程制图"课程（36~64 学时）教材，还可作为继续教育同类专业的教材。此外，本书还可供工程设计人员和机械设计与制造工程师参考使用。

本书针对基础课程的特点，继承传统内容的精华，融入编者多年积累的教学经验，着眼于新时期对人才培养的要求，以加强对学生综合素质及创新能力的培养为出发点，力求在不增加教师和学生负担的前提下，充分利用有限的教学资源，最大限度地调动学生的学习主动性和积极性，从而使工程图学教育从以"知识、技能"为主的教育向以"知识、技能、方法、能力、素质"综合培养的教育转化。本书具有如下特点：

1. 采用中、英文左右两栏同页对照编排的形式。

2. 以实用为主导，突出实用性和先进性。删减了图解法的内容，重点突出了与工程应用密切相关的投影基本理论、形体的表达方法及工程图样的阅读等内容，提供了典型例题及分析和解决问题的思路和方法，重视方法、能力、技能等综合能力素质的培养。

3. 强调徒手绘图的基本功训练。大量精心设计的习题集保证了充足的徒手画图练习和基本功训练，并注意正确处理徒手绘图、尺规绘图和计算机绘图三者之间的关系。

4. 全书采用了最新颁布的《技术制图》和《机械制图》的国家标准，根据需要选择并分别编排在正文或附录中，以培养学生贯彻最新国家标准的意识和查阅国家标准的能力。

5. 本书有配套有 CAI 多媒体电子教案（胡琳、程蓉等编），该电子教案覆盖教材的全部内容，充分运用各种软件功能，采用大量反映实物模型的动画演示，形象生动、逼真，启发性强，可大大激发学生的学习兴趣。该电子教案采用较易掌握的 PowerPoint 工具软件编制，为开放式课件。其最大的好处是：可由任课教师根据课程需要及教学习惯，方便地自行增加、删减或重组有关内容，或按自己的风格和特色进行改编，以满足个性化教学的要求。

6. 在与本书配套的《工程制图习题集》（英汉双语对照）（胡琳、程蓉主编）中附有一张练习光盘（学生版），在练习光盘中，有相应的立体模型动画演示等，便于学生自学或课外辅导，帮助学生掌握学习内容。练习题型博采众长，由浅入深，覆盖面宽，难度适宜，兼顾典型性和通用性。另外还配有习题解答和标准答案光盘（教师版），请选用该教材的院校与出版社联系索取。

本书由深圳大学的胡琳教授任主编，程蓉（深圳大学）、付芩（江汉大学）、孙炜（华南理工大学）任副主编。参加本书编写的还有王枫红（华南理工大学）、范雪、彭小波（深圳大学）、张明莉（北华航天工业学院）。本书由教育部工程图学教学指导委员会副主任委员、广东省工程图学学会理事长、华南理工大学的陈锦昌教授和香港城市大学的马维银教授担任主审。两位主审分别对书稿的中、英文进行了仔细的审阅，并提出了许多宝贵建议，在此表示衷心的感谢！

本书参考了国内外一些著述和教材，在此向有关作者致意！并感谢其他关心和帮助本书出版的人员。

本书得到深圳大学教材建设基金资助，在此特别表示衷心的感谢！

由于编者水平有限，书中难免有错误和疏漏，敬请各位专家及广大读者批评指正。

编　者

Contents 目 录

Foreword to the Third Edition　第 3 版前言
Foreword to the Second Edition　第 2 版前言
Foreword to the First Edition　第 1 版前言
Introduction　绪论 ·· 1
Chapter 1　Basic Knowledge of Drawings　第 1 章　制图基本知识 ·································· 4
　1.1　Drawing Tools and Their Utilization　绘图工具及其用法 ·· 4
　1.2　Related Provisions in National Standards　国家标准有关规定 ································ 7
　1.3　Geometric Construction　几何作图 ·· 17
Chapter 2　Basic Knowledge of Orthogonal Projection　第 2 章　正投影基础知识 ··········· 23
　2.1　Principles of Orthogonal Projection Method　正投影法原理 ·································· 23
　2.2　Projections of Points on a Solid　立体上点的投影 ·· 27
　2.3　Projections of Lines on a Solid　立体上直线的投影 ·· 29
　2.4　Projections of Planes on a Solid　立体上平面的投影 ·· 31
Chapter 3　Solids and Their Intersections　第 3 章　立体及其交线 ································ 34
　3.1　Projections of Polyhedral Solids　平面立体的投影 ·· 34
　3.2　Projections of Curved Solids　曲面立体的投影 ·· 38
　3.3　Intersection of Planes and Solids　平面与立体相交 ·· 41
　3.4　Intersection of Two Solids　两立体相交 ·· 49
Chapter 4　Composite Solids　第 4 章　组合体 ·· 58
　4.1　Analysis for Composite Solids　组合体的形体分析 ··· 58
　4.2　Drawing Views of Composite Solids　组合体三视图绘制 ······································ 60
　4.3　Reading Views of Composite Solids　组合体三视图的读图 ··································· 63
　4.4　Dimensioning Composite Solids　组合体的尺寸标注 ·· 70
Chapter 5　Axonometric Projections　第 5 章　轴测图 ··· 76
　5.1　Basic Knowledge of Axonometric Projections　轴测图基本知识 ···························· 76
　5.2　Isometric Projections　正等轴测图 ·· 77
　5.3　Cabinet Axonometry Projections　斜二轴测图 ·· 83
Chapter 6　General Principles of Representation　第 6 章　图样画法 ···························· 85
　6.1　Views　视图 ·· 85
　6.2　Sectional Views　剖视图 ·· 90
　6.3　Cuts　断面图 ··· 103
　6.4　Drawings of Partial Enlargement　局部放大图 ··· 107
　6.5　Simplified and Specified Representation　简化画法与规定画法 ··························· 108
Chapter 7　Expression for Commonly Used Machinery Components　第 7 章　常用机件的表达 ··· 113
　7.1　Screw Threads　螺纹 ·· 113
　7.2　Screw Fasteners　螺纹紧固件 ·· 122
　7.3　Gears　齿轮 ··· 129
　7.4　Pins　销 ··· 138
　7.5　Keys　键 ·· 139
　7.6　Springs　弹簧 ··· 142

7.7　Bearings　轴承 ·· 146

Chapter 8　Detail Drawings　第 8 章　零件图 ························ 151
8.1　Contents of Detail Drawings　零件图的内容 ······················· 151
8.2　Choosing Views of Detail Drawings　零件图的视图选择 ··········· 152
8.3　Representation Methods of Typical Parts　典型零件的表达方法 ······ 152
8.4　Manufacturing Processes of Parts　零件的结构工艺性 ············· 158
8.5　Dimensioning of Detail Drawings　零件图的尺寸标注 ············· 162
8.6　Technical Requirements in Detail Drawings　零件图的技术要求 ······ 165
8.7　Interpreting Detail Drawings　读零件图 ···························· 186
8.8　Mapping Parts　零件的测绘 ··· 197

Chapter 9　Assembly Drawings　第 9 章　装配图 ···················· 204
9.1　Summary of Assembly Drawings　装配图概述 ······················ 204
9.2　Representation Methods of Assembly Drawings　装配图的表达方法 ··· 209
9.3　Choosing Views of Assembly Drawings　装配图的视图选择 ········ 212
9.4　Dimensioning and Specifications for Assembly Drawings　装配图的标注 ··· 214
9.5　Rationality of Fitting Structures　装配结构的合理性 ················ 217
9.6　Mapping Units and Representation of Assembly Drawings　部件测绘及其装配图的画法 ··· 220
9.7　Interpreting Assembly Drawings and Extracting Detail Drawings　读装配图和拆画零件图 ··· 229

Chapter 10　Other Drawings　第 10 章　其他工程图 ················· 238
10.1　Development　展开图 ·· 238
10.2　Welding Drawings　焊接图 ·· 245
10.3　Chemical Drawings　化工制图 ···································· 253
10.4　Electric Diagrams　电气制图 ······································ 273

Appendices　附录 ··· 286
Appendix 1　Screw Threads　附录 1　螺纹 ································ 286
Appendix 2　Commonly Used Standard Parts　附录 2　常用标准件 ········ 289
Appendix 3　Commonly Used Elements of Parts　附录 3　常用的零件结构要素 ··· 305
Appendix 4　Surface Texture Parameter　附录 4　表面结构参数 ··········· 309
Appendix 5　Limits and Fits　附录 5　极限与配合 ························ 309
Appendix 6　Commonly Used Materials　附录 6　常用材料 ··············· 318
Appendix 7　Definitions for General Heat Treatment and Surface Treatment
　　　　　　附录 7　常用的热处理和表面处理名词解释 ················ 325
Appendix 8　Commonly Used Terminologies and Abbreviations in Engineering Drawing
　　　　　　附录 8　工程制图中常用的专业术语及缩略语 ············· 327

References　参考文献 ·· 332

Introduction

1. Subjects and features of this course

Engineering Drawing is an application oriented subject that introduces the preparation, representation and reading of engineering drawings. Similar to characters and numbers, it is one of the tools used by human for the expression, conception, analysis and communication of technical information. Considering scientific and technological level at the present time, engineering drawings are important documents in industry for design, manufacture, utilization and service, often called "the common technical language for engineers". Engineering Drawing is not only a foundation subject for a specific major, but also a part of the entire spectrum of engineering education. It provides theories and methodologies for all engineering talents to express their spatial and visual imagination.

The course mainly studies the basic theories and methods for the preparation and reading of engineering drawings as well as related national standards on *Mechanical Drawings* and *Technical Drawings*. The main features of the book are as follows:

(1) Foundation. It is a foundation subject for other engineering-related subjects and education and provides a basis for studying other subjects afterwards.

(2) Interdiscipline. It is a cross disciplinary subject that integrates geometry, projection theory, basic engineering knowledge, basic engineering specifications and standards, and advanced drawing techniques.

(3) Engineering. It is an engineering subject which has a close connection with body and shape construction, analysis and representation in engineering applications and requires continuous integration with engineering regulations and methodologies.

(4) Practicability. It is a widely used practical subject integrating both theoretical and engineering practices.

(5) Methodology. It provides a method covering engineering and visual imagination that could effectively train students with the ability in comprehensive spatial imagination and analysis.

(6) Universality. It is a common engineering language applicable to different regions, different disciplines, different languages, and whether for the past, present or the future. The representation of engineering drawings is always the same.

绪　　论

1. 本课程的研究对象及特征

工程图学是一门研究工程图样的绘制、表达和阅读的应用学科。工程图样和文字、数字一样，也是人类借以表达、构思、分析和进行技术交流的不可缺少的工具之一。就当代科学技术水平而言，工程图样仍是工业生产中进行设计、制造、使用、维修时的重要技术文件，有"工程界共同的技术语言"之称。因此，工程图学不仅是某个特定专业的基础课程，而且还是工程教育的一部分，为一切涉及工程领域的人才提供空间思维和形象思维表达的理论及方法。

本课程主要研究绘制、阅读工程图样的基本理论和方法，学习国家标准《机械制图》《技术制图》的相关内容。其主要特征体现为：

(1) 基础性　工程图学是作为一切工程和与之相关人才培养的工程基础课，并为后续的工程类专业课的学习提供基础。

(2) 交叉性　工程图学是几何学、投影理论、工程基础知识、工程基本规范和标准，以及现代绘图技术等多个学科相结合的产物。

(3) 工程性　工程图学与工程中的形体构成、分析及表达紧密相连，需要与工程规范、工程思想持续结合。

(4) 实用性　工程图学具有广泛的实际应用性，是理论与工程实践相结合的学科。

(5) 方法性　工程图学中所蕴涵的工程思维和形象思维的方法可以有效地培养学生的综合空间想象和分析能力。

(6) 通用性　工程图学作为工程界的通用语言，具有跨地域、跨行业性，无论古今中外，尽管语言不同，但是工程图样的表达方法都是相同的。

2. Nature and tasks of this course

Engineering Graphics is the study for the representation and communication of products and processes. An engineer drawing is a carrier of engineering and product information and the communication language in engineering and industry. The course is based on rigorous theoretical and highly practical materials that are important to train students with scientific thinking and innovation. It is an important technical foundation undergraduate course for general higher education in universities and colleges.

The tasks of this course are as following:

1) To study the basic theory and applications of orthographic projection and train students with abilities in design and innovation.

2) To train students with a balanced skill in hand drawing, instrument drawing and computer-aided drawing as well as the ability in reading mechanical drawings.

3) To train students with abilities in spatial and logical thinking, visual imagination, conceptualization, exploration and innovation.

4) To train students with serious working attitude, meticulous working style and their persistence in complying with related national standards.

3. Aims and topics of this course

This course has four major aims and topics:

1) Equip students with basic theory and methodologies and the ability of projection and spatial imagination.

2) Equip students with strong expression abilities to correctly, completely, clearly and reasonably represent parts and components.

3) Equip students with proficient drawing skills in producing skilled and qualified drawings meeting related standards.

4) Equip students with the ability in identifying and reading related national standards and proficiency in reading engineering drawings.

The study of the above topics enables students to develop their skills and abilities in spatial imagination, representation, and preparation and reading of engineering drawings. It can also help establish a solid foundation for future studies of technical subjects and produce a new generation of engineering talents with strong spatial imagination and creativity.

4. Study methodologies of this course

The course consists of projection theory and engineering practices. Different methodologies may be applied when studying different parts of this course.

2. 本课程的性质和任务

工程图学是研究工程与产品信息表达、交流与传递的学问。工程图是工程与产品信息的载体，是工程和工业领域交流的语言。本课程理论严谨，实践性强，对培养学生掌握科学的思维方法和创新意识有重要作用，是普通高等院校重要的技术基础课程。

本课程的主要任务是：

1）学习正投影法的基本理论及应用，培养创造性构型设计能力。

2）培养徒手绘图、尺规绘图、计算机绘图的综合能力及阅读简单机械图样的能力。

3）培养对物体的空间思维能力、逻辑思维能力、形象思维能力、构思造型能力和开拓创新精神。

4）培养严谨的工作态度、细致的工作作风、贯彻执行国家标准的意识。

3. 本课程的学习目的及内容

本课程的学习目的及内容可分为四个部分：

1）掌握正投影基本理论和方法，具有丰富的投影和空间想象能力。

2）能正确、完整、清晰、合理地表达机件，具有较强的表达能力。

3）能熟练、准确地绘制规范、合格的图样，具有扎实的绘图能力。

4）能查阅有关国家标准、看懂并正确理解工程图样，具有熟练的读图能力。

通过这四个部分的学习，培养学生的空间想象能力、表达能力、绘图能力、读图能力，为进一步的专业课学习和实践打下坚实的基础，塑造一代具有丰富空间想象力和创造性的现代工程技术人才。

4. 本课程的学习方法

本课程既有投影理论，又有较强的工程实践性，每一部分又有其各自的特点，故而学习方法也不尽相同。

1) When studying the projection theory, one should understand the basic concepts and basic rules. It should integrate projection analysis and geometry drawing techniques with their spatial imagination, logic reasoning and analytics judgment and establish the corresponding relationship between plane drawing and spatial shape. To improve one's spatial imagination is a step-by-step process and requires repeated studies from spatial shapes to planar representations and vise versa.

2) When studying the representation of parts and components, one should master the theory and methodologies of shape and body analysis and line and plane analysis through attending lectures or self-learning. One should be skilled in simplifying complicated problems such that the problem could be solved with minimum effort. One should also ensure careful observation, profound thinking, persistent trying and exercises and thus continuously improve his/her ability in spatial analysis and conceptualization.

3) When studying drawing techniques, attention should be paid to proper usage of tools and mastery of methods and skills in freehand drawing. All drawings must comply with related national standards. Skilled drawing using AutoCAD requires comprehensive exercise on computers and practice. Be patient in the course of drawing, always keep a precise, serious, responsible and meticulous attitude, and pay more attention on exploration, observation, and practice. Each drawing should be finished on schedule with all quality requirements satisfied.

4) When learning reading drawings, one should participate in real engineering work and accumulate knowledge through practice. One should try to apply the knowledge in practice and thus develop ability in making observation, thinking, conclusion, application and skills in problem solving. In the process of drawing reading and exercises, try to learn from duplication, revision, exploration and creative thinking. One must be able to complete this subject through typical examples and continued synthesis of exercises and practices.

In conclusion, one can master and consolidate the knowledge of this course through the integration of learning and practice, carefulness and seriousness, step-by-step learning approach and correct completion of related exercises in a timely manner.

1）在学习投影理论部分时，应注意对基本概念、基本规律的掌握，将投影分析、几何作图同空间想象、逻辑推理和分析判断结合起来，建立起平面图形与空间形体的对应关系。提高空间想象力是一个循序渐进的过程，需要从空间形体到平面图形、从平面图形到空间形体的反复学习才能实现。

2）在学习机件的表达方法时，通过听讲和自学，注意掌握形体分析法和线面分析法等构形分析的理论和方法，善于把复杂的问题转化为简单的问题，许多难题便可迎刃而解，也可收到事半功倍的效果。做到细心观察、勤于思考、不断尝试、认真练习，从而不断提高空间分析能力和构思能力。

3）在学习绘制图样时，应注意绘图工具的正确使用方法以及徒手绘图的方法和技巧，并自觉遵守国家标准的有关技术制图、机械制图中的规定。在学习AutoCAD绘图时，应多上机进行综合练习并实践。作图时要有耐心，多摸索、多看、多画、多练，以严谨、认真、负责、细致的态度，按时、优质地完成每一幅图样。

4）在学习读图时，注意联系工程实践并积累知识，注重将所学的理论知识在实践环节中加以运用，培养观察、思考、总结以及运用这些知识解决实际问题的能力。应在反复读图的练习过程中，注意"继承模仿、改造变换、联想创新"的运用，抓住典型例题，不断总结读图经验，就一定能完成本课程的学习任务。

总之，只要学与练相结合，发扬一丝不苟的精神，一步一个脚印地学习，保质保量地完成相应的练习题，就能使所学知识得以巩固，并真正学好本课程。

Chapter 1 Basic Knowledge of Drawings 第 1 章 制图基本知识

This chapter introduces basic knowledge of engineering drawing for further studies. Topics covered in this chapter include drawing tools and their utilization, related national standards, and drawing techniques based on geometric construction.

本章重点介绍绘图工具及其使用方法、国家标准的有关规定、几何作图方法等，其目的是为今后的学习打下必要的基础。

1.1 Drawing Tools and Their Utilization 1.1 绘图工具及其用法

In order to efficiently produce quality drawings, all tools must be used correctly.

为了提高绘图质量和效率，必须正确地使用绘图工具。

1.1.1 Drawing board 1.1.1 图板

A drawing board is used to fix drawing sheet and produce drawings on the sheet. The surface of the board should be flat and smooth. The left side is the lead side and should be straight for guiding the T-square as shown in Fig. 1-1a.

图板是用来固定图纸并进行绘图的。图板表面要求平整光滑，其左侧为导边，必须平直，如图 1-1a 所示。

1.1.2 T-square 1.1.2 丁字尺

A T-square is mainly used to draw horizontal lines as shown in Fig. 1-1a. It is also used to draw vertical lines and parallel lines in combination with a triangle as shown in Fig. 1-1b and Fig. 1-1c. It is made of blade and head. To drawing horizontal lines, place the head of the T-square in contact with the lead side of the drawing board with left hand and move the T-square to the desired position. Hold the pencil and draw the line from its left end to the right end. One should not directly use a T-square to draw vertical lines. One should not use the lower edge of a T-square to draw horizontal lines either.

丁字尺主要用来画水平线，还常与三角板配合画铅垂线和平行线，如图 1-1 所示。丁字尺由尺身和尺头构成，使用时，需用左手扶住尺头并使尺头内侧紧靠图板左侧导边，上下滑移到所需位置，然后沿丁字尺工作边自左向右画水平线。禁止直接用丁字尺画铅垂线，也不能用尺身下缘画水平线。

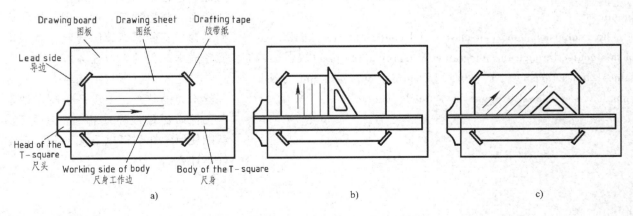

Fig. 1-1 Drawing board and T-square 图板及丁字尺的应用
a) Position the paper and draw horizontal lines 固定图纸及画水平线
b) Draw vertical lines 画竖直线 c) Draw parallel lines 画平行线

1.1.3 Triangle

A triangle is often used in combination with a T-square to draw vertical lines, lines with an inclination angle of 30°, 45° or 60° with horizontal lines as shown in Fig. 1-1 b and Fig. 1-1c. One can also use two triangles to draw parallel lines and orthogonal lines of any orientation as shown in Fig. 1-2.

1.1.3 三角板

三角板常与丁字尺配合使用,画水平线的垂直线,以及与水平成30°、45°或60°的斜线,如图1-1b、c所示。两块三角板配合使用,可画任意方向倾斜线的平行线和垂直线,如图1-2所示。

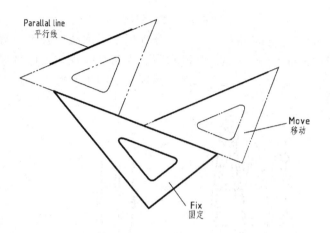
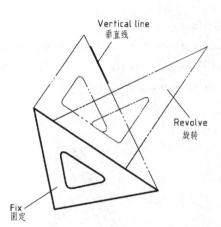

Fig. 1-2 Using two triangles to draw parallel lines and orthogonal lines of any orientation
用两块三角板配合作平行线或垂直线

1.1.4 Other drawing tools

1. Compass

A compass is used to draw circles and arcs. The leg of the compass with a step pin should be facing downward and the pencil tip on the other leg should have similar height to the pin step as shown in Fig. 1-3a. When drawing on paper, one can rotate the compass by revolving the handle clockwise and incline the compass slightly forward as shown in Fig. 1-3b. When drawing circles with different diameters, adjust the needle and the pencil of the compass as necessary, but they should always be adjusted perpendicular to the paper as shown in Fig. 1-3c and Fig. 1-4.

1.1.4 其他绘图仪器

1. 圆规

圆规用来画圆和圆弧。使用时,应将圆规钢针有台阶的一端朝下,并使台阶面与铅芯平齐,如图1-3a所示。画图时,按顺时针方向旋转圆规并使其稍向前倾斜,如图1-3b所示。画不同直径的圆时,要注意随时调整钢针和铅芯插腿,使其始终垂直于纸面,如图1-3c和图1-4所示。

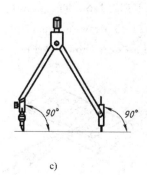

Fig. 1-3 The use of a compass 圆规的使用方法
a) Needle is slightly longer than the pencil lead 针尖略长于铅芯
b) Draw a circle in clockwise direction 按顺时针方向画圆
c) Steel needle and pencil lead straddle are perpendicular to paper 钢针和铅芯插腿垂直于纸面

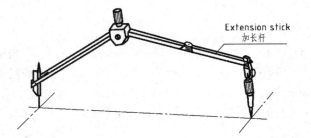

Fig. 1-4 Adding an extension stick to draw big circles　用加长杆画大圆

2. Dividers

Dividers are used for transferring distances and for equal subdivision of lines and circles. The two tips should meet together at the position as shown in Fig. 1-5.

2. 分规

分规用来量取线段长度及等分线段和圆。当两脚并拢时，两针尖应对齐，其用法如图1-5所示。

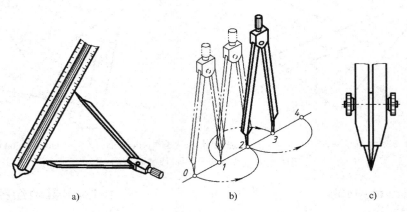

Fig. 1-5 The use of dividers　分规的用法
a）Size taking　量取尺寸
b）Size with equal subdivision　连续截取等长线段　c）Tip of needles　针尖并拢于一点

1.1.5 Measures for pencil sharpening

There are soft and hard drawing pencils marked with B or H. 2B pencil or B pencil is used to draw thick lines. H or 2H pencil is used to draw thin lines. HB pencil is used to write characters. Sand paper is usually used to sharpen pencil, as shown in Fig. 1-6a. There are two kinds of core forms of pencil as shown in Fig. 1-6b and Fig. 1-6c. The pencil with cone form shown in Fig. 1-6b is used to draw manuscript and to write, while the pencil with wedge form shown in Fig. 1-6c is used for deepening and thickening.

1.1.5 铅笔的削法

绘图铅笔的铅芯有软硬之分，用标号B或H表示。通常用2B或B铅笔画粗线；用H或2H铅笔画细线；用HB铅笔写字。通常用砂纸磨削铅笔的笔芯，如图1-6a所示。铅笔的铅芯有两种形状，如图1-6b、c所示。圆锥形笔芯的铅笔用来画底稿和写字，如图1-6b所示；楔形笔芯的铅笔用来加深加粗图线，如图1-6c所示。

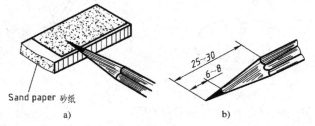 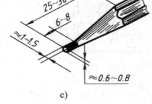

Fig. 1-6 Pencil sharpening　铅笔的磨削
a）Sharpening with sand paper　用砂纸磨削　b）Cone form　圆锥形　c）Wedge form　楔形

1.2 Related Provisions in National Standards

Engineering drawing is an important documentation used during the process of design and manufacturing. For the sake of convenience in technical communication, all drawings should comply with national standards. National standards are abbreviated as "GB".

1.2.1 Size and layout of drawing sheets (GB/T 14689—2008)

GB/T 14689—2008 is the standard for defining size and layout of drawing sheets. Here, GB/T is the abbreviation of the National Standard (GUO JIA BIAO ZHUN in Chinese). The code 14689 is a serial number of the standard. The code 2008 indicates the year when the standard was published.

1. Formats

While producing the drawing, one should usually adopt standard basic formats illustrated in Table 1-1. There are five standard formats, namely A0, A1, A2, A3 and A4. If it is necessary, one may also use extended formats of larger dimensions specified in the standard.

1.2 国家标准有关规定

工程图样是设计和制造过程中的重要技术资料,为了便于技术交流,所有工程图样都应遵守国家标准。国家标准简称"国标",用代号"GB"表示。

1.2.1 图纸幅面和格式（GB/T 14689—2008）

GB/T 14689—2008 是图纸幅面和格式的标准编号。其中"GB/T"是"国家标准（推荐性）"的汉语拼音字母缩写,"14689"是标准的顺序号,"2008"表示该标准颁布的年份。

1. 图纸幅面

绘制图样时,应优先采用表 1-1 所规定的基本幅面。幅面代号有 A0、A1、A2、A3、A4 五种,必要时,也允许选用国标规定的加长幅面。

Table 1-1 Basic formats 基本幅面 (mm)

Format codes 幅面代号	Dimensions 尺寸 B×L	Margin 周边 a	c	e
A0	841×1189	25	10	20
A1	594×841	25	10	20
A2	420×594	25	10	10
A3	297×420	25	5	10
A4	210×297	25	5	10

2. Border

One must use continuous thick lines to draw the border. There are two layouts to follow, i.e., with or without space for book binding (see Fig. 1-7 or Fig. 1-8). The dimensions for each of the layouts for the five standard formats are shown in Table 1-1. However, one may only use the same layout in case of the same product. For A4 format with booking space, one usually adopts a portrait orientation, while for A3 format with booking space, one usually adopts a landscape orientation.

3. Title block

One must draw a title block on each drawing. The title block should be located on the lower right corner of the sheet as shown in Fig. 1-7 and Fig. 1-8.

2. 图框

图框用粗实线画出,其格式分不留装订边和留装订边（见图 1-7 及图 1-8）两种,尺寸见表 1-1。但同一产品的图样只能采用一种格式。当图纸选用留装订边格式时,一般应采用 A4 幅面竖装和 A3 幅面横装。

3. 标题栏

每张图样上都必须画出标题栏。标题栏应位于图纸的右下角,如图 1-7 和图 1-8 所示。

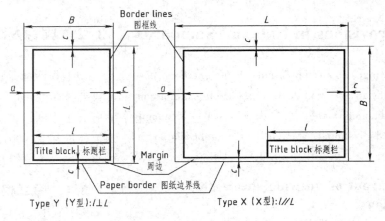

Fig. 1-7 Drawing sheets with binding area 有装订边的图框格式

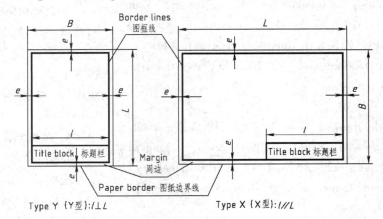

Fig. 1-8 Drawing sheets without binding area 无装订边的图框格式

The format and dimensions of the title block should also follow the national standard GB/T 10609.1—2008, illustrated in Fig. 1-9. The title block on exercise drawings used in this course can be simplified and a recommendation is illustrated in Fig. 1-10.

One should note the following: The outline of the title block must be drawn with continuous thick lines. The right-side and the base line should be in coincidence with the border.

标题栏的格式和尺寸遵守国家标准 GB/T 10609.1—2008 的规定，如图1-9 所示。学校的制图练习作业中常采用简化的标题栏，建议采用图1-10所示的标题栏尺寸与格式。

注意：标题栏的外框线一律用粗实线绘制，其右边和底边均与图框线重合。

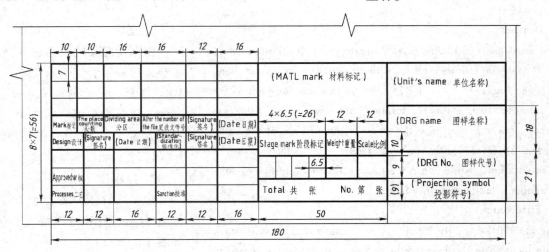

Fig 1-9 Title block of national standard 标准标题栏

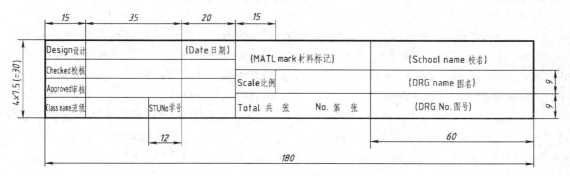

Fig. 1-10 Title block used for practice   练习用标题栏

1.2.2 Scales（GB/T 14690—1993）

The ratio between the dimensions on the drawing and the dimensions of corresponding features of the actual object is called the scale of the drawing. One should use an appropriate scale as suggested in Table 1-2.

1.2.2 比例（GB/T 14690—1993）

图中图形与其实物相应要素的线性尺寸之比，称为比例。绘图时，应从表1-2规定的系列中选取适当的比例。

Table 1-2 Scales 绘图的比例

Scale for actual dimensions 原值比例	1:1						
Down scale 缩小比例	(1:1.5) (1:1.5×10^n)	1:2 1:2×10^n	(1:2.5) (1:2.5×10^n)	1:3 (1:3×10^n)	(1:4) (1:4×10^n)	1:5 1:5×10^n	(1:6) (1:6×10^n)	1:10 1:1×10^n
Enlargement scale 放大比例	2:1 1×10^n:1	2×10^n:1	(2.5:1) (2.5×10^n:1)	(4:1) (4×10^n:1)	5:1 5×10^n:1		

Note：n is an integer and priority scales are those not written in a bracket. （注：n 为正整数；优先选用不带括号的比例值。）

For the same drawing, one should adopt the same scale and the scale should be marked in the title block. For dimensioning, one should always mark the actual dimensions no matter the drawing is enlarged or scaled down. In case of views with a different scale, it should be marked clearly in the drawing.

同一张图样上的各视图应采用相同的比例，并标注在标题栏中的"比例"栏内。图样无论放大或缩小，在标注尺寸时，应按机件的实际尺寸标注。当某个视图需要采用不同的比例时，必须另行标注。

1.2.3 Lettering（GB/T 14691—1993）

Lettering of characters or numbers on a drawing must follow related national standard：

1）The characters must be whole and clear. The distance between characters must be uniformly distributed, standing in a line.

2）The font for Chinese characters should be the "Fang Song" and the standard simplified Chinese characters should be used. The height of characters should not be less than 3.5mm and the width should be $h/\sqrt{2}$（h is the height）.

3）The number of characters is the same as the height of the characters, such as 1.8, 2.5, 3.5, 5, 7, 10, 14, 20mm.

1.2.3 字体（GB/T 14691—1993）

在图样中，书写文字或数字时，必须按国标规定书写，应做到：

1）字体工整、笔画清楚、间隔均匀、排列整齐。

2）汉字应写成长仿宋字，并采用国家正式公布推行的简化字。字高度不应小于3.5mm，字宽一般为$h/\sqrt{2}$（h表示字高）。

3）字体的号数，即字体的高度，其公称尺寸系列为：1.8, 2.5, 3.5, 5, 7, 10, 14, 20mm。

4) The font for all characters can be in italic font or normal/straight font. The width of the strokes should be about one tenth or one fourteenth of the character height. The italic font character should be inclined towards the right with an angel of 75° with respect to horizontal lines. In general, italic font is used for all types of drawings.

5) The number of characters for subscripts, superscripts, fractions, the limit symbol, and notes is usually one less compared with full size fonts used in the drawing.

Several examples of lettering are shown in Fig. 1-11.

4) 字母和数字可写成斜体或正体。其笔画宽度约为字高的 1/10 或 1/14。斜体字的字头向右倾斜,与水平线约成 75°角。图样上一般采用斜体字。

5) 用作指数、分数、极限偏差、注脚等的数字及字母,一般采用小一号字体。

字体示例如图 1-11 所示。

Fig. 1-11 Examples of lettering 字体示例

1.2.4 Lines (GB/T 4457.4—2002)

1. Line styles and their utilization

All line names, styles, line thicknesses and their applications are defined in national standard. There are nine line styles as shown in Table 1-3. When producing drawings, the width of lines should be 0.18, 0.25, 0.35, 0.5, 0.7, 1.0, 1.4, or 2mm. There are two line thicknesses, namely thick lines and thin lines. The thickness scale of the two lines is 2∶1. The width of thick lines is based on the dimension and structure of the drawing. Fig. 1-12 illustrates the applications of various line styles.

1.2.4 图线 (GB/T 4457.4—2002)

1. 图线的型式及其应用

国标规定了各种图线的名称、型式、代号及在图上的一般应用,常用 9 种线型见表 1-3。图线宽度 d 的推荐系列为:0.18,0.25,0.35,0.5,0.7,1.0,1.4,2mm。图线分粗、细两种。粗线的宽度 d 应按图形的大小和复杂程度选择。粗线与细线宽度之比为 2∶1。其应用举例如图 1-12 所示。

Table 1-3 Line styles 图线型式

Description 图线名称	Line styles 图线型式	Width 图线宽度	Usage 应用举例
Continuous thick line 粗实线	———————	d	Visible outlines and edges 可见轮廓线,可见棱边线

(Continued 续)

Description 图线名称	Line styles 图线型式	Width 图线宽度	Usage 应用举例
Continuous thin line 细实线	———————	d/2	Dimensions lines, extension lines, leader lines, section hatching lines, outlines of coincidence sections, root of threads and gears 尺寸线及尺寸界线、引出线剖面线、重合断面的轮廓线、螺纹的牙底线及齿轮的齿根线
Continuous thin irregular line 波浪线	～～～～	d/2	Boundary lines for broken views, boundary lines between a regular view and section views 断裂处的分界线，视图与剖视的分界线
Continuous thin straight line with intermittent zigzags 双折线	——⋀——⋀——	d/2	Boundary lines for broken views 断裂处的分界线
Thin long dashed line 细虚线	– – – – – 2~6 ≈1	d/2	hidden outlines, Hidden edges 不可见轮廓线，不可见棱边线
Thick long dashed line 粗虚线	— — — — —	d	Line style for surfaces with surface-treatment 允许表面处理的表示线
Thin long dashed dotted line 细点画线	—·—·— ≈20 ≈3	d/2	Axis lines, symmetrical center lines 轴线，对称中心线
Thick long dashed dotted line 粗点画线	—·—·— ≈10 ≈3	d	Cutting plane for section views 限定范围表示线
Thin long dashed double dotted line 细双点画线	—··—··— ≈15 ≈5	d/2	Outlines of adjacent alternative parts, extreme positions of moving parts, and imaginary termination lines 相邻辅助零件的轮廓线，可动零件极限位置的轮廓线，中断线

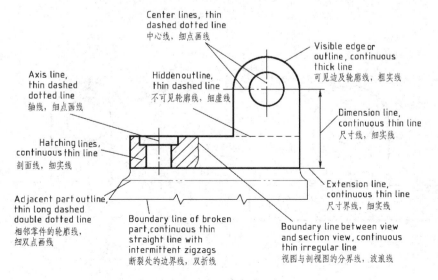

Fig. 1-12 Examples of line styles 图线应用示例

2. Notes on drawing lines

1) In the same drawing, one should use the same line width for the same type of lines. The length and the distance of different dashed lines, dashed dotted lines, long dashed double dotted lines should also be the same respectively.

2) The distance between two parallel lines should not be less than the width of two continuous thick lines. The limit distance must not be less than 0.7mm.

3) The center line should be longer than the drawing boundary from 3 to 5 mm. A longer dot, but not a shorter dot, should be used at the start and the end of the center line. The center of a circle should be the intersection point of long dots. When it is difficult to draw dotted lines or double dotted lines on a small drawing, it can be replaced by continuous thin lines, as shown in Fig. 1-13.

4) When two thin long dashed lines meet together, the line segments should meet (not space). When a thin dashed line is an extension of a continuous line, a little space should be kept at the intersection position, as shown in Fig. 1-14.

2. 图线的画法

1) 同一图样中，同类图线的宽度应一致。虚线、点画线及双点画线的线段长短和间隔应各自相等。

2) 两条平行线之间的距离应不小于粗实线的两倍宽度，其最小距离不得小于 0.7mm。

3) 绘制图的对称中心线时，应超出图外 3~5mm。首末两端应是长画，而不是短画。圆心应是长画的交点。在较小的图形上绘制细点画线或细双点画线有困难时，可用细实线代替，如图 1-13 所示。

4) 当两细虚线相交时，应线段相交（不可在空隙处相交）；细虚线是实线的延长线时，在连接处应留有间隙，如图 1-14 所示。

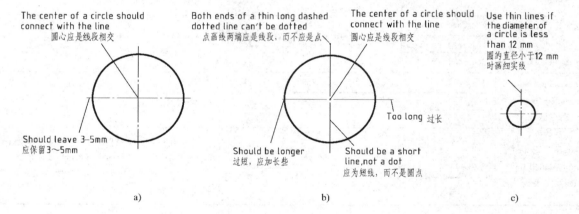

Fig. 1-13 Drawing of symmetrical center lines 对称中心线的画法
a) Correct 正确 b) Wrong 错误 c) Small circle 小圆形

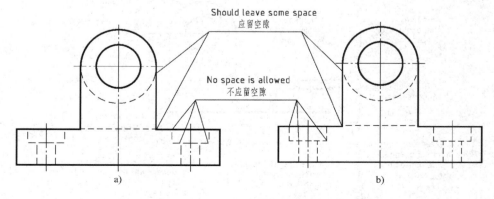

Fig. 1-14 Drawing of thin long dashed lines 细虚线的画法
a) Wrong 错误 b) Correct 正确

1.2.5 Dimensioning (GB/T 4458.4—2003)

1. Basic rules

1) The true sizes of parts are based on the dimensions marked on the drawings and have nothing to do with the sizes of the drawings and the accuracy of the drawings.

2) There is no need to mark the unit of dimensions if it is "mm". One only needs to mark the dimensions if the units are not "mm", such as "10cm" or "30°".

3) All dimensions marked in the drawing are dimensions of the final part. Otherwise one needs to clearly indicate on the drawing.

4) Each dimension of the part should be marked on the drawing only once. It should also be marked on a drawing that most clearly represents the corresponding constructs.

2. Dimension construction

A complete dimension is shown in Fig. 1-15 and is composed of an extension line, a dimension line, terminal features of a dimension line, and dimension text.

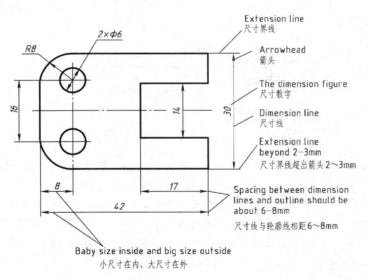

Fig. 1-15 Dimension elements with examples 尺寸的组成及标注示例

(1) Extension line An extension line is drawn with a continuous thin line and it illustrates the scope of the corresponding dimension. An extension line originates from a feature outline, an axis line, or a symmetrical center line. The outline, axis line and symmetrical center line can also be used as an extension line.

An extension line is usually drawn perpendicular to the dimension lines and the extension line should extend beyond the terminal of dimension line for about 2—3mm. If necessary, the extension line may also have a different inclination angle with the dimension line.

1.2.5 尺寸注法 (GB/T 4458.4—2003)

1. 基本规定

1) 机件的真实大小应以图样上所注的尺寸数值为依据，与图形的大小及绘图的准确度无关。

2) 图样中尺寸以毫米为单位时，不需标注计量单位的代号或名称。若采用其他单位，则必须注明相应计量单位的代号或名称，如30°、10cm等。

3) 图中所标注的尺寸数字，为该图样所示机件的最后完工的尺寸，否则应另加说明。

4) 机件的每一尺寸在图样上一般只标注一次，并应标注在最清晰反映该结构的图形上。

2. 尺寸的组成

一个完整的尺寸由尺寸界线、尺寸线、尺寸线终端以及尺寸数字组合表示，如图1-15所示。

(1) 尺寸界线 尺寸界线用细实线画出，以表示所注尺寸的界限范围。尺寸界线应由图形的轮廓线、轴线或对称中心线处引出，也可利用轮廓线、轴线或对称中心线作为尺寸界线。

尺寸界线一般应与尺寸线垂直，且超出尺寸线终端2~3mm。必要时也允许与尺寸线倾斜。

(2) Dimension line and its terminals

1) A dimension line must be drawn using a continuous thin line and cannot be replaced by existing lines in the drawing. It can not be in coincidence with any existing lines or drawn as extension of any existing lines.

2) For linear dimension, the dimension line must be parallel to the line segment being dimensioned. The distance between the dimension line and the nearest feature should be about 8—10mm. All distances between neighboring dimension lines should be as consistent as possible and are usually between 6—8mm. A baby size is usually placed inside and a big size is usually placed outside avoiding the intersection between dimension lines and extension lines or between dimension lines, as shown in Fig. 1-15.

3) A dimension line may have two alternative types of terminals, as shown in Fig. 1-16, i.e., arrowhead and oblique line. For the same drawing, one should use the same terminal and usually the arrowhead is used. It should be noticed that the tip of the arrowhead should just get in touch with the extension line, but not longer or shorter. When an oblique line is adopted for dimension line terminals, the dimension line and the extension line must be perpendicular from each other and the oblique line should be a thin solid line.

（2）尺寸线及尺寸线终端

1）尺寸线必须用细实线单独画出，不能用图上任何其他图线代替，也不能与图线重合或画在图线的延长线上。

2）标注线性尺寸时，尺寸线必须与所标注的线段平行。尺寸线与轮廓线的间隔为 8~10mm。当几条尺寸线互相平行时，其间隔尽量保持一致，一般间隔为 6~8mm，且小尺寸在内，大尺寸在外，以免尺寸线与尺寸界线相交或尺寸线与尺寸线之间相交，如图 1-15 所示。

3）尺寸线终端有两种形式：箭头和斜线，如图 1-16 所示。在同一张图样上，只能采用同一种尺寸线终端形式。通常使用箭头形式，注意箭头的尖端应正好接触尺寸界线，不要超出或不及。当采用斜线形式时，尺寸线和尺寸界线必须是互相垂直的直线，斜线用细实线绘制。

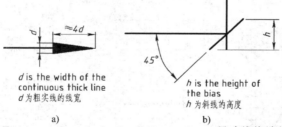

Fig. 1-16　Two types of terminals of a dimension line　尺寸线终端的两种形式
a) Arrow head　箭头　b) Oblique line　斜线

（3）Dimension text　Dimension text should usually be marked above the dimension line or in the breaking space of a dimension line. It should, be consistent in the same drawing. When there is not enough space, it can also be placed somewhere else using a leader line.

All linear dimensions should usually be marked heading upwards for horizontal dimensions, heading towards the left for vertical dimensions, or with and inclination heading upwards, as shown in Fig. 1-17a. Try to avoid placing the dimension line within the scope of 30°, and when inevitably, the dimension styles can be used according to Fig. 1-17b. If no confusion is created, dimension texts can also be entered in the breaking space of the dimension line, as shown in Fig. 1-17c. Dimension texts cannot be crossed by any drawing line, otherwise, the drawing line should be disconnected, as shown in Fig. 1-18.

（3）尺寸数字　线性尺寸的尺寸数字一般应注写在尺寸线的上方或中断处，但同一图样上最好保持一致。当位置不够时，也可引出标注。

线性尺寸数字，一般应按图1-17a所示的方向注写，即水平方向字头朝上，垂直方向字头朝左，倾斜方向字头保持朝上的趋势，并尽可能避免在图示30°范围内标注尺寸。当无法避免时，可按图 1-17b 所示的形式标注。在不致引起误解时，对于非水平方向的尺寸，其数字可水平地注写在尺寸线的中断处，如图 1-17c 所示。尺寸数字不可被任何图线所穿越通过，否则，需将该图线断开，如图1-18所示。

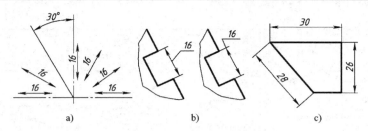

Fig. 1-17 Text orientations for dimensioning　尺寸数字注写方向

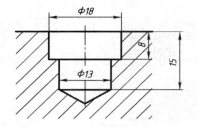

Fig. 1-18 Dimension text should not cover any drawing lines
尺寸数字不得有线穿插

(4) Examples in dimensioning　Table 1-4 provides some examples for various cases in dimensioning with dimension symbols.

（4）尺寸标注举例　表 1-4 给出了各类尺寸的标注方法。

Table 1-4　Dimension symbols　各类尺寸的注法

Dimension contents 标注内容	Description 说　明	Examples 示　例
Angle 角度	All dimension text should be marked with horizontal orientation. Extension lines should follow the radial direction and the dimension line should be an arc of which the center coincides with the apex point of the extension lines. All dimension text should be marked above the dimension line or in the breaking space of the dimension line. In case of insufficient space, dimension text may also be marked elsewhere with a leader 尺寸数字一律水平书写，尺寸界线应沿径向引出，尺寸线应画成圆弧，圆心在角的顶点。一般注在尺寸线的中断处，空间不够时允许引出标注	
Circle 圆	If the dimension is for the diameter of a circle, one should add the symbol "ϕ" before the dimension figure. The dimension line should go through the center of the circle 标注圆的尺寸时应在尺寸数字前加注符号"ϕ"。尺寸线应通过圆心	
Arc 圆弧	For arcs with radius, one should add the symbol "R" before the dimension figure. The dimension line should meet the center of the circle 标注半径尺寸时，应在尺寸数字前加注符号"R"。尺寸线应通过圆心	
Big arc 大圆弧	For arcs with large radius, one may draw dimension lines as shown in Fig. 1. When there is no need to mark the center position, one may draw dimension lines as shown in Fig. 2 在图纸范围内无法标出圆心位置时，可按图 1 标注；不需要标注圆心位置时，可按图 2 标注	
Baby size 小尺寸	In case of insufficient space, the arrowhead can be drawn outside the dimension line or replaced with a small dot. Dimension text can also be located outside the dimension line or marked with a leader line 没有足够的空间时，箭头可画在外面或用小圆点代替两个箭头，尺寸数字也可写在外面或引出标注	

(Continued 续)

Dimension contents 标注内容	Description 说　　明	Examples 示　　例
Surface of sphere 球面	One should add the symbol "S" before "φ" or "R", as shown in Fig. 1. However, the symbol "S" may be omitted for the end of shafts, screws, rivets and handles as long as it does not cause any misunderstanding, as shown in Fig. 2 应在 φ 和 R 前加注 "S"，如图 1 所示。对于轴、螺杆、铆钉及手柄端部，在不致引起误解时，可省略 "S"，如图 2 所示	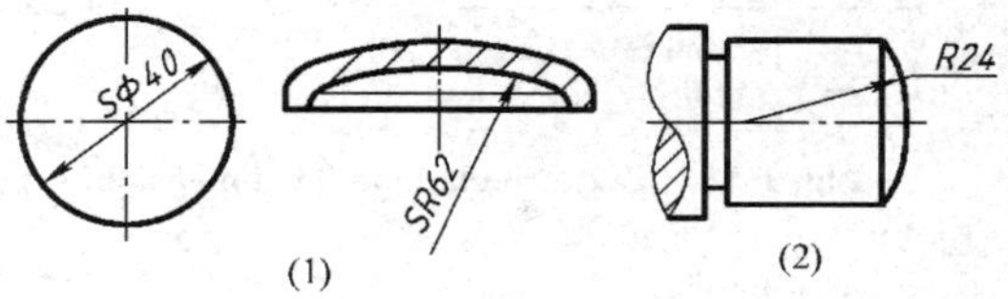 (1)　　　　(2)
The length of an arc and a chord 弧长和弦长	When dimensioning the length of a chord, the dimension line should run in parallel with the chord, and the extension lines should be paralles to the perpendicular bisector of the chord, as shown in Fig. 1. when dimensioning the length of an arc, dimensions line should be an arc, and the symbol "⌢" (Fig. 2) should be placed before the dimension text 标注弦长时，尺寸线应平行于该弦，尺寸界线应平行于该弦的垂直平分线，如图 1 所示；标注弧长尺寸时，尺寸线用圆弧，尺寸数字旁应加注符号 "⌢"，如图 2 所示	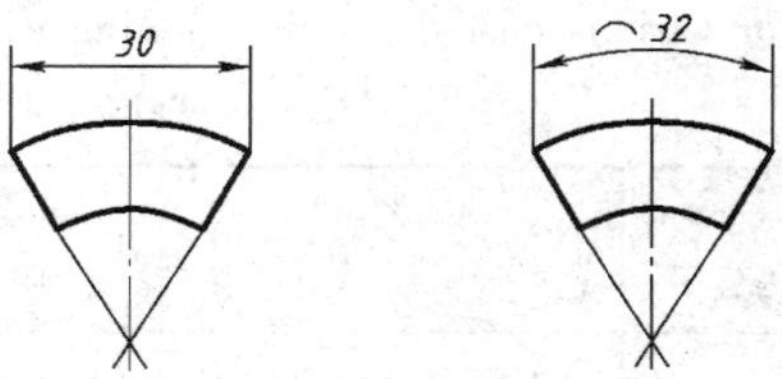 (1)　　　　(2)
Symmetry part 对称机件	The dimension line should be extended beyond the symmetry center line or broken boundary line. Only one arrowhead is required on the corresponding side of the extension line. In the drawing, the two pairs of parallel short thin lines perpendicular to the symmetry center line are the symbols of symmetry 对称机件只画出一半或大于一半时，尺寸线应略超出对称中心线或断裂处的边界线，仅在尺寸界线一端画出箭头。图中在对称中心线两端画出的与其垂直的平行细直线是对称符号	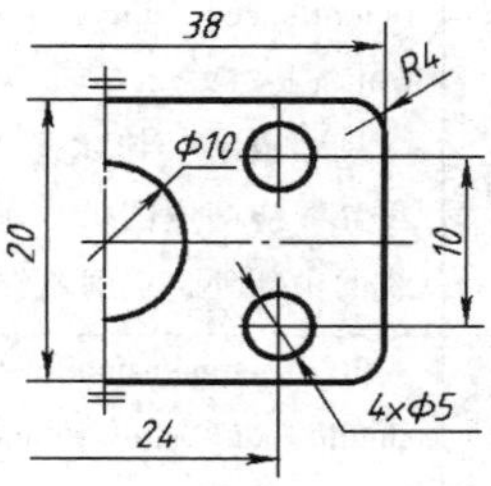
Smooth transition 光滑过渡处	Regarding to smooth transition, one may extend the outlines using continuous thin lines and find the intersection points. Extension lines for dimensioning can be drawn from the intersection points and may be inclined for clarity 在光滑过渡处，用细实线将轮廓线延长，并从它们的交点引出尺寸界线。尺寸界线如果垂直于尺寸线，则图线很不清晰，允许倾斜标注	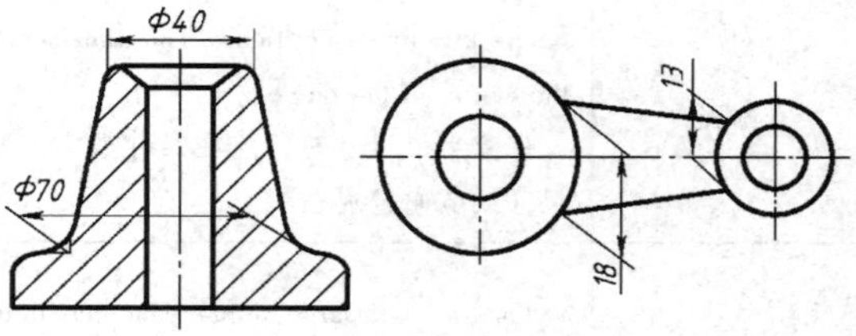
Square construction 正方形结构	Dimensions for square cross sections should be marked with the symbol "□" or marked in the form of 14×14. The crossing continuous thin lines on the drawing are the symbols of flat planes 标注剖面为正方形结构的尺寸时，可在边长尺寸数字前面加注符号 "□"，或用 14×14 的形式标注。图中相交的两细实线是平面符号	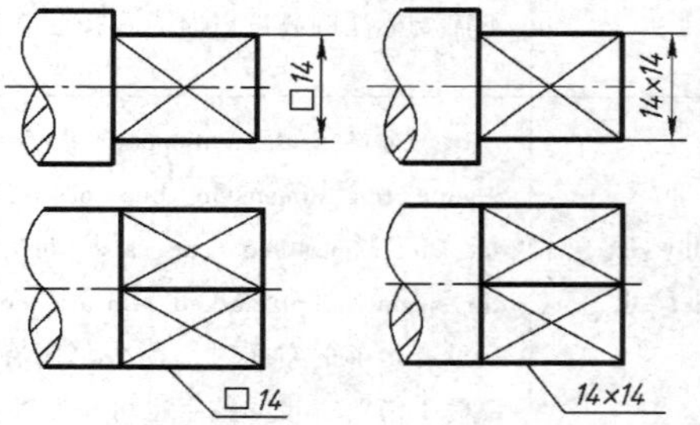

(Continued 续)

Dimension contents 标注内容	Description 说 明	Examples 示 例
Uniformly distributed holes 均布的孔	When positions and distributions of holes are obvious, as shown in Fig. 1, the location dimension and the mark with "EQS" can be omitted, as shown in Fig. 2 and Fig. 3 In Fig. 1, 8 × φ6 stands for 8 holes with the same diameter φ6 当孔的定位和分布情况在图中都已明确时（见图1），允许省略其位置尺寸和 EQS（均布）字样，如图2和图3所示 图1中 8×φ6，φ6 表示孔的直径，8 为孔的个数	

1.3 Geometric Construction

Outlines of machine components represent various geometric shapes. This section introduces some common methods of geometric construction that are often used for producing drawing.

1.3.1 Equilateral polygon

1. Regular hexagon

1) Using a compass to divide and draw a regular hexagon as shown in Fig. 1-19.

2) Using a triangle with a T-square to draw a regular hexagon as shown in Fig. 1-20.

1.3 几何作图

机件的轮廓形状一般都是几何图形，本节介绍一些常用的几何作图方法。

1.3.1 正多边形

1. 正六边形

1) 用圆规等分圆周作正六边形，如图 1-19 所示。

2) 用丁字尺和三角板配合作正六边形，如图 1-20 所示。

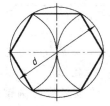

Fig. 1-19 Drawing a regular hexagon with a compass 用圆规画正六边形

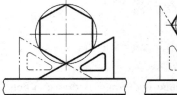

Fig. 1-20 Drawing a regular hexagon with a triangle and a T-square 用丁字尺和三角板画正六边形

2. Equilateral polygon

Fig. 1-21 illustrates how to draw a n (as shown in the figure, $n = 7$)-regular polygon with known diameter of the circum circle AE.

2. 正 n 边形

已知外接圆直径 AE，作正 n（图 1-21 中 $n=7$）边形，如图 1-21 所示。

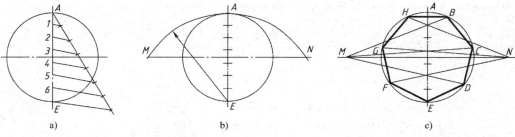

Fig. 1-21 The drawing of an n-regular polygon (n=7)　正n边形的画法（n=7）

a) Divide AE, the diameter of circum circle, by n parts (as shown in the figure n=7), and thus generate n-1 points in total (7-1=6). 将外接圆直径AE等分为n等份（图中n=7），即得n-1个点（7-1=6） b) Use E as center point, AE as radius, draw an arc that intersects the horizontal centre line of circum circle by M and N 以E为圆心，AE为半径画弧，与外接圆水平中心线交于M、N两点 c) Connect M and N respectively to every odd point (or even point) on AE with straight line, and then prolong each line in order to intersect with circum circle on points B, C, D, F, G, H, connect these points successively to generate final solution 分别由点M、N作直线与AE上每个奇数点（或偶数点）相连，并延长与外接圆交于B、C、D、F、G、H点，依次连接圆周上的各点即为所求

1.3.2 Slope and taper

1. Slope

The slope is an inclination of one line with respect to another line or the inclination of one plane with respect to another plane. The slope can be measured as the tangent of the angle between the two related lines or planes, i.e. slope = tanα = H/L as shown in Fig. 1-22a.

The slope symbol is shown in Fig. 1-22b. The slope is often represented in the form of 1 : n. When dimensioning, the oblique direction of the slope symbol should be in accordance with the actual slope direction. The drawing of slope are shown in Fig. 1-22c.

1.3.2 斜度和锥度

1. 斜度

斜度是指一条直线对另一条直线或一个平面对另一个平面的倾斜程度。斜度的大小可用两直线或两平面夹角的正切表示，即斜度 = tanα = H/L，如图1-22a所示。

斜度符号如图1-22b所示。斜度常以1∶n的形式标注，标注时要注意符号中的斜线方向与实际直线或平面的倾斜方向一致。斜度的作图方法如图1-22c所示。

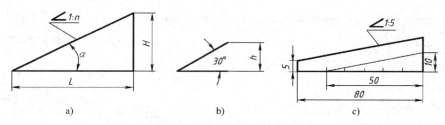

Fig. 1-22 Slope　斜度

a) The slope 斜度　b) The slope symbol 斜度符号　c) Drawing of slope 斜度的作图方法

2. Taper

The taper of a cone is defined as the ratio between the diameter of the bottom circle and the height. The taper can be measured as two times of the tangent of the half apex angle, i.e., taper = 2tan(α/2) = H/L as shown in Fig. 1-23a. The taper is often represented in the form of 1 : n. The taper symbol is shown in Fig. 1-23b. When dimensioning, the orientation of the taper symbol should be in accordance with the orientation of the actual cone. The drawing of taper are shown in Fig. 1-23c.

2. 锥度

锥度是指正圆锥的底圆直径与圆锥高度之比。锥度的大小可用圆锥素线与轴线夹角的正切的两倍表示，即锥度 = 2tan(α/2) = H/L，在图样中常以1∶n的形式标注，如图1-23a所示。锥度符号如图1-23b所示。锥度标注时要注意符号斜线方向与圆锥实际锥度方向一致。其作图方法如图1-23c所示。

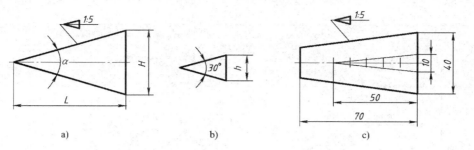

Fig. 1-23 Taper 锥度

a) The taper 锥度　　b) The taper symbol 锥度符号　　c) Drawing of taper 锥度的作图

1.3.3　Arc connection

The so-called arc connection is often used for smooth connection between lines and circles using arcs.

There are three cases, i.e.,

1) arc connection between two lines.
2) arc connection between one line and another piece of arc.
3) arc connection between two pieces of arcs.

The key in arc connection is to find the center of the connection arc and the tangent position on the corresponding lines, line and arc, and arcs respectively. Table 1-5 provides drawing techniques for four different cases of arc connection with known connection arc radius R.

1.3.3　圆弧连接

所谓圆弧连接，是指由一圆弧光滑地连接相邻两线段（或圆弧）的作图方法。

圆弧连接有三种基本形式：

1) 用圆弧连接两条已知直线。
2) 用圆弧连接已知直线和已知圆弧。
3) 用圆弧连接两个已知圆弧。

圆弧连接作图的关键在于，找出连接圆弧的圆心位置及连接圆弧与两条被连接线段相切的切点位置。表1-5给出已知连接圆弧半径为 R 的四种连接形式的作图方法。

Table 1-5　Arc connection　圆弧连接

Types of connection 连接要求	Drawing method and related procedures 作图方法和步骤		
	Finding the center O of a circle 求圆心 O	Finding the tangent points m, n 求切点 m、n	Drawing connection arc 画连接圆弧
Arc connection between two intersecting lines 连接相交两直线			
Arc connection between a line and a piece of arc 连接一直线和一圆弧			
External arc connection between two pieces of arcs 外接两圆弧			

Types of connection 连接要求	Drawing method and related procedures 作图方法和步骤		
	Finding the center O of a circle 求圆心 O	Finding the tangent points m, n 求切点 m、n	Drawing connection arc 画连接圆弧
Internal arc connection between two pieces of arcs 内接两圆弧			

(Continued 续)

1.3.4 Techniques for drawing 2D objects

Two-dimensional objects are drawn with line segments, arcs and curves. To produce accurate drawings, one must make a through analysis about known dimensions and the composition of the 2D objects. One should also make sure that which segment has full dimension and can be drawn directly and which segment does not have and needs to be produced graphically.

1. Analysis of 2D dimensions

All 2D dimensions can be classified into size demension and location dimension, respectively.

(1) Size dimension Size dimensions are used to determine the size of individual elements of the 2D objects, such as the length of a line segment, the diameter or radius of a circle or a piece of arc, and various angles. In Fig. 1-24, the dimensions $\phi20$, $\phi5$, $R15$, $R50$, $R10$, $\phi32$, $R12$ and 15 are size dimensions.

(2) Location dimension Location dimensions are used to determine the relative position of individual elements of the object. In Fig. 1-24, dimensions 8 and 75 are location dimensions.

1.3.4 平面图形的画法

平面图形都是由若干线段（直线、圆或曲线）连接而成的。要正确绘制一个平面图形，首先必须对平面图形进行尺寸分析和线段分析，弄清哪些线段尺寸齐全，可以直接画出来；哪些线段尺寸不全，需通过作图才能画出。

1. 平面图形尺寸分析

平面图形的尺寸可分为定形尺寸和定位尺寸两类。

（1）定形尺寸　定形尺寸是确定平面图形各部分形状大小的尺寸。如线段的长度、圆或圆弧的直径和半径以及角度大小等。图1-24中尺寸 $\phi20$、$\phi5$、$R15$、$R50$、$R10$、$\phi32$、$R12$、15等都是定形尺寸。

（2）定位尺寸　定位尺寸是确定平面图形各部分之间相对位置的尺寸。图1-24中尺寸8、75即为定位尺寸。

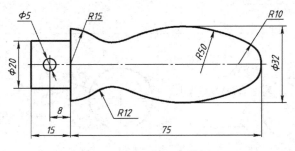

Fig. 1-24　Handle　手柄

2. Analysis of individual segments of the 2D objects

All segments (lines or arcs) can be classified as known segments, intermediate segments and connection segments.

2. 平面图形的线段分析

平面图形中的线段（直线或圆弧），可分为已知线段、中间线段和连接线段。

(1) Known segments Known segments are those with all dimensions for shape and position. In Fig. 1-24, the arcs of R15 and R10 are known segments.

(2) Intermediate segments Segments that have shaping dimensions and incomplete positioning dimensions are called intermediate segments. In Fig. 1-24, the arc of R50 is an intermediate segment.

(3) Connection segments Segments that have only size dimensions but without location dimensions are called connection segments that need to be determined geometrically through drawing techniques. In Fig. 1-24, the arc of R12 is a piece of connection arc.

When drawing 2D objects, one should first draw known segments, then draw intermediate segments, and finally draw connection segments.

3. General drawing procedures

(1) Preparation

1) Analyze dimensions and segments and plan for the drawing procedures.

2) Define the scale, select the size of drawing sheet, and fix the drawing sheet on the drawing board.

3) Draw the border and the title box.

(2) Rough sketches (use the handle of Fig. 1-24 as an example)

1) Procedure for rough sketch: First draw datum lines for drawing positioning, then draw all known segments, followed by intermediate segments and finally connection segments. See Fig. 1-25 for an illustration. Note that the arc of R50 should be produced before drawing the arc of R12.

2) Requirements: Use extra thin lines and the sketches should be accurate, clear and clean.

(3) Darken and make the line bolder

1) Before producing the final drawing, one should check the sketches carefully, correct all mistakes, and erase all unnecessary segments.

2) The procedure is as follows:

① First thick lines and then thin lines: One should first draw continuous thick lines, then draw dashed thin lines, long dashed dotted thin lines and continuous thin lines.

② First curve then straight: One should first produce curve segments and then straight lines.

③ Draw parallel lines from top downwards, then vertical lines from left to right, and finally inclined lines.

④ Dimensioning (as shown in Fig. 1-24).

⑤ Check and fill in the title block.

3) Requirements: Be consistent in drawing different types of segments. The drawing should be produced with smooth connection between elements, tidy lettering and be clean finally.

(1) 已知线段　定形尺寸和定位尺寸标注齐全的线段，称为已知线段。图1-24中R15和R10圆弧都是已知线段。

(2) 中间线段　只给出定形尺寸而定位尺寸不全的线段称为中间线段。图1-24中R50圆弧就是中间线段。

(3) 连接线段　只给出定形尺寸而没有定位尺寸的线段，称为连接线段。图1-24中R12圆弧就是连接线段。

画图时，应先画已知线段，再画中间线段，最后画连接线段。

3. 绘图的一般步骤

(1) 准备工作

1) 分析图形的尺寸与线段，拟订作图步骤。

2) 确定比例，选定图幅，固定图纸。

3) 画出图框和标题栏。

(2) 绘制底稿（以图1-24手柄为例）

1) 画底稿步骤：首先画出作图基准线以确定图形位置，然后依次画出已知线段和中间线段，最后画出连接线段，如图1-25所示。注意，图中两个连接弧中应先画R50圆弧，后画R12圆弧。

2) 要求：图线细淡，图形准确、清晰，图面整洁。

(3) 加深描粗

1) 加深描粗前，要检查底稿，修正错误，擦去多余图线。

2) 加深描粗步骤：

① 先粗后细：先加深全部粗实线，再加深全部细虚线、细点画线及细实线等。

② 先曲后直：描粗加深同一种线型时，应先画圆弧，后画直线。

③ 从上而下画水平线，从左至右画垂直线，最后画斜线。

④ 标注尺寸（见图1-24）。

⑤ 校对，填写标题栏。

3) 要求：同类图线粗细浓淡一致，连接光滑，字体工整，图面整洁。

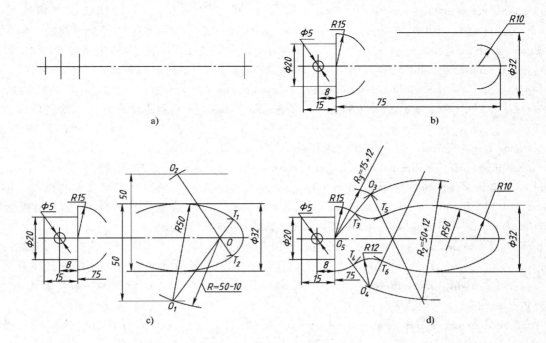

Fig. 1-25 Procedure for preparing rough sketches 画底稿步骤
a) Draw the datum lines 画基准线   b) Draw the known arcs and segments 画已知圆弧和线段
c) Draw the intermediate arcs 画中间圆弧   d) Draw the connection arcs 画连接圆弧

Chapter 2 Basic Knowledge of Orthogonal Projection

One of the important techniques for producing engineering drawings is orthogonal projection. Forms the basis for producing and reading method, which engineering drawings.

2.1 Principles of Orthogonal Projection Method

2.1.1 Construction of projection

In daily life, when the sunlight or lamplight irradiates an object, there will be a shadow of the object on the ground or wall. People abstract the natural phenomena with science to describe an object with projection. The source of light is called the projection center. The light ray is called the projection line. The pre-established plane is called the projection plane. The graphics on the plane is called the projection of the object (see Fig. 2-1). When the projection line passes through the object and the pre-established plane, there will be graphics produced on the plane. This method is called projection method.

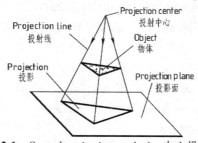

Fig. 2-1 Central projection method 中心投影法

2.1.2 Types of projection method

There are two types of projection methods, i.e. central projection method and parallel projection method.

1. Central projection method

The projection method, in which the projection lines meet at a point, is called central projection method (see Fig. 2-1).

The size of the central projection depends on the distance between the object and the projection plane. It can not show the exact dimension of the object and thus the central projection method is mainly used for producing the pictorial drawings of buildings.

2. Parallel projection method

The projection method, in which the projection lines are parallel with each other, is called parallel projection (see Fig. 2-2). Parallel projection method includes oblique projection method and orthogonal projection method.

第 2 章 正投影基础知识

正投影法是绘制工程图样的重要方法之一，是绘制和阅读工程图样的基础。

2.1 正投影法原理

2.1.1 投影的形成

在日常生活中，当阳光或灯光照射物体时，就会在地面或墙壁上呈现出该物体的影子。人们把这一自然现象加以科学的抽象，提出用投影的方法来表达物体。光源称为投射中心，光线称为投射线，预设的平面称为投影面，在投影面上所得到的图形称为该物体在此平面上的投影（见图2-1）。因此，投影法是投射线通过物体，向选定的平面投射，并在该面上得到图形的方法。

2.1.2 投影法分类

投影法分为两类：中心投影法和平行投影法。

1. 中心投影法

投射线汇交于一点的投影法称为中心投影法（见图2-1）。

中心投影法所得到物体的投影的大小随着物体与投影面距离的变化而变化，一般不能反映物体的实际大小，因此，中心投影法多用于绘制建筑物的直观图。

2. 平行投影法

投射线互相平行的投影法称为平行投影法（见图2-2）。平行投影法又分为斜投影法和正投影法。

（1）Oblique projection method　The projection method, in which the projection lines are inclined to the projection surfaces, is called oblique projection method (see Fig. 2-2a). The projection obtained by oblique method is called oblique projection.

（2）Orthogonal projection method　The projection method, in which the projection lines are orthogonal to the projection plane, is called orthogonal projection method (see Fig. 2-2b). The projection obtained by orthogonal projection method is called orthogonal projection.

Because the orthogonal projection can reflect the true size of the object, it is used in most engineering drawings. For convenience of description, "orthogonal projection" is called projection for short in this book.

（1）斜投影法　投射线倾斜于投影面的投影方法称为斜投影法，如图 2-2a 所示。由斜投影法得到的投影，称为斜投影。

（2）正投影法　投射线垂直于投影面的投影方法称为正投影法，如图 2-2b 所示。由正投影法得到的投影，称为正投影。

由于正投影能够真实地表达空间物体的形状和大小，因此，在工程技术上使用的图样多采用正投影法绘制。为了叙述简便，本书今后把"正投影"简称为投影。

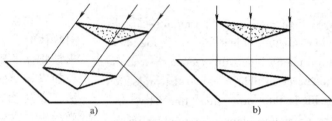

Fig. 2-2　Parallel projection method　平行投影法
a) Oblique projection method　斜投影法　b) Orthogonal projection method　正投影法

2.1.3　Features of orthogonal projection

1. Authenticity

When planar features or lines are in parallel with the projection plane, the projection shows the real shape or true length of the features being projected (see Fig. 2-3a).

2. Accumulation

When planar features or lines are orthogonal to the projection plane, the projection accumulates to a line or a point (see Fig. 2-3b).

3. Similarity

When planar features or lines are inclined to the projection plane in general, the projection has similar shape with the original feature. The length of the lines in the projection will be shorter than the true length of the corresponding lines being projected (see Fig. 2-3c). In similar shapes, the parallelism of lines before and after projection does not change and the connectivity of features does not change either after projection.

2.1.3　正投影的特性

1. 真实性

当平面图形或直线平行于投影面时，其投影反映实形或实长，如图 2-3a 所示。

2. 积聚性

当平面图形或直线垂直于投影面时，其投影积聚成一直线或一点，如图 2-3b 所示。

3. 类似性

当平面图形或直线倾斜于投影面时，平面图形的投影成类似形，线段的投影长度比实长短，如图 2-3c 所示。在类似形中，平面图形投影前后线段的平行性质不变，各边在投影中的连线顺序保持不变。

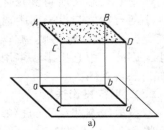 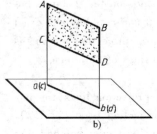 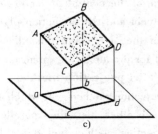

Fig. 2-3　Characteristics of orthogonal projection　正投影的特性
a) Authenticity　真实性　b) Accumulation　积聚性　c) Similarity　类似性

2.1.4 Three-view drawings

A drawing produced through orthogonal projection is called a view of the object. Because one view can only show the shape of the object from two sides, it can not represent the whole object. It is necessary to use multiple views in engineering drawings.

1. Construction of three-view drawings

First, put one object in a projection system with three projection planes that are orthogonal to each other. The object should be placed between the viewer and the projection planes respectively. Orthogonal projection method is used for producing three-view projections. In the three-view projection system, the three projection planes are orthogonal to each other and called H plane, V plane and W plane respectively. The H plane is a horizontal projection plane and is called H-Plane in short. The V plane is a frontal vertical plane and is called V-Plane in short. The W plane is a profile plane and is called W-Plane in short. The intersection lines of the three projection planes are called the projection axes, which are termed as X-axis, Y-axis and Z-axis respectively. These axes meet at a point which is called the origin. The drawing produced on the frontal plane (V-Plane) is called orthographic drawing, also called front view. The drawing produced on the horizontal projection plane (H-Plane) is called horizontal projection, also called top view. The drawing on the profile plane (W-Plane) is called profile projection, also called left-side view (see Fig. 2-4 a).

To draw three-view drawings on a drawing sheet, following the national standard (GB), the V-Plane is first fixed. The H-Plane turns down 90° around the OX-axis along the arrow direction and the W-Plane turns right 90° around the OZ-axis as shown in Fig. 2-4 b. As a result, the H-Plane and the W-Plane coincide with the V-Plane and the front view, the top view and the left view are on the same plane (see Fig. 2-4c). Since the size of the projection planes is not related to the views, it is not necessary to draw the border of the projection planes and the distance between individual views is decided by the breadth of the drawing sheet and the size of the views (see Fig. 2-4d).

2.1.4 物体的三视图

用正投影法绘制的物体正投影，也称为该物体的视图。由于物体的一个视图只能反映物体两个方向的形状，不能完整地表达物体，故在工程图样中多采用多面视图。

1. 三视图的形成

首先把物体放在三个互相垂直的投影面体系中，使其位于人与投影面之间，然后将物体分别向各投影面进行投影。在三个互相垂直的投影面体系中，H 面在水平位置，称为水平投影面（简称水平面）；V 面在正立位置，称为正立投影面（简称正面）；W 面在侧立位置，称为侧立投影面（简称侧面）。三个投影面的交线称为投影轴，分别记作 X 轴、Y 轴和 Z 轴，三个投影轴交于原点 O。在正立投影面（V 面）上所得的图形称为正面投影，亦称为主视图；在水平投影面（H 面）上所得的图形称为水平投影，亦称为俯视图；在侧立投影面（W 面）上所得的图形称为侧面投影，亦称为左视图，如图 2-4a 所示。

为了把三个视图画在一张图纸上，国家标准规定：V 面不动，将 H 面按图 2-4b 所示箭头方向，绕 OX 轴向下旋转 90°；将 W 面绕 OZ 轴向右旋转 90°，使它们都与 V 面重合，这样主视图、俯视图、左视图即可画在同一平面上（见图 2-4c）。由于投影面的大小与视图无关，因此，在画三视图时，可以不画出投影面的边界，视图之间的距离可根据图纸幅面和视图的大小来确定（见图 2-4d）。

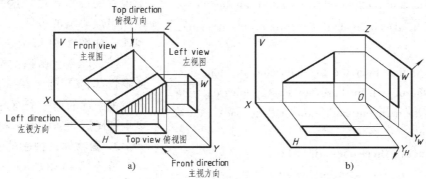

Fig. 2-4 Construction of three-view drawings 三视图的形成

a) The projection of an object in the three projection planes system 物体在三投影面体系中的投影

b) The system with three projection planes will be flattening 将三投影面体系展平

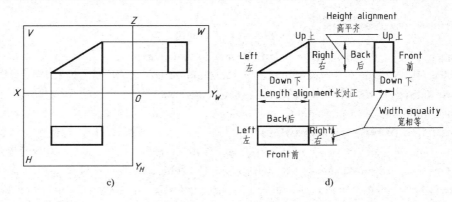

Fig. 2-4 Construction of three-view drawings   三视图的形成（续）
c) The relationship between the location of three-views   三视图的位置关系
d) Three-view drawings of the "three equality" and the relationship between the positions   三视图的"三等"关系及方位对应关系

2. View positions and rules for producing three-view drawings

The relative positions of three-view drawings are as follows: the top view is below the front view and the left view is on the right side of the front view. When the views are arranged following the national standard (GB) as mentioned above, there is no need to label the names of views.

Fig. 2-4 shows that:

The front view shows up/down and left/right relations of drawing features, and it reflects the height and length of the object.

The top view shows left/right and front/back relations, and it reflects the length and width of the object.

The left view shows up/down and front/back relations, and it reflects the height and width of the object.

Based on the above, the following rules can be concluded: The length of the front view and the top view should be aligned and is equal. The height of the front view and the left view should be aligned and is equal. The width of the top view and the left view should also be aligned and is equal. A basic rule that must be observed is therefore "length alignment, height alignment and width equality" when producing and reading drawings and it applies to not only the entire views, but also individual elements on the drawing, as shown in Fig. 2-4d.

It should be noticed that both the top view and the left view can show the width of the object. In these two views, the front side of the object is located far from the front view and the back side is most close to the front view.

3. Steps for producing three-view drawings

First put the object in the three-projection plane system as shown in Fig. 2-5 and, at the same time, choose a direction for front view projection. The front view should show most important features of the object.

2. 三视图的位置关系和投影规律

三视图的位置关系为：俯视图在主视图的正下方；左视图在主视图的正右方。国家标准规定，按这样位置配置视图时，一律不标注视图的名称。

从图 2-4 可以看出：

主视图反映物体上下、左右的位置关系，同时反映物体的高度和长度。

俯视图反映物体左右、前后的位置关系，同时反映物体的长度和宽度。

左视图反映物体上下、前后的位置关系，同时反映物体的高度和宽度。

由此可得出下列投影规律：主视图与俯视图长对正；主视图与左视图高平齐；俯视图与左视图宽相等。"长对正，高平齐，宽相等"是画图和看图必须遵循的投影规律，不仅整个视图的投影要符合这一规律，而且物体组成要素的投影也必须符合这一规律，如图 2-4d 所示。

应该注意，俯视图和左视图都能反映物体的宽度，在这两个视图中，离主视图远的那一个面是物体的前面，离主视图近的那一面是物体的后面。

3. 画物体三视图的步骤

首先将物体放在三投影面体系中，如图 2-5 所示，同时选择最能反映物体形状特征的方向作为主视图的投射方向。

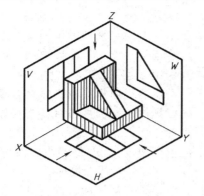

Fig. 2-5　Producing three-view drawings
　　　　三视图的产生

While keeping the position of the object fixed, project the object from three different directions and draw the corresponding views respectively based on the positions and relations of the drawer, the object and the final expected drawings. Further details of individual steps are highlighted in Fig. 2-6.

然后保持物体不动，按人、物、图的关系，从三个不同方向对物体进行投射，而后分别作三视图，具体步骤如图 2-6 所示。

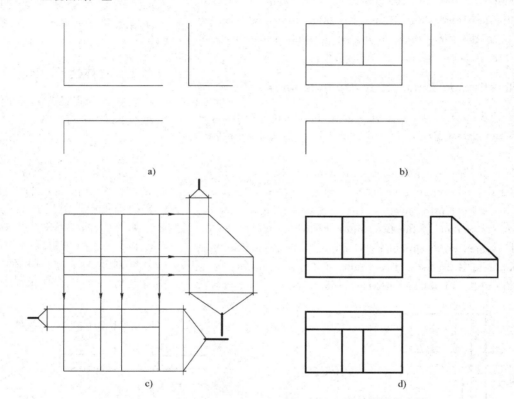

Fig. 2-6　Steps for producing three-view drawings　画三视图的步骤
a) Draw the datum line, noticing that each view is in proper distance　画基准线，注意各视图间距
b) Draw the front view　画主视图　c) Draw the top view, assuring the length alignment and equality; draw the left view, assuring the height alignment and the width equality　按长对正画俯视图，按高平齐、宽相等画左视图
d) Check and correct, then darken all lines according to the demand　检查修正，按要求描深图线

2.2　Projections of Points on a Solid

2.2.1　Projections of points

When drawing a perpendicular line through the point being projected to the three projection planes, there will be three intersection points corresponding to the three orthogonal projections and these intersection points are then the projected points for the corresponding views.

2.2　立体上点的投影

2.2.1　点的投影

点的三面投影就是过点分别向三个投影面所作垂线的垂足。

As an example, the projection of point A is shown in Fig. 2-7. The horizontal projection of the point is represented by its lower case character "a", the front and profile projection of the point are represented with a' and a" respectively.

The projection of a point should also follow the same principle for view projection, this rule is reflected in Fig. 2-7 as: $a'a \perp OX$, $a'a'' \perp OZ$, and the horizontal line going through "a" and the vertical line going through a" meet at a profile line that passes through the origin O with an inclination angle of 45° with the OY-axis. At the same time, orthogonal projection of geometrical element has a dependency property, that is the projection of a point on a line must be on the projected line and the projection of a point on a plane must be on the projected plane. If the point being projected is on the line which is at the same time on the plane, the projected point must be on the projected plane.

2.2.2 Relative positions of two points on a solid

The relative position of two points on a solid is the relative position of left and right (X), front and back (Y), up and down (Z) in the space.

As shown in Fig. 2-8, given projections of two points A and B, we want to analyze the relative positions of the projections of A and B. In the three projections, any two of the projections can show the relative position/location of the two points. Following the front projection, A is located on the upper right side of B. Following the horizontal projection, A is located in front of B. As a result, A is located on the upper, right and front position of B.

立体上点 A 的三面投影如图 2-7 所示。点的水平投影用相应的小写字母 a 表示，点的正面投影和侧面投影分别用 a' 和 a" 表示。

点的三面投影也必然符合视图的投影规律，这种投影规律在图 2-7 中表现为：$a'a \perp OX$；$a'a'' \perp OZ$；过 a 的水平线与过 a" 的铅直线交于过原点 O 的 45°斜线。同时，几何元素的正投影具有从属性，即：直线上的点的投影必在直线的同名投影上；平面上的点的投影必在平面的同面投影上。若点位于平面内的一条直线上，则点必在平面上。

2.2.2 立体上两点的相对位置

立体上两点的相对位置是指这两点在空间的左右（X）、前后（Y）、上下（Z）三个方向上的相对位置。

如图 2-8 所示，已知立体上 A、B 两点的三面投影，分析点 A 相对于点 B 的位置。因在点的三面投影中，任意两面投影均能反映两点的相对位置，如正面投影反映了点 A 在点 B 的右方和上方，水平投影反映了点 A 在点 B 的前方。因此，在空间上，点 A 在点 B 的右、上、前方相应位置处。

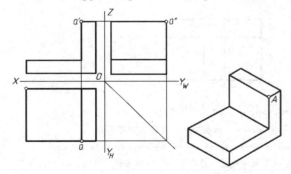

Fig. 2-7 Projection of points on a solid 立体上点的投影

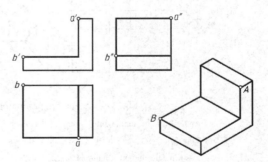

Fig. 2-8 Relative positions of two points on a solid 立体上两点的相对位置

2.2.3 Identification and labeling of coincident points

As shown in Fig. 2-9, when two points on the solid (such as A and D, B and C, C and D) are on the same vertical line of a projection plane, the two projection points will be the same point and the two points are called the coincident point of the projection plane. Therefore, A and D produce a coincident point on the H-Plane, B and C produce a coincident point on the V-Plane, C and D produce a coincident point on the W-Plane.

2.2.3 重影点的判别与标注

如图 2-9 所示，当立体上的两点（如点 A 与点 D，点 B 与点 C，点 C 与点 D）处于某一投影面的同一条垂线时，它们在该投影面上的投影重合为一点，则此两点称为该投影面的重影点。故此，A、D 两点是 H 面的重影点，B、C 两点是 V 面的重影点，C、D 两点是 W 面的重影点。

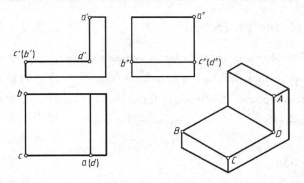

Fig. 2-9 Identification and labeling of coincident points　重影点的判别与标注

A coincident point can be either visible or invisible. The visibility is identified according to the "front covering back, left covering right, up covering down" principle. In Fig. 2-9, all invisible points are written in brackets.

For example, when A and D are projected to the H-Plane, A is above (up) D. A is thus visible, D is invisible. Their labels of the projections in H-Plane are a and (d). When B and C are projected to the V-Plane, C is in front of B. C is thus visible, B is invisible. Therefore their labels of the projections in V-Plane is c' and (b') respectively. When C and D are projected to the W-Plane, C is on the left side of D. C is thus visible, D is invisible. Their labels of the projections in the W-Plane is c'' and (d'').

重影点有可见与不可见之分，重影点的可见性根据"前遮后，左遮右，上遮下"判别，不可见的点用括号括起来，如图2-9所示。

如 A、D 两点在向 H 面投射时，点 A 在点 D 的上方，点 A 可见，点 D 不可见，故在 H 面投影的标记为 a (d)。B、C 两点在向 V 面投射时，点 C 在点 B 的前方，点 C 可见，点 B 不可见，故在 V 面投影的标记为 c' (b')。C、D 两点在向 W 面投射时点 C 在左，点 D 在右，点 C 可见，点 D 不可见，故在 W 面投影的标记为 c'' (d'')。

2.3　Projections of Lines on a Solid

2.3.1　Projections of lines

The projection of a line is a line on the drawing as shown in Fig. 2-10. It can be determined by the line connecting the two end points of the same line.

2.3　立体上直线的投影

2.3.1　直线的投影

直线的投影一般仍为直线，可由直线上两端点同面投影的连线来确定，如图2-10所示。

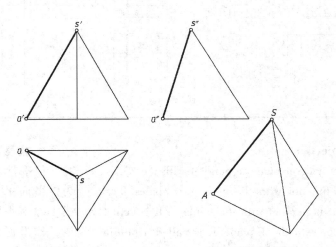

Fig. 2-10　Projection of a general line　直线的投影

2.3.2 Characteristics of line projection

Based on the analysis of the basic characteristics of orthogonal projection, we know that the projected line has three relative positions with respect to the projection plane. This leads to three classes and seven cases of relative positions in a three-view projection system.

1. Vertical lines projection

A vertical line is orthogonal to the projection plane. It must be in parallel with the other two projection planes. It is called a H-perpendicular line if it is vertical to the H-plane; it is called a V-perpendicular line if it is vertical to the V-plane; it is called a W-perpendicular line if it is vertical to the W-plane.

The projection analysis of the vertical lines in regular cases is shown in Table 2-1. The characteristics can be concluded as follows:

1) Its projection on its vertical plane converges to a point.

2) Its projections on the other two planes are both parallel with a projection axis, and show its true length.

2.3.2 直线的投影特性

从正投影的基本特性分析中可知,直线对一个投影面有三种相对位置。因此,直线在三投影面体系中对投影面的相对位置可分为三大类,共七种情况。

1. 投影面垂直线

垂直于一个投影面的直线称为投影面垂直线,它一定和另外两个投影面平行。垂直于 H 面的直线称为铅垂线;垂直于 V 面的直线称为正垂线;垂直于 W 面的直线称为侧垂线。

投影面垂直线的投影分析见表2-1,其投影特性可归纳为:

1)投影面垂直线在所垂直的面上的投影积聚为一点。

2)在另外两个投影面上的投影均平行于一个投影轴,且反映实长。

Table 2-1 Vertical lines projection 投影面垂直线的投影

Name 名称	H-perpendicular line 铅垂线	V-perpendicular line 正垂线	W-perpendicular line 侧垂线
Pictorial drawing 直观图			
Projection drawing 投影图			

2. Parallel lines projection

A parallel line of a projection plane is parallel with the corresponding projection plane but not with the other two planes. It is called a horizontal line if it is parallel with the H-plane. It is called a frontal line if it is parallel with the V-plane. It is called a profile line if it is parallel with the W-plane.

2. 投影面平行线

仅平行于一个投影面的直线,称为投影面平行线。平行于 H 面的直线,称为水平线;平行于 V 面的直线,称为正平线;平行于 W 面的直线,称为侧平线。

The projection analysis of the parallel lines is shown in Table 2-2 and the characteristics can be concluded as follows:

1) The projection of a parallel line shows its true length and the inclination angles formed with the projection axes show the oblique angles formed with the other two projection planes. In Table 2-2, α, β and γ are the oblique angles of the corresponding line formed with the horizontal plane, the frontal plane and the profile plane respectively.

2) The projections on the other two projection planes are shorter than its true length, and are all vertical to the projection axis.

投影面平行线的投影分析见表2-2,其投影特性可归纳为:

1)投影面平行线在所平行的面上的投影反映实长,它与投影轴的夹角反映直线对另外两个投影面的倾角。表2-2中α、β、γ分别为直线对水平面、正面及侧面的倾角。

2)在另外两个投影面上的投影长度比实长短,且共垂直于一个投影轴。

Table 2-2　Parallel lines projection　投影面平行线的投影

Name 名称	Horizontal line 水平线	Frontal line 正平线	Profile line 侧平线
Pictorial drawing 直观图	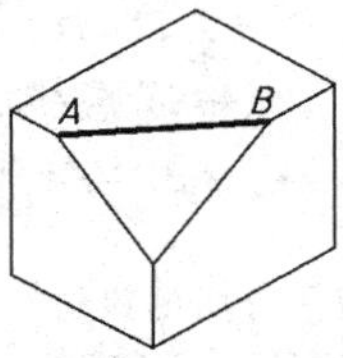	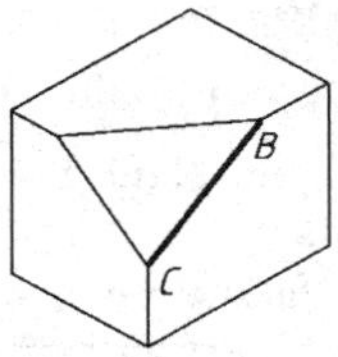	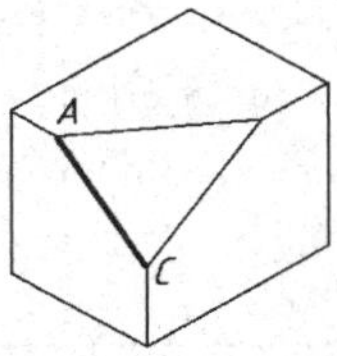
Projection drawing 投影图	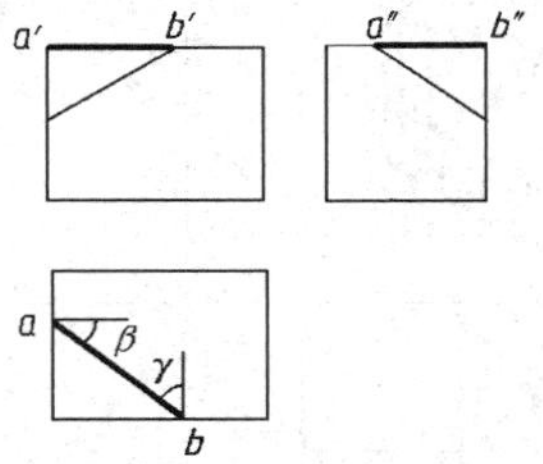	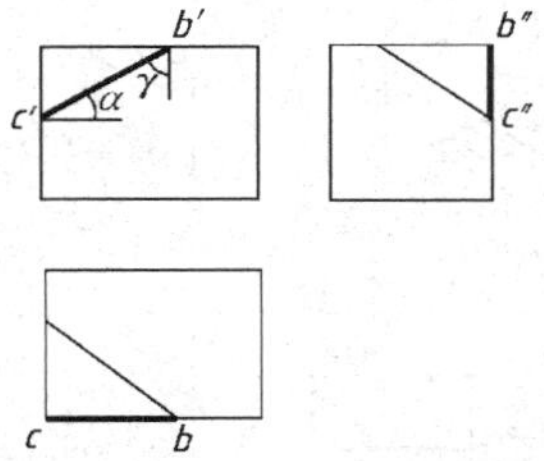	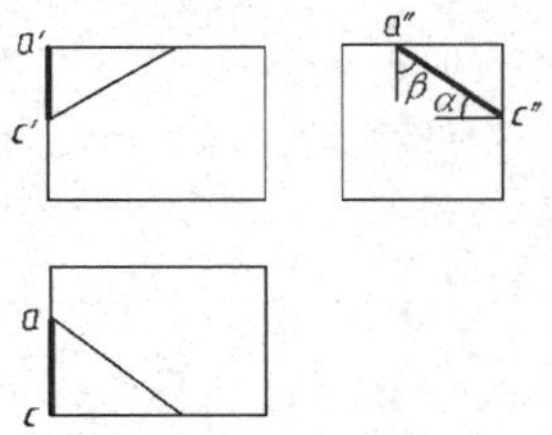

3. General position lines

In this case, the line being projected is oblique with all the three projection planes. In Fig. 2-10, edge SA of the regular triangular pyramid is oblique with the three projection planes. It is therefore a line in general position.

The characteristic of the line in general position is that the three projections are all oblique with the projection axes and shorter than its true length.

2.4　Projections of Planes on a Solid

2.4.1　Projections of planes

Planes of a solid are formed by a number of lines and therefore the projection of planes can be viewed as the projections of multiple lines.

3. 一般位置直线

与三个投影面都倾斜的直线称为一般位置直线。如图2-10所示,正三棱锥中的棱线SA相对于三个投影面都处于倾斜位置,即为一般位置直线。

一般位置直线的投影特性为:三个投影都倾斜于投影轴,且长度都比实长短。

2.4　立体上平面的投影

2.4.1　平面的投影

立体上的平面是由若干条线围成的平面图形,因此,立体上平面的投影就是这些线的投影。

2.4.2 Characteristics of plane projection

The relative positions of a plane in the three-view projection system can be classified into three classes and seven individual cases.

1. Parallel plane projection

In this case, the plane is parallel with the projection plane. It is called a horizontal plane if it is parallel with the H-plane. It is called a frontal plane if it is parallel with the V-plane. It is called a profile plane if it is parallel with the W-plane.

An analysis of parallel planes projection is shown in Table 2-3. The characteristics can be concluded as follows:

1) The projection on its parallel projection plane shows the true shape.

2) The projections on the other two projection planes both converge to a line, and are parallel with the corresponding projection axis that is itself parallel with the plane being projected.

2.4.2 平面的投影特性

平面在三投影面体系中的相对位置有以下三大类，共七种情况。

1. 投影面平行面

平行于投影面的面，称为投影面平行面。平行于 H 面的平面，称为水平面；平行于 V 面的平面，称为正平面；平行于 W 面的平面，称为侧平面。

投影面平行面的投影分析见表 2-3。

投影面平行面的投影特性可归纳为：

1) 投影面平行面在所平行的面上的投影反映实形。

2) 在另外两个投影面上的投影积聚成直线，且分别平行于所平行投影面相应的投影轴。

Table 2-3 Parallel plane projection 投影面平行面的投影

Name 名称	Horizontal plane 水平面	Frontal plane 正平面	Profile plane 侧平面
Pictorial drawing 直观图			
Projection drawing 投影图			

2. Vertical plane projection

In this case, the plane is vertical with one of the projection planes and oblique with the other two projection planes. The plane is called a H-perpendicular plane if it is only vertical with the H-plane. It is called a V-perpendicular plane if it is only vertical with the V-plane. It is called a W-perpendicular plane if it is only vertical with the W-plane.

The analysis of vertical planes projection is shown in Table 2-4. The characteristics of vertical planes projection can be concluded as follows:

1) The projection on its orthogonal plane converges to a line which is oblique with the projection axes. The inclination angles between the projected line and the projection axes show the oblique angles between the vertical plane and the two other projection planes.

2) The projections on the other two projection planes are similar to its true shape.

2. 投影面垂直面

仅垂直于一个投影面，相对其他两个投影面倾斜的平面，称为投影面垂直面。仅垂直于 H 面的平面称为铅垂面；仅垂直于 V 面的平面称为正垂面；仅垂直于 W 面的平面称为侧垂面。

投影面垂直面的投影分析见表 2-4。

投影面垂直面的投影特性可归纳为：

1) 投影面垂直面在所垂直的面上的投影积聚成与投影轴倾斜的直线段；该直线段与两投影轴的夹角反映该平面对相应两投影面的倾角。

2) 在另外两个投影面上的投影为类似形。

Table 2-4　Vertical plane projection　投影面垂直面的投影

Name 名称	H-perpendicular 铅垂面	V-perpendicular 正垂面	W-perpendicular 侧垂面
Pictorial drawing 直观图			
Projection drawing 投影图			

3. General position planes

In the views of a triangular pyramid (see Fig. 2-11), the plane △ASB is oblique with the three projection planes. Such a plane is called a general position plane.

The main characteristic of a general position plane is that the three projections are all similar to its true shape.

3. 一般位置平面

如图 2-11 所示，在三棱锥的视图中，棱面△ASB 与三个投影面都处于倾斜位置，这类平面称为一般位置平面。

一般位置平面的主要投影特性为：三面投影都为类似形。

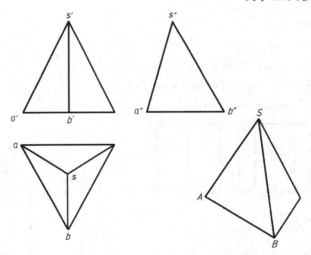

Fig. 2-11　The projection of a plane in general position　一般位置平面的投影

Chapter 3 Solids and Their Intersections 第 3 章 立体及其交线

Solids are surrounded by surfaces (planes and curved surfaces). The solids that are surrounded by planes are called polyhedral solids. The solids that are surrounded by curved surfaces and planes or curved surfaces are called curved or free-form solids.

立体由若干面（平面、曲面）所围成，完全由平面围成的立体称为平面立体；由曲面和平面或完全由曲面围成的立体称为曲面立体。

3.1 Projections of Polyhedral Solids

3.1 平面立体的投影

The most common polyhedral solids are prisms and pyramids. They are surrounded by side planes and bases. Intersection lines of side planes are called edge lines. Intersection lines of side planes and bases are called base edge lines. When drawing projections of polyhedral solids, we are actually drawing the projections of all edge lines and base edge lines followed by a visibility test of the projected edge lines and base edge lines.

最常见的平面立体有棱柱和棱锥，它们由棱面和底面所围成，各棱面的交线称为棱线，棱面与底面的交线称为底边。画平面立体的投影图，实质是画出所有棱线和底边的投影，并判别其可见性。

3.1.1 Prism

3.1.1 棱柱

1. Projection of a prism

1. 棱柱的投影

A regular hexagonal prism, as shown in Fig. 3-1, is surrounded by both upper and lower hexagon bases and six side planes (rectangles). Both the upper and lower bases are horizontal planes. Both the front and back planes are frontal planes.

如图 3-1 所示，正六棱柱由上、下正六边形和六个棱面（矩形）所构成，上下两平面为水平面，前后两平面为正平面。

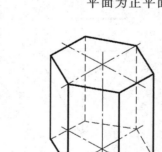
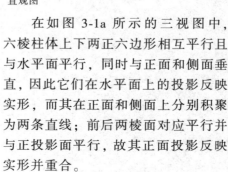

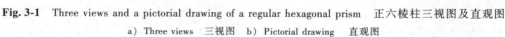

Fig. 3-1 Three views and a pictorial drawing of a regular hexagonal prism   正六棱柱三视图及直观图
a) Three views 三视图 b) Pictorial drawing 直观图

As shown in Fig. 3-1a, since the upper and lower hexagon bases of the regular hexagonal prism are parallel to the H-plane and perpendicular to the V-plane and W-plane, the projections show true shapes (right hexagon) and coincide on the top view and the projections are two lines (accumulation lines) on the front and left views. Since the front and back side planes of the regular hexagonal prism are parallel to the V-plane and perpendicular to the H-plane and W-plane, the projections show true shapes (rectangle) and coincide on the front view.

在如图 3-1a 所示的三视图中，六棱柱体上下两正六边形相互平行且与水平面平行，同时与正面和侧面垂直，因此它们在水平面上的投影反映实形，而其在正面和侧面上分别积聚为两条直线；前后两棱面对应平行并与正投影面平行，故其正面投影反映实形并重合。

Since the other side planes of the regular hexagonal prism are perpendicular to the H-plane and inclined to the other two, their projections show similar shapes (rectangles) on the front and left views and are lines (accumulation lines) on the top view. The six-side planes of the regular hexagonal prism accumulate as the six lines of the right hexagon on the top view.

Drawing steps for producing the three basic views of a regular hexagonal prism are shown in Fig. 3-2.

The projection characteristics of prisms are as follows: If the edge lines of the prisms are perpendicular to a projection plane, they will show a polygon and true shapes of the upper and lower bases on that view. The view on the projection plane that is parallel to the edge lines of the prisms shows a rectangle or a combination of several rectangles, and the projections of the edge lines are the common edge lines of these rectangles.

而六个棱面与水平投影面垂直，则其水平投影积聚为六条直线重叠在六边形的六条边上。同理在侧面投影上，前后棱面积聚为直线，其他棱面均为类似形。

正六棱柱三视图的画图步骤如图 3-2 所示。

正棱柱的投影特点是：在垂直于棱线的投影面上的投影为多边形，反映上底面和下底面的实形。在平行于棱线的投影面上的投影为一个矩形或多个矩形的组合，且棱线的投影为组合矩形的公共边。

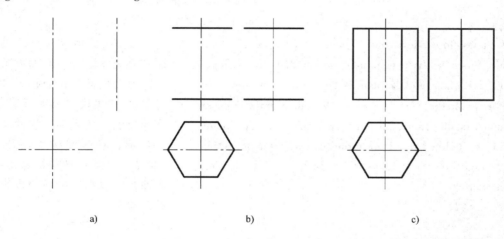

Fig. 3-2 Drawing procedure of three views of a regular hexagonal prism 正六棱柱三视图的作图过程
a) Add centerlines and symmetry lines 画出中心线和基准线 b) Add lines for the upper and lower faces 画出顶面和底面的实形俯视图
c) Complete the three views 完成三视图

2. Getting points on the surfaces of a prism

Getting points on the surfaces of a prism can be based on the accumulation characteristics of plane projection. As shown in Fig. 3-3, point A is on a regular triangular prism. Since projection a' of point A on the front view is known, its projections a and a'' on the top view and left view respectively, could be obtained. One may first draw the projection a of point A on the top view on basis of accumulation. According to the projections a and a' of point A on the top and front views, one may finally draw the projection a'' of point A on the left view. Since the side plane containing point A is not visible on the left view, the projection a'' of point A is also invisible.

2. 棱柱表面上取点

在棱柱的表面上取点时，可利用表面的积聚性进行作图。如在图 3-3 所示的正三棱柱的表面上有一点 A（位于右侧棱面上），已知其正面投影 a'，需求它的水平投影和侧面投影。利用棱面的水平投影有积聚性，求出点 A 的水平投影 a，再根据 a 和 a' 求出侧面投影 a''。由于右侧棱面的侧面投影为不可见，所以 a'' 也不可见。

3.1.2 Pyramid

1. Projection of a pyramid

Pyramids are surrounded by one base and several side planes. Each side plane is a triangle. All edge lines intersect at the same point, called the vertex of a pyramid.

3.1.2 棱锥

1. 棱锥的投影

棱锥由底面和棱面所围成，各棱面都是三角形，各棱线均相交于同一点，此点即棱锥的顶点。

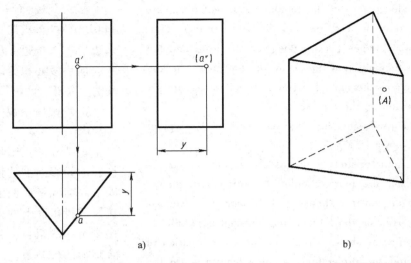

Fig. 3-3 Getting a point on the surface of a regular triangular prism 正三棱柱表面上取点的方法
a) Three views 三视图 b) Pictorial drawing 直观图

Fig. 3-4 shows three views and a pictorial drawing of a right square pyramid. Since its base is parallel to the H-plane and perpendicular to the V-plane and W-plane, The projection shows true shape (a square) on the top view while the other projections are two line segments on the front and left views. Since the side planes of the pyramid are inclined to all the three planes, their projections show similar triangles on these views. All edge lines of the pyramid intersect at the vertex.

图 3-4 所示是一正四棱锥的投影图和直观图。它的底面放置成水平位置，所以底面的正面投影和侧面投影都是水平线段；而水平投影是正方形，反映它的实形。它的各棱面都是三角形，且各棱线均相交于同一点（顶点）；由于各棱面都是一般位置平面，所以它们的各个投影都是类似的三角形。

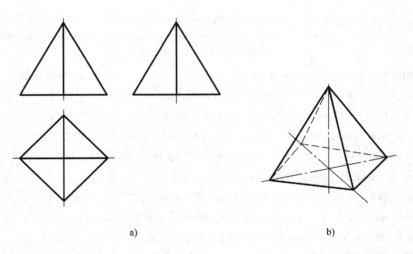

Fig. 3-4 Three views and a pictorial drawing of a right square pyramid 正四棱锥三视图及直观图
a) Three views 三视图 b) Pictorial drawing 直观图

Drawing steps are shown in Fig. 3-5: Draw the projection of the base of the pyramid (usually first draw the projection that shows the true shape) and then the projection of the vertex and edge lines. Finally, test the visibility of the edge lines.

画棱锥投影图的步骤如图 3-5 所示：一般也是先画底面的投影（先画反映实形的那个投影），然后画顶点的投影和各棱线的投影，最后判别其可见性。

Projection characteristics of pyramids: When bases of pyramids are parallel to the *H*-plane, their projections show true shapes (polygons) on the top view and the other projections are horizontal line segments on the front and left views. Projections of all edge lines intersect the projection of the vertex on the same view.

棱锥的投影特点是：当底面平行于水平面时，底面在该投影面上的投影是多边形，反映实形，在其他投影面上的投影均积聚成直线段。各棱线的同面投影均相交于顶点的相应投影。

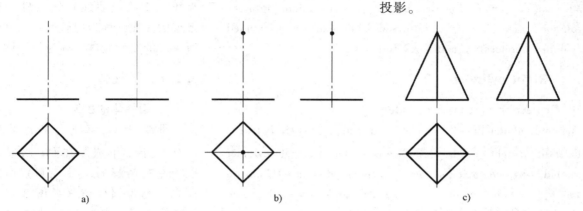

Fig. 3-5 Producing three views of a right square pyramid 正四棱锥三视图的作图过程
a) Draw symmetry lines and the base of the pyramid 画出对称线、基准线和锥的底面投影
b) Add the vertex of the pyramid 确定棱锥顶点的投影
c) Complete three views after adding lines for the side surfaces 画出各棱线的投影并完成全图

2. Getting points on the surfaces of a pyramid

As shown in Fig. 3-6, point *B* is on the back side plane. Projection *b* of the point *B* is known on the top view. Draw an auxiliary line *S*Ⅰ going through the point *B* on the pyramid.

Drawing steps: Draw projection *s*1 of the auxiliary line *S*Ⅰ going through the point *b* on the top view. Draw projections *s'*1' and *s"*1" of the auxiliary line and then projection *b'* of point *B* could be obtained on the basis of projection principle. Following the projection *b'* of point *B* on the front view, projection *b"* of point *B* could also be obtained. Since the point *B* is on the back plane of the quadrangular pyramid, the projection *b'* of point *B* is invisible, shown as (*b'*).

2. 棱锥表面取点

如图 3-6 所示，已知后棱面上点 *B* 的水平投影，补画其余投影。在后棱面上过锥顶作一条包含点 *B* 的辅助线 *S*Ⅰ。

作图步骤是：过水平投影 *b* 作辅助线 *S*Ⅰ 的水平投影 *s*1，然后作出该辅助线的正面投影 *s'*1' 和侧面投影 *s"*1"，根据线上求点的原理，分别按点的投影规律求出点 *B* 的正面投影 *b'* 和侧面投影 *b"*。因点 *B* 在后棱面上，故 *b'* 为不可见，用 (*b'*) 表示。

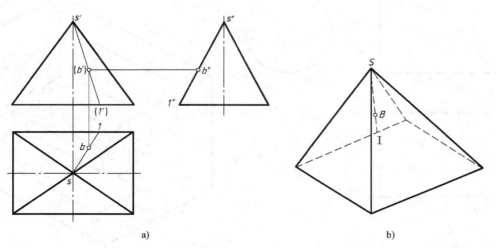

Fig. 3-6 Methods of getting a point on the surface of a quadrangular pyramid 正四棱锥表面上取点的方法
a) Three views 三视图 b) Pictorial drawing 直观图

3.2 Projections of Curved Solids

In engineering applications, commonly used curved solids are produced by revolution, such as cylinder, cone, sphere and so on. When an element (a line or a curve) revolves around an axis, a revolution surface is produced. Curved solids that are surrounded by revolution surfaces are called revolution solids.

3.2.1 Right cylinder

1. Projection of a right cylinder

When generator line A_1A_2 revolves around an axis O_1O_2 that is parallel to it, a right cylindrical surface is produced. A right cylinder is surrounded by one curved surface (a cylindrical surface) and two end bases, As shown in Fig. 3-7. When its axis is perpendicular to the W-plane, it will be projected as a circle on the projection plane. The profile of the circle is the profile projection of the curved surface and has accumulation. The circle is also the profile projection of the cylinder bases. The cylinder bases are perpendicular to the vertical (V) and horizontal (H) projection planes and are projected as line segments on both the front and top views. The right cylinder is projected as a rectangle with the same size and shape on both the front and top views.

As shown in Fig. 3-7, when the right cylinder is projected to the H-plane, elements A_1A_2 and B_1B_2 (generator lines at different positions are called elements of the cylindrical surface) define the upper and lower limits of horizontal projection of the right cylinder. The front and back lines A_1A_2 and B_1B_2 are called limit elements for the top view. The limit elements are called the outlines of the top view.

3.2 曲面立体的投影

工程中常见的曲面立体如圆柱体、圆锥体、圆球等是回转体。一条母线（直线或曲线）绕一轴线回转形成的曲面，称为回转面；由回转面所围成的曲面立体，称为回转体。

3.2.1 圆柱体

1. 圆柱体的投影

母线 A_1A_2 绕与它平行的轴线 O_1O_2 旋转，形成圆柱面。圆柱面及位于左右两端的底面（圆）所围成的立体称为圆柱体（见图3-7）。由于轴线垂直于侧面，所以圆柱体的侧面投影为一个圆，这个圆的圆周是圆柱面的侧面投影，有积聚性；该圆也是圆柱底面和顶面的侧面投影。圆柱底面的正面投影和水平投影均积聚成铅垂方向的直线段。圆柱体的正面投影和水平投影是形状和大小相同的矩形。

如图3-7所示，将圆柱体向 H 面投影时，圆柱面上最前和最后的素线 A_1A_2、B_1B_2（母线的各个不同位置称为圆柱面的素线）的投影确定了圆柱面水平投影的前后范围，素线 A_1A_2、B_1B_2 称为对 H 面的转向轮廓线，简称对 H 面的轮廓线。

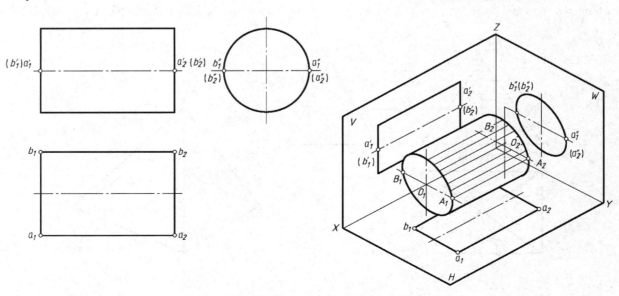

Fig. 3-7 Three views of a cylinder   圆柱的三视图

Limit elements of a revolution surface have the following characteristics:

1) Limit elements are associated with the direction of projection. The elements A_1A_2 and B_1B_2, e.g., are limit elements for the top view. The frontal projections $a'_1a'_2$ and $b'_1b'_2$ of the elements A_1A_2 and B_1B_2, respectively, coincide with the frontal projection of the axis. The profile projections $a''_1a''_2$ and $b''_1b''_2$ of the elements A_1A_2 and B_1B_2 respectively, converge to two points. Since projections $a'_1a'_2$ and $a''_1a''_2$ of the element A_1A_2 do not define projective limits of the frontal and profile projections, A_1A_2 is not limit element for the front view and left view. Therefore, the frontal and profile projections of A_1A_2 should not be drawn.

2) Limit elements are usually form the boundary between visible and invisible regions of the corresponding projection direction. The elements A_1A_2 and B_1B_2, e.g., are visible and invisible demarcation lines of the right cylinder on the top view.

2. Getting points on the surfaces of a right cylinder

As shown in Fig. 3-8, there are three points A, B and C on the surface of the right cylinder. Given the projections a' and b' of points A and B respectively, on the front view and the projection c of point C on the top view. We want to find out the other projections a and b on the top view, a'', b'' and c'' on the left view, and c' on the front view.

Since the cylinder is put vertically, the axis is perpendicular to the V-plane. The horizontal projection of the right cylinder has accumulation on the top view. The horizontal projections a and b of points A and B could be obtained on the circle following the projection principle and accumulation. The profile projections a'' and b'' of points A and B could then be obtained. Since point A is on the left half cylinder, the profile projection a'' of point A is visible. Also since point B is on the right half cylinder, the profile projection b'' of point B is invisible, shown as (b'') in brackets. Point C is located on the upper base of the cylinder. The front projection c' and profile projection c'' could be easily obtained following the projection principle.

回转面的轮廓有以下特点：

1）转向轮廓线是相对于某投射方向而言。例如 A_1A_2 和 B_1B_2 是对 H 面的轮廓线，它们的正面投影 $a'_1a'_2$、$b'_1b'_2$分别与轴线的正面投影重合，它们的侧面投影 $a''_1a''_2$、$b''_1b''_2$ 分别积聚成一点。由于$a'_1a'_2$、$a''_1a''_2$并不确定圆柱体正面投影和侧面投影的范围，所以对正面和侧面而言，它们不是轮廓线。因此 A_1A_2 的正面投影和侧面投影均不必画出。

2）转向轮廓线通常是回转面上对某投射方向可见部分与不可见部分的分界线。例如素线 A_1A_2 和 B_1B_2 就是圆柱面水平投影的可见部分（上半部）与不可见部分（下半部）的两条分界线。

2. 圆柱体表面取点

如图 3-8 所示，圆柱体表面上有三点 A、B、C，已知点 A、B 的正面投影和点 C 的水平投影已知，求作它们的另两面投影。

由于圆柱体为竖立放置，轴线垂直于水平面，圆柱体的水平投影具有积聚性，点 A、B 的水平投影可直接按点的投影规律和积聚性求得，再由此求得其侧面投影。因点 A 位于左半个圆柱体，其侧面投影是可见的，而点 B 位于右半个圆柱体，其侧面投影为不可见，用（b''）表示。点 C 位于圆柱体上底面，直接按点的投影规律求得其正面投影 c' 和侧面投影 c''。

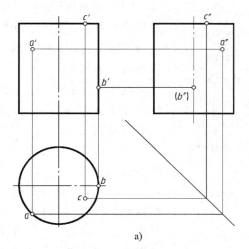
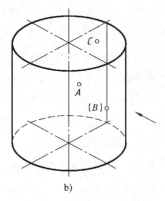

a)   b)

Fig. 3-8 Method of getting points on the surface of a cylinder   圆柱表面上取点的方法
a) Three views   三视图   b) Pictorial drawing   直观图

3.2.2 Cone

1. Projection of a cone

When an element (a line) SA revolves around an axis SO that intersects with the element (or line) SA, we define the curved surface as the revolution surface of a cone. As shown in Fig. 3-9, a cone is surrounded by one curved surface and a lower base. When the axis is perpendicular to the H-plane, the cone is projected as a circle on the top view. This circle is the horizontal projection of the cone and the horizontal projection of the base plane. The frontal and profile projections of the base accumulate to two horizontal line segments. The frontal and profile projections of the cone are isosceles triangles with the same shape and size. The sides of the isosceles triangles are limit elements for the front and left views.

3.2.2 圆锥体

1. 圆锥体的投影

母线 SA 绕与它相交的轴线 SO 旋转，形成圆锥面，它与底面（圆）围成圆锥体（见图 3-9）。当轴线垂直于 H 面时，该圆锥体的水平投影是一个圆，这个圆既是圆锥体的水平投影，也是底面的水平投影。底面的正面投影和侧面投影均积聚成水平直线段。圆锥体的正面投影与侧面投影是形状和大小相同的等腰三角形，该等腰三角形的腰就是圆锥体正面投影和侧面投影的转向轮廓线。

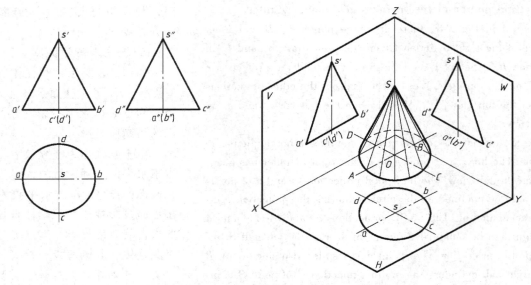

Fig. 3-9 Three views of a cone　圆锥的三视图

2. Getting points on the surfaces of a cone

（1）Auxiliary element method　First draw an auxiliary element containing the projected point on the cone surface. Then draw the three views of the auxiliary element. The required projection of the given point can be obtained by drawing the three views of the given point on the auxiliary element (see Fig. 3-10a).

As shown in Fig. 3-10a, given the frontal projection m' of point M on the cone surface, we want to draw projections m, m'' of point M in other views.

Drawing steps: Draw three views of an auxiliary element SA going through the point M and the apex of the cone (see Fig. 3-10a). Draw the three views of the point M on the auxiliary element following the projection principle.

2. 圆锥体表面取点

（1）辅助素线法　在圆锥上通过所求点作一条素线，先求该素线的三面投影，再求出该素线上此点的投影即可，如图 3-10a 所示。

在图 3-10a 中，已知圆锥上一点 M 的正面投影 m'，求其水平投影 m 和侧面投影 m''。

作图步骤：过锥顶作一条包含点 M 的辅助素线 SA，作出其三面投影 $s'a'$、sa、$s''a''$，根据求线上一点的投影原理，点 M 在 SA 上，必定也在此圆锥面上，故可由 m' 在 sa、$s''a''$ 上分别按点的投影规律求出 m、m''，如图 3-10a 所示。

(2) Auxiliary circle method Given the frontal projection R' of point K, first draw a horizontal plane containing point K on the cone surface. The plane will intersect with the cone and the intersection is an auxiliary circle containing point K. Then draw three views of the auxiliary circle. The projections of the point K can be obtained by drawing the three views of the point K (see Fig. 3-10b).

Drawing steps: Draw a horizontal plane containing point K and intersecting with the cone. The intersection will be an auxiliary circle containing point K on the cone surface. This auxiliary circle is parallel to the base. Draw a horizontal line a'b' through the projection k', and the line a'b' is the frontal projection of the auxiliary circle. Draw the horizontal projection of the auxiliary circle at the radius of a'b'/2 centered around the point s (see Fig. 3-10b). Complete the three views of the auxiliary circle and then point K on the cone surface. Since point K is on the right half surface of the cone, projection k'' of the point K is invisible, shown as (k'') in brackets.

（2）辅助圆法 在圆锥面上，平行于底面的任意圆称为纬圆。点 K 的正面投影 k' 已知，过点 K 作一辅助纬圆，先求纬圆的投影，再求点 K 其他两面投影，如图 3-10b 所示。

作图步骤：过点 K 在圆锥面上作一平行于底面的纬圆，先在正面投影过 k' 作一水平线 a' b' 即得到辅助纬圆的正面投影，再作此纬圆的水平投影，即以 s 为圆心以 a' b'/2 长为半径作圆（见图 3-10b）。根据可见性，由 k' 求得 k，再由 k' 和 k 求得 k''。由于点 K 在圆锥面的右半部分上，所以 k'' 不可见，用（k''）表示。

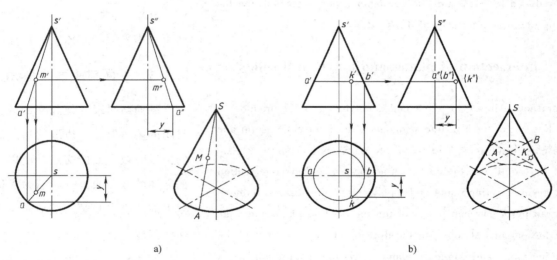

a)　　　　　　　　　　　　b)

Fig. 3-10 Method of getting points on the surface of a cone　圆锥表面上取点的方法
a) Element method 素线法　b) Circle method 纬圆法

3.3 Intersection of Planes and Solids

When a plane intersects surfaces of a solid, a definite edge is formed. The definite edge is called the intersection of the plane and the solid. The plane is called an intersecting plane. The planar area surrounded by the intersection (loops of lines or curves) is called a section (see Fig. 3-11). Common seen engineering cut-solids are shown in Figure 3-12.

3.3.1 Characteristics of intersection line

1) Segments of an intersection line are common segments to both the intersecting plane and surfaces of the solid. Every point on the segments of the intersection line is a common point of the intersecting plane and surfaces of the solid.

3.3 平面与立体相交

平面与立体表面相交时的交线，称为截交线，这个平面称为截平面，由截交线围成的平面图形称为断面，如图 3-11 所示。工程上常见的截切体如图 3-12 所示。

3.3.1 截交线的特性

1) 截交线是截平面和立体表面的共有线，截交线上任一点都是截平面和立体表面的共有点。

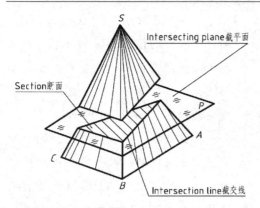

Fig. 3-11　Intersection of planes and solids
平面与立体相交

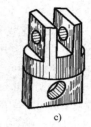

Fig. 3-12　Cut-solids　截切体
a) Vee block　V形块　b) Ball bearing　球轴承　c) Cross joint　十字接头

2) Since surfaces of a solid form a closed volume, segments of the intersection line also form a closed planar area.

Based on the above properties, drawing of an intersection line can be reduced to a task for finding common points on both the intersecting plane and surfaces of the solid.

3.3.2　Intersection of planes and polyhedral solids

Intersection line of a plane and a polyhedral solid produces a loop of line segments. The line segments of an intersection form one or more closed polygons. Vertices of the polygons are points of edge lines of the solid and the intersecting plane. To draw intersection lines of an intersecting plane and a polyhedral solid can be reduced to the drawing of common points of the intersecting plane and edge lines of the polyhedral solid, or the drawing of common lines of the intersecting plane and boundary planes of the polyhedral solid.

Example 3-1　As shown in Fig. 3-13 a, a triangular pyramid is cut by a V-perpendicular plane P. Draw projections of the intersection line of the triangular pyramid and the cutting plane P.

Analysis：Since the triangular pyramid is cut by a V-perpendicular plane P, the intersection segment forms a triangle. To draw the triangle, one can draw three common points Ⅰ, Ⅱ, Ⅲ on the three edge lines of the triangular pyramid and the intersecting plane.

Drawing steps：

1) According to the accumulation of plane p' which intersects with the three edges of the triangular pyramid, figure out the projections $1'$, $2'$, $3'$ of the intersected points Ⅰ, Ⅱ, Ⅲ on the front view (see Fig. 3-13b).

2) Draw projections 1, 2, 3 and $1''$, $2''$, $3''$ of points Ⅰ, Ⅱ, Ⅲ on the top and left views respectively.

2) 由于立体表面占有一定的空间范围，所以截交线是封闭的平面图形。

根据以上特性，求作截交线可归结为求作平面与立体表面共有点。

3.3.2　平面与平面立体相交

平面与平面立体相交的截交线是封闭的多边形，多边形的顶点是立体的棱线与截平面的交点。所以，求平面立体的截交线，可归结为求棱线和截平面的交点问题，或平面与平面的交线问题。

例3-1　已知三棱锥被正垂面所截切，求截交线的投影（见图3-13a）。

分析：由于三棱锥被正垂面P截切，其截交线为一三角形，要求此三角形的投影，只需求出截平面与三棱锥的三条棱线的交点即可。

作图步骤：

1) 利用正垂面的积聚性投影p'，求得p'与三棱锥三条棱线交点Ⅰ、Ⅱ、Ⅲ的正面投影$1'$、$2'$、$3'$（见图3-13b）。

2) 然后求得水平投影1、2、3，以及侧面投影$1''$、$2''$、$3''$。

3) Join the projections 1, 2, 3 and 1″, 2″, 3″ of points Ⅰ, Ⅱ, Ⅲ at the intersection line on the top view and left view with straight lines (see Fig. 3-13c).

3）分别依次连接 1、2、3 和 1″、2″、3″，即可得到截交线上 Ⅰ、Ⅱ、Ⅲ 点在水平投影面和侧面投影面上的投影（见图 3-13c）。

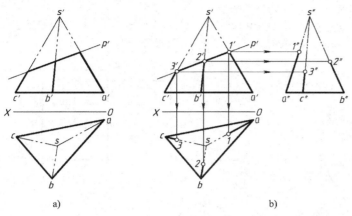
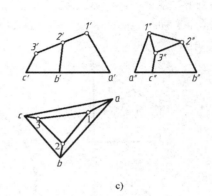

Fig. 3-13 Intersecting a triangular pyramid 截切三棱锥

3.3.3 Intersection of planes and curved solids

Intersection line of a plane and a curved solid usually produces a closed planar curve. In special cases, the intersection is a plane polygon. To draw the intersection line of a plane and a curved solid can be reduced to draw a series of common points on the intersecting plane and surfaces of the curved solid.

1. Intersection of planes and right cylinders

While a right cylinder is intersected by a plane, there are three possible situations depending on the relative position of the intersecting plane and the axis of the right cylinder (see Table 3-1).

3.3.3 平面与曲面立体相交

平面与曲面立体相交的截交线一般是封闭的平面曲线，特殊情况下为平面多边形。求平面与曲面立体截交线可归结为求平面与曲面立体上一系列素线或纬圆与截平面的交点问题。

1. 平面与圆柱体相交

平面截切圆柱体时，根据截平面与圆柱轴线的相对位置可分为三种情况（见表 3-1）。

Table 3-1 Intersection lines of a plane and a right cylinder 平面与圆柱体相交的截交线

Description 说明	Parallel intersecting plane with respect to the axis 截平面平行于轴线	Perpendicular intersecting plane with respect to the axis 截平面垂直于轴线	Inclined intersecting plane with respect to the axis 截平面倾斜于轴线
Pictorial drawing 直观图			
Drawing projection 投影图			

Example 3-2 Draw three views with projections of the intersection line between a right cylinder and a V-perpendicular plane (see Fig. 3-14).

Analysis: When the right cylinder is intersected by the plane, according to Table 3-1, the profile projection of the intersection line is an ellipse in general on the cylindrical surface. The ellipse is projected as a straight line on the front view and a circle on the top view. The main task for this example is to draw the projection of the ellipse on the left view.

例 3-2 求一正垂面截切圆柱体后截交线的投影（见图 3-14）。

分析：圆柱表面被正垂面斜切后，对照表 3-1 可知，其截交线一般为一个椭圆，该截交线的正面投影积聚为一条直线，水平投影积聚在圆周上，本例题主要是求出椭圆截交线的侧面投影。

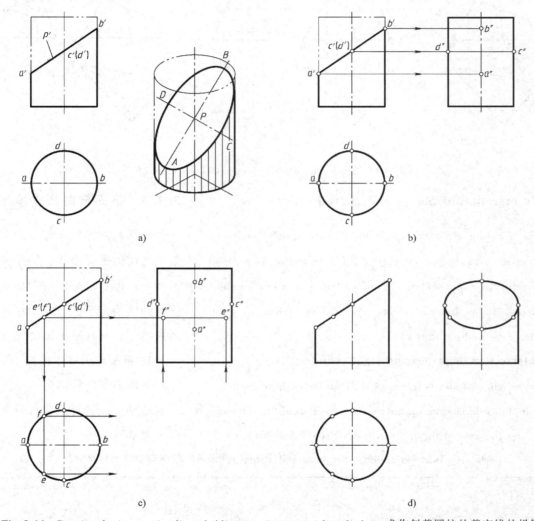

Fig. 3-14 Drawing the intersection line of oblique section on a right cylinder　求作斜截圆柱的截交线的投影
a) Analyze the inter section line　分析截交线　b) Draw special points　找特殊位置的点
c) Draw general points　找一般位置的点　d) Complete the projection　完成投影

Drawing steps:

1) Draw special points. Special points A and B are on the limit elements of the cylinder for the front view. Special points C and D are on the limit elements of the cylinder for the left view. The point A and B are also the end points of the major axis of the ellipse. The point C and D are the end points of the minor axis of the ellipse (see Fig. 3-14a and b). Draw projections a'', b'', c'', d'' of the special points A, B, C, D on the left view, following the projections a, b, c, d and a', b', c', d' of the special points A, B, C, D on the front and top views.

作图步骤：

1) 作特殊点。特殊点 A 和 B 是相对于主视图的转向轮廓线上的共有点，同时也是椭圆截交线长轴端点。特殊点 C 和 D 是相对左视图的转向轮廓线上的共有点，同时也是椭圆截交线短轴端点，如图 3-14a、b 所示。根据正面投影 a'、b'、c'、d' 和水平投影 a、b、c、d，可以求得侧面投影 a''、b''、c''、d''。

2) Draw general points. For producing an accurate projection of the intersection line, one may increase the number of general points. To get projections of general points *E*, *F* on the left view (Fig. 3-14c), first mark the coincident projections *e'* (*f'*) of *E*, *F* on the front view, and then draw projections *e*, *f* and *e"*, *f"* of *E*, *F* on the top and left views.

3) Complete the projection line of the ellipse on the left view by smoothly connecting the projected points after finding enough common points projections on the left view (see Fig. 3-14d).

Example 3-3 Draw three views with incision projections of the locating shaft, shown in Fig. 3-15a.

2）作一般位置点。为了尽可能保证截交线的准确性，可以增加一般点的数量。如图 3-14c 所示，求作一般点 *E* 和 *F* 的侧面投影 *e"*、*f"*，先在截交线的正面投影上任取一对重影点的正面投影 *e'*(*f'*)，然后再作出点 *E*、*F* 在俯视图和左视图上的投影 *e*、*f* 及 *e"*、*f"*。

3）完成椭圆截交线的侧面投影。在画出足够的共有点的侧面投影后，光滑连接各点，即完成作图（见图 3-14d）。

例 3-3 画出图 3-15a 所示定位轴切口的三视图。

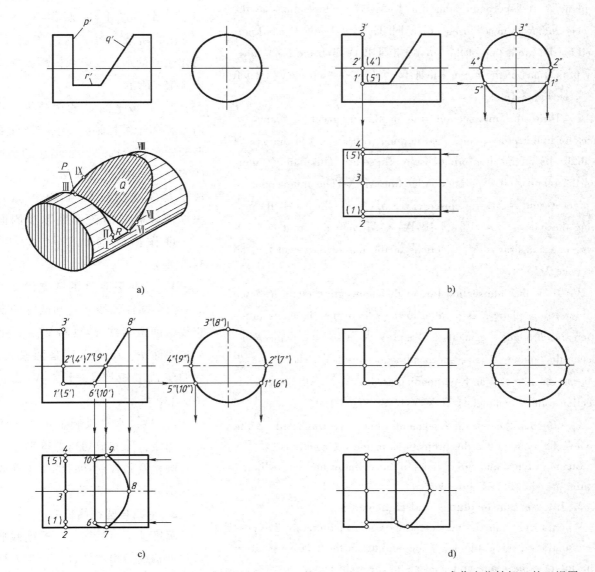

Fig. 3-15 Solution of drawing three views with incision projections of the locating shaft 求作定位轴切口的三视图
a) Analyze the intersection line 分析截交线 b) Draw the intersection line of the intersecting plane *P* 作 *P* 平面的截交线
c) Draw the intersection line of the intersecting plane *Q* 作 *Q* 平面的截交线
d) Draw the intersection line of the intersecting plane *R* and complete the drawing 作 *R* 平面的截交线并完成全图

Analysis: As shown in Fig. 3-15a, the cylindrical surface of the shaft is intersected by a V-perpendicular plane Q, a horizontal plane R and a profile plane P. The plane R intersects the plane P at line Ⅰ-Ⅴ. The plane R intersects the plane Q at line Ⅵ-Ⅹ. Since projections of planes P, Q and R, are of accumulation on the front view, projections of the corresponding intersection line coincide with projections p′, q′ and r′respectively. The main task for this example is to draw projections of the intersection lines of the shaft and three planes on the top view and left view.

Drawing steps:

1) Draw the intersection line of the intersecting plane P. Since the intersecting plane P is perpendicular to the axis of the locating shaft, the intersection line is a piece of arc. The plane P intersects the plane R at V-perpendicular line Ⅰ-Ⅴ. The intersections of the plane P and R form a closed curve Ⅰ-Ⅱ-Ⅲ-Ⅳ-Ⅴ-Ⅰ(see Fig. 3-15a). Based on the projection of the curve Ⅰ-Ⅱ-Ⅲ-Ⅳ-Ⅴ-Ⅰ on the front view, draw projections of the curve Ⅰ-Ⅱ-Ⅲ-Ⅳ-Ⅴ-Ⅰ on the top view and left view (see Fig. 3-15b).

2) Draw the intersection line of the intersecting plane Q. Since the intersecting plane Q is inclined with respect to the axis of the shaft, its intersection produces an ellipse arc. The plane Q intersects the plane R at V-perpendicular line Ⅵ-Ⅹ. The intersection of the plane Q and R forms a closed curve Ⅵ-Ⅶ-Ⅷ-Ⅸ-Ⅹ-Ⅵ. By projecting the closed curve Ⅵ-Ⅶ-Ⅷ-Ⅸ-Ⅹ-Ⅵ onto the front and left views, one can draw the projections of the closed curve on the top view (see 3-15c).

3) Draw the intersection line of the intersecting plane R. Since the intersecting plane R is parallel to the axis of the shaft, its intersection line produces segments Ⅰ-Ⅵ and Ⅴ-Ⅹ. Also, the plane R intersects the plane P at V-perpendicular line Ⅰ-Ⅴ. The plane R intersects the plane Q at V-perpendicular line Ⅴ-Ⅹ. All lines of intersection form a rectangle Ⅰ-Ⅵ-Ⅹ-Ⅴ-Ⅰ(see Fig. 3-15d).

4) Remove the two line segments that were truncated on the top view. This concludes the projections of the intersections.

Common types and three views of cut cylinder and cut cylindrical tube are shown in Figure 3-16.

2. Intersection of planes and right cones

When a right cone is intersected by a plane, there are five possible situations (see Table 3-2) depending on the relative position of the intersecting plane and the axis of the right cone.

Example 3-4 Draw three views with projections of the intersection line between a right cone and a V-perpendicular plane (see Fig. 3-17a).

分析：如图 3-15a 所示，切口由侧平面P、正垂面Q 和水平面R 截切圆柱形成，截平面R 与P、Q 两截平面分别相交于线段Ⅰ-Ⅴ、Ⅵ-Ⅹ。因各截平面的正面投影都具有积聚性，所以各条截交线的正面投影分别和 p′、q′、r′重合，本例题主要是求出切口的水平投影和侧面投影。

作图步骤：

1) 作截平面P 的截交线。截平面P 垂直于圆柱轴线，它和圆柱面的截交线是一段圆弧，平面P 和平面R 的交线是一段正垂线Ⅰ-Ⅴ，它们组成了一个弓形Ⅰ-Ⅱ-Ⅲ-Ⅳ-Ⅴ-Ⅰ，如图 3-15a 所示。根据它的正面投影，可作出它的水平投影和侧面投影，如图 3-15b 所示。

2) 作截平面Q 的截交线。截平面Q 倾斜于圆柱轴线，它与圆柱面的截交线是部分椭圆，平面Q 和平面R 的交线是一段正垂线Ⅵ-Ⅹ，它们组成了封闭截交线Ⅵ-Ⅶ-Ⅷ-Ⅸ-Ⅹ-Ⅵ。根据部分椭圆的正面投影和侧面投影，可作出它的水平投影，如图 3-15c所示。

3) 作截平面R 的截交线。平面R 平行于圆柱轴线，它和圆柱面的截交线是两段素线Ⅰ-Ⅵ和Ⅴ-Ⅹ，平面R 与平面P、Q 的交线是两段直线，它们组成了矩形Ⅰ-Ⅵ-Ⅹ-Ⅴ-Ⅰ，如图 3-15d 所示。

4) 擦去水平投影上被切去的两段轮廓线，即完成切口的投影。

圆柱及圆柱筒切口的常见形式及三视图如图 3-16 所示。

2. 平面与圆锥体相交

圆锥被平面截切时，根据截平面与圆锥轴线的相对位置，分为五种情况（见表 3-2）。

例 3-4 求作正垂面与圆锥的截交线（见图 3-17a）。

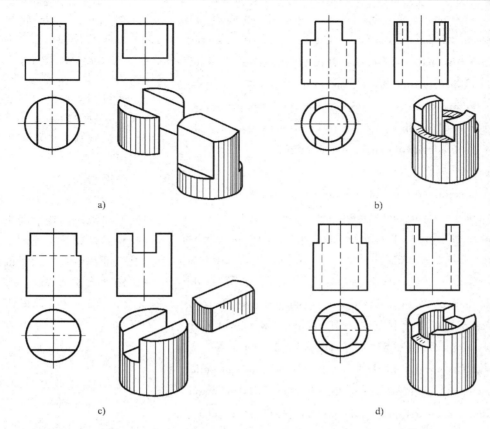

Fig. 3-16 Three views with incision projections of the cylinder and cylindrical tube
有切口的圆柱及圆筒的三视图

Table 3-2 Intersection line of a cone and a plane 平面截切圆锥体的截交线

Intersecting plane position 截平面位置	Plane P is perpendicular to the axis 平面 P 垂直于轴线 $\theta = 90°$	Plane P is tilted to the axis 平面 P 倾斜于轴线 $\theta > \alpha$	Plane P is parallel with a contour line 平面 P 平行于一条轮廓线 $\theta = \alpha$	Plane P is parallel with the axis 平面 P 平行于轴线 $\theta = 0$ $\theta < \alpha$	Plane P is across the apex point P 过锥顶
Projection of intersection lines 截交线的投影	Circle 圆	Ellipse 椭圆	Parabola + straight line 抛物线+直线	Hyperbola + straight line 双曲线+直线	Isosceles triangle 等腰三角形
Pictorial drawings 直观图					
Projection drawings 投影图					

Analysis: As shown in Fig. 3-17a, since *V*-perpendicular plane *P* intersects all elements of the right cone, the intersection line is an ellipse. The major axis of the ellipse is line Ⅰ-Ⅱ. The minor axis of the ellipse is line Ⅲ-Ⅳ. The projections of the ellipse and the plane *P* coincide in a line on the front view. The projections of the ellipse are still an ellipse on the top and left views, but the projections do not produce true shape of the ellipse.

Drawing steps:

1) Draw special points. Special points Ⅰ and Ⅱ are on the limit elements of the right cone for the front view. Special points Ⅴ and Ⅵ are on the limit elements of the right cone for the left view. Following the accumulation of the plane *P*, draw the projections 1′, 2′, 5′, (6′) of the points Ⅰ, Ⅱ, Ⅴ, Ⅵ on the front view and then down to give the corresponding projections 1, 2, 5, 6 of the points Ⅰ, Ⅱ, Ⅴ, Ⅵ on the top view. Finally draw the projections 1″, 2″, 5″, 6″ of the points Ⅰ, Ⅱ, Ⅴ, Ⅵ on the left view. The special points Ⅰ and Ⅱ are also the end points of the major axis of the ellipse. Since the line Ⅰ-Ⅱ is the major axis of the ellipse, draw a *V*-perpendicular line Ⅲ-Ⅳ on the center of the line Ⅰ-Ⅱ. The *V*-perpendicular line Ⅲ-Ⅳ is the minor axis of the ellipse. Draw the projections 3′, (4′) on the front view, then draw the projections 3, 4 and 3″, 4″on the top and left views using the auxiliary circle method (see Fig. 3-17b).

2) Draw general points. Draw the projections of the general points Ⅶ, Ⅷ using the auxiliary circle method (see Fig. 3-17c).

3) Join the projections of the points Ⅰ-Ⅴ-Ⅲ-Ⅶ-Ⅱ-Ⅷ-Ⅳ-Ⅵ-Ⅰ with a smooth curve on the top view and left view. Namely, the completion of the projections of the intersection line.

分析：从图 3-17a 可以看出，由于正垂面 *P* 与圆锥的所有素线都相交（θ>α），所以截交线为一个椭圆，其长轴为Ⅰ-Ⅱ，短轴为Ⅲ-Ⅳ。截交线的正面投影与平面 *P* 的正面投影 *p*′重合成一条直线段，其水平投影和侧面投影仍为椭圆，但都不反映空间椭圆的实形。

作图步骤：

1) 作特殊点。特殊点Ⅰ和Ⅱ是圆锥主视图转向轮廓线上的点，特殊点Ⅴ和Ⅵ是圆锥左视图转向轮廓线上的点。可利用截平面正面投影的积聚性，先求出这些特殊点的正面投影 1′、2′、5′、(6′)，再作出它们的水平投影 1、2、5、6 和侧面投影 1″、2″、5″、6″。特殊点Ⅰ、Ⅱ又是椭圆截交线长轴的端点。过线段Ⅰ-Ⅱ的中点作其垂直平分线Ⅲ-Ⅳ，即为椭圆截交线的短轴。该短轴为正垂线，可在主视图上绘出其正面投影 3′(4′)，过点 3′(4′)作辅助纬圆，就可作出其他两个投影 3、4 和 3″、4″，如图 3-17b 所示。

2) 作一般点。利用辅助纬圆法，作出椭圆截交线上点Ⅶ和Ⅷ的三面投影，如图 3-17c 所示。

3) 依次连接Ⅰ-Ⅴ-Ⅲ-Ⅶ-Ⅱ-Ⅷ-Ⅳ-Ⅵ-Ⅰ各点的水平投影和侧面投影，即完成该截交线的投影。

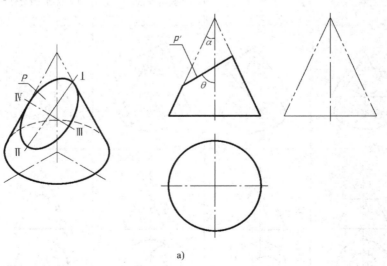

Fig. 3-17 Drawing three views of frustum of a cone 求作截头圆锥的三视图

a) Analyze the intersection line 分析截交线

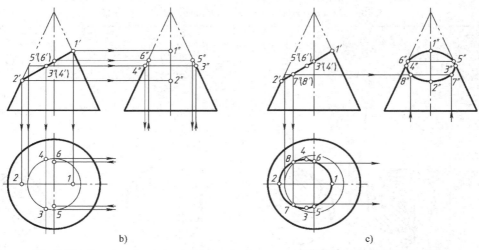

Fig. 3-17 Drawing three views of frustum of a cone　求作截头圆锥的三视图（续）
b）Draw special feature points　作特殊位置的点　c）Draw general points and complete the projection　作一般位置点的投影并完成全图

3. Intersection of planes and spheres

As a ball is intersected by a plane, their intersection line can be addressed in two situations according to the relative position of the intersecting plane and the pivot of sphere (see Table 3-3).

If the intersecting plane P is parallel or perpendicular to the pivot, the projection of intersection line is a circle, which can be drawn with auxiliary circle method; else if the intersecting plane P is tilted to the pivot, the projection of intersection line is an ellipse, which can be drawn with auxiliary circle by figuring out the intersection points on the sphere.

3.4　Intersection of Two Solids

If two solids Intersect each other, it is interpenetration (or intersection). such combination is called intersection of solids, on which one or more boundaries produced are called intersection curves. When the two solids intersect, there are three cases in terms of the geometric characteristics of solids:

1) Intersection of two polyhedral solids (see Fig. 3-18a).
2) Intersection of a polyhedral solid and a curved solid (see Fig. 3-18b).
3) Intersection of two curved solids (see Fig. 3-18c).

In the following section of this chapter, we mainly discuss the intersection of a polyhedral solid and a curved solid and the intersection of two curved solids.

3. 平面与圆球相交

圆球被平面截切时，根据截平面与圆球轴线的相对位置，其截交线分为两种情况（见表3-3）。

当截平面P平行或垂直于圆球轴线时，截交线投影为圆形，可利用辅助纬圆法求其投影。当截平面P倾斜于圆球轴线时，截交线投影为椭圆，可采用纬圆法在球面上找点来求其投影。

3.4　两立体相交

两立体彼此相交即为相贯，这样结合的立体称为相贯体，在其表面产生的一条或多条分界线称为相贯线。根据立体的几何性质，两立体相交可分为三类：

1）两平面立体相交（见图3-18a）。
2）平面立体与曲面立体相交（见图3-18b）。
3）两曲面立体相交（见图3-18c）。

本节主要介绍平面立体与曲面立体相交和两曲面立体相交。

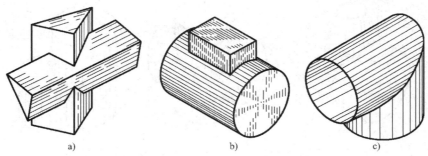

Fig. 3-18　Intersection of solids　相贯体
a）Intersection of two polyhedral solids　两平面立体相交　b）Intersection of a polyhedral solid and a curved solid　平面立体与曲面立体相交
c）Intersection of two curved solids　两曲面立体相交

Table 3-3 Intersection lines of sphere   圆球的截交线

Intersecting plane position 截平面位置	Plane P is parallel or perpendicular to the pivot P 平行或垂直于圆球轴线			Plane P is tilted to the pivot P 倾斜于圆球轴线
Projection of intersection lines 截交线的投影	Circle 圆			Ellipse 椭圆
Pictorial drawing 直观图				
Projection drawing 投影图				

3.4.1　Characteristics of intersection curves

(1) Closed loop　When two curved solids meet, the intersection curve is usually a closed space curve (see Fig. 3-19a). The intersection curve may be either a general curve (see Fig. 3-19b) or a straight line segment in special cases (see Fig. 3-19c).

(2) Commonality　The intersection curve is a common line of both intersecting solids, which is also their boundary. Points on the intersection curve are common points of both intersecting bodies.

3.4.1　相贯线的性质

（1）封闭性　两曲面立体的相贯线一般是封闭的空间曲线（见图3-19a），特殊情况下可以是平面曲线（见图3-19b）或直线（见图3-19c）。

（2）共有性　相贯线是两相交立体表面的共有线，也是它们的分界线。相贯线上的点是两立体表面的共有点。

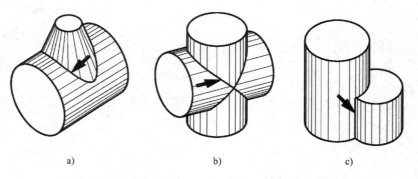

Fig. 3-19　Types of intersection curve   相贯线的形式
a) Closed space curve　封闭的空间曲线   b) General curve   平面曲线   c) Straight line   直线

3.4.2 Views of intersection curves

Drawing of the three views of the intersection curve can be accomplished by taking selected points on the curved surfaces.

Example 3-5 Draw the front view of the intersection of a quadrangular prism and a cylinder (see Fig. 3-20a).

Analysis:

1) Shape analysis. The intersection curve is formed by intersecting the curved surface of a cylinder with the four faces of a quadrangular prism, in which the front and back surfaces are parallel to the cylinder's pivot, therefore their intersection lines of surfaces are two line segments that are parallel to the cylinder's pivot; the left and right surfaces are perpendicular to the cylinder's pivot, therefore their intersection lines of surfaces are two arcs. Hence, the intersection curve is formed by connecting all these intersection segments.

2) Projection analysis. The horizontal projection of the intersection curve accumulates on the rectangle 2-3-5-6, and the profile projection of the intersection curve accumulates on the arc (5″)-1″-(3″) and 6″-(4″)-2″.

3.4.2 相贯线的画法

可以利用在立体表面上取点的方法来画相贯线的投影。

例 3-5 正四棱柱与圆柱相交，求作其主视图（见图 3-20a）。

分析：

1) 形体分析。相贯线由正四棱柱的四个侧面与圆柱面的交线组成，其中前后两个棱面与圆柱的轴线平行，交线为两段与圆柱轴线平行的直线；左右两个棱面与圆柱的轴线垂直，交线为两段圆弧，将这些交线连接起来即为相贯线。

2) 投影分析。相贯线的水平投影积聚在四边形 2-3-5-6 上，相贯线的侧面投影积聚在圆弧（5″）-1″-(3)″和 6″-（4″）-2″上。

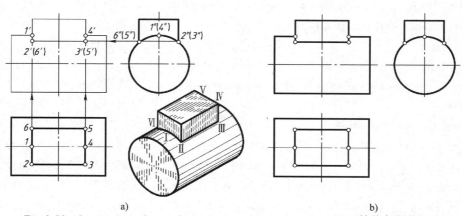

Fig. 3-20　Intersection of a quadrangular prism and a cylinder　正四棱柱与圆柱相交

Drawing steps:

1) Draw special points. The special points of intersection curve are mainly addressed in terms of common points and limitation points that are located on the limiting element. The horizontal and profile projections of the intersection curve are given. According to the given projections 1, 2, 3, 4, 5, 6 and projections 1″, 2″, (3″), 4″, (5″), 6″ of points, find and locate points on surface in order to figure out the projections 1′, 2′, 3′, 4′, (5′), (6′) of points on the front view (see Fig. 3-20a).

2) Connect each point successively to complete the front view of intersection curve (see Fig. 3-20b).

The common forms of intersection of quadrangular prism and cylinder are shown in Table 3-4.

作图步骤：

1) 作特殊点。相贯线上的特殊点主要是转向轮廓线上的共有点和极限位置点。其水平投影和侧面投影都是已知的，利用面上取点的方法，由已知投影 1、2、3、4、5、6 和 1″、2″、（3″）、4″、（5″）、6″，进而求得 1′、2′、3′、4′、（5′）、（6′），如图 3-20a 所示。

2) 将各点依次连接起来，即得到相贯线的正面投影，如图 3-20b 所示。

四棱柱与圆柱相贯的常见形式见表 3-4。

Table 3-4 Intersection of quadrangular prism and cylinder 四棱柱与圆柱相贯

Description 说明	A quadrangular prism intersects with a cylinder 四棱柱与圆柱相贯	Cut a cylinder with a quadrangular prism 圆柱切四棱柱孔	Cut a cylindrical tube with a quadrangular prism 圆筒切四棱柱孔
Pictorial drawing 直观图		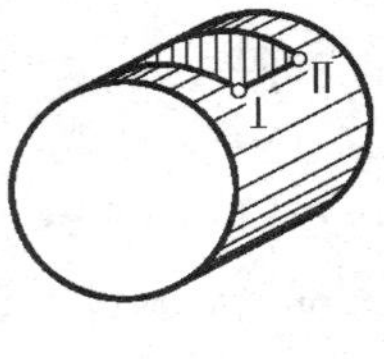	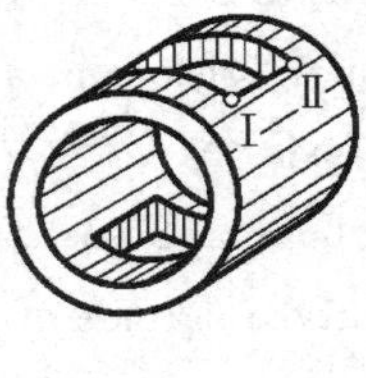
Projection drawing 投影图	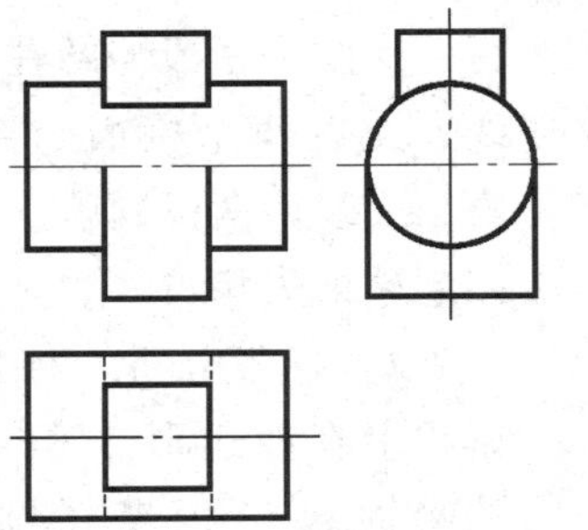	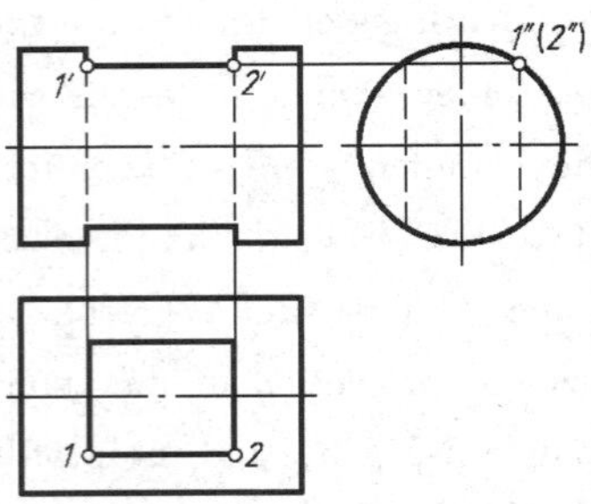	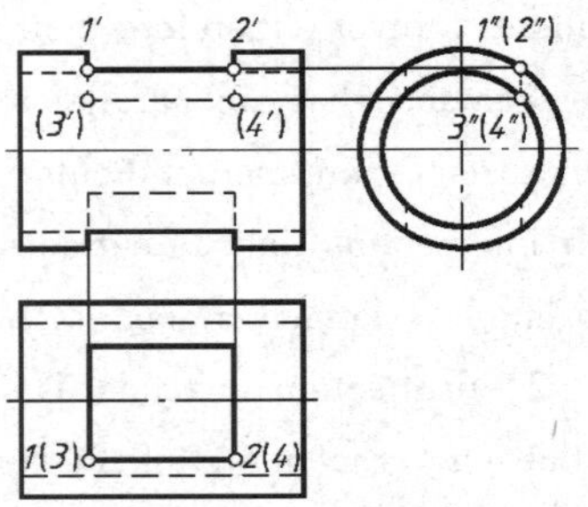

Example 3-6 Draw three views with projections of orthogonal intersection between two cylinders, shown in Fig. 3-21a.

Analysis:

1) Shape analysis. Fig. 3-21a shows that the two cylinders have different diameters. The intersection curve of the two cylinders is a closed space curve.

2) Projection analysis. The projection of the intersection curve of the two cylinders coincides in the projection of the big cylinder on the top view as an arc on the circle (the projection of the big cylinder is a circle on the top view). The projection of the intersection curve coincides in the projection of the small cylinder on the left view as a circle. The main task is to draw the projection of the intersection curve on the front view.

Drawing steps:

1) Draw special points. The special points of intersection curve are mainly addressed in terms of common points and limitation points that are located on the limiting element. The left outline of the larger cylinder intersects with the smaller cylinder at points Ⅰ and Ⅲ. The top, bottom, front, and back outlines of the smaller cylinder are respectively intersected with the larger cylinder at points Ⅰ, Ⅲ, Ⅱ and Ⅳ. Therefore, there are four points Ⅰ, Ⅲ, Ⅱ and Ⅳ of intersection curve that are located on the outline, such points are also limitation points and their horizontal projections are given already. According to the given projections 1, 2, (3), 4 and the projections 1″, 2″, 3″, 4″ of the points, find and locate points on the surface in order to figure out the projections 1′, 2′, 3′, (4′) of the points Ⅰ, Ⅱ, Ⅲ, Ⅳ on the front view (see Fig. 3-21a).

例 3-6 求两圆柱正交的相贯线（图 3-21a）。

分析：

1) 形体分析。由图 3-21a 可知，这两个正交圆柱体直径不同，其相贯线为一封闭的空间曲线。

2) 投影分析。相贯线的水平投影和大圆柱的水平投影重合，为一段圆弧，相贯线的侧面投影和小圆柱的侧面投影重合，为一个圆。本例题主要是求出相贯线的正面投影。

作图步骤：

1) 作特殊点。相贯线上的特殊点主要是转向轮廓线上的共有点和极限位置点。大圆柱左侧的轮廓线和小圆柱相交于Ⅰ、Ⅲ两点，小圆柱的上、下、前、后四条轮廓线和大圆柱交于Ⅰ、Ⅲ、Ⅱ、Ⅳ四点，因此，相贯线在轮廓线上的共有点有Ⅰ、Ⅲ、Ⅱ、Ⅳ四个，它们也是极限位置点，其水平投影和侧面投影都是已知的。由此根据已知投影1、2、(3)、4 和 1″、2″、3″、4″，可以求得1′、2′、3′、(4′)，如图 3-21a 所示。

2) Draw general points. Firstly, select coincident points 5 (6) on the horizontal projection of the intersection curve. Secondly, according to the projection rule—equal width, figure out the side projections 5″ and 6″. Finally, figure out the projections 5′ and 6′ on the front view (see Fig. 3-21b).

3) Connect each point successively with a smooth line, and thus to complete the front projection of the intersection curve (see Fig. 3-21c).

2) 作一般点。先在相贯线的已知水平投影中取重影点 5（6），根据"宽相等"投影规律求出侧面投影 5″、6″，然后再求出主视图上的投影 5′、6′（见图 3-21b）。

3) 将各点依次光滑连接起来，即得到相贯线的正面投影，如图 3-21c 所示。

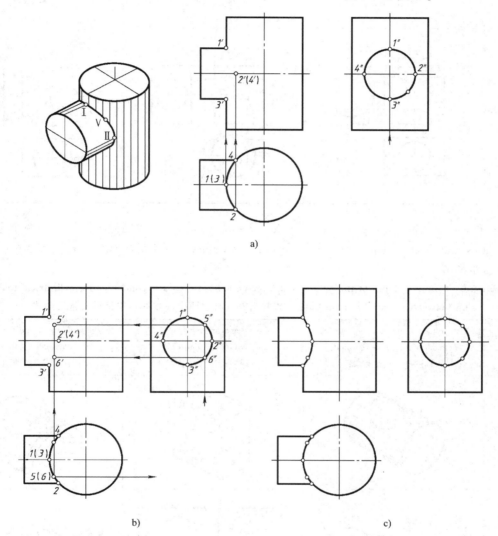

Fig. 3-21 Drawing intersection curve of two cylinders 求作两圆柱的相贯线
a) Analyze and mark special points 分析并找出特殊点 b) Draw general points 求作一般位置的点
c) Complete the projection 完成全图

Common types of orthogonal intersection of two cylinders are shown in Table 3-5. As given in Table 3-5, if the diameters of the two intersecting cylinders are unequal, the curved direction of the intersection curve is always oriented to the pivot axis of larger cylinder; if the diameters of intersecting cylinders are both equal, then the intersection curve will be shaped into two plane curves (a pair of perpendicular ellipses), and their relevant projections are accumulated into two intersecting straight line segments.

两圆柱正交相贯的常见形式见表 3-5。由表 3-5 可知，当正交相贯的两圆柱直径不等时，相贯线的弯曲方向总是朝向直径大的圆柱的轴线；当正交相贯的两圆柱直径相等（即公切于一个球面）时，相贯线变为两条平面曲线（两个相互垂直的椭圆），其投影积聚为两条相交直线段。

Table 3-5　Common types of orthognal intersection of two cylinders　两圆柱正交相贯的常见形式

Description 说明	Diameters are unequal 直径不等		Diameters are equal 直径相等
	$D > d$	$D < d$	$D = d$
Pictorial drawing 直观图	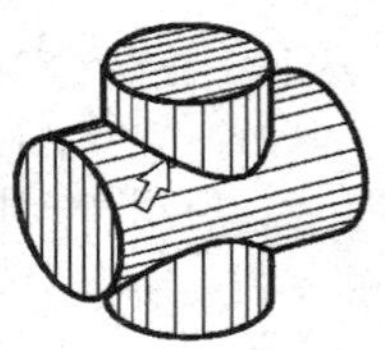	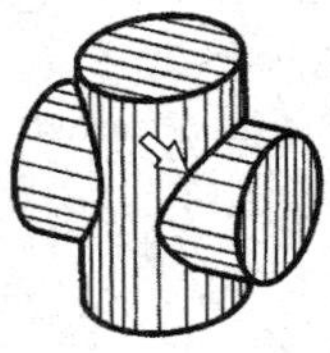	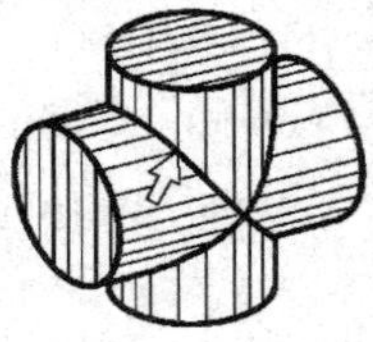
Projection drawing 投影图	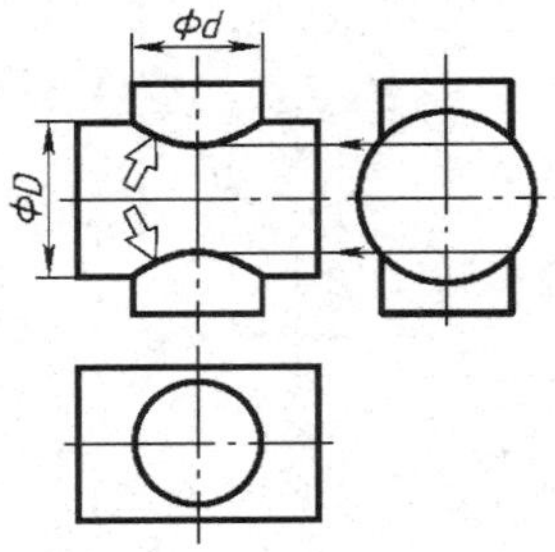	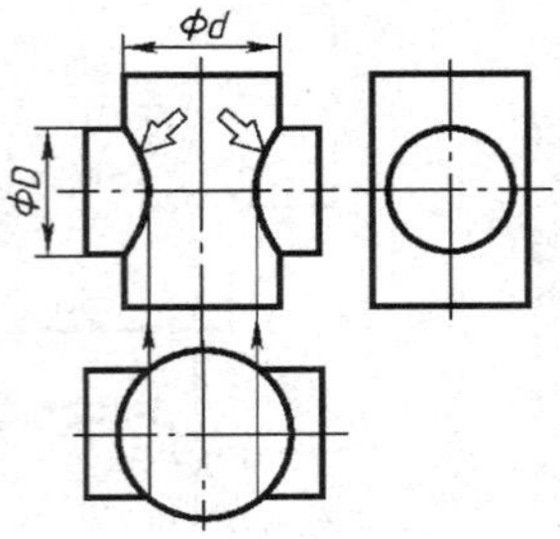	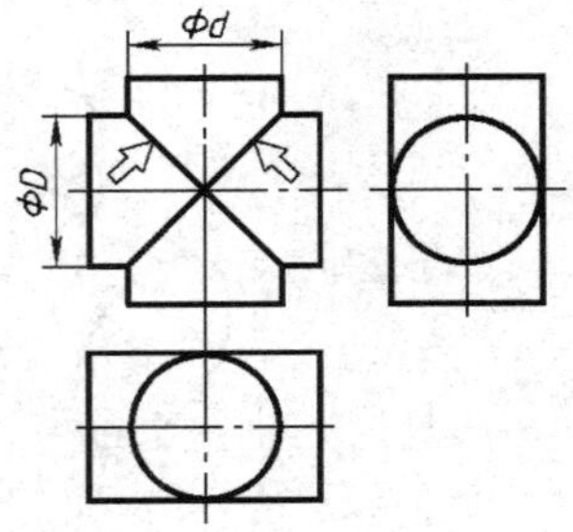

Common types of the orthogonal intersection between a cyclinder and a cylindrical hole and between two cyclindrical holes are shown in table 3-6.

圆柱与圆柱孔、两个圆柱孔正交相贯的常见形式见表3-6。

Table 3-6　Common types of the orthogonal intersection between a cyclinder and a cylindrical hole and between two cyclindrical holes　圆柱与圆柱孔、两个圆柱孔正交相贯的常见形式

Description 说明	Orthognal intersection of a cylinder and a cylindrical hole 圆柱与孔正交相贯	Orthognal intersection of cylindrical holes 圆柱孔正交相贯	Orthognal intersection of cylindrical tubes 圆筒正交相贯
Pictorial drawing 直观图	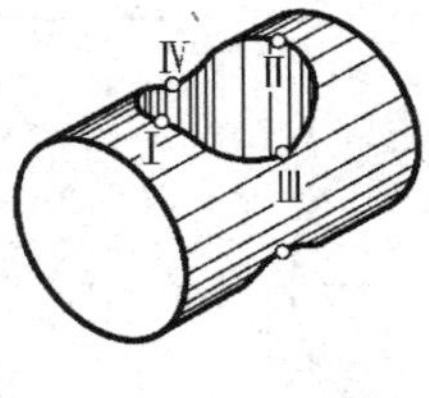	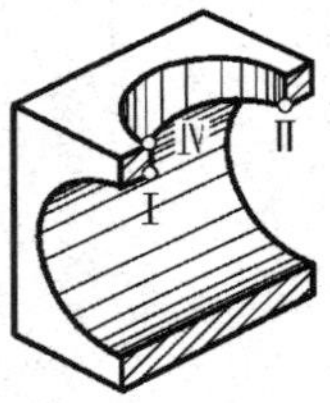	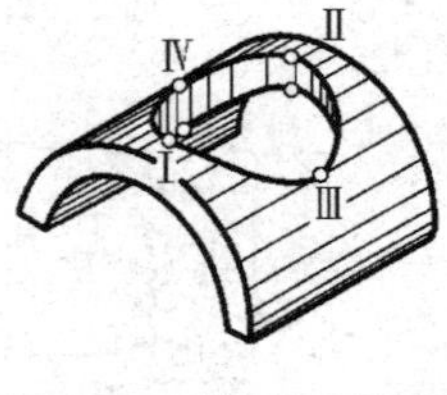
Projection drawing 投影图	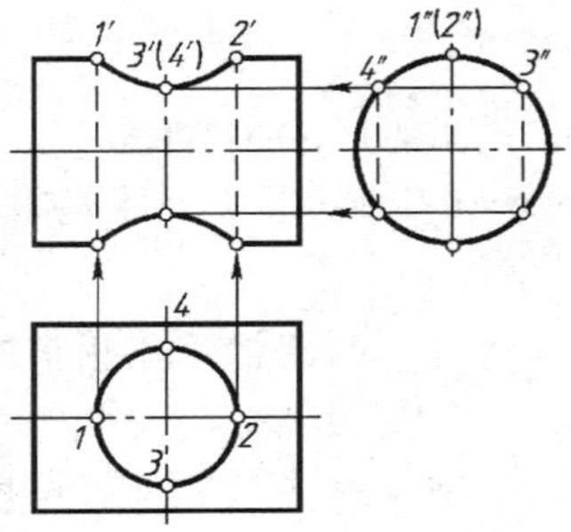	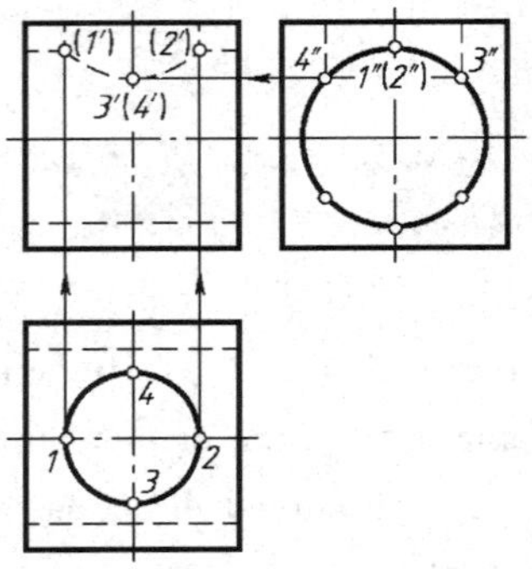	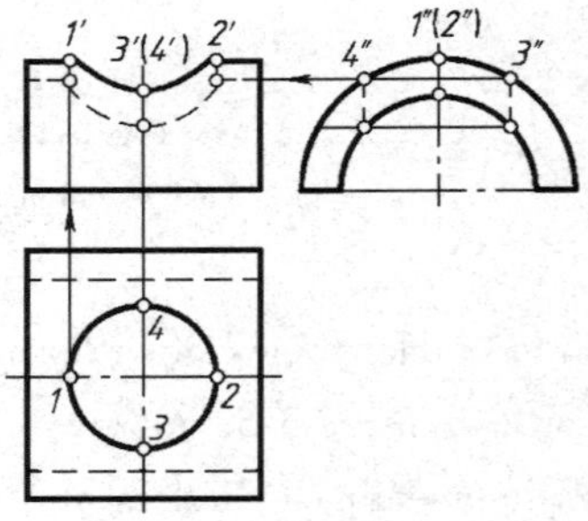

Example 3-7　Draw the projections of the intersecting curve on V and H planes for the orthogonal intersection of a cylinder and a cone, shown in Fig. 3-22a.

例 3-7　求圆柱与圆锥正交相贯的 V 面和 H 面投影（图3-22a）。

Analysis:

As given in 3-22a, the axis of the cylinder is orthogonal to the axis of the cone, the axis of the cylinder is perpendicular to the plane W, the projection of the cylinder n the plane W is accumulated into a circle, and the projection of the intersecting line on the plane W is overlapped on the circle, so the projection of the intersecting line on the plane W is known, only two other projections are required. Because the intersection curve is located on cone's surface, it can be drawn by finding intersecting points on the cone's surface.

Drawing steps:

1) Draw special points. The special points of the intersection curve are mainly addressed in terms of common points and limit points that are located on the limiting element. The top and bottom outlines of cylinder intersect with the front outline of cone at point Ⅰ and Ⅱ, and thus are the uppermost and lowermost points of the intersection curve, which can be drawn by finding interesting points on surface such as 1″, 2″→1′, 2′→1, 2. The front and rear outlines of cylinder interest with the cone at Ⅲ and Ⅳ, which are the foremost and backmost points of the intersection curve, respeceively. And their side projection are already given. Draw an auxiliary horizontal plane Q across the pivot of the cylinder, which has the horizontal projection intersects with the cone at a horizontal circle q. Extend the front and rear limiting elements of the horizontal projection of the cylinder until intersect with the circle q, and thus figure out 3, 4, and 3′, 4′, which are the forefront and the last points. The limit points Ⅴ (5″) and Ⅵ (6″) are found by using the side vertical surface S which has been tapered and tangent to the cylinder. Then the horizontal surface P is obtained by crossing the point of 5″ and 6″ respectively, and the horizontal plane projections 5 and 6 are obtained according to the width equality rule, then the 5′ and 6′ are obtained, as shown in Fig. 3-22b.

分析：

由图 3-22a 可知，圆柱与圆锥的轴线正交，圆柱的轴线垂直于 W 面，其 W 面投影积聚为圆，相贯线的 W 面投影重合在该圆上，所以，实际上相贯线的 W 面投影是已知的，只要求其他两个投影即可。由于相贯线又在锥表面上，因此可利用圆锥表面取点的方法，求出相贯线。

作图步骤：

1) 作特殊点：相贯线上的特殊点主要是转向轮廓线上的共有点和极限位置点。圆柱的上下两条轮廓线和圆锥的正面轮廓线相交于Ⅰ、Ⅱ两点，即为相贯线上的最高、最低点，利用面上取点的方法求得 1″、2″→1′、2′→1、2。圆柱的前、后两条轮廓线与圆锥交于Ⅲ、Ⅳ两点，即为相贯线上的最前、最后点，其侧面投影已知。过圆柱轴线作辅助水平面 Q，其水平投影与圆锥交于一水平位置的圆 q，延伸圆柱水平投影的前后两条转向轮廓线与该圆相交，求得 3、4 和 3′、4′，为最前、最后点。用过锥顶且与圆柱相切的侧垂面 S，找到极限点Ⅴ (5″) 和Ⅵ (6″)，再分别过 5″、6″点作辅助水平面 P，根据"宽相等"分别求出水平面投影 5、6，进而求出 5′、6′，如图 3-22b 所示。

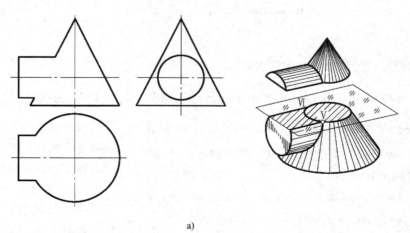

a)

Fig. 3-22 Drawing the projections of the intersecting curve on the V, H planes for the orth ogonal intersection of a cylinder and a cone　求作圆柱与圆锥正交相贯的 V、H 面投影

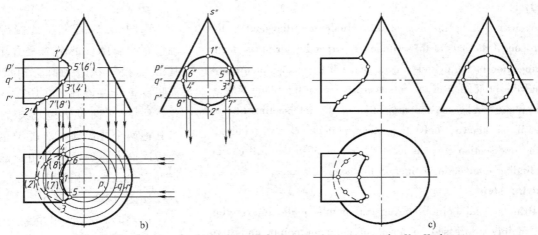

Fig. 3-22 Drawing the projections of the intersecting curve on the V, H planes for the orthogonal intersection of a cylinder and a cone   求作圆柱与圆锥正交相贯的 V、H 面投影（续）

2) General points: Firstly, select points $7''$, $8''$ as shown in the given side view of the intersection curve. Secondly, draw auxiliary horizontal plane P and R across $7''$, $8''$, and then figure out the horizontal projection 7, 8 and 9, 10 according to the width equality rule, and then figure work out the points $7'$, $8'$, as shown in Fig. 3-22b.

3) After distinguishing the visibility of each point, connect the points successively with smoothed line, and thus complete the front view projection of the intersection curve (see Fig. 3-22c).

Common types of orthogonal intersection of a cylinder and a cone are shown in Fig. 3-23.

2）作一般点：先在相贯线的已知侧面投影中取点 $7''$、$8''$，再分别过 $7''$、$8''$ 点作辅助水平面 R，根据"宽相等"分别求出水平面投影 7、8，进而求出 $7'$、$8'$，如图 3-22b 所示。

3）判别可见性后，将各点依次光滑连接起来，即得到相贯线的正面和水平投影，如图 3-22c 所示。

圆柱与圆锥正交相贯的常见形式见图 3-23。

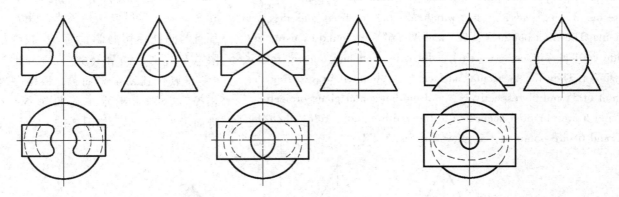

Fig. 3-23 Common types of orthogonal intersection of a cylinder and a cone   圆柱与圆锥正交相贯的常见形式

3.4.3 Special cases of intersection of two curved solids

When two curved solids intersect, the intersection curve may be either a straight line or a planar curve, depending on the surfaces and their relative positions.

1) When two curved solids are tangent with a common sphere, the intersection curve is an ellipse (see Fig. 3-24a).

2) When two curved solids share a common axis, the intersection curve is a circle perpendicular to their axis (see Fig. 3-24b).

3.4.3 曲面立体相交的特殊情况

两曲面立体相交时，相贯线可能是平面曲线或直线段，这取决于两立体的表面和其相对位置。

1）若两个曲面立体公切于圆球，则这两个回转体的相贯线为平面曲线，如图 3-24a 所示。

2）若两个曲面立体具有公共轴线时，相贯线为垂直于轴线上的圆，如图 3-24b 所示。

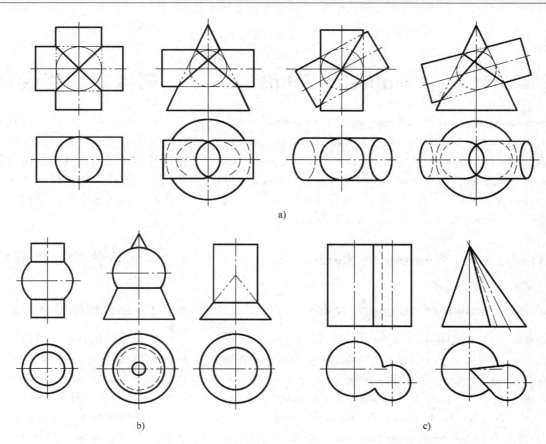

Fig. 3-24 Special cases of intersection curves　相贯线的特殊情况

3) When the axes of two cylinders are parallel to each other, the intersection curve produces lines (see Fig. 3-24c).

4) When two cones have a common vertex, the intersection curve also produces lines (see Fig. 3-24c).

3) 当两个圆柱轴线平行时，相贯线为直线，如图 3-24c 所示。

4) 当两圆锥共顶相交时，相贯线也为直线，如图 3-24c 所示。

Chapter 4　Composite Solids

From the point of geometry, all mechanical parts are composed of some basic geometrical bodies, such as columns, cones, spheres, etc. Generally, such mechanical parts are called composite solids. Preparing and reading the drawings of composite solids are important contents of this book, at the same time, they have great significance in developing the space imagination for students.

4.1　Analysis for Composite Solids

4.1.1　Configuration of composite solids

Composite solids can be constructed through various Boolean operations, such as union (superposition), subtraction (cutting), and integration (hybrid construction). The integration approach is most commonly used. In Fig. 4-1, the basic bodies are a cylinder and a hexagonal prism and the final object is produced through union operation. In Fig. 4-2, the final object is formed by cutting off two small vaulted poles, a cylinder and a semi-cylindrical from the base cuboid using subtraction operation. In Fig. 4-3, the final object is produced through a combination of union and subtraction operations.

第4章　组合体

从几何学的角度看，所有机械零件都是由若干柱、锥、球等基本几何体组成的，我们将其称为组合体。因此，画、读组合体视图是本课程的重要内容，对培养学生的空间想象能力意义重大。

4.1　组合体的形体分析

4.1.1　组合体的组合形式

组合体的组合形式可由布尔运算的方法分为几种：叠加式（堆积）、切割式（切除）及综合式。其中综合式最为常见。图4-1所示为叠加式，构成该组合体的基本形体是圆柱体和六棱柱；图4-2所示为切割式，从较大长方体中切去两个拱形柱、一个圆柱和一个半圆柱；图4-3为综合式，该形体既有叠加，也有切割。

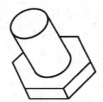

Fig. 4-1　Union　叠加式

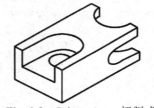

Fig. 4-2　Subtraction　切割式

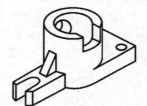

Fig. 4-3　Integration　综合式

4.1.2　Relative positions of surfaces on a composite solid

There are usually four types of relative positions between the adjacent surfaces: alignment, non-alignment, tangency and intersection. When preparing drawings, one must correctly express their relative positions, and can not miss any lines or draw excessive lines. When reading a drawing, one should keep in mind about these relationships to clearly imagine the overall shape of the solid.

1. Alignment

When two neighboring parallel planes are aligned, there is no line between their projections in a view.

Analysis: As the front planes and the back planes of the two cuboids shown in Fig. 4-4a are aligned, there is no line between their projections in the front view. In the case shown in Fig. 4-4b, only the front planes of the two cuboids are aligned, and thus there is no line between their projections in the front view.

4.1.2　形体表面间的相对位置关系

组合体相邻表面的相对位置关系通常有四种情况：平齐、不平齐、相切、相交。画图时要正确表示表面间的相对位置关系，不多画线也不漏画线。看图时注意这些关系，才能想清楚组合体的整体结构形状。

1. 平齐

当相邻平行表面平齐时，在视图中两表面的投影之间不画线。

分析：图4-4a中两个长方体的前后两面都平齐，故在主视图上不画线；图4-4b中两个长方体只有前表面平齐，则在主视图中这两表面间不画线。

2. Non-alignment

When two neighboring parallel planes are not aligned, there should be a line to separate their projections. The two side planes shown with arrows in Fig. 4-4a and Fig. 4-4 b illustrate this case that there is a line in the left view between their projections. The dashed line in the front view of Fig. 4-4b indicates another case that the two back planes are not aligned and there should be a line between their projections.

2. 不平齐

当相邻平行表面不平齐时,在视图中两表面的投影之间应有线隔开。如图 4-4a、b 所示的立体图和左视图箭头所指处。图 4-4b 所示的主视图中所画出的虚线,是表示其后表面不平齐,它们的正面投影之间应有线。

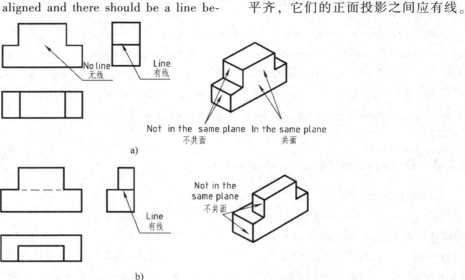

Fig. 4-4　Planes with or without alignment　表面平齐与不平齐

3. Tangency

When two neighboring surfaces are in tangent condition, there is no line between their projections.

Analysis: The cylinder and the side plane of the supporting board shown in Fig. 4-5 are in tangent condition. As the plane and the cylinder are connected smoothly, there is no visual boundary between the two surfaces and there is no line between their projections. But one should notice the tangents positions of the projections in the front and left views.

4. Intersection

When two neighboring planes intersect, one should draw the projection of the intersection line as shown in Fig. 4-5.

3. 相切

当相邻表面相切时,在相切处不画线。

分析：图 4-5 中圆柱体和底板两立体是相切关系。由于平面光滑地与圆柱面相切连接,两表面之间不再有分界线存在,故其投影在相切处无线。但应注意底板在主视图和左视图中的切点投影位置。

4. 相交

当相邻表面相交时,在相交处应画出交线的投影,如图 4-5 所示。

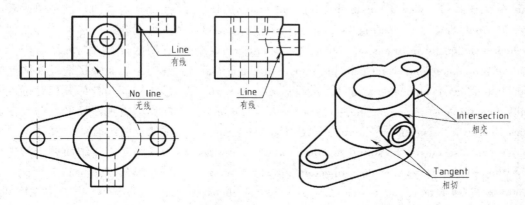

Fig. 4-5　Two surfaces meet in tangent or intersection conditions　两表面相切与相交

4.2 Drawing Views of Composite Solids

Now we use the support in Fig. 4-6 to explain the methods and steps of drawing various views of the composite solids.

4.2.1 Shape analysis method

Since a composite solid is composed of basic solids, we imagine that the composite solid is disassembled to a few of individual basic solids. We then think how these individual solids are used to form final object and identify the relative positions of neighboring surfaces. Shape analysis is a basic method in preparing and reading drawings.

4.2.2 Drawing methods and procedure

1. Shape analysis

The support shown in Fig. 4-6 is composed of the left bottom-board 1, the right bottom-board 2, the rib 3 and the cylinder 4. There is a U groove in the left bottom-board 1, two circular holes in the right bottom-board 2, a quadrate groove in the front of the cylinder 4 and a circular hole in its back. The left bottom-board 1 and the rib 3 intersect cylinder 4, the right bottom-board 2 is in tangent condition with cylinder 4, and the rib 3 is superposed on the left bottom-board 1.

4.2 组合体三视图绘制

下面以图 4-6 所示的支架为例，说明画组合体三视图的方法和步骤。

4.2.1 形体分析法

由于组合体是由若干基本体组合而成的，所以形体分析就是假想地把组合体分解为若干基本体，并确定它们的组合形式以及相邻表面间的相互位置。形体分析法是画图和看图的基本方法。

4.2.2 画图方法和步骤

1. 形体分析

如图 4-6 所示，支架由左底板 1、右底板 2、肋板 3、直立圆筒 4 组成。左底板 1 上开有 U 形槽，右底板 2 上有两个小圆孔，直立圆筒 4 前面有方槽，后面有圆孔。左底板 1、肋板 3 与直立圆筒 4 相交，右底板 2 与直立圆筒 4 相切，左底板 1 与肋板 3 叠加。

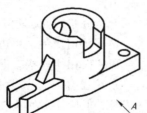

Fig. 4-6　Shape analysis of support　支架的形体分析

2. Selection of the front view

The front view is the most important one in the three views since it usually shows the main shape of the composite solid. We usually start from the front view for both drawing and reading. The selection of the front view is to confirm its projection direction and its placement in relation to the projection plane. We usually choose the direction of projection from which can the important shape and positions of individual components can be mostly reflected. The views should show the most important shape characteristics and most of the planes should be in special orientation relative to the projection plane. We can also choose a natural location. Based on the analysis, we choose the arrow A as the projection direction toward the front view in Fig. 4-6 and, in this case, it is better to choose the natural location. At the same time, the number of dotted lines on other views is as little as possiple.

2. 确定主视图

三视图中主视图是最重要的视图，因为主视图通常是反映组合体主要形状的视图，且画图或看图大都从主视图开始。选择主视图就是确定主视图的投射方向和相对于投影面的放置问题。通常是选择最能反映其主要形状特征和各部分相对位置的投射方向作为主视图投射方向；安放位置应最能反映组合体位置特征，并使多数表面相对于投影面处于特殊位置，也可选择其自然位置。因此，图 4-6 中以 A 所指方向作为主视图投射方向，按自然位置放置较好。同时也兼顾使其他视图的虚线数量尽量少。

3. Drawing procedure

1) Select the scale and the size of drafting paper based on the size of the composite solid.

2) Draw the datum lines on each view. The datum can be symmetry planes, bottom plane, end faces, axes of revolution, and symmetry center lines, whose projections have accumulation characteristic (Fig. 4-7a).

3) Draw the erect cylinder, and first draw the top view (Fig. 4-7b).

4) Draw the right bottom-board. Notice that it is tangent with the erect cylinder and there is no line in the front view (Fig. 4-7c).

5) Draw the left bottom-board. Notice that there is an intersection lines between the left bottom-board and the erect cylinder (Fig. 4-7d).

6) Draw the rib. Notice that there is an intersection line between the rib and the erect cylinder (Fig. 4-7e).

7) Check and draw with regular lines. After the draft, especially notice to check whether the connection, tangency and intersection between individual components follow the principles of projection. After careful check, erase all redundant lines, and then draw all regular lines following the sketches. The whole drawing is completed (Fig. 4-7f).

3. 画图步骤

1) 根据组合体大小，选比例、定图幅。

2) 画各视图作图基准线。通常选组合体中投影有积聚性的对称面、底面、端面、回转轴线、对称中心线等作为基准（见图4-7a）。

3) 画圆筒，并先画俯视图（见图4-7b）。

4) 画右底板，注意右底板与圆筒是相切关系，主视图中无线（见图4-7c）。

5) 画左底板，注意其与圆筒之间的相贯线（见图4-7d）。

6) 画肋板，注意其与圆筒之间的相贯线（见图4-7e）。

7) 检查描深。完成底稿后，特别要注意检查各个基本体之间连接、相切、相交等关系的处理是否符合投影原理。经全面检查后，擦去多余图线，再描深、加粗，完成全图（见图4-7f）。

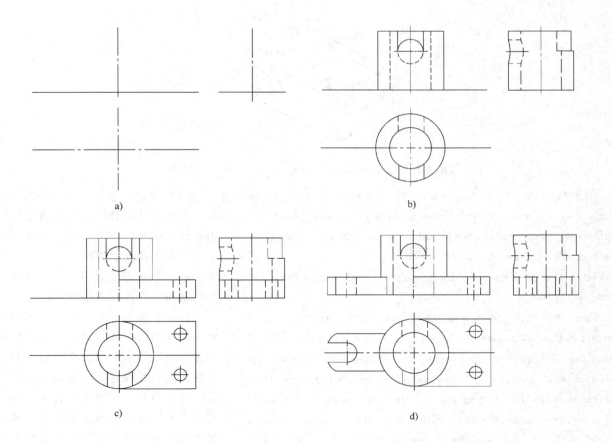

Fig. 4-7 Steps for producing the three-view drawing 画三视图的步骤

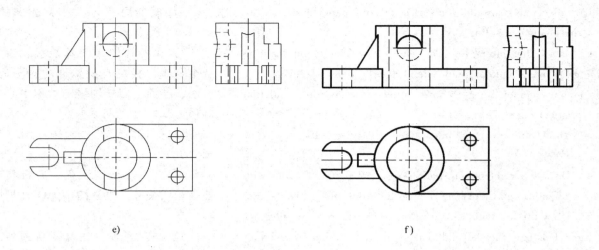

e)　　　　　　　　　　　　　f)

Fig. 4-7　Steps for producing the three-view drawing　画三视图的步骤（续）

Example 4-1　Draw the three views of the composite solid shown in Fig. 4-8.

（1）Shape analysis　The composite solid is an incised solid. We imagine that the composite solid is produced from a cuboid after cutting off four parts as shown in Fig. 4-8.

例 4-1　画图 4-8 所示组合体的三视图。

（1）形体分析　该组合体为切割体，可以看作是从长方体上切去四部分形成的，如图 4-8 所示。

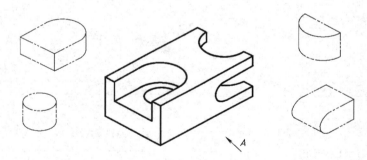

Fig. 4-8　Shape analysis of Example 4-1　例 4-1 形体分析

（2）Selection of the front view　The direction of arrow A can clearly show the relative positions of the composite solid itself and the individual components. We therefore select this direction as the direction of projection for the front view.

（3）The drawing procedures　The drawing procedures are shown in Fig. 4-9.

From the above two examples, we know that the shape analysis method is mostly suited for union type as well as hybrid type solids. We may first draw the view of individual components and then deal with the intersection lines of surfaces and their relative positions. For cutting type solids, we should first analyze the shape, determine the basic parts, draw the views of the basic parts, and finally draw the three views of individual components.

（2）确定主视图　如图 4-8 中 A 所示方向反映组合体及各组成部分间相互位置关系较清楚，故应选此方向作为主视图的投射方向。

（3）画图基本步骤　图 4-9 所示为绘制该组合体三视图的步骤。

由以上两例可以看出，对于叠加型和综合型组合体，画三视图宜用形体分析法，分别画出每个组成部分的视图，然后处理它们的表面交线及相对位置；而对于切割型组合体，则宜先进行形体分析，确定基本体，画出基本体的三视图，然后，再画出各个切去部分三视图。

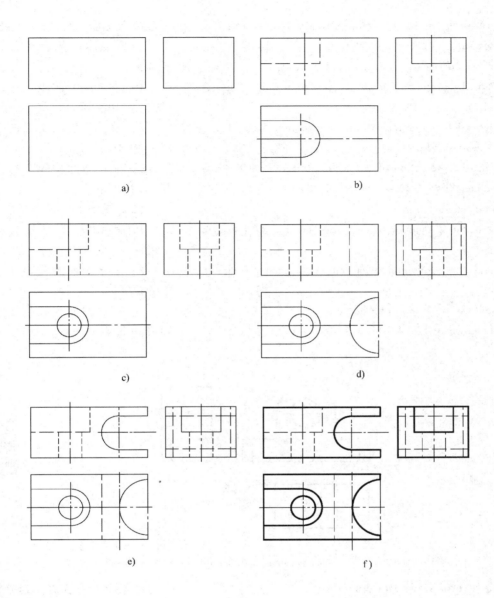

Fig. 4-9 Steps for drawing three views 画三视图的步骤
a) Draw three views of the basic cube 画基本立体的三视图
b) Draw three views of the left U groove cutted H-perpendicularly 画左侧垂直于 H 面切得的 U 形槽的三视图
c) Draw three views of the left cyclindrical hole cutted H-perpendicularly 画左侧垂直于 H 面切得的圆柱孔的三视图
d) Draw three views of the right semi-column slot cutted H-perpendicularly 画右侧垂直于 H 面切得的半圆柱的三视图
e) Draw three views of the right U groove cutted V-perpendicular 画右侧垂直于 V 面切得的 U 形槽的三视图
f) Finally check and draw with appropriate line styles 检查并描深

4.3 Reading Views of Composite Solids

From above discussions, we know that drawing views is to draw 3D solids on 2D drawing sheets using orthogronal projection method. In this section, we introduce techniques for reading 2D drawings by imagining the shape of 3D solids based on the principle of orthogonal projection.

4.3 组合体三视图的读图

由前述可知,画图是把空间的组合体用正投影法画成平面图形。本节要研究的读图则是根据平面图形,运用正投影的投影原理想象出空间组合体的形状。

4.3.1 Outline of reading views

1. Understanding the meaning of individual lines and line frames of the views

(1) Lines and curves (including circular arcs)

1) Represent the projection of an accumulation plane (line A shown in Fig. 4-10).

2) Represent the projection of an intersection line of two planes (line B shown in Fig. 4-10).

3) Represent the projection of the outline of a cylindrical surface or ruled surface (line C shown in Fig. 4-10).

(2) Each closed line frame

1) Represent the projection of a plane (frame Ⅰ shown in Fig. 4-10).

2) Represent the projection of a curved surface (frame Ⅱ shown in Fig. 4-10).

3) Represent the projection of a curved surface and its tangent plane (frame Ⅲ shown in Fig. 4-10).

4.3.1 读图要点

1. 要读懂视图中图线、线框的含义

(1) 视图上的直线和曲线(包括圆弧)

1) 表示具有积聚性表面的投影(如图4-10中线A)。

2) 表示两表面交线的投影(如图4-10中线B)。

3) 表示曲面轮廓线的投影(如图4-10中线C)。

(2) 视图上的封闭线框

1) 表示平面的投影(如图4-10中线框Ⅰ)。

2) 表示曲面的投影(如图4-10中线框Ⅱ)。

3) 表示曲面与其相切平面的投影(如图4-10中线框Ⅲ)。

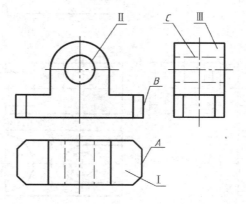

Fig. 4-10 The meaning of lines and line frames on a drawing   图线及线框的含义

2. Reading several views simultaneously

It may not be possible to reveal the general spatial shape of a composite solid just by one view. We must read all views simultaneously and analyze them using the projection rules. In Fig. 4-11a and Fig. 4-11b, the front views and left views are the same. However, the shapes of the corres ponding bodies are different as the top views are different. In Fig. 4-11c and Fig. 4-11d, the front views and top views are the same. However, the shapes of the corres ponding bodies. are not the same as the left views are different.

When reading drawings, we should not make a conclusion after reading only one or two views. We must carefully compare all views and make a conclusion until they can be fully explained following the rules of projection. The reading is a process of comparing the possible spatial shapes with the actual drawings. This is an iterative process and we should try again and again until the correct shape is obtained.

2. 要几个视图联系起来读

一般仅凭一个视图不能完全确定组合体的空间形状,读图时要将各视图联系起来读,根据投影规律进行分析比较。如图4-11a、b所示,虽然两立体的主、左视图相同,但俯视图不同,两者形状也不同。而图4-11c、d中的主、俯视图相同,但左视图不同,两立体形状也不同。

因此在读图过程中切忌看了一两个视图就下结论。要反复对照各视图,直至确认其均符合投影规律时,才能最后下结论。读图的过程是不断地把空间形状与各视图反复对照、反复修改的过程。只有不断修正,才能想象出正确的形体。

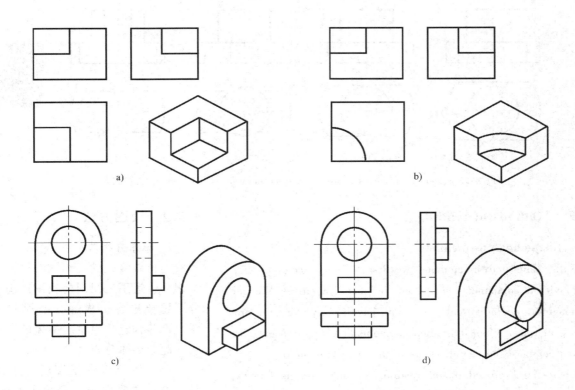

Fig. 4-11 Reading several views simultaneously　将各视图联系起来读

3. Starting from the view that most reflect shape characteristics

In Fig. 4-11a and Fig. 4-11b, we can see that the top views most represent the shape characteristics of the composite solids. Thus the top views are the character views of these solids. In Fig. 4-11 c and d, the left views are the character views. While reading views, we must pay special attention to the character view for identifying the basic shape of the composite solid.

4. Analyzing the relative positions and intersection lines of neighboring surfaces

While reading views, we should also pay attention to analyzing the positons of adjacent surfaces, such as front and back, top and bottom as wellas intersection. In Fig. 4-12a, the diameter of the cylinder is smaller than the width of the bottom-board. One thus sees an intersection curve (a circle) in the top view. In Fig. 4-12b, the diameter of the cylinder is equal to the width of the bottom-board, thus there is an intersection curve (a circle) in a tangent condition with the two side planes of the board on the top view, In Fig. 4-12c, the diameter of the cylinder is larger than the width of the bottom-board. Thus there are two line segments in the front view representing the intersection lines of the cylinder and bottom-board. as shown in fig. 4-12c. In Fig. 4-12d, both the cylinder and the bottom-board are cut by a plane. It produces intersections between the projected circle and lines on the top view.

3. 要从最能反映形状特征的视图读起

从图 4-11a、b 可看出，俯视图最能代表相应组合体的形状特点，因而为特征视图；图 4-11c、d 中，特征视图都为左视图。所以，看图时，应特别注意抓住特征视图，以确定组合体基本形状。

4. 分析相邻表面间的相互位置和交线

读图时，要注意分析相邻面的前后、高低和相交等相互位置关系。图 4-12a 中圆柱直径小于底板的宽度，因而有交线（圆）产生；图 4-12b 中圆柱直径与底板宽度相同，俯视图中的线和圆相切；图 4-12c 中圆柱直径大于底板宽度，其交线在主视图上显示为两条线段；图 4-12d 中圆柱与底板一同被平面所截，产生交线（圆+直线）。

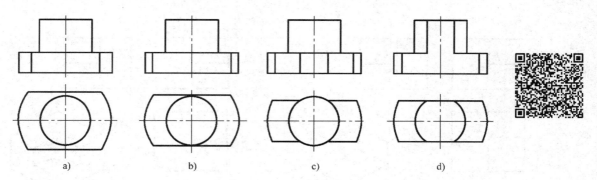

Fig. 4-12 Intersection curves between two adjacent surfaces 相邻表面间的交线

4.3.2 Methods of reading

1. Shape analysis method

With shape analysis method, we first divide the complex view into individual components based on the loops of curves. We can then imagine their individual shapes and locations by using the projection principles. When individual components are figured out, we can then compose the overall shape of the composite solid.

The general procedures of reading drawings are as follows: first read the primary components, then read the other components; first read easy components, then read difficult components; first read the overall shape for each component, then read its shape details. Now, we use the plank shown in Fig. 4-13 as an example to illustrate the steps of the shape analysis method.

(1) Partition of drawing elements Usually begin with the front view. It can be divided into four parts 1′, 2′, 3′, 4′ as shown in Fig. 4-13.

(2) Comparison of drawing elements Find their corresponding projections in other views based on the projection relationship with rulers and compasses as shown in Fig. 4-13.

(3) Understanding the shapes from the drawing elements

1) Firstly read the line frame Ⅱ. Obviously, it is a half column as shown in Fig. 4-14.

4.3.2 读图方法

1. 形体分析法

把比较复杂的视图按线框分成几个部分，运用三视图的投影原理，先分别想象出各组成部分的形状和位置，再综合起来想象出整体的结构形状，此为形体分析法。

用形体分析法读图的一般顺序是：先看主要部分，后看次要部分；先看容易确定的部分，后看难以确定的部分；先看某一组成部分的整体形状，后看其细节部分形状。下面，以图4-13所示支架为例，说明形体分析法读图的具体步骤。

（1）分线框 一般从主视图上分，该图可分成四个线框，如图4-13中线框1′、2′、3′、4′。

（2）对线框 用尺和圆规，根据投影关系，找到三个线框的另两个投影，如图4-13所示。

（3）看线框识形体

1）一看立体Ⅱ的线框。如图4-14所示，很明显，这个立体是一个半圆筒。

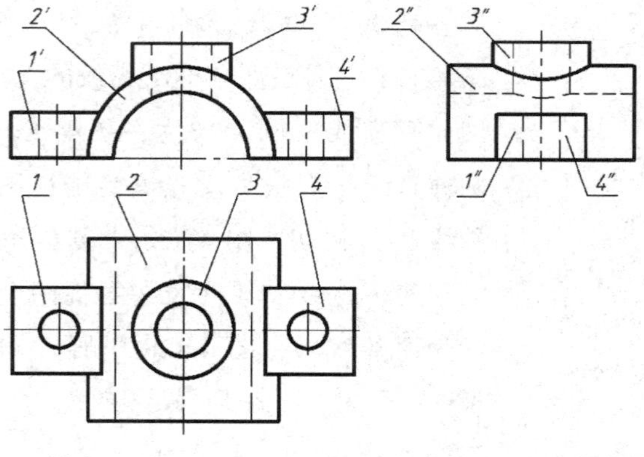

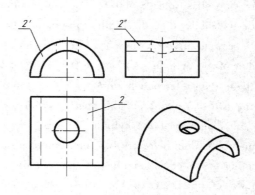

Fig. 4-13 Three-view drawing of a plank 支架的三视图

Fig. 4-14 Three-view drawing and pictorial clrauing of Ⅱ
立体Ⅱ的三视图和立体图

2) Secondly read the line frame Ⅲ. It is a cylindrical column as shown in Fig. 4-15. Its lower side is tangent to the half column Ⅱ and thus is cut off for attaching to the half column Ⅱ.

3) Thirdly read the line frames Ⅰ and Ⅳ. They have the same shape as shown in Fig. 4-16. The three views show the body is a square prism. Seeing from the front view, the right part of the square prism is cut off by a cylinder for the same objective as for that Ⅲ attaches to Ⅱ.

(4) Overall imagination Based on the above analysis, we know the shapes of individual components. We can then analyze their relations and relative locations from the three views (see Fig. 4-14): Ⅰ, Ⅲ and Ⅳ intersect with Ⅱ, Ⅲ is on the top of Ⅱ, Ⅰ and Ⅳ are on the two sides of Ⅱ, respectively. Thus, the space shape of the plank is concluded as shown in Fig. 4-17.

Example 4-2 Given the front and the top views shown in Fig. 4-18, draw the left view of the solid.

Analysis: From the given views, we can conclude that the solid is made up of three parts.

2）二看Ⅲ的线框。如图4-15所示，从三视图看，立体Ⅲ应为一个圆筒。其下侧为了与半圆筒Ⅱ相贯，所以切去了一部分形成圆柱面，因为这部分要贴在圆柱Ⅱ上。

3）三看Ⅰ、Ⅳ的线框。两者的形状相同，如图4-16所示，表示一个四棱柱。从主视图可以看出，四棱柱右侧切去一部分形成圆柱面，其目的同Ⅲ一样是为了贴在Ⅱ上。

（4）综合想象 通过以上分析，可得知各部分的形状，再读已知的组合体三视图，分析各个部分的相对位置关系：Ⅰ、Ⅲ、Ⅳ与Ⅱ是相交关系，Ⅲ在Ⅱ的上面，Ⅰ、Ⅳ在Ⅱ的两侧。至此，可以得到支架的形状，如图4-17所示。

例4-2 如图4-18所示，已知主、俯视图，补画左视图。

分析：从已知视图看，该组合体由三部分组成。

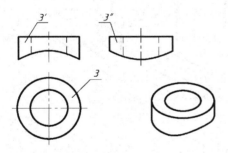

Fig. 4-15 Three-view drawing and pictorial drawing of Ⅲ
立体Ⅲ的三视图和立体图

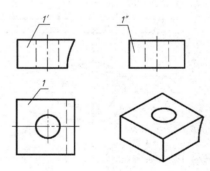

Fig. 4-16 Three-view drawing and pictorial drawing of Ⅰ and Ⅳ
立体Ⅰ、Ⅳ的三视图和立体图

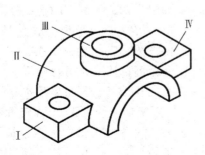

Fig. 4-17 Pictorial drawing 立体图

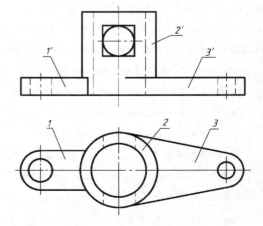

Fig. 4-18 The front and top views of Example 4-2
例4-2图形

(1) Partition of line frames　The top view can be divided into three line frames 1, 2 and 3, as shown in Fig. 4-18.

(2) Comparring the line frames　By shape comparing, the line frames 1′, 2′ and 3′, corresponding to the line frames 1, 2 and 3 on the top view respectively, are obtained on the front view, as shown in Fig. 4-18.

(3) Understanding the shape　Since the front view of Ⅱ is a rectangle and its top view is a circle, it is therefore a column. From the visibility of the rectangle and the circle in the front view, the front hatch is quadrate and the back hatch is circular. Also, the front view of Ⅰ is a rectangle, the top view is a half-circle and a rectangle. It is therefore a vaulted column. From the two views of Ⅲ, the basic shape is a cuboid. For their relationships, Ⅰ intersects with Ⅱ and Ⅲ is tangent to Ⅱ.

(4) Overall imagination　Based on the above analysis, we have clearly identified the space shape. We could then draw its left view as shown in Fig. 4-19. Be sure to check whether the three views meet the projection rules.

(1) 分线框　如图 4-18 俯视图所示,可分为 1、2、3 个线框。

(2) 对线框　由俯视图 1、2、3 各线框形状,对应找出主视图的线框 1′、2′、3′,如图 4-18 所示。

(3) 识形体　Ⅱ 的主视图为矩形,其俯视图为圆形,故形体应为圆柱体。又因为 Ⅱ 的主视图上矩形和圆形投影的可见性,故其上开孔应为前方后圆;Ⅰ 的主视图为矩形,俯视图为半圆加矩形,故该形体应为拱形柱;从 Ⅲ 的两视图看,其基本体应为长方体。从相对位置来看,Ⅰ 与 Ⅱ 为相交关系,Ⅲ 与 Ⅱ 为相切关系。

(4) 综合想象　通过以上分析,即可得出组合体形状,进而可以补画出左视图,如图 4-19 所示。注意检查三视图之间的投影是否符合投影规律。

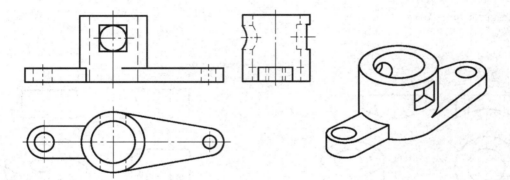

Fig. 4-19　The resulting three-view drawing and a pictorial drawing of Example 4-2　例 4-2 的三视图及立体图

Example 4-3　Draw the missing lines shown in Fig. 4-20a.

Analysis: From the three views, its main shape is a cuboid which is cut and piled. In Fig. 4-20b, the front view of Ⅰ is a triangle. Its top view is a rectangle. It is therefore a triangular prism. We can then fill a rectangle in the left view. From the three views of Ⅱ, it is a cuboid on the right side. Ⅴ is nearly the same as Ⅱ, but Ⅴ is only cut by Ⅲ. Since the front plane of Ⅴ and Ⅱ belong to the same plane, there is no line in the front view. Furthermore, Ⅲ is a cuboid cut from Ⅴ and its three views are rectangles. We should thus add a line in the front view. The left view can also be processed similarly. Finally, Ⅳ is obtained from Ⅱ by cutting off a quarter column at the top right corner and it appears as a rectangle in the other two views. We should therefore add a line in the front view and the top view.

例 4-3　补齐图 4-20a 所示视图中的缺线。

分析:综合三视图看,该形体主结构为长方体经过切割与叠加而成。如图 4-20b 所示,Ⅰ 的主视图为三角形,俯视图为矩形,故它是三棱柱,俯视图中应补上一条线,左视图应补一矩形;从 Ⅱ 的三视图看,应为在右侧的长方体;Ⅴ 与 Ⅱ 基本相同,只是在 Ⅴ 上切去的是 Ⅲ,因 Ⅴ 与 Ⅱ 前表面平齐,故在主视图中无线;Ⅲ 是在 Ⅴ 上被切去的一个长方体,其三视图应为矩形,故主视图应补上一条线,左视图同理;Ⅳ 是在 Ⅱ 的右上角切去了 1/4 圆柱而成,其另两视图均应为矩形,故在主、俯视图上应各画上一条直线。

From the above analysis, we can then add the missing lines. We can also produce the pictorial drawing after reading and add the missing lines afterwards until the drawing represents the object shown in Fig. 4-21.

根据以上分析，即可补画出所缺图线。也可以在看图时画出形体的直观草图（图4-21），对照所补画的图线，直至图物相符。

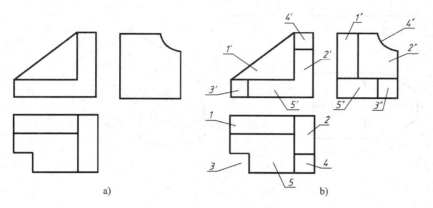
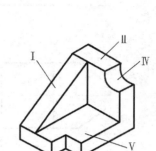

Fig. 4-20　Three-view drawings of Example 4-3　例4-3 三视图
a) Incomplete drawing with missing lines　补画漏线　b) Analysis　分析

Fig. 4-21　Axonometric drawing of Example 4-3　例4-3 立体图

2. Analysis of lines and planes

A composite solid may be considered as an enclosed volume by several surfaces (planes and curves). There are often intersection curves/lines between the surfaces/planes. With this method, we partition the composite solid into individual boundary surfaces. We analyze and identify the space shapes of the individual surfaces and their relative locations following the rules of projection. We can further determine intersection curves/lines and their relative locations and finally imagine the overall space shape of the composite solid.

In the following, we use the drawing of a board as an example to illuminate its application.

Example 4-4　The front view and the top view of the board are shown in Fig. 4-22a. Add the missing left view.

Analysis：From the given two views, the basic body is a cuboid. Its left view is therefore a rectangle as shown in Fig. 4-22b. The front view can be divided into frames 1′ and 2′ and they show two planes of the board respectively. Based on the projection, frame 1′ and the two symmetrical oblique lines fore-and-aft in the top view satisfy projection properties. We can then conclude that I is are two H-perpendicular planes and their side projections are similar to the drawings in the front view respectively. For the same reason, we can conclude that II are two frontal planes which are symmetrical fore-and-aft, and the side projections are two lines. The projection of frame 3 on the top view should be an oblique line in the front view following projection analysis. Obviously, III is a V-perpendicular plane. Its side projection is similar to the drawing in the top view. The projection of IV is a horizontal line in the front view. It is therefore a horizontal plane and the side projection should be a line. The lower side VI can be identified with the same principle. The projections of V in the front view and the top view are lines with different lengths and are perpendicular to the OX-axis. The space shape of V should therefore be a profile plane and its side projection is a rectangle which shows the true shape. The analysis of the right plane is the same as that for V. The overall shape is thus figured out. The three-view drawing is shown in Fig. 4-22b and the pictorial drawing is shown in Fig. 4-22c.

2. 线面分析法

组合体也可以看成是由若干个面（平面和曲面）围成，面与面间常存在交线。线面分析法就是把组合体分成若干个面，根据其投影规律确定其空间形状和相对位置，并判别表面的空间形状和相对位置，进一步确定交线及其相对位置，从而想象出组合体的形状。

下面以压板图为例，说明线面分析法的应用。

例4-4　如图4-22a所示，已知压板的主、俯视图，补画左视图。

分析：从两视图看，该立体基本体为长方体，因此，其左视图主轮廓为矩形，如图4-22b所示。主视图可分成1′、2′两个线框，分别表示压板的两个表面。根据投影，线框1′与俯视图中前后对称的两条斜线符合投影关系，进而可以确定Ⅰ为两个铅垂面，其侧面投影分别为与其主视图相类似的图形；同理，可确定Ⅱ为前后对称的两个正平面，其侧面投影应为两条直线；俯视图中的线框3，按投影分析，对应在主视图的投影为斜线，显然，面Ⅲ为正垂面，侧面投影与俯视图中的投影图形类似；面Ⅳ的主视图为一条水平线，故为水平面，其侧面投影应为一直线；底面Ⅵ同理。面Ⅴ在主、俯视图中的投影为不等长直线，且垂直于OX轴，故面Ⅴ应为侧平面，形状为矩形，其侧面投影反映实形；压板右侧面与面Ⅴ同。至此可得压板的整体形状，其三视图如图4-22b所示，直观图如图4-22c所示。

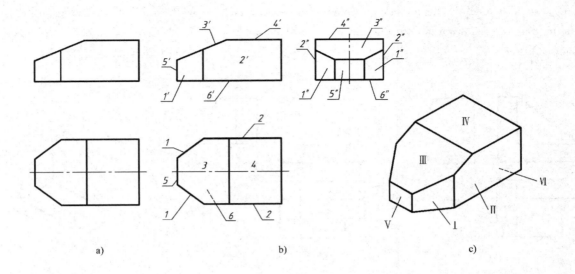

Fig. 4-22 Drawings of Example 4-4 例 4-4 图形
a) The two given views 已知二视图 b) Analyze and get the left view 分析及求得左视图
c) Pictorial drawing 立体图

It should be pointed out that the analysis of lines and planes is complementary to the shape analysis method. When analyzing composite solids, we can use the two methods together and first analyze the shape followed by the analysis of lines and planes if needed. For example, during the analysis of the board drawings, shape analysis method is first used to identify the main shape. Further details are identified through the analysis of lines and planes. We should avoid using one of the methods without considering the other one.

4.4 Dimensioning Composite Solids

The three-view drawing can only show the shape of a composite solid. Further information is supplemented by dimensions, which is very important for determining the size of composite solids.

4.4.1 Basic requirements for dimensioning

Correctness: There should be no mistake with all dimensions and the national standard (GB) should be observed.

Integrity: The dimensions must fully define the shape and size of the part, and relative positions of individual components. It can not be omitted, or be repeated.

应该指出，对于相当数量的组合体而言，线面分析法是形体分析法的补充。分析组合体时，可先进行形体分析，再根据需要进行线面分析，两种方法结合使用。例如在压板的视图分析中，首先进行的是形体分析，确定主体结构，而后对具体结构进行线面分析。切不可将两种方法分割开来使用。

4.4 组合体的尺寸标注

三视图只能表达组合体的形状，组合体的大小则要根据视图上所标注的尺寸来确定。因此，标注组合体尺寸，对确切表达组合体大小非常重要。

4.4.1 标注尺寸的基本要求

正确：尺寸无误，标注方法符合国标规定。

完整：所注尺寸必须能完全确定零件的形状、大小及其构成要素间的相对位置，不遗漏，不重复。

Clarity: The arrangement of the dimensions should be clear and clean.

Reasonableness: The dimensions must be marked according to the requirements of design, manufacture and inspection.

4.4.2 Dimensioning of basic bodies, cutting bodies and interpenetration bodies

1. Dimensioning of basic bodies

One usually needs to mark all dimensions including length, width and height. Demensionins methods for some common basic solids are shown in Fig. 4-23. For body of revolution, it only needs to mark the radial dimension (the diameter symbol "ϕ" is added before the dimension figure) and the axial dimension.

In Fig. 4-23, $S\phi$ represents the diameter of the sphere and in Fig. 4-24, SR represents the radius of the sphere.

The dimensions mentioned above are mainly used for defining the shape of basic boidies.

4.4.2 常见基本体、截切体及相贯体的尺寸注法

1. 基本体的尺寸标注方法

标注基本体的尺寸,一般要注出长、宽、高三个方向的尺寸。图 4-23 所示为几种常见基本体的尺寸标注方法。对于回转体,通常只要注出径向尺寸(直径尺寸数字前加符号"ϕ")和轴向尺寸。

图 4-23 中,$S\phi$ 表示球体直径,图 4-24 中 SR 表示球体半径。

这里所述的尺寸,主要是用来确定基本体自身形状大小的定形尺寸。

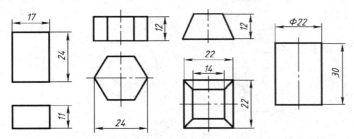 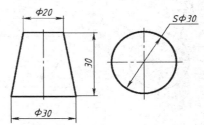

Fig. 4-23 Dimensioning methods for some common basic solids 一些常见基本体的尺寸标注方法

2. Dimensioning of subtraction/cutting type solids

When dimensioning cutting type solids, there are also dimensions for defining location of individual shape elements such as cutting planes, in addition to the above shape dimensions. As the intersection curve is uniquely defined after the relative position of the intersecting plane is determined, there is no need to mark dimensions of intersection curves. Fig. 4-24 shows several examples for dimensioning several cutting type solids.

3. Dimensioning of interpenetration type solids

Similar to the demensions of cutting type solids, there are also location dimensions in addition to size dimensions for interpenetration type bodies. With size dimensions and location dimensions, the interpenetration curves are uniquely defined, and there is no need to mark dimensions of interpenetration lines. Fig. 4-25 illustrates several examples for dimensioning several interpenetration type solids.

2. 截切体的尺寸标注方法

在标注截切体的尺寸时,除应标注上述定形尺寸外,还应注出确定截平面位置的定位尺寸。由于截平面在形体上的相对位置确定后,截交线即被唯一确定,因此对截交线不应再标注尺寸。图 4-24 所示为几种常见截切体的尺寸标注示例。

3. 相贯体的尺寸标注方法

与截切体的尺寸注法一样,相贯体除了应注出两相贯基本体的定形尺寸外,还应注出确定两相贯基本体相对位置的定位尺寸。当定形尺寸和定位尺寸注全后,两相贯基本体的相贯线即被唯一确定,因此对相贯线本身不应再标注尺寸。图 4-25 所示为几种常见相贯体的尺寸标注示例。

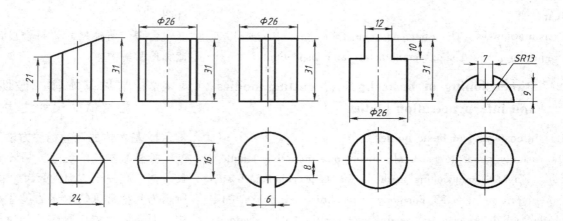

Fig. 4-24 Dimensioning examples of some common cutting solids 几种常见截切体的尺寸标注示例

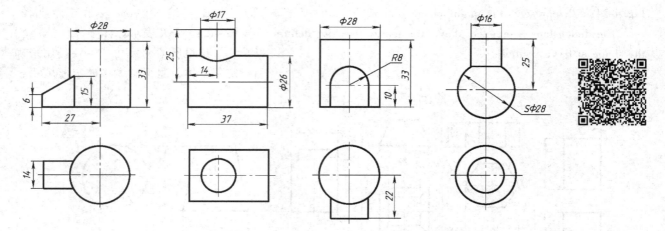

Fig. 4-25 Dimensioning examples of some common interpenetration solids 几种常见相贯体的尺寸标注示例

4. Dimensioning of supporting board elements

Fig. 4-26 illustrates several examples for dimensioning commonly used board elements. All the center-to-center spacing of holes, including general holes and threaded holes, and grooves on the board should be dimensioned. Since the basic shape of the board and the distribution of holes and grooves are different, the formats of location dimensions for dimensioning center-to-center spacing are also different.

4. 机件上常见平板的尺寸标注方法

图 4-26 所示为几种机件上常见平板的尺寸标注示例。板上孔（包括一般孔和螺孔）、槽等的中心距都应注出。而且由于板的基本形状和孔、槽的分布形式不同，其中心距定位尺寸的标注形式也不一样。

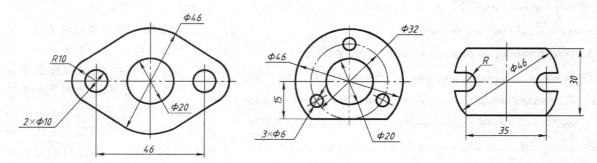

Fig. 4-26 Dimensioning examples of some common flat plates of solid parts 一些机件上常见平板的尺寸标注示例

4.4.3 Dimensioning general composite solids

1. Selection of dimension datums

Before dimensioning, we must select dimension datums. A dimension datum indicates common starting position of related dimensions. There should be at least one dimension datum respectively for length, width and height of a composite solid. Some geometrical elements, such as symmetry planes, bottom surfaces of supporting boards, end surfaces, axes, centers of circles are usually selected as dimension datums. In Fig. 4-27, the axis of the cylinder is the dimension datum for length; the front-back symmetry plane of the part is the dimension datum for width; the bottom surface of the supporting board is the dimension datum for height.

4.4.3 组合体的尺寸标注方法

1. 选定尺寸基准

在视图上标注尺寸，要先确定尺寸基准。尺寸基准是指尺寸的起始位置，是度量尺寸的起点。组合体在长、宽、高三个方向上至少分别有一个尺寸基准。通常选用组合体的对称面、底面、端面、轴线、圆心等几何要素作为尺寸基准。如图4-27所示，支架上圆筒的轴线为长度方向的尺寸基准；前后对称面为宽度方向的尺寸基准；底板的底面为高度方向的尺寸基准。

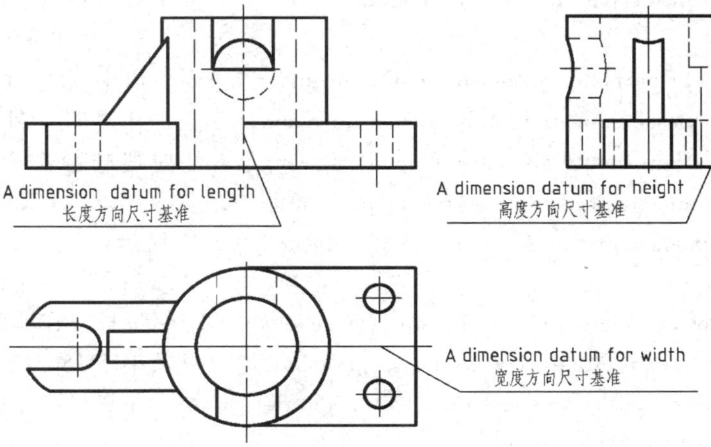

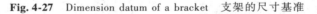

Fig. 4-27 Dimension datum of a bracket 支架的尺寸基准

2. Integrity in dimensioning

Dimensions should be fully and consistently marked without any missing or redundant dimensions. The following summarizes three general types of dimensions:

1) Shaping dimensions show sizes, such as the length, width and height of the basic parts.

2) Positioning dimensions show the relative positions between individual basic bodies.

3) Overall dimensions show the overall length, width and height of a composite solid.

To dimension completely, an effective method is to first analyze the shape of the composite solid and then mark all types of dimensions as mentioned above based on the individual shape elements and their relative locations. In Fig. 4-28, the bracket is composed of the left bottom-board, the right bottom-board, the rib and the cylinder. In the views, we should include not only all size dimensions, but also all location dimensions of the shape elements. The steps are shown in Fig. 4-29.

2. 标注尺寸要完整

标注尺寸要完整，既无遗漏，又不重复或多余。尺寸有三类：

1) 定形尺寸是表示各基本体长、宽、高三个方向大小的尺寸。

2) 定位尺寸是表示各基本体之间相对位置的尺寸。

3) 总体尺寸是表示组合体外形的总长、总宽、总高的尺寸。

要做到尺寸标注完整，有效的方法是先对组合体进行形体分析，然后根据各基本体形状及其相对位置注全上述三类尺寸。如图4-28所示，支架由左底板、右底板、肋板和圆筒组成，在它的各视图上，不但要注全各组成部分大小的尺寸，即定形尺寸，还应将各部分的定位尺寸标注出来。支架尺寸标注的步骤如图4-29所示。

3. Clarity in dimensioning

The dimensions should be properly distributed for easy reading. The following should be noticed:

1) Dimensions should be marked in the view that most reflects the characteristics of the related shapes and locations. In Fig. 4-28, the dimensions of 26 and 16 of the square-groove in the cylinder should be better marked in the front view which can clearly show its shape and location characteristics. The external dimensions $R6$ and $R14$ of the left bottom-board should be marked in the top view, not in the front view or in the left views.

2) The location dimensions and shaping dimensions of the same shape element should be marked in the same view as far as possible. In Fig. 4-28, the shaping dimension of $2\times\phi 8$, and the location dimensions of 26, 38 of the two circular holes in the right bottom-board are all marked in the top view for the reading convenience.

3) All dimensions should be better arranged outside the individual view in order to avoid possible overlap of dimension lines and feature lines of the drawing. They should also be arranged following the principle of "big outside and small inside". An example can be found in Fig. 4-28 for the dimensioning of 26, (42), 38, 44 and 92 in the top view.

4) For the diameters of cylinders and cones, the dimensions should be better marked in a non-circle view as far as possible, while for the semi-circular arcs and the arcs less than 180°, the radial dimension should be marked in the circle view.

3. 标注尺寸要清晰

尺寸标注布局要恰当，便于看图。所以尺寸标注应注意：

1）尺寸应尽可能标注在最能反映形体形状特征、位置特征的视图上。如图 4-28 所示，圆筒上方槽的定位尺寸 26 和宽度尺寸 16 应注在反映形状特征明显的主视图上较好，左底板外形尺寸 $R6$、$R14$ 应注在俯视图上，不应注在主、左视图上。

2）同一基本体的定位尺寸和定形尺寸应尽可能集中标注在同一视图上。例如图 4-28 中右底板上两圆孔的定形尺寸 $2\times\phi 8$ 和定位尺寸 26、38 集中标注在俯视图上，以便于看图。

3）尺寸应尽量配置在视图的外面，以避免尺寸线和轮廓线交错重叠，并遵照"外大里小"的尺寸标注原则进行。例如图 4-28 中俯视图上的尺寸 26 和 42，以及 38、44 和 92 所示。

4）标注圆柱、圆锥的直径时，应尽量标注在非圆视图上；而半圆弧以及小于 180°的圆弧，其半径尺寸必须标注在投影为圆弧的视图上。

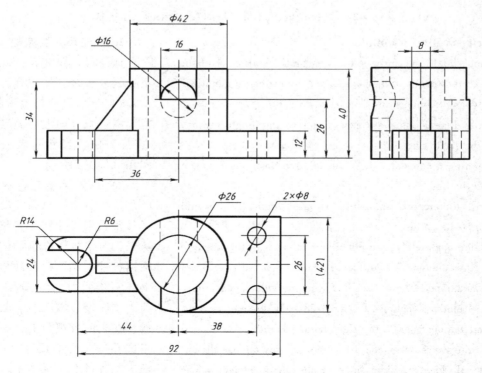

Fig. 4-28 Dimensions of a bracket 支架的尺寸

Several examples can be found in Fig. 4-28, such as the outside diameter of φ42 of the cylinder in the front view, and the external dimensions of R6, R14 of the left bottom-board in the top view.

5) There should be no dimensions for intersection lines and the dimensions should not be marked on dashed lines as far as possible.

6) When adjusting the overall size, if there is an arc structure, the overall size is usually marked to the arc center, and thus it is determined indirectly, as shown in Fig. 4-29c, in which the size of length is 106 (14+44+48). If the diameter (or radius) of the circle (or the arc) and the length of the corresponding direction are same, the size of length is employed as a reference size and is marked with brackets in order to avoid a closed dimension chain, as shown in Fig. 4-29d, the brackets are needed to be added to the size of the bottom-board 42 (the same as the outer diameter φ42 of the cylinder).

Fig. 4-29 is an example illustrating the steps for marking dimensions.

如图 4-28 中主视图上圆筒外径尺寸 φ42 和俯视图上左底板外形尺寸 R6、R14 所示。

5) 组合体的交线上不能标注尺寸，且尽量不在虚线上标注尺寸。

6) 当有圆弧结构需要调整总体尺寸时，通常将总体尺寸标注到圆弧中心，此时该方向总体尺寸将间接得到，如图 4-29c 中的长度尺寸为 106（14+44+48）。当圆形（圆弧）的直（半）径与该方向长度相同时，为避免封闭尺寸链，应将长度尺寸作为参考尺寸并加括弧，如图 4-29d 中底板宽度尺寸 42（与圆筒外径尺寸 φ42 相同）应加括弧。

图 4-29 为尺寸标注步骤举例。

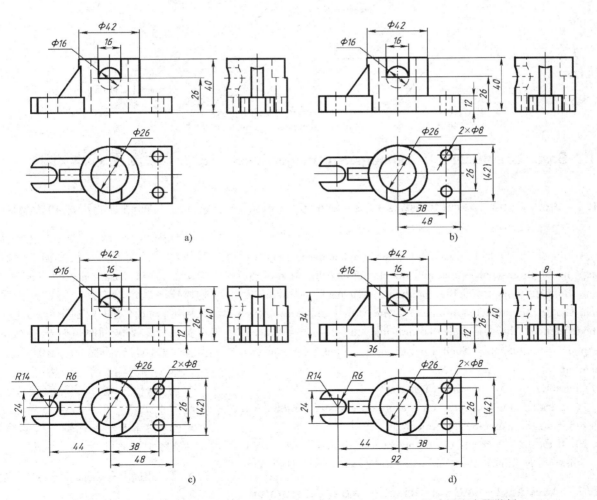

Fig. 4-29 An example illustrating the steps for marking dimensions 尺寸标注步骤举例
a) Mark the dimensions of the cylinder, and the hole, square slot on it 标注圆筒及其上圆孔和方槽的尺寸
b) Mark the dimensions of the right bottom-board and the hole on it 标注右侧底板及其上圆孔的尺寸
c) Mark the dimensions of the left bottom-board and the U slot on it 标注左侧底板及其上 U 形槽的尺寸
d) Mark the dimensions of left rib plate and adjust the overall size 标注左侧肋板的尺寸并调整总体尺寸

Chapter 5　Axonometric Projections

An axonometric projection is the one in which the object is viewed in such a position that several faces appear in a single view (see Fig. 5-1). Axonometric drawings are three-dimensional but poorly measurable and it is difficult to draw. Therefore, axonometric drawings are often used to supplement multiview drawings.

第5章　轴　测　图

轴测图是一种能同时反映物体三维空间形状的单面投影图（见图5-1）。这种图富有立体感，但度量性差，作图困难，在工程应用中一般只作为辅助图样。

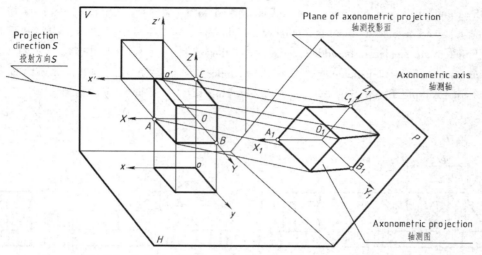

Fig. 5-1　Construction of an axonometric projection　轴测图的形成

5.1　Basic Knowledge of Axonometric Projections

5.1　轴测图基本知识

5.1.1　Form and characteristics of an axonometric projection

As shown in Fig. 5-1, place a P-plane (axonometric projection plane) in the three-view system and select a proper direction of projection. Both the object and the coordinate system of the three-view system should be projected to the P-plane with parallel projection method. Hence, the new projected drawing can reflect three coordinate planes simultaneously. Such a projection is called an axonometric projection.

Characteristics of axonometric projections are:

1) Parallel line segments on the object are still parallel with each other on the axonometric projection.

2) If the line segments on the object are parallel to the coordinate, they are parallel to the coordinate on the axonometric projection.

5.1.2　Axes angles and coefficient of axial deformation

As shown in Fig. 5-1, the projections (O_1X_1-axis, O_1Y_1-axis and O_1Z_1-axis) of coordinate axes (OX-axis, OY-axis and OZ-axis) on the P-plane are called axonometric axes. The resulting angles ($\angle X_1O_1Y_1$, $\angle X_1O_1Z_1$ and $\angle Y_1O_1Z_1$) are called axes angles.

5.1.1　轴测图的形成和投影特性

如图5-1所示，在原三投影面体系适当位置设置一个投影面 P（轴测投影面），选取适当的投射方向 S，将物体连同确定其空间位置的直角坐标系用平行投影法投射到 P 平面上，所得到的投影图能同时反映出三个坐标面，这样的投影图称为轴测图。

轴测图具有如下投影特性：

1) 物体上互相平行的线段，在轴测图上仍然互相平行。

2) 物体上与坐标轴平行的线段，在轴测图中仍平行于相应轴测轴。

5.1.2　轴间角和轴向伸缩系数

如图5-1所示，空间坐标轴（OX、OY、OZ）在轴测投影面 P 上的投影（O_1X_1、O_1Y_1、O_1Z_1）称为轴测轴，轴测轴之间的夹角（$\angle X_1O_1Y_1$、$\angle X_1O_1Z_1$、$\angle Y_1O_1Z_1$）称为轴间角。

The length ratio between line segments on the axonometric axis and that on the corresponding space coordinate axis is called a coefficient of axial deformation. The coefficient of axial deformation of the X-axis is $p = O_1A_1/OA$. The coefficient of axial deformation of the Y-axis is $q = O_1B_1/OB$ and the coefficient of axial deformation of the Z-axis is $r = O_1C_1/OC$.

When preparing an axonometric drawing, one could determine the length of line segments according to the corresponding coefficient of axial deformation. In this chapter, we introduce the methods of two commonly used axonometric projections, i.e., isometric projections and cabinet axonometry projections.

5.2 Isometric Projections

5.2.1 Construction of an isometric projection, axes angles and coefficients of axial deformation

1. Construction of an isometric projection

To produce an isometric projection, it is necessary to view an object that its principal edges are equally inclined to the viewer and hence are foreshortened equally.

2. Axes angles and coefficients of axial deformation

In an isometric projection, coordinate axes (O_1X_1-axis, O_1Y_1-axis and O_1Z_1-axis) are equally inclined to the P-plane and the isometric axes form three axes angles of 120° from each other (see Fig. 5-2). Due to the same axes angles, it can be shown that the coefficients of axial deformation for the three axes are $p_1 = q_1 = r_1 = 0.82$. It means that the length, width and height of an object should be reduced to 0.82 times in the drawing. However, as the utilization of such a ratio is quite complicated for drawing preparation, we often use a simplified ratio with $p = q = r = 1$. It results in a view that the size are (1/0.82 =) 1.22 times larger than that produced with the isometric projection, but conveys the same pictorial presentation of the objects. Fig. 5-3b and Fig. 5-3c show different isometric views with the different coefficients of axial deformation, respectively.

轴测轴上的线段与空间坐标轴上对应线段的长度比称为轴向伸缩系数。X轴的轴向伸缩系数$p = O_1A_1/OA$,Y轴的轴向伸缩系数$q = O_1B_1/OB$,Z轴的轴向伸缩系数$r = O_1C_1/OC$。

在画轴测图时,可按相应的轴向伸缩系数直接量取有关线段的尺寸。下面分别介绍两种常用轴测图的画法:正等轴测图画法和斜二等测图画法。

5.2 正等轴测图

5.2.1 正等轴测图的形成、轴间角和轴向伸缩系数

1. 正等轴测图的形成

当空间直角坐标轴向轴测投影面倾斜的角度相同时,用正投影法得到的投影图称为正等轴测图。

2. 正等轴测图轴间角和轴向伸缩系数

在正等轴测图中,三个坐标轴与轴测投影面倾斜的角度相同,其三个轴间角相等,都是120°(见图5-2),可以证明,轴向伸缩系数$p_1 = q_1 = r_1 = 0.82$。也就是在画图时物体在长、宽、高三个方向的尺寸均要缩小0.82倍。由于按这一轴向伸缩系数作图烦琐,在制图中常采用简化的轴向伸缩系数,即取$p = q = r = 1$。也就是在作图时,沿各轴方向取相应的实长,这种正等轴测图较采用简化前轴向伸缩系数的正等轴测图其放大倍数为(1/0.82 =)1.22倍,但不影响对物体形状的理解。图5-3b、c所示分别为根据两种轴向伸缩系数画出的轴测图。

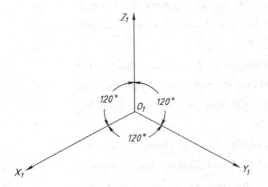

Fig. 5-2 Axes angles of isometric projection 正等轴测图的轴间角

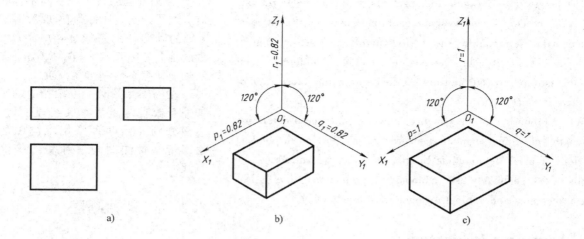

Fig. 5-3 Isometric projections with different coefficients of axial deformation 用不同轴向伸缩系数绘制的正等轴测图
a) Orthographic views 正投影图 b) Isometric projection 正等轴测图 ($p_1=q_1=r_1=0.82$) c) Isometric projection 正等轴测图 ($p=q=r=1$)

5.2.2 Isometric projections of polyhedral solids

The main purpose of an isometric view is to provide a pictorial presentation which reveals as many details of solids as possible, In order to draw the isometric projection, one should set up the axis coordinate system firstly, then draw line segments with length values of the original figure in the corresponding axial (length, width and height).

Example 5-1 Produce an isometric projection of a regular hexagonal prism.

Analysis: The lower base of this regular hexagonal prism is a right hexagon which is parallel to the horizontal plane. For convenience, the center of the right hexagon may be selected as the origin O of the isometric coordinate system (see Fig. 5-4a).

Drawing steps:

1) Draw the isometric axes. Set up an orthogonal coordinate system in the three-view system as shown in Fig. 5-4b.

2) Draw the lower base of right hexagon. Measure the line segments O_1M and O_1N on O_1X_1, and make $O_1M\ O_1M=OM$ and $O_1N=ON$. The points A, B, C and D are determined by the length b on Y_1-axis and the length of AB and CD along X_1-axis. Draw the lower base hexagon of the regular hexagonal prism by connecting the six points A, B, N, C, D and M.

3) Draw six parallel line segments to O_1Z_1 with length h starting from points A, B, N, C, D and M respectively. The six parallel line segments are defined as the six edge lines of the regular hexagonal prism.

4) Connect the end points of the six line segments above to form the upper regular hexagon of the prism. Erase the unnecessary lines and complete the isometric projection of the hexagonal prism (see Fig. 5-4c).

5.2.2 平面立体正等轴测图的画法

绘制正等轴测图的主要目的是用立体图尽可能详细地表示一个立体，选择正等轴测轴时必须牢记这一点。

例5-1 画出正六棱柱的正等轴测图。

分析：正六棱柱的底面是处于水平位置的正六边形。为使作图方便，可以把坐标原点定在六棱柱底面的中心 O 处，如图5-4a所示。

作图步骤：

1) 画出轴测轴。在原图上建立正交坐标系，如图5-4b所示。

2) 画出底面的正六边形。在 O_1X_1 轴上截取 O_1M、O_1N，使 $O_1M=OM$、$O_1N=ON$；在 O_1Y_1 轴上以尺寸 b 和线段 AB、CD 的长度来确定 A、B、C、D 各点。按顺序连接 A、B、N、C、D 和 M 六点得底面正六边形。

3) 由六边形各顶点 A、B、N、C、D 和 M 分别作 O_1Z_1 的平行线，截取各平行线长为 h，即得到六条棱线。

4) 依次连接各点，即得上顶面正六边形。擦去多余线段，即得到该正六棱柱的正等轴测图，如图5-4c所示。

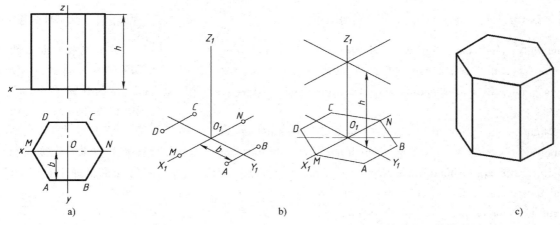

Fig. 5-4 Isometric projection of a regular hexagonal prism 六棱柱的正等轴测图

Example 5-2 Produce an isometric projection of a polyhedral solid shown in Fig. 5-5a.

Analysis: This polyhedral solid is formed by cutting off a quadrangular prism with a trapeziform base and a regular triangular prism from a cuboid.

Drawing steps:

1) Set up an isometric coordinate system as shown in Fig. 5-5b.

2) Draw the isometric axes and complete the isometric projection of the integrated cuboid before cutting/trimming (see Fig. 5-5b).

3) Cut off the left upper quadrangular prism as shown in Fig. 5-5c, measure the sizes along the directions of the corresponding isometric axes and position the corresponding points I, II, III, IV, and VII. Draw related edges according to the axonometric characteristics as shown in Fig. 5-5c.

4) Cut off the lower left corner as shown in Fig. 5-5d in a similar way. The complete drawing is shown in Fig. 5-5e.

例 5-2 求图 5-5a 所示平面立体的正等轴测图。

分析：该立体可以看成是从长方体上先后切去一个以梯形为底面的四棱柱和一个底角（三棱柱）后所形成的平面立体。

作图步骤：

1）建立正等轴测坐标系，如图 5-5b 所示。

2）画出轴测轴，并作未切前完整长方体的轴测图，如图 5-5b 所示。

3）切去左上方的四棱柱，沿相应轴测轴方向量取尺寸，找出相应的 I、II、III、IV、VII 各点，根据平行线段在轴测图上相应平行的性质，画出图形，如图 5-5c 所示。

4）切去左前下角的三棱柱，同理，截取各点后，画出图形，如图 5-5d 所示。图 5-5e 所示为完成的正等轴测图。

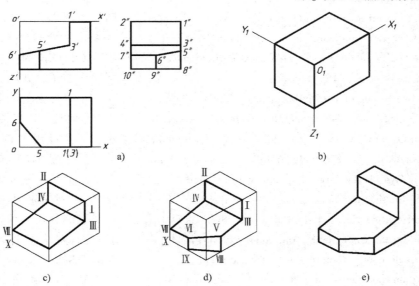

Fig. 5-5 Isometric projection with cutting method 用切割法作正等轴测图

5.2.3 Isometric projections of revolution bodies

Generally, the base plane or section of all bodies of revolution is a circle. When the circular base plane is parallel to an isometric projection plane, its corresponding isometric projection appears as an ellipse.

A "four-center approximate ellipse" method is adopted to get the isometric projection of a circle, in which four arcs with different centers are drawn approximately and connected orderly to form the elliptical projection of the circle. Steps of drawing such a "four-center approximate ellipse" on the horizontal plane are shown in Fig. 5-6.

Drawing steps:

1) Draw a rhombus EFGH with center O_1 and let the dimensions of line segments AB and CD equal to ϕ (the diameter of the circle shown in Fig. 5-6a), as shown in Fig. 5-6b.

5.2.3 回转体正等轴测图的画法

回转体的截面或底面一般都呈圆形。当圆所在的平面平行投影面时，其轴测投影均为椭圆。

在正等轴测图中采用"四心近似椭圆"法，即以四个圆心为中心近似地画四条圆弧并顺序连接成椭圆。图5-6 所示为在水平面上作"四心近似椭圆"的步骤。

作图步骤：

1) 以圆心 O_1 为坐标原点，过圆心 O_1 在轴测轴 X_1Y_1 上量取 AB、CD 等于 ϕ（图 5-6a 中圆的直径），得到菱形 EFGH，如图 5-6b 所示。

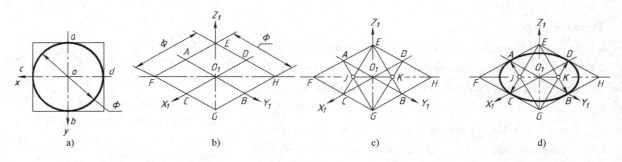

Fig. 5-6 Steps of drawing "four-center approximate ellipse"　"四心近似椭圆"的作图步骤

2) Connect GA and mark the intersection point between GA and the large diagonal FH as J. Connect GD and mark the intersection point between GD and the large diagonal FH as K. Similarly, connect EC and EB produces the same intersection points J and K respectively, as shown in Fig. 5-6c.

3) Take E, G, J and K as the centers of the four arcs respectively. The radius of the two small arcs is defined by $R_1 = JA = JC = KB = KD$, while the radius of the two large arcs is defined by $R_2 = GA = GD = EC = EB$, as shown in Fig. 5-6d.

The isometric projection of a circular face on the front plane and the side plane appears are also appeared as ellipse and can be constructed in a similar way by using the "four-center approximate ellipse method". As shown in Fig. 5-7, it follows that the major axis of the oval is always on the large diagonal of the rhombus.

1) For a circle on the horizontal plane, the major axis of the ellipse on the isometric projection is perpendicular to the axis O_1Z_1.

2) For a circle on the front plane, the major axis of the ellipse on the isometric projection is perpendicular to the axis O_1Y_1.

3) For a circle on the side plane, the major axis of the ellipse on the isometric projection is perpendicular to the axis O_1X_1.

2) 连接 GA 与长对角线 FH 交于点 J，连接 GD 与长对角线交于点 K，同样，连接 EC 和 EB 与长对角线也分别交于点 J、K，如图 5-6c 所示。

3) 以点 E、G、J 和 K 作为四段圆弧的圆心，以 $R_1 = JA = JC = KB = KD$ 为半径画两段小圆弧，以 $R_2 = GA = GD = EC = EB$ 为半径画两段大圆弧，即完成作图，如图 5-6d 所示。

正平面上的圆和侧平面上的圆的正等轴测图也为椭圆，如图 5-7 所示，同样可用"四心近似椭圆"法绘制。由图 5-7 可知，椭圆的长轴总在菱形的长对角线上。

1) 对于水平面上的圆，其轴测投影椭圆的长轴垂直于 O_1Z_1 轴。

2) 对于正平面上的圆，其轴测投影椭圆的长轴垂直于 O_1Y_1 轴。

3) 对于侧平面上的圆，其轴测投影椭圆的长轴垂直于 O_1X_1 轴。

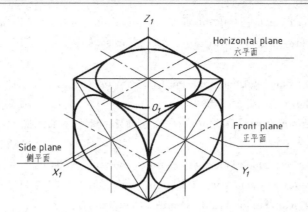

Fig. 5-7 The major axis of the oval is always on the large diagonal of the rhombus 椭圆形的长轴总在菱形的长对角线上

Example 5-3 Produce an isometric projection of the cylinder according to the two given views of acylinder, as shownin Fig. 5-8a.

例 5-3 已知圆柱的两个视图（见图 5-8a），作出该圆柱的正等轴测图。

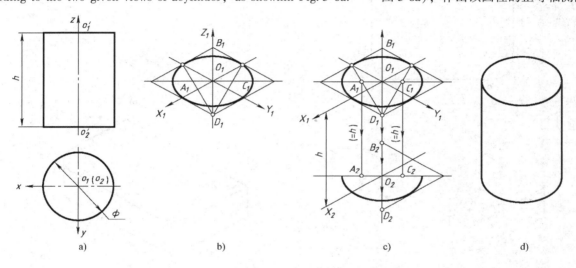

Fig. 5-8 Drawing an isometric projection of a cylinder 画圆柱的正等轴测图

Analysis: Both the upper and the lower faces are circles on the H-plane and are ellipses on the horizontal plane. Produce the two iso-circles firstly with. the major axes perpendicular to the vertical iso-axis. Then, complete the iso-view by connecting the corresponding tangent points as shown in Fig. 5-8.

Drawing steps:

1) Define the upper isometric coordinate system as shown in Fig. 5-8a.

2) Draw the isometric axes, and produce the rhombus with center O_1 by measuring line segments equal to ϕ (the diameter of the cylinder). Then draw the isometric projection of the circle of the upper face by the "four-center approximate ellipse" method, as shown in Fig. 5-8b.

3) Then drop the upper isometric coordinate $X_1O_1Y_1$ and centers A_1, B_1, C_1 and D_1 down a distance equal to h (the height of the cylinder), adopt the same method and draw the ellipse of the half bottom face, as shown in Fig. 5-8c.

分析：圆柱的顶面和底面均为水平面上的圆，其正等轴测投影皆为椭圆，且长轴应垂直于轴测轴 O_1Z_1。可先作上、下两椭圆，再作其外公切线，即可求得该圆柱的正等轴测图。

作图步骤：

1) 在原视图上确定顶面坐标系，如图 5-8a 所示。

2) 画出轴测轴，通过圆心 O_1 在轴测轴上量取线段等于 ϕ（圆柱的直径）作菱形，然后用"四心近似画椭圆法"画出顶圆的正等轴测图，如图 5-8b 所示。

3) 将顶圆的坐标面 $X_1O_1Y_1$ 及四个圆心 A_1、B_1、C_1 和 D_1 下移 h（圆柱的高度），并采用相同的方法画出椭圆在底面的正等测投影，如图 5-8c 所示。

4) Complete the drawing by connecting the corresponding tangent points of the upper and lower ellipse with two vertical line segments respectively.

5) Erase all unnecessary lines and complete the isometric projection of the cylinder, as shown in Fig. 5-8d.

Example 5-4 Complete the isometric projection of a block with round corners (see Fig. 5-9).

Analysis: A round corner of a block is a quarter of a cylindrical column. For producing the required isometric projection, one can combine the drawing methods of plane solid and cylindrical column.

4) 分别作上、下两椭圆的垂直外公切线。

5) 擦去多余的线条，完成圆柱的正等轴测图，如图 5-8d 所示。

例 5-4 完成带圆角平板的正等轴测图（见图 5-9）。

分析：平板的圆角可看成是圆柱体的 1/4，作图时应将平面立体的画法与圆柱体的画法结合起来。

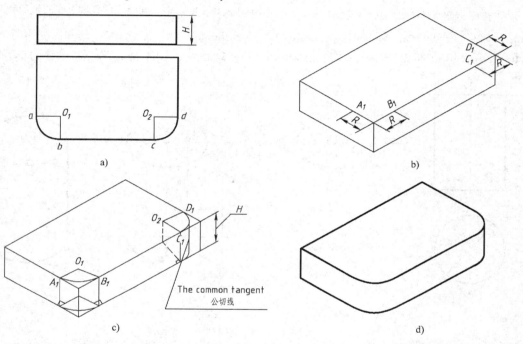

Fig. 5-9 Drawing isometric projection of a block with round corners　画带圆角平板的正等轴测图

Drawing steps:

1) Set up an isometric coordinate system and find the tangent positions a, b, c and d for the two corners as shown in Fig. 5-9a.

2) Produce the isometric projection of the base block and intercept radius R along the two sides of the cylindrical corners, to find the tangent positions A_1, B_1, C_1, D_1 as shown in Fig. 5-9b.

3) Draw the lines perpendicular to the edges of the block through A_1, B_1, C_1, D_1, and the intersection points of the two perpendicular lines are O_1 and O_2, which are the centre points of the approximate arcs. Draw two arcs with radii of distance between the centre points to its tangent point O_1A_1, O_2C_1. Then offset both the centre points and the tangent points a distance with H (the height of the block) downwards to the lower face, and draw arcs out by the same method of the upper face. In the end, draw a common tangent for the upper and lower arcs, thus get the isometric projections of the round corners, as shown in Fig. 5-9c.

4) Erase all unnecessary lines and deepen the visible lines (see Fig. 5-9d).

作图步骤：

1) 在原视图中确定坐标原点和切点 a、b、c、d，如图 5-9a 所示。

2) 画出平板的正等轴测图，并沿圆角两边分别截取半径 R 得到切点 A_1、B_1、C_1、D_1，如图 5-9b 所示。

3) 过切点 A_1、B_1、C_1、D_1 分别作平板各对应边线的垂线，两垂线的交点分别为 O_1、O_2 点，即为近似圆弧的圆心。然后分别以各自的圆心到顶点的距离 O_1A_1、O_2C_1 为半径画圆弧与对应边相切，再将切点、圆心都平行下移 H 距离（平板的高度），以与顶圆相同的半径画圆弧与对应边相切，然后再作上、下圆弧的公切线，即完成圆角的正等轴测图，如图 5-9c 所示。

4) 擦去多余线条，并将可见轮廓线描深，完成作图（见图 5-9d）。

5.3 Cabinet Axonometry Projections

5.3.1 Construction of a cabinet axonometry projection, axes angles and coefficients of axial deformation

1. Construction of a cabinet axonometry projection

When an axonometric projection plane is parallel to a coordinate plane, and the coefficients of axial deformation of two axes that parallel to the coordinate plane are equal, a cabinet axonometry is produced, as shown in Fig. 5-10.

5.3 斜二轴测图

5.3.1 斜二轴测图的形成、轴间角和轴向伸缩系数

1. 斜二轴测图的形成

使轴测投影面平行于一个坐标平面，且平行于坐标平面的那两个轴的轴向伸缩系数相等，即可得到斜二轴测投影，如图 5-10 所示。

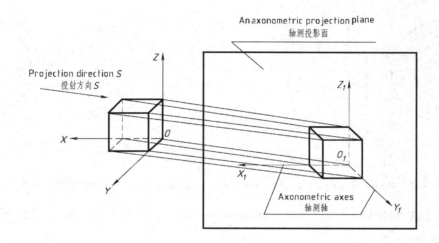

Fig. 5-10 Construction of a cabinet axonometric projection 斜二轴测图的形成

2. Axes angles and coefficients of axial deformation of a cabinet axonometry projection

Since the frontal plane (XOZ-plane) is parallel to the axonometric projection plane, the axonometric projection appears in true size and shape. The angle between axonometric axes is $\angle X_1 O_1 Z_1 = 90°$. The coefficients of axial deformation for $O_1 X_1$-axis and $O_1 Z_1$-axis are the same with $p_1 = r_1 = 1$. The angle between $O_1 Y_1$-axis and a horizontal line is $45°$. The coefficient of axial deformation for $O_1 Y_1$-axis is $q_1 = 0.5$, as shown in Fig. 5-11.

Based on the above, we know that the oblique projections of circles that are parallel to the XOZ projection plane appear in true sizes and shapes. For example, in Fig. 5-12, all circular faces of the wheel that are parallel to the frontal plane, appear as perfect circles on the oblique drawing. For circular faces that are parallel to other planes (XOY、YOZ), the cabinet axonometry projections are ellipses. One should note the followings: In oblique drawings, the faces containing circles should be parallel to the XOZ projection plane, so that the projections of the circles appear in real sizes and shapes, which make the drawing task more simple and convenient.

2. 斜二轴测图轴间角和轴向伸缩系数

由于正平面（即坐标面 XOZ）平行于轴测投影面，这个坐标面的轴测投影反映实形，其轴间角 $\angle X_1 O_1 Z_1 = 90°$，这两根轴（$O_1 X_1$ 及 $O_1 Z_1$）的轴向伸缩系数 $p_1 = r_1 = 1$。$O_1 Y_1$ 轴与水平线成 $45°$ 角，其轴向伸缩系数为 $q_1 = 0.5$，如图 5-11 所示。

由此可知，平行于 XOZ 坐标面的圆，其斜二轴测投影反映实形。例如图 5-12 中，轮子上面的圆形在斜二测投影中仍为正圆形，而轮子上平行于另外两个坐标平面（XOY、YOZ）的圆形的斜二测投影为椭圆。注意：在作斜二轴测图时，应尽量使圆所在的面平行于 XOZ 面，其投影反映圆的真实尺寸和形状，从而使作图简单、方便。

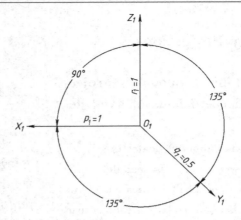

Fig. 5-11 Angles between oblique axes and coefficients of axial deformation 轴间角和轴向伸缩系数

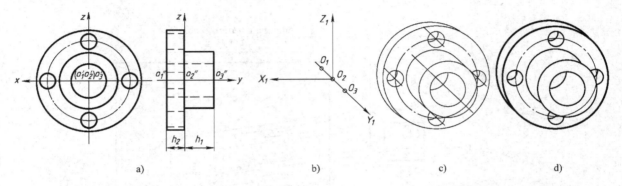

Fig. 5-12 Procedure for producing a cabinet axonometry projection 斜二轴测图的画图步骤

5.3.2 Drawing procedure for a cabinet axonometry projection

The following example illustrates steps of drawing a cabinet axonometry projection as shown in Fig. 5-12.

Example 5-5 Produce a cabinet axonometry projection of the "Flange" shown in Fig. 5-12a.

Analysis: The "Flange" consists of several bodies of revolution. It has three planes with circles. The cabinet axonometry projection will be produced by selecting the three planes with circles are parallel to the front plane ($X_1O_1Z_1$).

Drawing steps:

1) Define the origin of coordinate and axes as shown in Fig. 5-12a.

2) Draw the oblique axes and position origins O_1, O_2 and O_3 of the three planes. Let $O_1O_2 = h_2/2$ and $O_2O_3 = h_1/2$ as shown in Fig. 5-12b.

3) Draw all circles on the three planes as shown in Fig. 5-12c.

4) Draw the four small circles on related planes as shown in Fig. 5-12c.

5) Erase all unnecessary lines and complete the view with regular line styles as shown in Fig. 5-12d.

5.3.2 斜二轴测图的画法

下面举例说明画斜二轴测图的方法与步骤，如图 5-12 所示。

例 5-5 画出图 5-12a 中法兰盘的斜二轴测图。

分析：法兰盘由若干回转体组成，它在三个平面上都有圆。为作图方便，选择三个带圆的平面平行于 $X_1O_1Z_1$ 面，因此它们的斜二轴测图都反映实形。

作图步骤：

1) 在视图上确定坐标原点和坐标轴，如图 5-12a 所示。

2) 画出轴测轴，分别确定三个平面的轴测坐标系，依次确定三个轴测坐标原点 O_1、O_2、O_3，使 $O_1O_2 = h_2/2$、$O_2O_3 = h_1/2$，如图 5-12b 所示。

3) 分别画出三个平面上的圆，如图 5-12c 所示。

4) 确定法兰盘中四个小孔的圆心，并画出小圆孔，如图 5-12c 所示。

5) 擦去多余线条，描深可见轮廓线，即完成法兰盘的斜二轴测图，如图 5-12d 所示。

Chapter 6　General Principles of Representation

In real engineering applications, a machine component or part is usually composed of primitive features with various basic shapes and structures. Therefore, how to draw such machine components' drawings correctly and rapidly is an important issue. The chapter introduces methods for representing commonly used machine components using views, sections, sectional drawings and so on following related national standards on technical and engineering drawing. We also introduce how to select appropriate representations in order to produce complete, clear and concise drawings based on the structure and form characteristics of the given machine components.

6.1　Views

A graphical representation of a machine component produced using an orthogonal projection method following related national standards is called a view of the component. A view usually mainly shows visible faces of a part, with invisible faces only when necessary. Views can be divided into principal views, directional views, partial views and oblique views.

6.1.1　Principal views and reference arrow views

Sometimes the shapes and structures of a part are so complicated that they can not be represented clearly only by the front view, the top view and the left side view. To express all the faces parallel to the six principal projection planes, six principal views are required. The way to create the principal views of a part is specified in the national standard: First put the part inside a cube whose six faces are the principal projection planes (as shown in Fig. 6-1), then project the part on these six planes and one principal view is obtained from each projection plane (as shown in Fig. 6-2). The six principal views include the front view which is obtained by the projection from front to rear, the top view from up to down and the left view from left to right as mentioned before, and they also include the right view from right to left, the bottom view from down to up and the rear view for rear to front.

第6章　图样画法

在实际工程应用中，零件是由若干基本体组成的组合体，它们的结构多种多样，如何正确、快速地绘制零件图样是一个重要的问题。本章主要介绍按照技术制图与机械制图国家标准规定的视图、剖视图、断面图及其他表达方法绘制常用零件图样。并讲述如何根据零件的结构特点，选用适当的表达方法，在完整、清晰地表达零件各部分形状结构的前提下，力求绘图简便。

6.1　视图

根据有关标准和规定，用正投影法所绘制的机件的图形称为视图。视图中一般只画出零件的可见部分，必要时才画出不可见部分。视图分基本视图、向视图、局部视图和斜视图四种。

6.1.1　基本视图及向视图

当零件外形复杂时，只用主视图、俯视图和左视图还不能清楚地表达机件的后面、底面及右面的形状。为了把零件上平行于基本投影面的六面都表示清楚，可画出零件的六个基本视图。国家标准中规定：以正六面体的六个面为基本投影面，将零件放在正六面体当中，如图6-1所示；分别向六个基本投影面投射，便得到六个基本视图，如图6-2所示。这六个基本视图除前面学过的主视图（由前向后投射所得的视图）、俯视图（由上向下投射所得的视图）和左视图（由左向右投射所得的视图）外，还有右视图（由右向左投射所得的视图）、仰视图（由下向上投射所得的视图）和后视图（由后向前投射所得的视图）。

After unfolding the projection cube and the six principal views are arranged on a drawing sheet in the order as shown in Fig. 6-2, which follows the projection rules. In this case, there is no need to indicate the view names.

六个投影面展开后,各视图之间位置应符合投影规律,如图 6-2 所示。此时,一律不需要标注视图的名称。

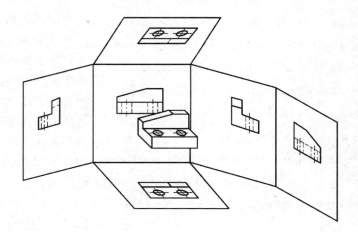

Fig. 6-1 Organization of the six principal projection planes on a drawing sheet 六个基本投影面

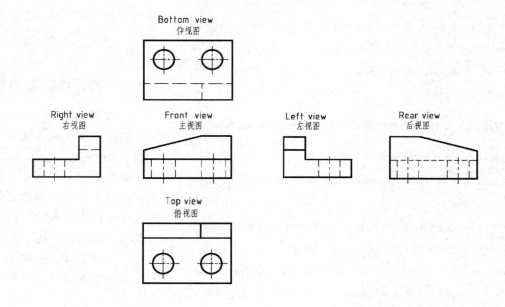

Fig. 6-2 Positions of the six principal views 六个基本视图的位置

If any of the views shown in Fig. 6-2 is not arranged on the proposed default position, or if any of the views does not follow orthogonal projection, the name of the view should be marked above the view and the projection direction should also be clearly indicated at the viewing position with the same symbol. In the later case, the view is called a reference arrow view as shown in Fig. 6-3.

如果某个视图的位置不符合投影规律,或者不属于基本视图,则应在上方标出视图的字母名称,并在相应的视图附近用箭头指明投射方向,注上同样的字母。这样的视图称为向视图,如图 6-3 所示。

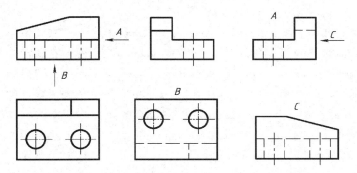

Fig. 6-3 Reference arrow views and their marks　向视图及其标注

6.1.2　Brief introduction of the third-angle projection

In China, the first angle projection is preferably used following related national standards. In some other countries, the third angle projection may also be used. This subsection provides a brief introduction of the third angle projection.

As shown in Fig. 6-4, when the third angle projection is adopted, a part is placed in the third quadrant. By orthogonal projection method, its projection views are created on six projection planes which can be assumed to constitute a transparent projection cube, with the part inside it while the observer outside. Therefore the position relationship between the part and the observer is always kept as observer— projection plane—part.

The main differences between the third angle projection and the first angle projection are the followings:

(1) Names and positions of the views are different

As shown in Fig. 6-4, the projected view on the vertical projection plane V is called the front view (elevation), the projected view on the horizontal plane H is called the top view (plane), and the view projected on the right projection plane W is called the right view. The relative projection position among each view is maintained. When the views are positioned on a drawing sheet, the top view is placed on top of the front view and the right view is placed on the right of the front view.

6.1.2　第三角画法简介

我国优先采用第一角画法并按照相关国家标准来绘制零件图样。有些国家还采用第三角画法。下面简要介绍第三角画法。

如图 6-4 所示，使用第三角画法绘制图样时将所画零件放在第三分角中，使投影面（假想为透明的）置于观察者与零件之间，按照观察者—投影面—零件的相互位置关系，然后用正投影法绘制零件图样。

第三角画法与第一角画法绘制的图样的主要区别是：

（1）视图的名称和配置不同

在图 6-4 中，V 面上所得的投影称为主视图，H 面上所得的投影称为俯视图，W 面上所得的投影称为右视图。各视图之间保持投影关系。投影展开后，俯视图在主视图的上方，右视图在主视图的右方。

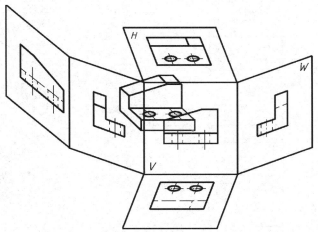

Fig. 6-4 Unfolding of six principal views in the third angle projection
第三角画法中六个基本投影面的展开

(2) Relative position of the views is different

As shown in Fig. 6-4, when positioning the views on the same drawing sheet, the H-plane is rotated upwards and the W-plane is rotated towards the right side so that both of the planes are aligned on the same plane with the V-plane. As a result, the lower side in the top view and the left side in the right view indicate the front side of the object. The upper side in the top view and the right side in the right view indicate the back side of the object.

In the third angle projection, there are also six principal views whose positions are shown in Fig. 6-5. When the third angle projection is adopted, one must draw the identification symbol of the third angle projection in the title box as shown in Fig. 6-6.

Although there are some differences between the two orthogonal projection systems, it is not difficult to master the third angle projection if one is familiar with the first angle projection.

6.1.3 Partial views

When some parts of an object are not clearly shown, but there is no need to add a compalte principal view, a partial view may be used to show the local details. A partial view produces a local projection of an object as shown in Fig. 6-7. The true shape of left and right bosses are not presented in the front view and top view, however, they are completely and clearly shown with the addition of two partial views.

The following provides some general guidelines for drawing and marking partial views:

1) An arrow marked with a capital letter is used to show the projection direction near the corresponding view and the same letter should be marked above the partial view, such as the "A" direction partial view shown in Fig. 6-7. When the partial view is placed according to the projection rules and there are no other views between

(2) 视图中反映前后关系不同

在图 6-4 中，H 面向上旋转，W 面向右旋转，与 V 面展开成一个平面。因此，俯视图的下方和右视图的左方都表示零件的前面；俯视图的上方和右视图的右方都表示零件的后面。

第三角画法也有六个基本视图。这六个基本视图的配置形式如图 6-5 所示。采用第三角画法绘制的图样中，必须在标题栏中画出第三角画法的识别符号，如图 6-6 所示。

尽管第三角画法和第一角画法有所不同，但如果熟练掌握第一角画法，并能触类旁通，就不难掌握第三角画法。

6.1.3 局部视图

当零件的某一部分外形没有表达清楚，又没有必要画出整个基本视图时，可只将零件的某一部分向基本投影面投射，所得到的视图称为局部视图，如图 6-7 所示。图中，零件的主、俯视图未能反映出零件左、右凸合的真实形状，但配合两个局部视图，就表达得很完整、清晰。

一般情况下，局部视图的画法和标注有如下规定：

1) 一般应在相应视图的投射部位附近用带大写拉丁字母的箭头指明投射方向，并在局部视图的上方标注相应的字母，如图 6-7 中的"A"向局部视图。当局部视图按投影关系配

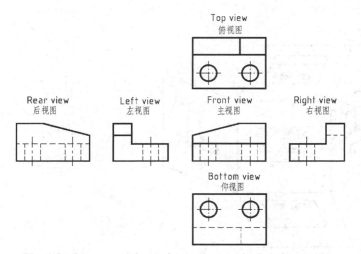

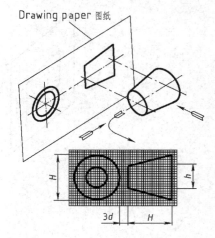

Fig. 6-5 Positions of the six basic views in third angle projection
第三角画法的六个基本视图的位置

Fig. 6-6 Identification symbol for the third angle projection 第三角画法的识别符号

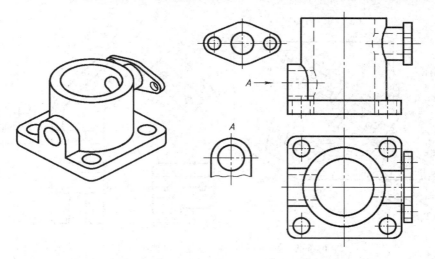

Fig. 6-7 Partial view and its marking   局部视图的画法和标注

the principal view and the partial view, both the projection direction and the marking letter can be omitted as shown with the other partial view in Fig. 6-7.

2) Continuous thin irregular lines are used to illustrate the rupture boundary of partial views, such as the partial view "A" shown in Fig. 6-7. When the complete partial structure is shown and its outline forms a closed drawing region, the continuous thin irregular line does not need to be drawn, such as the other partial view in Fig. 6-7.

In addition, the partial view of a local symmetrical structure on a part can be drawn near the structure, like the keyway structure shown in Fig. 6-8.

置，中间又没有其他图形隔开时，箭头和字母均可省略标注，如图6-7中的另一局部视图所示。

2) 局部视图的断裂边界应以波浪线表示，如图6-7中"A"向局部视图。当所表示的局部结构是完整的，且外轮廓线又呈封闭状时，波浪线可省略不画，如图6-7中未作标注的局部视图。

另外，零件上局部对称结构（如键槽）的局部视图，可按图6-8所示的方法就近绘制。

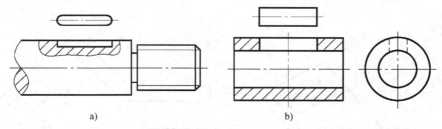

Fig. 6-8 Partial views of local symmetric structures   局部对称结构的局部视图
a) partial view of a keyway 键槽的局部视图 b) partial view of a symmetric hole 对称孔的局部视图

6.1.4 Oblique views

When the principal views cannot show the desired true size of an inclined surface, an auxiliary view can be projected to a reference plane parallel to the oblique surface. The auxilary view is called an oblique view as shown in Fig. 6-9.

The following provides some general guidelines for drawing and marking oblique views:

1) An arrow marked with a capital letter is used to show the projective direction near the corresponding view (the arrow perpendicular to the inclined plane), and the same letter should be marked above the oblique view as shown in Fig. 6-9.

6.1.4 斜视图

基本视图不能反映零件倾斜表面的实际形状时，可增加平行于倾斜表面的平面，然后将倾斜部分向该平面投射，所得到的视图称为斜视图，如图6-9所示。

斜视图的画法和规定如下：

1) 画斜视图时，必须在相应视图的附近，用带大写字母的箭头指明投射方向（箭头应垂直于倾斜表面），并在斜视图上方标注相同的字母，如图6-9所示。

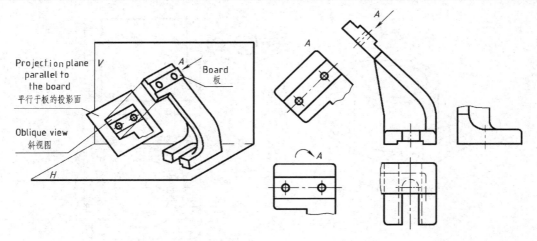

Fig. 6-9 Oblique views (Ⅰ) 斜视图(Ⅰ)

2) Fig. 6-9 shows that oblique views should be placed following the projection rules. It is permitted to rotate oblique views in order to avoid misunderstanding, but it is required to mark the name with a capital letter, such as "A", beside the arrow and the symbol of the rotation direction "⌒" or "⌒", whose radius is the same as the height of the letter as shown in Fig. 6-9 and Fig. 6-10.

3) Oblique views are usually drawn as partial oblique views to illustrate the oblique part shape only, and the rupture boundary is indicated by a continuous thin irregular line as shown in Fig. 6-9 and Fig. 6-10. When an oblique structure is completely shown on the partial view and its outline forms a closed drawing region, the continuous thin irregular line can be omitted.

2) 斜视图一般按投影关系配置，如图6-9所示，必要时也可配置在其他适当位置。在不致引起误解时，允许将图形旋转，在斜视图上方注明"A"，同时标注旋转方向的符号"⌒"或"⌒"，其半径大小与字高相同，字母写在箭头附近，如图6-9、图6-10所示。

3) 斜视图一般只需表达零件倾斜部分的形状，常画成局部的斜视图，其断裂边界应以波浪线表示，如图6-9、图6-10所示。如果所表示的倾斜结构是完整的，且外轮廓线封闭时，波浪线可省略不画。

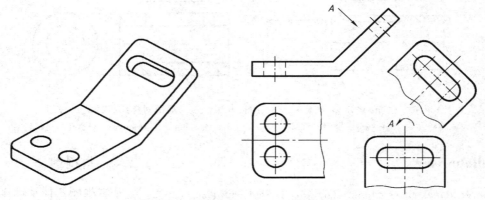

Fig. 6-10 Oblique views (Ⅱ) 斜视图(Ⅱ)

6.2 Sectional Views

In case of complex internal structures or complex structures behind the view, it might cause problems with too many hidden lines on the standard orthogonal views. For clarity and easy comprehension, sectional views are adopted.

6.2 剖视图

为了清楚地表达零件内部或被遮盖部分的结构形状，以免画内部结构复杂的零件时，图形上出现过多的虚线，导致层次不清，影响图形清晰，给读图、画图带来困难，绘图时可采用剖视图画法。

6.2.1 An overview of sectional views

The construction of a sectional view is shown in Fig. 6-11b, where the object is cut open with an imaginary plane. The standard front view and bottom view are shown in Fig. 6-11a, and the method of drawing a sectional view is shown in Fig. 6-11c and Fig. 6-11d, where the front view has been converted to a full sectional view and the cut portion is cross-hatched. Hidden lines have been omitted since they are not needed. The position of the cutting plane is shown with two short heavy line segments. The main objective of drawing a sectional view is to show complex internal structures with thick lines such that it will be easier to understand and it will also be convenient for dimensioning.

The sectional views produced by a cutting plane should be represented with appropriate hatching style. Different hatching styles should be used for different types of materials, as shown in Table 6-1.

Hatching style for metal material, e.g., is usually drawn with thin lines slanted at 45°. Section lines of the same part should be in parallel with the same inclined direction and spacing, as shown in Fig. 6-12.

6.2.1 剖视图概述

剖视图的形成如图 6-11b 所示，用一个假想的剖切面剖开物体，以显示物体内部的结构。标准主视图和俯视图如图 6-11a 所示，画剖视图的方法如图 6-11c、d 所示：主视图为全剖视图，剖面区域应画出剖面线，虚线一般不再画出，用粗短线表示剖切平面的位置。采用剖视图的目的，是使机件上一些原来看不见的内部结构能用粗实线来表示。这样便于看图和标注尺寸。

画剖视图时，应在剖切面剖到的零件实体部分画上剖面符号。剖面符号随零件所用材料类别的不同而有所不同，详见表 6-1。

一般金属材料的剖面符号是与水平成 45°的细实线。同一零件各视图中剖面线应画成方向相同、间隔相等的平行线，如图 6-12 所示。

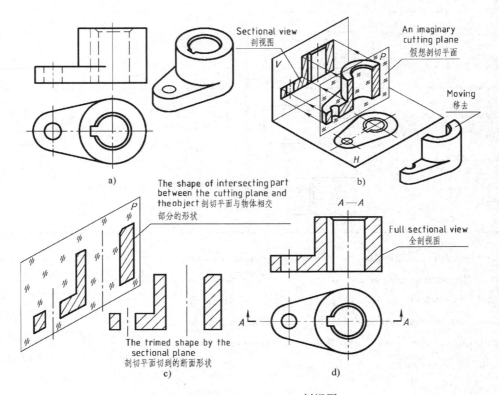

Fig. 6-11 Sectional views 剖视图

a) Two orthogonal views and a pictorial view of an object 物体的两正投影图和立体图

b) Cut the object with an imaginary plane P and remove the front part of the object 用假想剖切面剖切机件后，移去前半部分

c) The cross section showing areas with shading cut by the imaginary cutting plane 假想剖切面剖切到的区域

d) A sectional view (the front view) projected from the remaining part of the object 将剩下部分向正投影面投射后得到的剖视图（主视图）

Table 6-1 Symbols of sections 剖面符号

Material 材料	Symbol 符号	Material 材料	Symbol 符号
Metal material (excluding materials specified otherwise) 金属材料（已有规定剖面符号者除外）	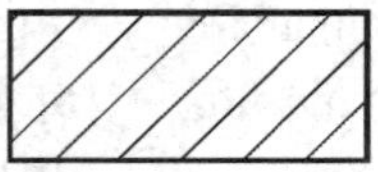	Non-metal material (excluding materials specified otherwise) 非金属材料（已有规定剖面符号者除外）	
Coil winding components 线圈绕组元件	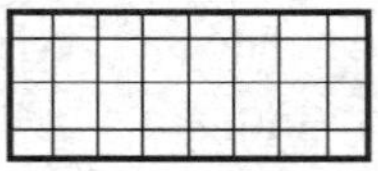	Clay around groundwork 基础周围的泥土	
Overlap armor plates of rotor, armature, transformer, reactor, etc. 转子、电枢、变压器、电抗器等的叠钢片	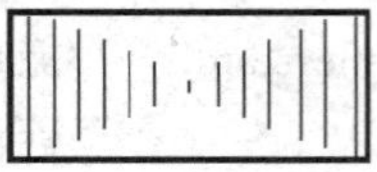	Concrete 混凝土	
Molding sand, filling sand, powder metallurgy, grinding wheel, ceramic blade, carbide blade, etc. 型砂、填砂、粉末冶金、砂轮、陶瓷刀片、硬质合金刀片等	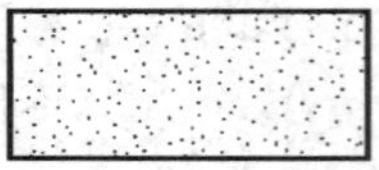	Reinforced concrete 钢筋混凝土	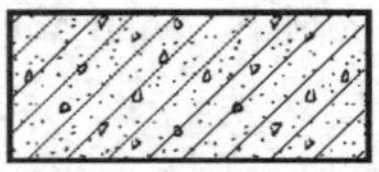
Glass and other transparent material for observation 玻璃及供观察用的其他透明材料	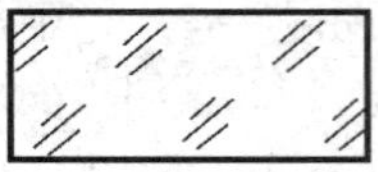	Grid (screen, filter screen, etc.) 网格（筛网、过滤网等）	
Plywood (without distinction of number of plies) 木质胶合板（不分层数）	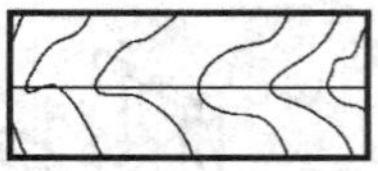	Brick 砖	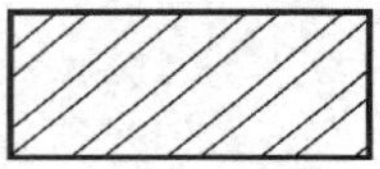
Wood 木材 — Longitudinal section 纵断面		Liquid 液体	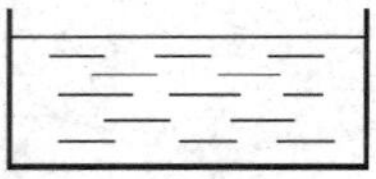
Wood 木材 — Radial section 横断面	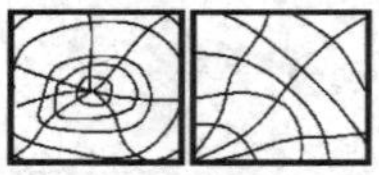		

If the section line is parallel to the main outline or the symmetry line of the section area of a graphics, it shall be drawn as oblique parallel lines which angled 30° or 60° with the symmetry line or the main outline as shown in Fig. 6-13a. For graphics that do not have section lines drawn in 45°, the oblique direction of section lines should be in accordance with other graphics as shown in Fig. 6-13b.

6.2.2 Drawing methods of sectional views

The following provides some general guidelines for drawing and marking sectional views:

当画出的剖面线与图形的主要轮廓线或剖面区域的对称线平行时，可将剖面线画成与主要轮廓线或剖面区域的对称线成 30°或 60°的平行线，如图 6-13a 所示。剖面线不画成 45°的图形中，剖面线的倾斜方向仍与其他图形上剖面线方向相同，如图 6-13b 所示。

6.2.2 剖视图画法

画剖视图时应注意下面几个问题：

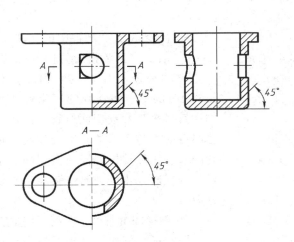

Fig. 6-12 The section lines of the same part on different views should be consistent
同一零件各视图中的剖面线应一致

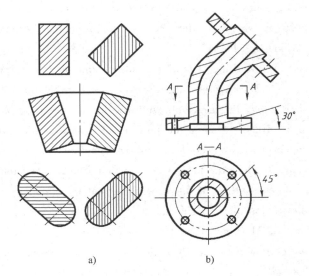

Fig. 6-13 Oblique directions of section lines
剖面线的倾斜方向
a）Section lines angled 30° or 60° with the main outline or the symmetry line
与主要轮廓线或对称线成 30° 或 60°
b）Directions of section lines in all views should be the same
剖面线倾斜方向应相同

1）A cutting plane passes through a symmetry plane or an axis of an object, such as the graphics shown in Fig. 6-11.

2）Because section cutting is imaginary, apart from the sectional views, full views should be produced for other views, such as the top view shown in Fig. 6-11d.

3）For clarity, hidden lines should be omitted in a sectional view as shown in Fig. 6-14b. However, if it helps with the hidden lines, they can still be drawn as shown in Fig. 6-15.

1）剖切面一般应通过零件的对称面或轴线，如图 6-11 中剖切面与俯视图的对称线重合。

2）由于剖切是假想的，所以一个视图取剖视后，其他视图仍应完整画出，如图 6-11d 中的俯视图。

3）为了使图形更加清晰，剖视图中应省略不必要的虚线，如图 6-14b 所示。但如果画出某一虚线有助于读图时，也可画出虚线，如图 6-15 所示。

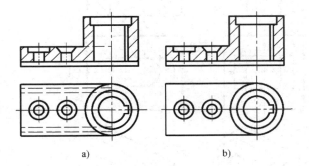

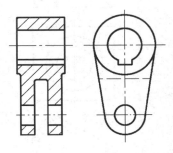

Fig. 6-14 Usually no hidden lines for sectional views
剖视图一般省略不必要的虚线
a）Cases with hidden lines 画出其虚线
b）Cases without hidden lines 省略不必要的虚线

Fig. 6-15 Sectional views with hidden lines for better understanding
虚线有助于读图时应画出

4) It is necessary to carefully analyze the structures of holes and grooves being cut in order to avoid mistakes and omissions. Fig. 6-16 shows the differences when producing sectional views of various step holes.

5) Marking of cutting positions and sectional views.

① Marking of cutting position and projection direction. Use short heavy line segments with a width of (1—1.5) b and a length of 5—10mm to show the position of a section plane in the corresponding views. Arrows are used in combination with the line segments to show the projection direction. The line segments should not touch the outline of the view. A capital letter can be used to mark the section plane beside the arrows.

② Marking of sectional views. Use the same letter as that of the cutting plane to show the name of a sectional view, such as "×—×", at the top position of the sectional view.

③ Marking omission. Arrows could be omitted when there are no other views between sectional views and standard views, and the sectional views are placed following the projection rules. All markings could be omitted when there is only a single cutting plane that goes through the object's symmetry plane or one of the basic symmetry planes, the section views are produced in appropriate positions following the projection rules, and not separated by other views.

6) The positions of sectional views are usually the same as that of standard views. They could be placed at other appropriate positions when necessary.

6.2.3 Types of sectional views

Commonly used sectional views include full sectional views, half-sectional views and local sectional views.

4）要仔细分析被剖切孔、槽的结构形状，以免错漏。图 6-16 所示为四种不同结构阶梯孔的投影。

5）剖切位置和剖视图的标注。

① 剖切位置及投射方向的标注。在相应的视图上用宽（1~1.5）b、长5~10mm、中间断开的短粗实线为剖切符号来表示剖切面位置。剖切符号尽可能不与图形的轮廓线相交，在它的起、止和转折处标注上相同的字母，并在起、止处画出箭头表示投射方向。

② 剖视图的标注。在剖视图的上方，用与标注剖切位置相同的字母标出剖视图的名称"×—×"。

③ 标注的省略。当剖视图按投影关系配置，中间又没有其他图形隔开时，可省略箭头；当单一剖切面通过机件的对称平面或基本对称平面，且剖视图按投影关系配置，中间又没有其他图形隔开时，可省略标注。

6）剖视图的配置与基本视图的配置规定相同，必要时允许配置在其他适当位置。

6.2.3 剖视图的种类

剖视图分为全剖视图、半剖视图和局部剖视图三种。

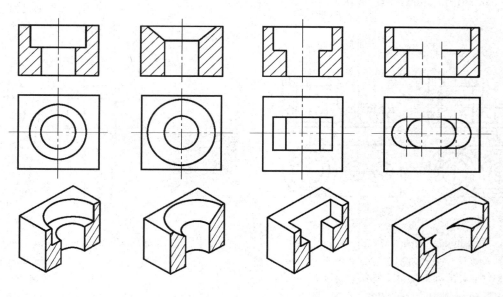

Fig. 6-16 Projections of four types of two-step holes with different structures 四种不同结构阶梯孔的投影

1. Full sectional views

A full sectional view is a view formed by passing a cutting plane fully through an object and removing the part of the front half. It is often used to show a non-symmetrical part with a simple shape and complex inner structure such as that shown in Fig. 6-17. In this figure, the front view, top view and left view are all full sectional views obtained by passing a cutting plane parallel to the projection plane fully through the object. A full sectional view shall be marked following the respective guidelines. In Fig. 6-17, the front view is a full section view through the symmetrical plane between the front and the back portion, and there are no other views to separate them, so all the labeling can be omitted. The top and left views are full sectional views through non-symmetrical cutting planes (through the hole center or the axis), only the arrows could be omitted, and the section symbols, the letters and the names of views, such as "A—A", "B—B" must be marked. Fig. 6-18 shows the marking method for different projection directions using the same cutting plane.

1. 全剖视图

用剖切面完全剖开零件后所得到的剖视图称为全剖视图。全剖视图主要用于外形简单、内部结构复杂且不对称的零件。如图 6-17 所示，图中的主、俯、左视图都是用一个平行于相应投影面的剖切平面完全地剖开零件后所得到的全剖视图。全剖视图应按规定标注。图 6-17 中的主视图是从零件前后对称平面进行全剖的，且处于基本视图位置，可不加标注。俯视图和左视图均从非对称平面（通过孔中心或轴线）进行剖切，只符合省略箭头的规定，必须用剖切符号、字母表明剖切位置，并在俯视图和左视图上方标注相应剖视图的名称，如 "A—A" "B—B"。图 6-18 给出了在同一剖切位置向不同方向进行投射的剖视图标注方法。

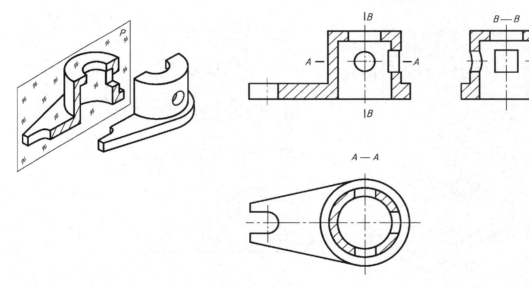

Fig. 6-17 Full sectional views（Ⅰ） 全剖视图（Ⅰ）

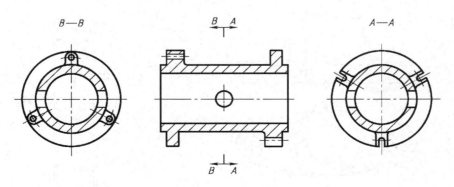

Fig. 6-18 Full sectional views（Ⅱ） 全剖视图（Ⅱ）

2. Half-sectional views

For symmetric objects, their views on the projection plane which is perpendicular to the symmetric plane can be drawn. half-original and half-sectional across the symmetric line, Such views are called half-section views. The method of drawing an orthographic half-section view is shown in Fig. 6-19, where both the internal and external features can be seen. Hidden lines are omitted in the sectional view.

The marking method of a half-sectional view is the same as that of full sectional views. If a half-section is symmetrical, the marking can be omitted, such as the front and left views shown in Fig. 6-19. Otherwise, one must mark a cutting plane symbol with arrows, and mark "×—×" above the section view, as "$A-A$" shown in Fig. 6-19.

When an object shape is nearly symmetrical and the non-symmetrical part has been shown clearly by other views, it could be drawn by the half-sectional views as shown in Figs. 6-20 and Fig. 6-21.

2. 半剖视图

当零件具有对称平面时，向垂直于对称平面的投影面上投射所得的图形，可以对称中心线为界，一半画成剖视图，另一半画成视图。用这种表达方法获得的剖视图称为半剖视图。半剖视图一方面可表达零件的内部结构，另一方面可表达零件的外部形状，其中不可见轮廓线省略不画，如图 6-19 所示。

半剖视图的标注方法与全剖视图相同。剖视图的剖切位置若处于对称平面上，则不必标注，如图 6-19 中的主、左视图的剖切位置；剖视图的剖切位置若不是对称平面（如剖切平面 A）所处位置，则需注明剖切面位置符号和字母，并在剖视图上方注明"×—×"，如图 6-19 中的"$A-A$"。

当零件形状接近于对称，且其不对称部分已另有视图表达清楚时，也可以画成半剖视图，如图 6-20、图 6-21 所示。

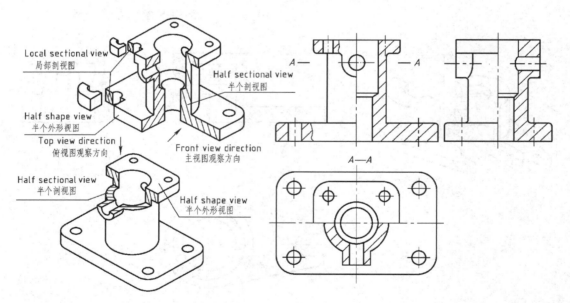

Fig. 6-19 Half-sectional views（Ⅰ） 半剖视图（Ⅰ）

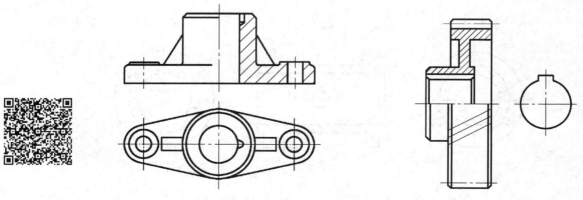

Fig. 6-20 Half-sectional views（Ⅱ） 半剖视图（Ⅱ） **Fig. 6-21** Half-sectional views（Ⅲ） 半剖视图（Ⅲ）

Attention should be paid to the following when drawing a half-sectional view:

1) Centerlines separating the two halves of a view with and without section respectively, should not be heavy lines, but should be a thin long dashed dotted line.

2) Hidden lines are omitted unless it is necessary to show internal structures, such as that of Fig. 6-20.

3) Half (or a quarter) of several symmetrical section views in the same projection direction could be integrated into a new sectional view and the corresponding names must be marked beside the respective sectional views as shown in Fig. 6-22.

画半剖视图时应注意：

1) 剖视图与视图的分界处为中心线时，应画成细点画线，切不可画成粗实线。

2) 内部形状已表达清楚的，虚线不再画出，如图 6-20 所示。

3) 可将投射方向一致的几个对称图形各取一半（或四分之一）合并成一个图形。此时应在剖视图附近标出相应的剖视图名称，如图 6-22 所示。

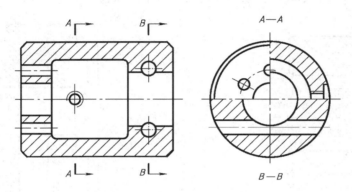

Fig. 6-22　Half-sectional views（Ⅳ）　半剖视图（Ⅳ）

3. Local sectional views

A local (broken-out) sectional view is used to show interior features by breaking away a portion of a part. In Fig. 6-23, a portion of the part is broken out from the front view to reveal inner details of the part, while the other portions are remained to show the external shape and position of the flange whose inner hole shapes are revealed in the top view with another local sectional view.

Attention should be paid to the following when drawing a local sectional view:

1) The dividing lines between local sectional views and standard views must be continuous thin irregular lines which cannot superpose outlines, as shown in Fig. 6-24 and Fig. 6-25.

2) When the structure being cut is symmetrical, the centerline of it can be treated as the dividing line, as the cutting view of the cylindrical structure in the top view shown in Fig. 6-26.

3) Continuous thin irregular lines can not be drawn within holes, grooves and outside the outlines, because there are no broken trails. See Fig. 6-26 and Fig. 6-27 for a comparison with correct and incorrect sections.

4) If the cutting position is obvious, there is no need to mark it on the drawing.

3. 局部剖视图

用剖切面局部地剖开零件所得的剖视图称为局部剖视图。如图 6-23 所示，主视图剖切一部分，用来表达零件的内部结构，保留的局部外形则用来表达凸缘形状及其位置。俯视图剖切局部，表达凸缘的内孔结构。

画局部剖视图时应注意：

1) 局部剖视与视图应以波浪线分界，波浪线不可与图形轮廓线重合，如图 6-24、图 6-25 所示。

2) 当被剖切的结构对称时，可以该结构对称中心线作为局部剖视与视图分界线，如图 6-26 的俯视图中圆筒结构的剖视。

3) 波浪线不应画在通孔、通槽内或画在轮廓线外，因为这些地方没有断裂痕迹。图 6-26 和图 6-27 给出了波浪线在通孔处的正误对比画法。

4) 剖切位置明显的局部剖视图可以不标注。

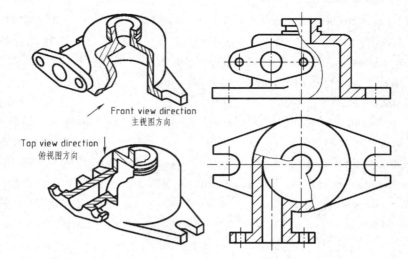

Fig. 6-23　Local sectional views（Ⅰ）　局部剖视图（Ⅰ）

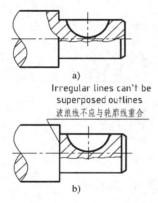

Fig. 6-24　Local sectional views（Ⅱ）　局部剖视图（Ⅱ）
　　a）Right　正确　b）Wrong　错误

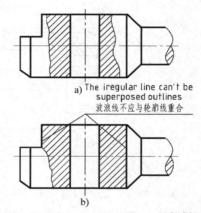

Fig. 6-25　Local sectional views（Ⅲ）　局部剖视图（Ⅲ）

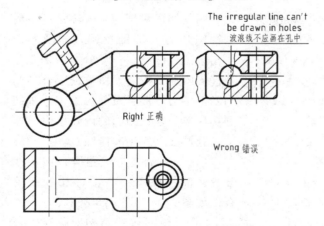

Fig. 6-26　Local sectional views（Ⅳ）　局部剖视图（Ⅳ）
　　a）Right　正确　b）Wrong　错误

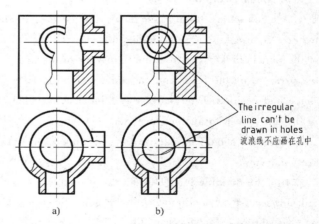

Fig. 6-27　Local sectional views（Ⅴ）　局部剖视图（Ⅴ）
　　a）Right　正确　b）Wrong　错误

6.2.4　Types of cutting planes

As the shapes and structures of mechanical parts vary a lot, different forms of cutting planes can be selected according to their structural characteristics to create their sectional views. Commonly used forms of cutting planes are specified in the national standard as follows：

6.2.4　剖切面种类

由于机械零件的结构形状各不相同，画剖视图时可根据其结构特点选用不同形式的剖切面。国家标准规定常用的剖切面有以下几种：

1. Single cutting planes

1) A cutting plane parallel to a principal projection plane. Full sectional views, half-sectional views and local sectional views all have a single cutting plane.

2) An inclined cutting plane that is not parallel to any principal projection plane. Fig. 6-28 gives a section view with an inclined cutting plane that is not parallel to any principal projection planes. The proposed cutting plane can show the real inner structure of the inclined part. Its drawing method, rules and markings are the same as oblique views besides drawing the section lines as shown in Fig. 6-29 and Fig. 6-30.

1. 单一剖切面

1) 用平行于某一基本投影面的平面剖切。前面介绍的全剖视图、半剖视图和局部剖视图都是单一剖切面剖切的图例。

2) 用不平行于任何基本投影面的（斜）剖切面剖切。为了表达机件上倾斜部分的内部结构，用不平行于任何基本投影面的（斜）剖切面剖开零件，如图 6-28 所示。这种表达方法可以真实地反映零件倾斜部分的内部结构，除应画出剖面线外，其画法、图形的配置及标注与斜视图相同（见图 6-29 及图 6-30）。

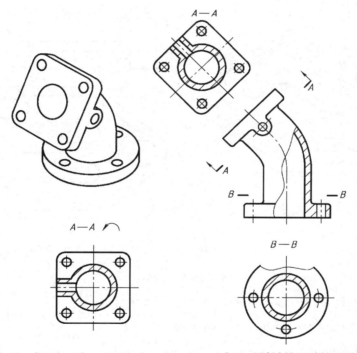

Fig. 6-28 Sectional view with an inclined cutting plane（Ⅰ） 用斜剖切面剖切的剖视图（Ⅰ）

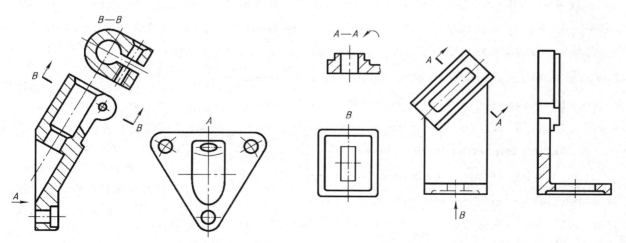

Fig. 6-29 Sectional view with an inclined cutting plane（Ⅱ）
用斜剖切面剖切的剖视图（Ⅱ）

Fig. 6-30 Sectional view with an inclined cutting plane（Ⅲ）
用斜剖切面剖切的剖视图（Ⅲ）

2. Multiple parallel cutting planes

A view produced with multiple parallel cutting planes is a full sectional view in which the cutting plane is offset to pass through important features as shown in Fig. 6-31. In this case, a single cutting plane cannot cut all the holes of the object. Fig. 6-31 shows that three parallel cutting planes are used to cut at the same time (one cutting the cylindrical hole on the left, one cutting the annular hole in the middle and the third cutting the cylindrical hole on the right) to get a section made up by section views of three portions, and then all the holes can be clearly shown in the same sectional view.

2. 几个平行的剖切平面

当需要表达分布在几个相互平行平面上的零件内部结构时，可用几个平行的剖切面沿这些重要结构剖开零件来表达，如图 6-31 所示。图 6-31 中零件上的孔只用一个剖切面不能都剖切到，假想用三个互相平行的剖切面来剖切（一个剖左端圆柱孔，一个剖中间的非圆孔，最后一个剖右端的圆柱孔），所得到的三部分剖视图合画成一个由三个平行剖切平面剖切的剖视图，则可表达清楚孔的内部结构。

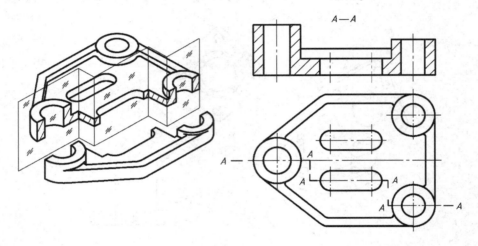

Fig. 6-31 Sectional view with multiple parallel cutting planes (Ⅰ)　用几个平行的剖切平面剖切的剖视图 (Ⅰ)

No lines are needed at the turn of the sectional views, and the turn cannot be overlapped by the component outlines, and the sectional views shall not be overlapped. Incomplete elements (such as hole, groove, boss club and rib, etc.) shall not appear on the sectional views. Only when two elements have the same symmetrical center-lines or axes, can they be drawn half divided by the center-lines, such as Fig. 6-32 and Fig. 6-33. The starting ending and turning of cutting planes position should be marked with the same cutting symbols. Cutting symbols (the label letter *A*) and the corresponding letter "*A—A*" should be placed above the section view.

3. Multiple intersecting cutting planes

(1) Two intersecting cutting planes　When representing the inner holes and grooves on rotational objects, such as wheels, plates and covers, one can adopt a full sectional view formed by the intersected cutting planes of which the intersecting line is perpendicular to a principal projection plane, as shown in Fig. 6-34.

剖视图中剖切平面转折处不画任何图线。转折部位不应与机件轮廓线重合。剖切面不得相互重叠，以免剖视图形紊乱。剖视图内不应出现不完整要素（如孔、槽、凸台、肋板等结构不应一部分剖去，一部分保留）。仅当两个要素具有公共对称中心线或轴线时，可以各画一半，并以中心线作为分界，如图 6-32、图 6-33 所示。剖切面起、讫、转折处画剖切符号并加注相同的字母（如字母 *A*），剖视图上方注明相应字母（如 "*A—A*"）。

3. 几个相交的剖切面

（1）用两个相交的剖切面剖切　当需要表达具有公共回转轴线的机件，如轮、盘、盖等零件上的孔、槽等内部结构时，可采用两个相交且交线垂直于某一基本投影面的剖切面剖开机件的表达方法，如图 6-34 所示。

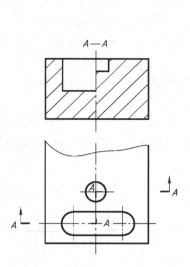

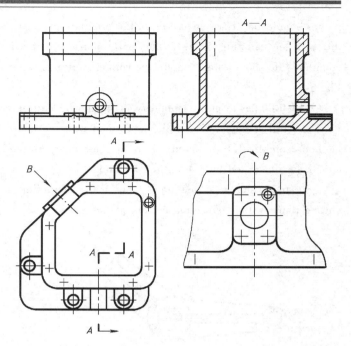

Fig. 6-32 Sectional view with multiple parallel cutting planes (Ⅱ)
用几个平行剖切平面剖切得到的剖视图（Ⅱ）

Fig. 6-33 Sectional view with multiple parallel cutting planes (Ⅲ)
用几个平行剖切平面剖切得到的剖视图（Ⅲ）

In this case, one cutting plane cannot cut the small hole, middle hole and bass club at the same time, so two intersecting cutting planes whose intersecting line being the axis of the middle hole are used to cut the three holes. As shown in Fig. 6-34, the cutting planes are perpendicular to the V-plane. Please remember that when creating the left view, first rotate the cutting plane not parallel to the W-plane around the intersection line so that all cutting planes are parallel to the projection plane and then do the projection job. At the start, end and turn of cutting planes, cutting symbols, the marking letter "A" should be marked correctly, and the corresponding letter "A—A" should be placed above the section view.

The intersecting line of the cutting planes shall overlap the center axis of a hole to avoid the deformation of the projection.

在该图例中，用一个剖切面不能同时剖到零件的小孔、中间孔及凸台。现用两个剖切面剖切零件，交线是大孔轴线，垂直于正立投影面。这样可同时剖到三个孔。画图时要注意将正垂面剖切到的结构绕交线（轴线）旋转到与侧立投影面平行后进行投射，如图 6-34 所示。还应在剖切平面起、讫、转折处画上剖切符号，注写字母"A"，并在剖视图上方注明"A—A"。

剖切面相交线应与零件上某孔中心轴线重合，以免投影变形。

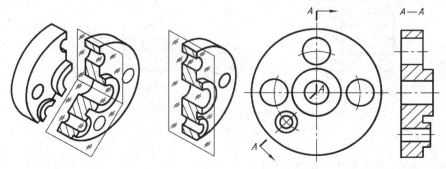

Fig. 6-34 Sectional view with multiple intersecting cutting planes (Ⅰ)　用几个相交的剖切面剖切的剖视图（Ⅰ）

Inclined cutting planes should be rotated to horizontal or vertical orientation, where the original structures (overlap the structures of the inclined cutting planes after rotation) should not be drawn, as shown in Fig. 6-35. The structures behind the cutting

倾斜的剖切平面旋转后，转平位置上原有结构（与斜切平面上的结构重叠）不再画出，如图 6-35 所示。剖切面后的其他结构一般仍按原来的

planes should be drawn according to the original positions, such as the oil hole in Fig. 6-35. If the part being cut becomes not complete, it should then be drawn without cutting, as shown in Fig. 6-36.

（2） Composite cutting planes A composite cutting sectional view is a full sectional view which is formed with the above cutting methods combined. It is mainly used to show many structures that cannot be illustrated by one cutting method.

Fig. 6-37 shows a sectional view with combining a single cutting plane with intersecting cutting planes.

位置投影，如图 6-35 的油孔。当剖切后产生不完整要素时，此部分按不剖绘制，如图 6-36 所示。

（2）采用几个剖切平面剖切 当采用上述某一种剖视方法都不能同时表达出零件的多处结构时，可采用几个剖切面剖切。

图 6-37 所示为将单一剖切平面与相交剖切平面结合起来的情况。

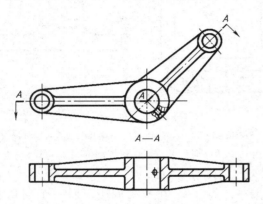

Fig. 6-35 Sectional view with multiple intersecting cutting planes （Ⅱ） 用几个相交剖切面剖切得到的剖视图（Ⅱ）

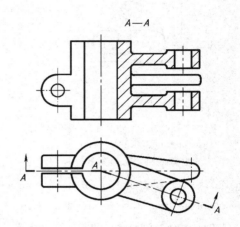

Fig. 6-36 Sectional view with multiple intersecting cutting planes （Ⅲ） 用几个相交剖切面剖切得到的剖视图（Ⅲ）

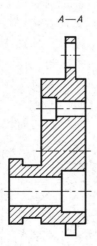
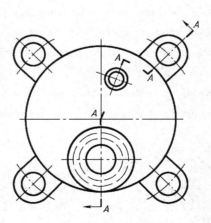

Fig. 6-37 Sectional view with composite cutting planes （Ⅰ） 用几个剖切平面剖切得到的剖视图（Ⅰ）

Fig. 6-38 shows an A—A composite sectional view which uses two parallel planes in the upside and two intersecting planes in the lowerside.

4. Others

1） A sectional view being cut with cylindrical faces must be drawn after rotational alignment, as shown in Fig. 6-39.

如图 6-38 所示，上部采用两个平行的剖切平面，下部采用两个相交的剖切平面，结合起来得到 A—A 剖视图。

4. 其他

1） 用柱面剖切零件结构所得的剖视图应按展开绘制，如图 6-39 所示。

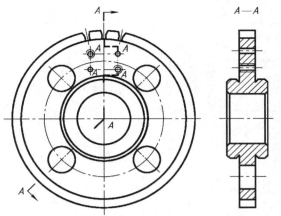

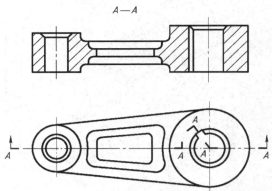

Fig. 6-38 Sectional view with composite cutting planes (Ⅱ)
用几个剖切平面剖切的剖视图（Ⅱ）

Fig. 6-39 Sectional view with cylinders
用柱面剖切的剖视图

2) In a sectional view, one more partial cutting can be done. When using this method, the section lines in the two cutting areas should have the same direction and spacing but should be staggered and the name should be labelled. The drawing method is shown in Fig. 6-40. When the cutting position is obvious, the marking can be omitted.

3) When it is necessary to show the structures in front of a cutting plane, they must be drawn according to imaginary projection outlines, as shown in Fig. 6-41.

2）在剖视图中，还可再进行一次局部剖切。采用这种表达方法时，两个剖面区域的剖面线应同方向、同间隔，但要互相错开，并用引出线标注其名称，如图 6-40 所示。当剖切位置明显时，也可省略标注。

3）在需要表示位于剖切面前的结构时，这些结构按假想投影轮廓线绘制，如图 6-41 所示。

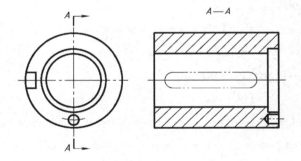

Fig. 6-40 One more partial cutting in a sectional view
在剖视图中再进行一次局部剖切

Fig. 6-41 Illustration of structures in front of a cutting plane
剖切面前的结构按假想投影的轮廓线绘制

6.3 Cuts

6.3.1 Concept of cuts

When an imaginary cutting plane passing through an object, the cross section is drawn, which is called a cut, as shown in Fig. 6-42a. Generally, cutting plane symbol should be marked at cutting position. It is different from a sectional view, in which besides drawing cross-sectional shape at the cutting position, one also needs to draw features behind the cutting plane, as shown in Fig. 6-42b.

6.3 断面图

6.3.1 断面图的概念

当用假想的剖切面将零件某部分切断，仅画出该剖切断面处的图形，即为断面图，如图 6-42a 所示。断面图上一般应画出剖面符号。与断面图不同，剖视图除画出断面形状外，还需画出剖切面后方结构的投影，如图 6-42b 所示。

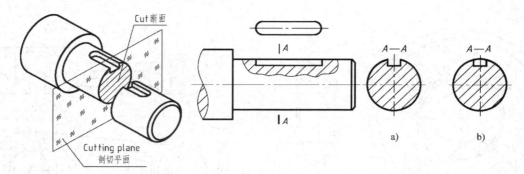

Fig. 6-42 Differences between a cut and a sectional view 断面图与剖视图的区别
a) Cut view 断面图 b) Sectional view 剖视图

A cut view is usually used to illustrate the shape of section area of an object or the holes and grooves in the solid shaft. In order to obtain the actual shape of an object, the cutting plane must be perpendicular to the main outlines or axis of the object.

6.3.2 Types of cuts

Cuts can be divided into removed cuts and superposed cuts.

1. Removed cuts

A removed cut is a cross section that is drawn outside the outline of the original structure.

One must pay attention to the following when drawing removed cuts:

1) Heavy lines are used to draw outlines of a removed cut and the cut is usually placed on the extension line of the cutting line, such as that shown in Fig. 6-43.

2) Fig. 6-44 shows that removed cuts can be placed at other appropriate positions as well. It is necessary to draw cutting symbols and arrows as shown in Fig. 6-43 when the cut is not symmetric; on the contrary, it can be drawn in the middle of the view as shown in Fig. 6-45.

断面图一般用于表达零件某一部分的切断表面形状，或轴及实心杆上孔、槽等的结构形状。为获得零件结构实形，剖切面必须垂直零件的主要轮廓线或轴线。

6.3.2 断面图的种类

断面图可分为移出断面图和重合断面图两种。

1. 移出断面图

画在视图轮廓外的断面图称为移出断面图。

画移出断面图时，要注意以下几点：

1) 移出断面的轮廓线用粗实线绘制，并且应尽量配置在剖切线的延长线上，如图6-43。

2) 如图6-44所示，为了合理布置图面，移出断面图也可以配置在其他适当位置。当断面图形不对称时，需画剖切符号及箭头，如图6-43所示；当断面图形对称时，还可以将移出断面画在视图的中断处，如图6-45所示。

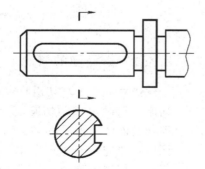

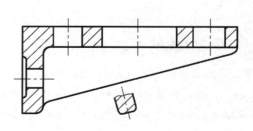

Fig. 6-43 Drawing method of removed cuts (Ⅰ)
移出断面图的画法（Ⅰ）

Fig. 6-44 Drawing method of removed cuts (Ⅱ)
移出断面图的画法（Ⅱ）

3) When the cutting plane passes through a non-round hole, two fully separate sections will appear. In this case, the structures should be drawn as a sectional view in stead of a cut, as shown in Fig. 6-46.

3）当剖切面通过非圆孔，会导致出现完全分离的两个断面时，则这些结构应按剖视图绘制，如图 6-46 所示。

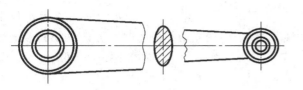

Fig. 6-45 Drawing method of removed cuts (Ⅲ) 移出断面图的画法（Ⅲ）

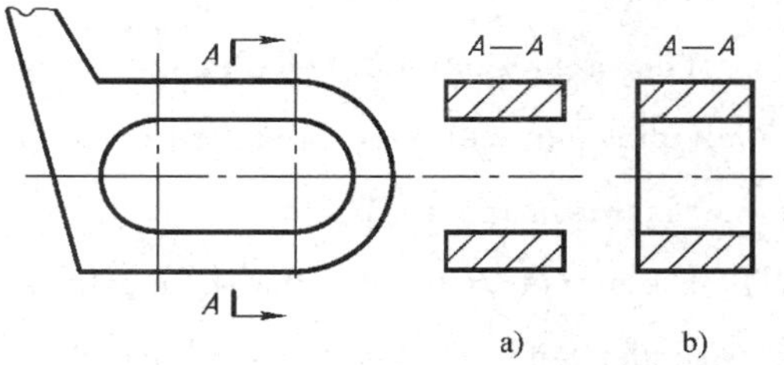

Fig. 6-46 Drawing method of removed cuts (Ⅳ) 移出断面图的画法（Ⅳ）
a) Wrong 错误 b) Right 正确

4) When necessary, a removed cut can be positioned in other places. It can be rotated if it would not be misunderstood, as shown in Fig. 6-47.

5) The removed cut by two intersecting planes must be disconnected in the middle, as shown in Fig. 6-48.

6) When the cutting plane passes through the axes of holes or hollows, those features must be drawn as sections. If the cut is symmetrical, it only needs to be labeled with letters and cutting symbols in the cutting plane positions, and be marked "A—A" above the cut, as shown in Fig. 6-49.

4）必要时可将移出断面图配置在其他适当的位置。在不引起误解时，允许将图形旋转，具体的标注形式如图 6-47 所示。

5）由两个相交剖切面切出的移出断面，中间应断开，如图 6-48 所示。

6）剖切面通过圆孔、圆坑的轴线时，断面图中这些结构按剖视图画出（见图 6-49）。断面对称时，只需用字母、剖切符号注明断面位置，并在断面图上方加注"A—A"，如图 6-49 所示。

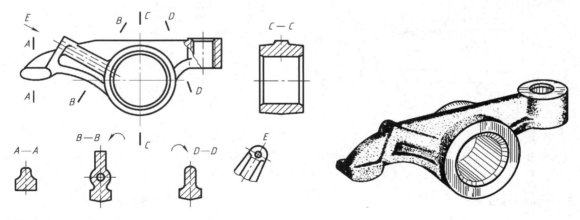

Fig. 6-47 Drawing method of removed cuts (Ⅴ) 移出断面图的画法（Ⅴ）

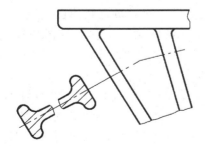

Fig. 6-48 Drawing method of removed cuts (Ⅵ) 移出断面图的画法（Ⅵ）

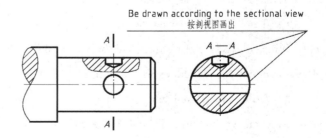

Fig. 6-49 Drawing method of removed cuts (Ⅶ) 移出断面图的画法（Ⅶ）

7) If the shape of the cut is non-symmetrical, it must be labeled with arrows to show the projection direction. If the cut is not on the extension line of the cutting line, it must be marked with letters, as shown in Fig. 6-50.

8) When representing a complex curved face with a series of cuts, only draw outlines of each cut and place them at the same position, as shown in Fig. 6-51.

9) It is permitted to omit the section lines when there would be no misunderstanding, but the marking of the cutting positions and cross sections should follow the respective guidelines. The drawing method is shown in Fig. 6-52.

7) 断面图形状不对称时，剖切符号还需加画箭头，表示断面图的投射方向（见图 6-50）。断面图不在剖切线延长线上时，还需加注字母，如图 6-50 所示。

8) 用一系列断面图表示机件上较复杂的曲面时，可以只画出断面轮廓，并配置在同一个位置上，如图 6-51 所示。

9) 在不致引起误解时，零件图中的移出断面图允许省略剖面线，但剖切位置和断面图的标注必须遵守规定，如图 6-52 所示。

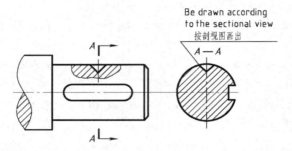

Fig. 6-50 Drawing method of removed cuts（Ⅷ） 移出断面图的画法（Ⅷ）

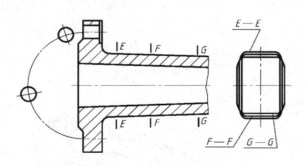

Fig. 6-51 Drawing method of removed cuts（Ⅸ）
移出断面图的画法（Ⅸ）

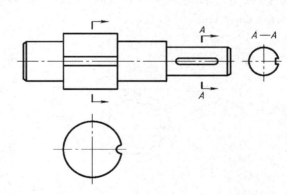

Fig. 6-52 Drawing method of removed cuts（Ⅹ）
移出断面图的画法（Ⅹ）

2. Superposed cuts

A superposed cut is a cross section drawn inside the outline of the original structure, as shown Fig. 6-53.

1) Thin lines are used to draw the outlines of superposed cuts. When it overlaps the outlines of the view, the outlines must be drawn completely, as shown in Fig. 6-53.

2) If the superposed cut is symmetrical, the marking can be omitted as shown in Fig. 6-53 and Fig. 6-54. Otherwise, it must be marked with a cutting symbol with arrows, and the letters may be omitted as shown in Fig. 6-55.

2. 重合断面图

画在视图内的断面图称为重合断面图，如图 6-53 所示。

1) 重合断面图的轮廓线用细实线画出。当它与视图中轮廓线重叠时，视图中的轮廓线仍需完整画出而不中断，如图 6-53 所示。

2) 重合断面图为对称图形时，不加标注，如图 6-53、图 6-54 所示。重合断面图为不对称图形时，应标出剖切符号及箭头，并且可以省略字母，如图 6-55 所示。

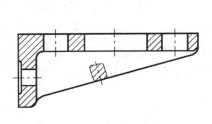
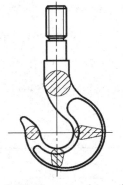
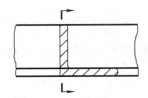

Fig. 6-53 Superposed cuts (Ⅰ)
重合断面图（Ⅰ）

Fig. 6-54 Superposed cuts (Ⅱ)
重合断面图（Ⅱ）

Fig. 6-55 Superposed cuts (Ⅲ)
重合断面图（Ⅲ）

6.4 Drawings of Partial Enlargement

A drawing of partial enlargement is a drawing that presents a partial local structure of an object with a scale-up local view. The drawing of partial enlargement can be represented by regular views, sections, or cuts, hence it is not confined by any particular drawing methods of the enlarged part, as shown in Fig. 6-56. The drawing of partial enlargement should be positioned nearby the original part. The scale-up portion should be circled out with continuous thin line, and then the scale should be written above the view. As when necessary, the same part can be represented with multiple drawings of partial scale-up, as shown in Fig. 6-57.

6.4 局部放大图

将零件的部分结构用大于原图形的比例画出的图形，称为局部放大图。局部放大图可画成视图，也可画成剖视图、断面图，它与被放大部分的表达方式无关，如图 6-56 所示。局部放大图应尽量配置在被放大部位附近。绘制局部放大图时，用细实线圈出被放大部位，并在图形上方注明比例大小。必要时可用几个图形来表示同一个被放大部分的结构，如图 6-57所示。

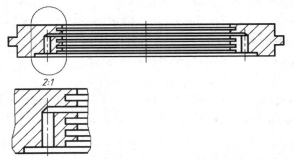

Fig. 6-56 Drawing of partial enlargement to represent the part 机件的局部放大图

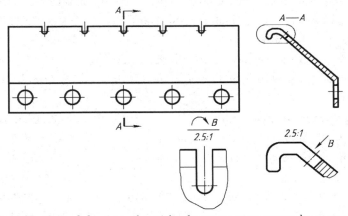

Fig. 6-57 Several drawings of partial enlargement to represent the same part
用几个局部放大图表达机件的同一部位的画法

When there are several scale-up parts on the same object, they must be labeled with Roman numerals in turn, the corresponding number and scale should aslo be marked above each of the scale-up views. The concrete drawing method is shown in Fig. 6-58.

当同一零件上有几个被放大的部分时，应用罗马数字依次标明被放大部位，并在局部放大图的上方标注出相应的罗马数字和所采用的比例，如图 6-58 所示。

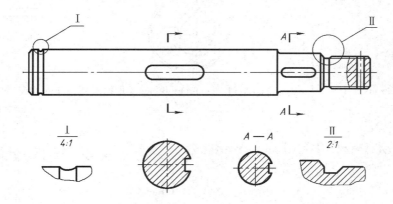

Fig. 6-58 Drawings of partial enlargement for different parts of the same object
同一机件的几个不同部位的局部放大图

6.5 Simplified and Specified Representation

Simplified representation is a drawing method for representing drawings in simplified but comprehensible format, on the premise of avoiding distortions in terms of intrinsic scales and structures. The following are some common uses of simplified representation.

1. Simplified representation of the same features

1）If a part contains several repetitive features（such as gears, grooves, etc.）which are distributed following certain rules, it is only needed to draw a portion of such features and use continuous thin lines to represent the rest, however, the number of all features must be labeled on the part drawing, as shown in Fig. 6-59.

If the repetitive features are a particular set of holes (such as round holes, screw holes, counter bored holes, etc.) with an unified diameter, only draw one or portion of them for representation, and represent the rest holes with dashed dotted lines to indicate the center position. It is necessary to mark the number of the holes on the detail drawing, as shown in Fig. 6-60.

2) Fig. 6-61 shows the drawing method of holes, which are distributed equably around the axis of a column inform flange or similar parts.

2. Simplified representation of projection

1) Structures with a small pitch in a part could be drawn according to its small end if it is represented in one view, as shown in Fig. 6-62.

2) If a circle or an arc has the included angle of 30° or less than 30° to projective plane, its projection can be represented with a circle or an arc directly, as shown in Fig. 6-63.

6.5 简化画法与规定画法

简化画法是在不妨碍将零件的形状和结构表达完整、清晰的前提下，力求制图简便、看图方便的一些简化表达方法。下面介绍一些比较常用的简化画法。

1. 相同结构的简化画法

1）当零件具有若干相同结构（如齿、槽等）并且这些结构按一定规律分布时，只需画出几个完整的结构，其余用细实线连接，但在零件图中则必须注明该结构的总数，如图 6-59 所示。

当这些相同结构是直径相同的孔（如圆孔、螺孔、沉孔等）时，可以仅画出一个或几个，其余只需用点画线表示其中心位置，并在零件图中注明孔的总数，如图 6-60 所示。

2）圆柱形法兰和类似零件上均匀分布的孔可按图 6-61 所示的方法绘制。

2. 图形中投影的简化画法

1）零件上斜度不大的结构，如在一个图形中已表达清楚，其他图形中可按小端画出，如图 6-62 所示。

2）与投影面倾斜角度小于或等于 30°的圆或圆弧，其投影可用圆或圆弧代替，如图 6-63 所示。

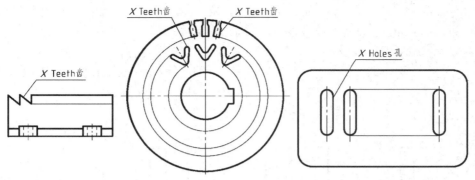

Fig. 6-59 Simplified representation of the same features 相同结构的简化画法

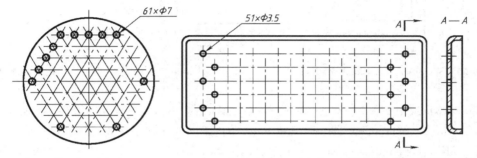

Fig. 6-60 Simplified representation of regularly distributed holes 规律分布的孔的简化画法

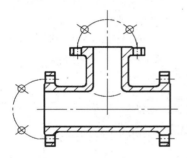

Fig. 6-61 Simplified representation of holes distributed regularly in flanges 法兰上均匀分布的孔的简化画法

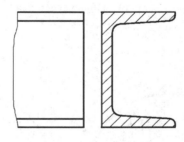

Fig. 6-62 Simplified representation of small pitch structure 小斜度结构的简化画法

3) Intersection line of small structures on a part can be simplified or omitted on other views if it has been expressed clearly in one view, as shown in Fig. 6-64.

3）零件上较小的结构所产生的交线，如果在一个图形中已表达清楚，则在其他图形中可简化或省略，如图 6-64 所示。

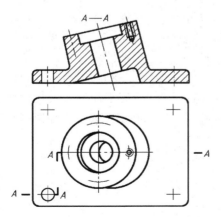

Fig. 6-63 Simplified representation of an circle at 30° 角度小于或等于30°圆的简化画法

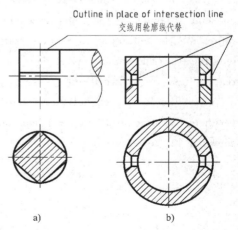

Fig. 6-64 Simplified representation of intersection line between small features 较小结构交线的简化画法

4) Small fillets or 45°chamfers is permitted to omit, but they should be dimensioned or illuminated in engineering requirements, as shown in Fig. 6-65.

4) 在不致引起误解时, 零件图中的小圆角、锐边的小倒圆或45°小倒角允许省略不画, 但必须注明尺寸或在技术要求中加以说明, 如图6-65所示。

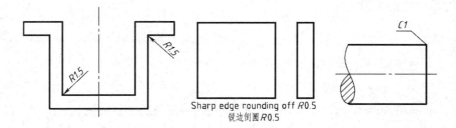

Fig. 6-65 Simplified representation of small fillets and 45°chamfers　　小倒圆或45°小倒角的简化画法

5) The view of symmetrical part can be drawn with just a half or a quarter of it, and two continuous parallel thin lines should be drawn at the two ends of center lines perpendicularly, as shown in Fig. 6-66.

6) Partial views of symmetrical structures can be drawn as shown in Fig. 6-67.

5) 在不致引起误解时, 对于对称零件的视图可只画一半或四分之一, 并在对称中心线的两端画出两条与其垂直的平行细实线, 如图6-66所示。

6) 对称结构的局部视图可按图6-67所示方法绘制。

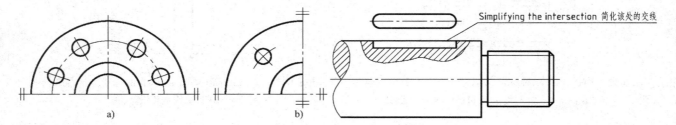

Fig. 6-66 Simplified drawing method for a symmetric part 对称零件的简化画法　　**Fig. 6-67** Simplified drawing method for a symmetric structure 对称结构的简化画法

7) Long parts (such as axes, poles, and so on), which has constant or regular profile along its length direction, can be simply represented by cutting off the middle portion of the part, just keep only the two ends and represent their cuts with continuous thin irregular lines, as shown in Fig. 6-68.

8) When a view cannot clearly show a plane, a plane symbol (with two continuous thin lines) can be drawn on the view, as shown in Fig. 6-69.

7) 较长的零件(轴、杆等)沿长度方向的形状一致或按一定的规律变化时, 可断开后缩短绘制, 断裂处用波浪线表示, 如图6-68所示。

8) 当图形不能充分表达平面时, 可用平面符号(用两条细实线画出对角线)表示, 如图6-69所示。

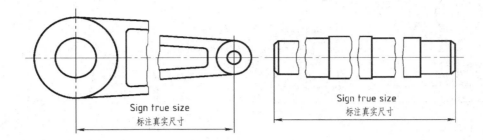

Fig. 6-68 Simplified drawing method of long parts 较长的零件的简化画法

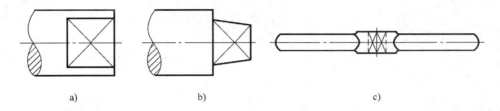

Fig. 6-69 Simplified drawing method using plane symbols 平面的简化画法

9) Reticulation material, weaving material, and diamond knurling on a part should be represented in simplified forms by drawing partially with continuous thick lines which attach to outlines, also they can be marked with the default relevant annotation instead, as shown in Fig. 6-70.

9）网状物、编织物及零件上的滚花部分，一般采用在轮廓线附近用粗实线局部画出的方法表示，如图6-70所示，也可以图中标注表达，省略不画花纹。

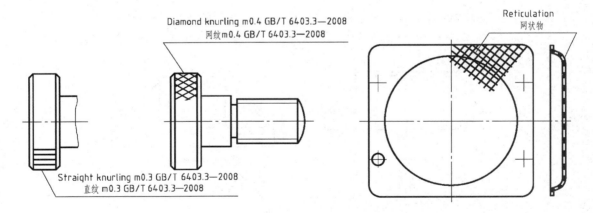

Fig. 6-70 Simplified drawing method of reticulation or knurling structure 网状物及滚花的简化画法

3. Specified representations

1) For ribs, spokes, and thin-walled structures, section symbols need not to be drawn if they are cut lengthwise. However, they should be distinguished from the other structures with continuous thick lines, as shown in Fig. 6-71.

2) When features lilce ribs, spokes and holes are uniformly located on a revolved body but not at cutting plane, they should be rotated to the cutting plane, as shown in Fig. 6-72.

3. 规定画法

1）对于零件上的肋板、轮辐及薄壁等，如按纵向剖切，这些结构都不画剖面符号，而用粗实线将它与其邻接部分分开，如图6-71所示。

2）当零件回转体上均匀分布的肋板、轮辐、孔等结构不处于剖切面上时，应将这些结构旋转到剖切面上画出，如图6-72所示。

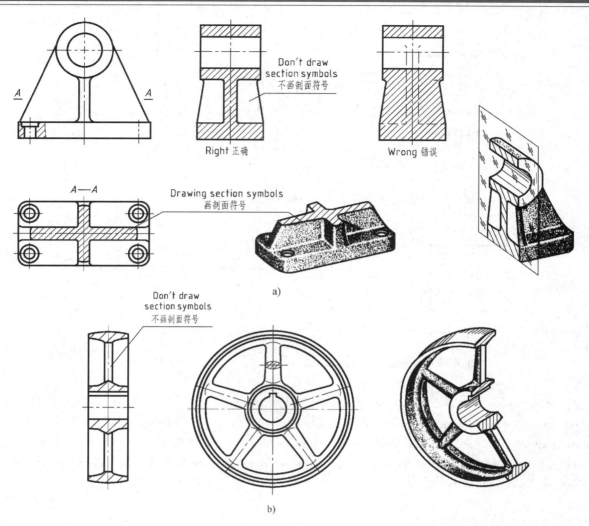

Fig. 6-71 Specified representation of ribs or spokes 肋板、轮辐的规定画法
a) Sectional view of rib 肋板的剖视图　b) Sectional view of spoke 轮辐的剖视图

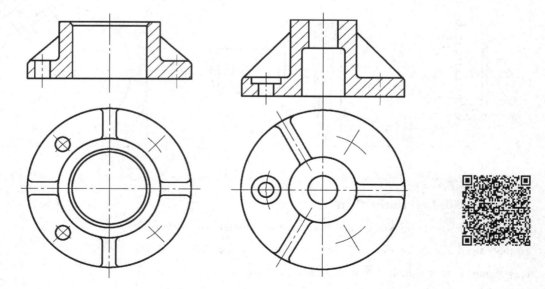

Fig. 6-72 Specified representation of symmetrically features on a revolved body 回转体上均匀分布结构的规定画法

Chapter 7 Expression for Commonly Used Machinery Components

Commonly used machinery components are divided into standard parts and commonly used parts. Standard parts are those whose structures, shapes, dimensions, marks and specifications are standardized. Typical examples are bolts, studs, screws, nuts, washers, keys, pins, and rolling bearings. Commonly used parts are those whose structure and shape elements are partially standardized, such as gears and springs. In engineering, standard parts and commonly used parts are the most widely used components. In addition to structures, styles and dimensions, national standards stipulate a series of signs, symbols and marks for drawing expression in order to improve production efficiency, reduce production costs and make use of specialized production. All such standardization must be observed when drawings. The emphasis of this chapter will be on standardized drawing methods, symbols and their indication methods.

7.1 Screw Threads

7.1.1 Overview

1. Screw threads manufacturing

Screw threads are continuous protrusions or grooves fabricated along a helix on a revolving component with the same cross sections. Screw threads on an external surface are called external threads, and those on an internal surface are called internal threads, as shown in Fig. 7-1.

第7章 常用机件的表达

常用机件包括标准件和常用件。标准件是指对结构、形状、尺寸、标记及技术要求都标准化的零件或部件，如螺栓、双头螺柱、螺钉、螺母、垫圈、键、销、滚动轴承等。常用件是指对结构和形状要素部分标准化的零件或部件，如齿轮、弹簧等。在工程上，标准件和常用件是机器、设备中使用最广泛的零件。为了提高生产效率，降低产品成本，便于专业化生产，国家标准除了对它们的结构、形式和尺寸予以规定外，还规定了一系列的画法符号、代号和标记，在图样表达时应遵守。本章重点介绍它们的规定画法、代号和标注方法。

7.1 螺纹

7.1.1 概述

1. 螺纹的形成

在回转表面上沿螺旋线所形成的具有相同断面的连续凸起和沟槽称为螺纹。在外表面上形成的螺纹称为外螺纹，在内表面上形成的螺纹称为内螺纹，如图7-1所示。

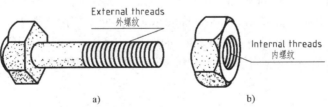

Fig. 7-1 Screw threads 螺纹
a) External threads 外螺纹 b) Internal threads 内螺纹

Various methods are employed in thread manufacturing, such as turning, rolling, and using tools like thread chasers and spiral taps, as shown inFig. 7-2. In the process of lathe manufacturing, the workpiece clamped with a chuck, rotates at constant speed. After reaching to a given distance along the radial direction, the turning tool moves with constant speed along the axial direction and thread curves are thus manufactured on the surface of the workpiece. For workpieces of small diameters, external threads and internal threads can be manufactured manually with the help of a thread chaser or a spiral tap.

螺纹的加工方法很多，如车削、滚压及用圆板牙、丝锥等工具加工，如图7-2所示。在车床上加工时，用卡盘夹持工件做等速旋转。当车刀沿径向进刀后，沿圆柱轴线方向匀速移动，便在工件表面加工出螺纹线。对直径较小的工件，可用板牙及丝锥手工加工内、外螺纹。

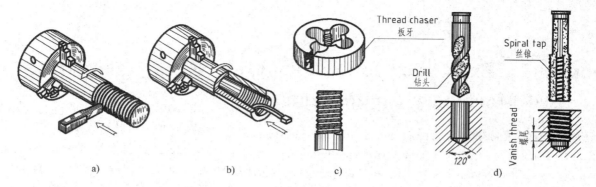

Fig. 7-2 Manufacturing methods of threads 螺纹的加工方法
a) External thread lathing 车削外螺纹 b) Internal thread lathing 车削内螺纹
c) Covering external thread 套外螺纹 d) Drilling 与 tapping internal thread 钻孔与攻内螺纹

2. Thread structure

(1) Thread head For installation purpose, preventing damage and providing operational safety, thread heads are shaped with certain features and the commonly used features are chamfer, round head and flat head as shown in Fig. 7-3.

2. 螺纹结构

(1) 螺纹端部 为了便于安装、防止损坏和安全操作，通常将螺纹端部加工成一定的形状，常见的形式有倒角、圆头及平头，如图7-3所示。

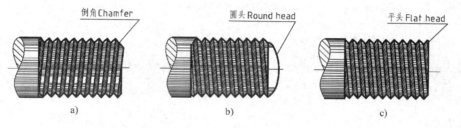

Fig. 7-3 Screw thread heads 螺纹端部
a) Chamfer 倒角 b) Round head 圆头 c) Flat head 平头

(2) Vanish thread and undercut A vanish thread is the incomplete segment at the end of a thread, originated when the tooth shape gradually becomes low for the lathe blade and gradually getting out of the workpiece. In order to avoid the formation of vanish thread and to make it easier for the lathe blade to complete its work, a small groove, called undercut, can be lathed in advance at the end of the screw termination as shown in Fig. 7-4.

(2) 螺尾及退刀槽 当车削螺纹终了时，在车刀逐渐离开工件处出现的一段牙型逐渐变浅的不完整螺纹，称为螺尾。为了避免产生螺尾和便于退刀，可在螺纹终止处预先车出一个小槽，称为退刀槽，如图7-4所示。

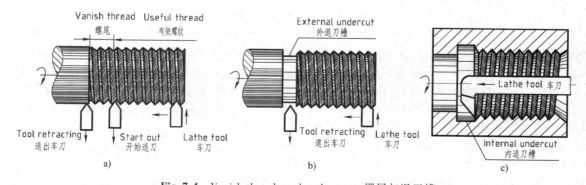

Fig. 7-4 Vanish threads and undercuts 螺尾与退刀槽
a) Vanish thread 螺尾 b) External undercut 外退刀槽 c) Internal undercut 内退刀槽

3. Thread elements

(1) Form of threads   The outline of the thread section cutting through the thread axis shows the form of threads. The common form of threads are triangular, trapezoidal, buttress and rectangular profiles as shown in Fig. 7-5.

3. 螺纹要素

(1) 螺纹牙型   通过螺纹轴线剖切时，螺纹断面的轮廓形状称为螺纹的牙型。常用螺纹的牙型有三角形、梯形、锯齿形和矩形，如图7-5所示。

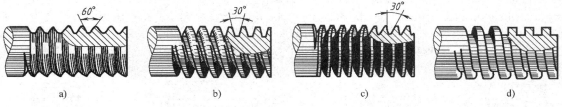

Fig. 7-5   Form of threads   螺纹牙型
a) Triangular (metric) thread   三角形（普通）螺纹   b) Trapezoidal (acme) thread   梯形螺纹
c) Buttress thread   锯齿形螺纹   d) Rectangular square thread   矩形（方）螺纹

(2) Thread diameters   Thread diameters include the major diameter (d, D), the minor diameters (d_1, D_1) and the pitch diameter (d_2, D_2). External thread diameters are indicated with lowercase letters, while internal thread diameters are indicated with capital letters, as shown in Fig. 7-6.

(2) 螺纹直径   螺纹的直径包括大径（d, D）、小径（d_1, D_1）、中径（d_2, D_2）。外螺纹的直径用小写字母表示，内螺纹的直径用大写字母表示，如图7-6所示。

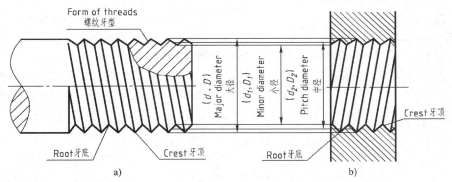

Fig. 7-6   Thread diameters   螺纹的直径
a) External thread   外螺纹   b) Internal thread   内螺纹

The thread's nominal diameter is generally the major diameter of the thread. The major diameter is the diameter of the column or taper which is tangential to crest of the external thread or the root of the internal thread. The minor diameter is the diameter of the column or taper which is tangential to the root of the external thread or the crest of the internal thread. The pitch diameter is the diameter of the column or taper where the thread groove width is equal to the thread ridge thickness.

(3) Number of threads (n)   A screw thread can be a single-start thread or a multi-start thread. If a screw thread has only one helix, it is called a single-start thread. While if it has two or more helices (equally distributed along the axis), it is called a double-start thread or multistart thread. The number of threads as shown in Fig. 7-7 is specified with letter "n".

螺纹的公称直径一般是指螺纹的大径。螺纹大径指与外螺纹牙顶或内螺纹牙底相切的假想圆柱或圆锥的直径。螺纹小径指与外螺纹牙底或内螺纹牙顶相切的假想圆柱或圆锥的直径。螺纹中径指通过圆柱或圆锥螺纹上牙厚和牙槽宽相等的圆柱或圆锥的直径。

(3) 螺纹线数（n）   螺纹有单线螺纹和多线螺纹之分。沿一条螺旋线所形成的螺纹，称为单线螺纹。沿两条或两条以上（在轴向等距分布的）螺旋线所形成的螺纹，称为双线螺纹或多线螺纹。螺纹线数用n表示，如图7-7所示。

(4) Pitch (P)　The axial distance between two points of adjacent teeth on the pitch diameter line is called thread pitch, symbolized as P (see Fig. 7-7).

(5) Lead (P_h)　The axial distance between two points of adjacent teeth of the same spiral line on the pitch diameter line is called lead, denoted as P_h ($P_h = nP$), as shown in Fig. 7-7.

(6) Revolving direction　A thread is called a right-hand thread if it follows clockwise turning when entering, otherwise it is known as a left-hand thread. A thread can be classified following the right or left spire rule, according to the methods shown in Fig. 7-8. Left-hand threads are generally used for devices with special requirement, such as a gas bottle switch.

(4) 螺距（P）　相邻两牙在中径线上同名牙侧间的轴向距离，称为螺距，用 P 表示（见图 7-7）。

(5) 导程（P_h）　同一条螺旋线上相邻两牙在中径线上同名牙侧间的轴向距离，称为导程，用 P_h（$P_h = nP$）表示（见图 7-7）。

(6) 旋向　顺时针旋转时旋入的螺纹称为右旋螺纹，反之为左旋螺纹。可用右手或左手螺旋规则，按图 7-8 所示的方法判断。左旋螺纹一般用于有特殊要求的零件，如煤气罐开关等。

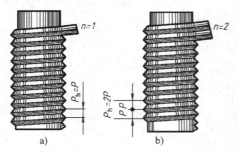

Fig. 7-7　Number of threads, pitch and lead
　　螺纹的线数、螺距和导程
　　a）Single-start threads　单线螺纹
　　b）Multi-start threads　多线螺纹

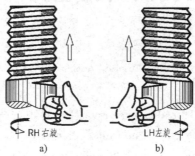

Fig. 7-8　Revolving directions of threads　螺纹旋向
　　a）Right-hand threads　右旋螺纹
　　b）Left-hand threads　左旋螺纹

4. Types of threads

(1) Classification following the degree of thread standardization　According to the degree of thread standardization, there are three types of threads.

1) Standard thread. The thread profile, diameters and pitch are standardized.

2) Special thread. The thread profiles follow national standards, but the diameter and pitch do not necessarily have to meet such standards.

3) Nonstandard thread. The thread profiles are not specified in national standards.

(2) Classification according to application　There are four types of threads.

1) Thread for fastening joining, or fastening thread for short. Typical examples include general thread, mini thread, transition fitting thread and interference fitting thread.

2) Thread for transmission, or transmission thread for short. Typical examples include trapezoidal thread, buttress thread and retangular thread.

3) Thread for pipes, or pipe threads for short. Typical examples include: pipe threads with 55 degree thread angle where pressure-tight joints are not made on the threads, pipe threads with 55 degree thread angle where pressure-tight joints are made on the threads, pipe threads with 60 degree thread angle where pressure-tight joints are made on the threads, and metric taper pipe thread.

4) Special application thread, or special threads for short. Typical examples include self-drilling screw thread, wooden screw thread screw thread for gas tanks.

Commonly used standard threads are shown in Table 7-1.

4. 螺纹的种类

(1) 按标准化程度分类　可分为三大类。

1）标准螺纹。螺纹牙型、直径和螺距均符合国家标准。

2）特殊螺纹。螺纹牙型符合国家标准，而直径和螺距不符合国家标准。

3）非标准螺纹。螺纹牙型不符合国家标准。

(2) 按螺纹用途分类　可分为四大类。

1）紧固联接用螺纹，简称紧固螺纹。典型的紧固螺纹有普通螺纹、小螺纹、过渡配合螺纹和过盈配合螺纹。

2）传动用螺纹，简称传动螺纹。典型的传动用螺纹有梯形螺纹、锯齿形螺纹和矩形螺纹。

3）管用螺纹，简称管螺纹。典型的管螺纹有55°非密封管螺纹、55°密封管螺纹、60°密封管螺纹和米制圆锥管螺纹。

4）专门用途螺纹，简称专用螺纹。典型的专用螺纹有自攻螺钉螺纹、木螺钉螺纹和气瓶专用螺纹。

常用标准螺纹的分类见表 7-1。

Table 7-1　Classification of commonly used standard threads　常用标准螺纹的分类

Type of thread 螺纹分类			Profiles or angles 牙型及牙型角	Symbol 特征代号	Number of national standards 国家标准代号	Application 用途说明
Fastening thread 紧固螺纹	Metric thread 普通螺纹	Coarse plain thread 粗牙普通螺纹	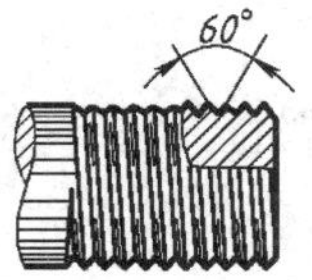 60°	M	GB/T 197—2003	Used in general links 用于一般零件的联接
		Fine plain thread 细牙普通螺纹				Used in small and thin parts 用于细、薄零件等
	Mini thread 小螺纹			S	GB/T 15054.4—1994	
Transmission thread 传动螺纹	Metric trapezoidal screw thread 梯形螺纹		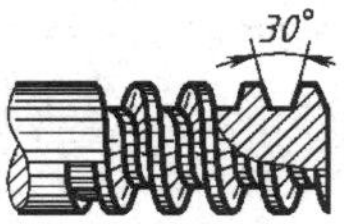 30°	Tr	GB/T 5796.1~5796.4—2005	It can transfer power and movement in two directions (the machine screw rod, and so on) 可传递双向动力和运动（如机床上的丝杠等）
	Buttress thread 锯齿形螺纹		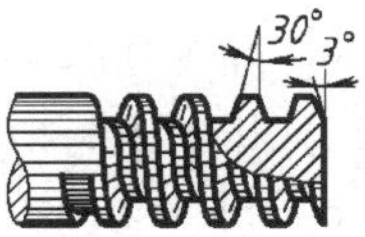 30° 3°	B	GB/T 13576.4—2008	It can transfer power in only one direction (the drive screw rod of screw press) 只能传递单向动力（如螺旋压力机的传动丝杠）
Pipe thread 管螺纹	Pipe threads with 55 degree thread angle where pressure-tight joints are not made on the threads 55°非密封管螺纹		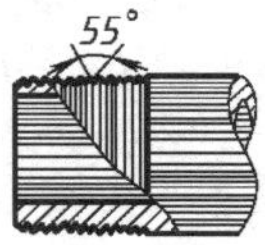 55°	G	GB/T 7307—2001	Used in low pressure tube linkage 用于低压管路的联接
	Pipe threads with 55 degree thread angle where pressure-tight joints are made on the threads 55°密封管螺纹	Taper external thread 圆锥外螺纹	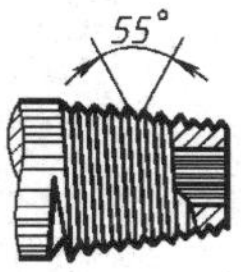 55°	R₁	GB/T 7306.1~7306.2—2000	Used in medium and high pressure tube (water pipes, oil pipes, gas pipes, etc.) linkage 用于中、高压管路（水管、油管、煤气管等）的联接
				R₂		
		Taper internal thread 圆锥内螺纹	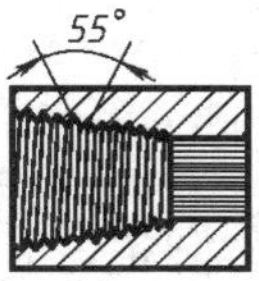 55°	Rc		
		Parallel internal thread 圆柱内螺纹	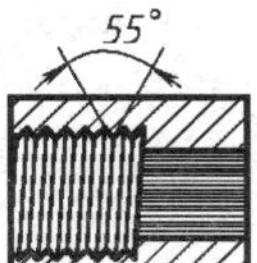 55°	Rp		
	Pipe threads with 60 degree thread angle where pressure-tight joints are made on the threads 60°密封管螺纹	Taper thread (external and internal) 圆锥管螺纹（内、外）	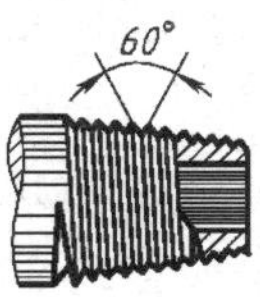 60°	NPT	GB/T 12716—2011	
		Parallel internal thread 圆柱内螺纹	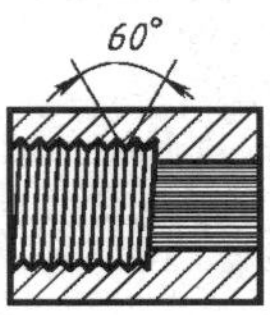 60°	NPSC		

7.1.2 Methods of drawing threads

1. Description of external threads (see Fig. 7-9)

7.1.2 螺纹的规定画法

1. 外螺纹的画法（见图7-9）

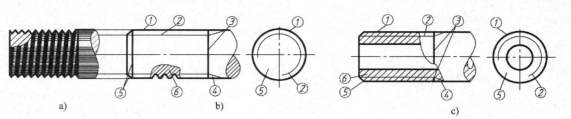

Fig. 7-9 Description of external threads 外螺纹的画法
a) Metric threads 普通螺纹 b) Description of metric threads 普通螺纹的画法
c) Description of pipe threads 管螺纹的画法

① Major diameter d should be draw with continuous thick lines, which indicate the thread's crest.

② Minor diameter d_1 should be drawn with continuous thin lines, which indicate the thread's root. Only draw a 3/4 circle on the circular view.

③ The thread end line should be drawn with continuous thick lines.

④ Vanish thread can be drawn with continuous thin lines which have an angle of 30° against the axis, whereas the vanish thread usually can be omitted.

⑤ Chamfer shouldbe drawn with continuous thick lines which have an angle of 45° against the axis. The chamfer is not required to be drawn on the circular view.

⑥ Draw section lines until the continuous thick lines.

2. Description of internal thread (see Fig. 7-10)

① 大径 d 用粗实线表示，对应于螺纹牙顶。

② 小径 d_1 用细实线表示，对应于螺纹牙底，并且在圆形的视图中只画3/4圈。

③ 螺纹终止线画粗实线。

④ 螺尾可画成与轴线成30°夹角的细实线，常省略不画。

⑤ 倒角画成粗实线，斜边与轴线成45°夹角，在圆形的视图中不画倒角圆。

⑥ 剖面线画到粗实线处。

2. 内螺纹的画法（见图7-10）

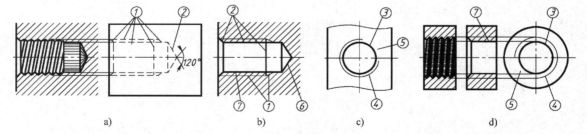

Fig. 7-10 Description of internal threads 内螺纹的画法
a) Non-cutting description of the blind hole 螺纹不通孔的不剖画法 b) Cutting description of the blind hole 螺纹不通孔的剖开画法
c) Circular view of the internal thread 在圆形视图中螺孔的画法 d) Drawing of the through hole of thread 螺纹通孔的画法

（1）Non-cutting (see Fig. 7-10 a and Fig. c)

① In the noncircular view, the major diameter D, the minor diameter D_1, the thread end line, the vanish thread, and the chamfer are denoted with dashed lines.

② The conic portion of the threaded blind hole should be drawn with dashed lines which form an angle of 120°.

③ In the circular view, the minor diameter D_1 is drawn with continuous thick lines.

④ In the circular view, only 3/4 circle of the major diameter D is drawn with continuous thin lines.

⑤ In the circular view, the chamfer is not drawn.

（1）不剖画法（见图7-10a、c）

① 非圆视图中，大径 D、小径 D_1、螺纹终止线、螺尾、倒角均画虚线。

② 螺纹不通孔的锥坑画成120°夹角的虚线。

③ 在圆形视图中，小径 D_1 画粗实线。

④ 在圆形视图中，大径 D 只画3/4圈细实线。

⑤ 在圆形视图中不画倒角圆。

(2) Cutting (Fig. 7-10b and Fig. d)

① In the noncircular view, the major diameter D and the vanish thread are drawn with continuous thin lines. The vanish thread usually be omitted.

② In the noncircular view, the minor diameter D_1, the thread end line, and the chamfer are drawn with continuous thick lines.

③ In the circular view, the minor diameter D_1 is drawn with continuous thick lines.

④ In the circular view, only 3/4 circle of the major diameter D is drawn with continuous thin lines.

⑤ In the circular view. the chamfer is not drawn.

⑥ The conic portion of the threaded blind hole can be drawn with continuous thick lines which form an angle of 120°.

⑦ Draw section lines until the continuous thick lines.

3. Assembly drawing of internal and external thread joints

In the section view of a thread joint, the joint portion is drawn according to the description of the external thread, and the rest is drawn following their respective descriptions, as shown in Fig. 7-11.

Notes: ①The major and minor diameter lines of internal and external threads must be aligned accordingly. ②When the section passes the axis of the solid pole, the thread is not cut and only its outline is drawn.

(2) 剖开画法（见图 7-10 b、d）

① 非圆视图中，大径 D、螺尾可画成细实线。螺尾常可省略不画。

② 非圆视图中，小径 D_1、螺纹终止线、倒角均画粗实线。

③ 在圆形视图中，小径 D_1 画粗实线。

④ 在圆形视图中，大径 D 只画 3/4 圈细实线。

⑤ 在圆形视图中不画倒角圆。

⑥ 螺纹不通孔的锥坑画成 120° 夹角的粗实线。

⑦ 剖面线画到粗实线处。

3. 内、外螺纹联接的画法

在绘制螺纹联接的剖视图时，其联接部分应按外螺纹的画法绘制，其余部分仍按各自的画法绘制，如图 7-11 所示。

注意：①内、外螺纹的大径线、小径线应分别对齐；②当通过实心杆件的轴线剖开时按不剖处理，即只画外形。

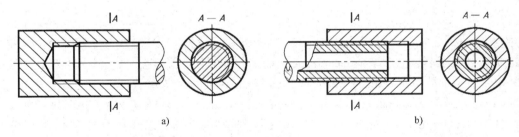

Fig. 7-11 Description of internal and external thread joints 内、外螺纹联接的画法
a) Metric thread joints 普通螺纹联接 b) Pipe thread joints 管螺纹联接

7.1.3 Thread marks

Since standard drawing methods have been adopted, in order to better identify thread types and elements, it is necessary to mark threads on drawings according to established formats. There are two kinds of the thread marks, namely standard thread marks and non-standard thread marks, explained as follows.

1. Standard thread marks

According to the national standard (GB/T 4459.1—1995), standard threads must be labeled in drawings with their corresponding symbols, as shown in Table 7-1.

7.1.3 螺纹的标注

由于螺纹采用统一规定的画法，为了便于识别螺纹的种类及其要素，对螺纹必须按规定格式在图上进行标注。螺纹的标注方法分标准螺纹的标注和非标准螺纹的标注两种，下面分别进行说明。

1. 标准螺纹的标注

按国家标准（GB/T 4459.1—1995）规定，标准螺纹应在图上注出相应的特征代号，见表 7-1。

(1) Marks of metric threads The marking format of the single-start thread is the following:

| Type symbol 特征代号 | × | Nominal diameter 公称直径 | Pitch(blank for coarse thread) 螺距(粗牙省略) | — | Tolerance zone symbol 公差带代号 | — | Symbol of joint length (blank for medium) 旋合长度代号 (中等省略) | — | Revolving direction (blank for right-hand thread) 旋向(右旋省略) |

The marking format of the multiple-start thread is:

| Type symbol 特征代号 | × | Nominal diameter 公称直径 | Ph Lead P Pitch Ph导程 P 螺距 | — | Tolerance zone symbol 公差带代号 | — | Symbol of joint length (blank for medium) 旋合长度代号 (中等省略) | — | Revolving direction (blank for right-hand thread) 旋向(右旋省略) |

1) Thread tolerance zone symbols. It explains the thread's permissible tolerance, which includes pitch diameter tolerance and major diameter tolerance, shown with figures and letters. Figures denote the degree of tolerance, while letters denote basic deviation. If the tolerance of the pitch diameter is equal to that of the major diameter, only one of them should be marked.

2) Joint length symbols. The length of thread joints are divided into three groups, including short joints, normal joints, and long joints, denoted with S, N and L, respectively. Generally, the joint length symbol is not marked, and considered as normal. Table 7-2 lists the marking of commonly used standard threads.

(2) Marks of piple threads

| Type symbol 特征代号 | Dimension symbol 尺寸代号 | Tolerance grade 公差等级 |

Notes: ①The pipe thread must be marked with a leading line, elicited from the major diameter; ②Tolerance grade symbol: For the Pipe threads with 55 degree thread angle where pressure-thght joints are not made on the threads, external threads are marked with two degrees (A and B), but the internal threads are not marked; For the other types of pipe threads, tolerance grade is not marked.

(3) Marks of trapezoidal (acme) threads

| Type symbol 特征代号 | × | Nominal diameter 公称直径 | Pitch(single start) 螺距(单线) / lead(pitch) 导程(P螺距) | Revolving direction (blank for right-hand thread) 旋向(右旋省略) | — | Tolerance zone symbol 公差带代号 | Joint length (blank for medium) 旋合长度 (中等省略) |

2. Marks of special and nonstandard threads

(1) Mark of special threads The word "special" must be added before the thread profile symbol, and mark the major diameter and the pitch, as shown in Fig. 7-12a.

(2) Marks of nonstandard threads The values of the major and minor diameters, the pitch and the thread profile must be marked, as shown in Fig. 7-12b.

（1）普通螺纹的标注 单线普通螺纹的一般标注格式为：

多线普通螺纹的一般标注格式为：

1) 公差带代号。螺纹的公差带代号表示螺纹允许的尺寸公差（包括中径公差和顶径公差两种），由数字和字母组成。其数字为公差等级，字母为基本偏差。若中径、顶径公差相同，则只标注一个代号。

2) 旋合长度代号。螺纹的旋合长度分为短、中、长三组，分别用S、N、L（即Short、Normal、Long的第一个字母）表示。一般情况下可不加标注，按中等长度考虑。在表7-2中列出了常用标准螺纹的标注示例。

（2）管螺纹的标注

注意：①管螺纹必须采用指引线标注，指引线从大径线引出；②公差等级代号：对于55°非密封管螺纹，外螺纹分A、B两级标注，内螺纹则不标注；对于其他类型管螺纹，不标准。

（3）梯形螺纹标注

2. 特殊螺纹与非标准螺纹的标注

（1）特殊螺纹的标注 应在螺纹特征代号前加注"特"字，并标注出大径和螺距，如图7-12a所示。

（2）非标准螺纹的标注 应标注出螺纹的大径、小径、螺距和牙型，如图7-12b所示。

Table 7-2 The demonstration of standard threads in common use 常用标准螺纹的标注示例

Classification 螺纹类别		Marking example 标注图例	Meaning of symbols 代号的意义	Explanation 说明
Fastening thread 紧固螺纹	Coarse plain thread 粗牙普通螺纹	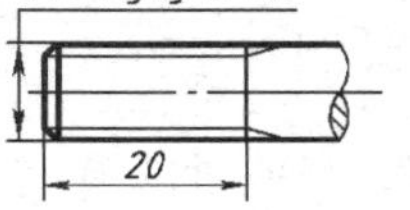 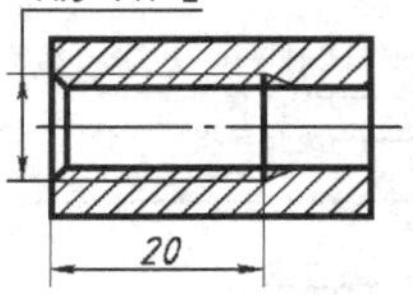 M10-5g6g-S-LH (length 20) M10-7H-L (length 20)	M10 – 5g 6g – S – LH — Left-hand thread 左旋 — Short Joint length 短旋合长度 — Tolerance zone of major diameter 顶径公差带 — Tolerance zone of pitch diameter 中径公差带 — Major diameter 螺纹大径 M10 – 7H – L — Long joint length 长旋合长度 — Tolerance zone of pitch diameter and major diameter (equal) 中径和顶径公差带(相同)	1. The pitch of a coarse plain thread, the number of thread for a single-start thread and the revolving direction of a right-hand thread are not marked, but for multiple-start threads and left-hand threads they must be marked 2. If the tolerance zone of the pitch diameter is equal to that of the major diameter, only one of them should be marked, e.g. 7H 3. If the joint length is normal, it is not marked 4. The marked length does not include the vanish thread 1. 粗牙螺纹不注螺距，单线、右旋螺纹不注线数和旋向，多线和左旋螺纹要标注 2. 中径和顶径公差带相同时，只注一个代号，如 7H 3. 中等旋合长度不标注 4. 图中所注长度不包括螺尾
	Fine plain thread 细牙普通螺纹	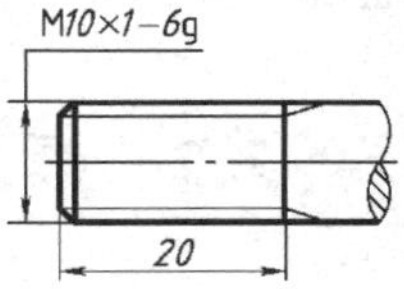 M10×1-6g (length 20)	M 10×1 — Pitch 螺距	1. The pitch of fine thread must be marked 2. Description used for coarse plain thread 1. 细牙要注螺距 2. 其他规定同粗牙普通螺纹
Pipe thread 管螺纹	Pipe threads with 55 degree thread angle where pressure-tight joints are not made on the threads 55°非密封管螺纹	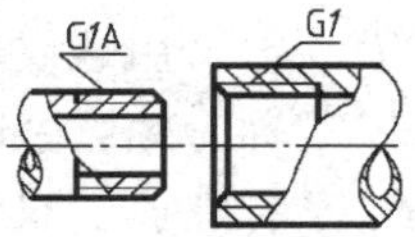 G1A G1	G1 A — Tolerance grade 公差等级 — Dimension symbol 尺寸代号	1. The dimension symbol of the pipe thread is not the major diameter value, which is established accordingly when drawing 2. Mark with a note. 3. The right-hand thread is omitted 1. 管螺纹尺寸代号不是螺纹大径，作图时应据此查出螺纹大径 2. 只能以旁注的方式引出标注 3. 右旋省略不注
	Pipe threads with 55 degree thread angle where pressure—tight joints are made on the threads—Parallel internal threads 55°密封圆柱内螺纹	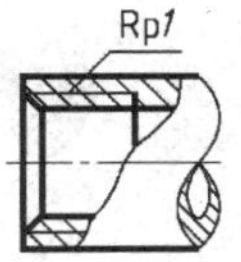 Rp1	Rp1 — Dimension symbol 尺寸代号 — Parallel internal thread 圆柱内螺纹	
	Pipe threads with 55 degree thread angle where pressure-tight joints are made on the threads—Taper internal and external threads 55°密封圆锥内螺纹与圆锥外螺纹	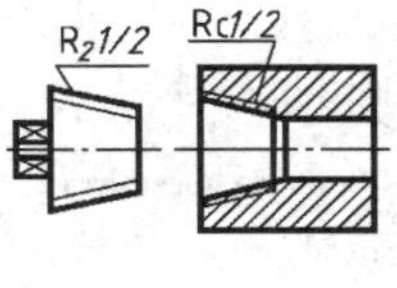 R₂1/2 Rc1/2	R₂1/2 — Dimension symbol 尺寸代号 — Taper external thread to match the taper internal thread 与圆锥内螺纹相配的圆锥外螺纹 Rc1/2 — Dimension symbol 尺寸代号 — Taper internal thread 圆锥内螺纹	

(Continued 续)

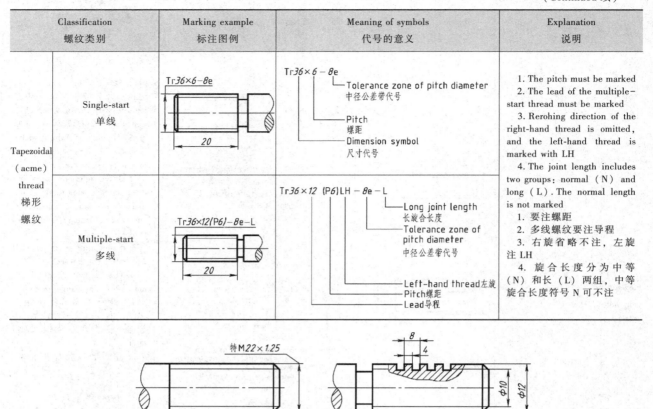

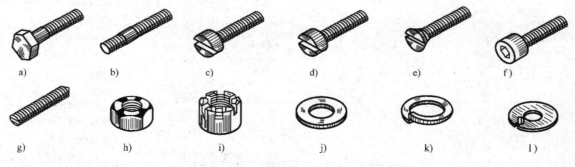

Fig. 7-12 Marks of special and nonstandard threads 特殊螺纹与非标准螺纹的标注
a) Special thread 特殊螺纹 b) Nonstandard thread 非标准螺纹

7.2 Screw Fasteners

7.2.1 Types of screw fasteners

Parts with threads used for joining and fastening are called screw fasteners. The most widely used components as screw fasteners are bolts, studs, screws, nuts, washers, as shown in Fig. 7-13.

7.2 螺纹紧固件

7.2.1 螺纹紧固件的种类

用螺纹联接并起紧固作用的零件称为螺纹紧固件。常用的螺纹紧固件有螺栓、双头螺柱、螺钉、螺母、垫圈等，如图 7-13 所示。

Fig. 7-13 Screw fasteners 螺纹紧固件
a) Hexagon head bolt 六角头螺栓 b) Double end stud 双头螺柱 c) Slotted cheese head screw 开槽圆柱头螺钉
d) Slotted pan head screw 开槽盘头螺钉 e) Slotted countersunk flat head screw 开槽沉头螺钉 f) Hexagon socket cap screw 圆柱头内六角螺钉 g) Slotted set screw with cone point 开槽锥端紧定螺钉 h) Hexagon nut 六角螺母
i) Hexagon slotted nut 六角开槽螺母 j) Plain washer 平垫圈 k) Single coil spring lock washer 弹簧垫圈 l) Lock washer 止动垫圈

7.2.2 Symbols of screw fasteners

The structures and dimensions of screw fasteners have been standardized, and they should be represented as follows:

| Name 名称 | Standard number 标准号 | Specification, precision 规格、精度 | Other requirements for the style and the dimension 型式与尺寸的其他要求 | — | Performance grade or material and heat treatment 性能等级或材料及热处理 | — | Surface treatment 表面处理 |

Example: A hexagon head bolt, the parameters of which are thread specification d = M12, nominal length l = 80mm, performance grade is 8.8, grade of products is A, and oxidization as surface treatment, is marked as:

Bolt GB/T 5782—2016   M12×80-10.9-A-O. Its simple mark is: Bolt GB/T 5782   M12×80.

Table 7-3 lists various commonly used screw fasteners, and their respective symbols.

7.2.2 螺纹紧固件的标记

螺纹紧固件的结构形式及尺寸均已标准化，其完整的标记格式为：

举例：螺纹规格 d = M12、公称长度 l = 80mm、性能等级为8.8级、产品等级为A、表面氧化处理的六角头螺栓，完整标记为：

螺栓 GB/T 5782—2016   M12×80-10.9-A-O；其简化标记为：螺栓 GB/T 5782   M12×80。

表7-3列出了一些常用的螺纹紧固件及其规定标记。

Table 7-3 Widely used screw fasteners and their respective symbols 常用螺纹紧固件及其规定标记

Name and standard number 名称及标准代号	Illustration 图例	Descriptive mark and explanation 规定标记及说明
Hexagon head bolts Product grades A and B 六角头螺栓　A和B级 GB/T 5782—2016	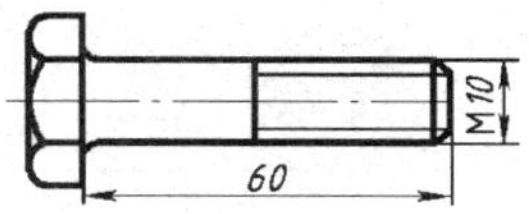	The hexagon head bolt (thread specification d = M10, nominal length l = 60mm, product grade A) is marked as: Bolt GB/T 5782 M10×60 螺纹规格 d = M10、公称长度 l = 60mm 的 A 级六角头螺栓标记： 螺栓　GB/T 5782 M10×60
Studs　螺柱 Double end studs 双头螺柱 b_m = 1d （GB 897—88） Double end studs 双头螺柱 b_m = 1.25d （GB 898—88） Double end studs 双头螺柱 b_m = 1.5d （GB 899—88） Double end studs 双头螺柱 b_m = 2d （GB 900—88）	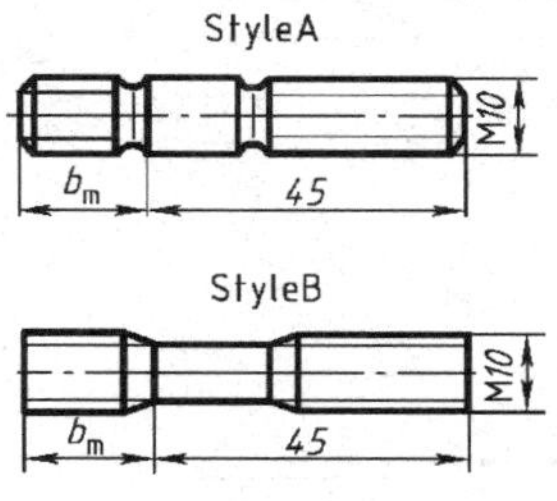	Double end stud (two ends are coarse threads, d = M10, l = 45mm, level of performance is 4.8, no surface treatment, style B, b_m = 1d) is marked as: Stud GB 897 M10×45 两端均为粗牙普通螺纹、d = M10、l = 45mm、性能等级为4.8级、不经表面处理、B型、b_m = 1d 的双头螺柱标记： 螺柱　GB 897 M10×45
Slotted cheese head screws 开槽圆柱头螺钉 GB/T 65—2016	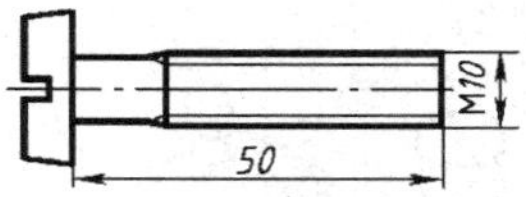	The slotted cheese head screw (d = M10, l = 50mm, level of performance is 4.8, and no surface treatment, product grade A) is marked as: Screw GB/T 65 M10×50 螺纹规格 d = M10、公称长度 l = 50mm、性能等级为4.8级、不经表面处理的A级开槽圆柱头螺钉的标记： 螺钉　GB/T 65 M10×50

(Continued 续)

Name and standard number 名称及标准代号	Illustration 图例	Descriptive mark and explanation 规定标记及说明
Slotted pan head screws 开槽盘头螺钉 GB/T 67—2016	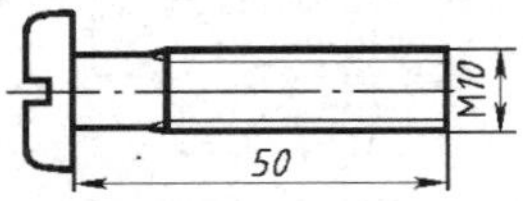	The slotted pan head screw (d = M10, l = 50mm, level of performance grade is 4.8, and no surface treatment, product grade A) is marked as: Screw GB/T 67 M10×50 螺纹规格 d = M10、公称长度 l = 50mm、性能等级为 4.8 级、不经表面处理的 A 级开槽盘头螺钉的标记： 螺钉 GB/T 67 M10×50
Slotted countersunk flat head screws 开槽沉头螺钉 GB/T 68—2016	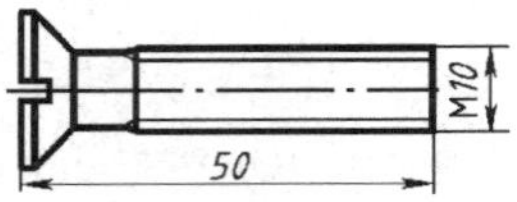	The slotted countersunk flat head screw (d = M10, l = 50mm, level of performance is 4.8, and no surface treatment, product grade A) is marked as: Screw GB/T 68 M10×50 螺纹规格 d = M10、公称长度 l = 50mm、性能等级为 4.8 级、不经表面处理的 A 级开槽沉头螺钉的标记： 螺钉 GB/T 68 M10×50
Countersunk flat head screws with cross recess 十字槽沉头螺钉 GB/T 819.1—2016	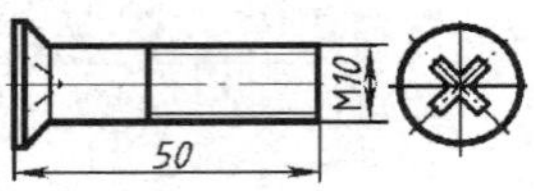	The countersunk flat head screw with cross recess (d = M10, l = 50mm, level of performance is 4.8, style H, and no surface treatment, product grade A) is marked as: Screw GB/T 819.1 M10×50 螺纹规格 d = M10、公称长度 l = 50mm、性能等级为 4.8 级、H 型不经表面处理的 A 级十字槽沉头螺钉的标记： 螺钉 GB/T 819.1 M10×50
slotted set screws with cone point 开槽锥端紧定螺钉 GB 71— 85	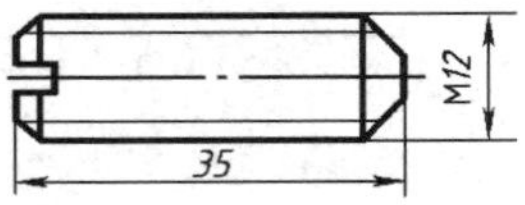	The slotted set screw with cone point (d = M12, l = 35mm, performance grade is 14H, and oxidization as surface treatment) is marked as: Screw GB 71—85 M12×35 螺纹规格 d = M12、公称长度 l = 35mm、性能等级为 14H 级、表面氧化的开槽锥端紧定螺钉的标记： 螺钉 GB 71—85 M12×35
Slotted set screws with long dog point 开槽长圆柱端紧定螺钉 GB 75— 85	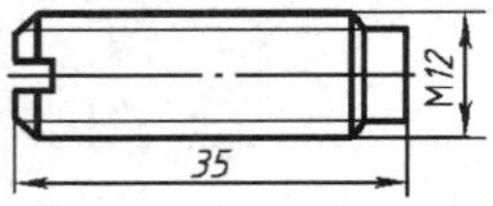	The slotted set screw with long dog point (d = M12, l = 35mm, performance grade is 14H, and oxidization as surface treatment) is marked as: Screw GB 75—85 M12×35 螺纹规格 d = M12、公称长度 l = 35mm、性能等级为 14H 级、表面氧化的开槽长圆柱端紧定螺钉的标记： 螺钉 GB 75—85 M12×35
Hexagon nuts, style 1 Product grades A and B 1 型六角螺母 A 级和 B 级 GB/T 6170—2015	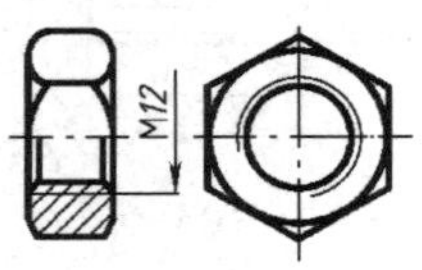	The hexagon nut (D = M12, performance grade is 8, no surface treatment, product grades A) is marked as: Nut GB/T 6170 M12 螺纹规格 D = M12、性能等级为 8 级、不经表面处理的 A 级 1 型六角螺母的标记： 螺母 GB/T 6170 M12

7.2.3 Assembly drawing methods of screw fasteners

1. Common joining methods

There are three kinds of joining methods for screw fasteners as shown in Fig. 7-14. Their utilization is to fasten two parts together. Different joining methods can be selected based on the requirement and thickness of parts to be fastened.

The drawing method that one uses dimensions obtained from respective tables (see Table 2-1 to Table 2-11 in appendices) to draw screw fasteners is called looking-up table method. On the other hand, one may also use dimensions proportional to the nominal diameter (d, D) when drawing screw fasteners. This method is called the proportional drawing method.

Examples shown in Fig. 7-15a are the screw fasteners drawn with the proportional drawing method, and their simple descriptions are shown in Fig. 7-15b.

7.2.3 螺纹紧固件的装配画法

1. 常见联接方式

螺纹紧固件有三种联接方式，如图 7-14 所示。它们的作用是将两个零件紧固在一起。根据零件被紧固处的厚度和使用要求选用不同的联接方式。

画图时，通过查表（参见附录表 2-1～表 2-11）获得螺纹紧固件各个部位参数的方法称为查表法；将螺纹紧固件各部位的尺寸用公称直径（d, D）的不同比例画出的方法，称为比例法。

采用比例法绘制螺纹紧固件，如图 7-15a 所示，其简化画法如图 7-15b 所示。

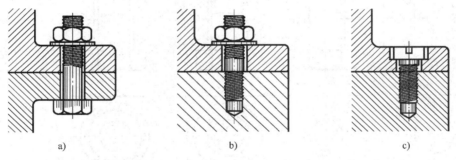

Fig. 7-14 Joining methods of screw fasteners　螺纹紧固件的联接方式
a) Bolt joint　螺栓联接　b) Stud joint　双头螺柱联接　c) Screw joint　螺钉联接

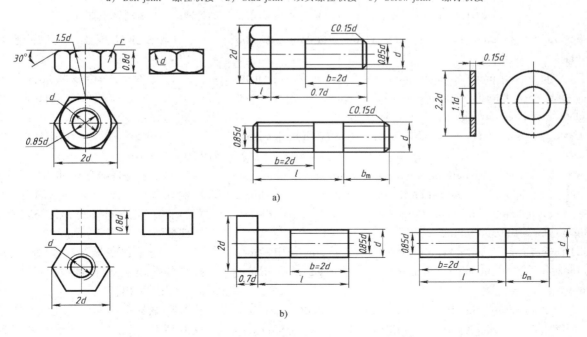

Fig. 7-15 Proportional drawing methods of screw fasteners　用比例法绘制螺纹紧固件
a) Original drawings　原形图　b) Simplified descriptions　简化图

2. Drawing methods of a bolt assembly

A bolt assembly is composed of a bolt, a nut, and a washer. The bolt assembly is used when joint parts are not too thick to drill through holes as shown in Fig. 7-16. The proportional drawing method and steps of bolt assemblies are shown in Fig. 7-17.

2. 螺栓装配图的画法

螺栓联接由螺栓、螺母、垫圈组成。螺栓联接用于被联接的两零件厚度不大,可钻成通孔的情况,如图7-16所示。螺栓装配图的比例画法和作图步骤如图7-17 所示。

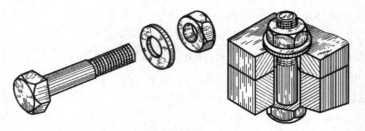

Fig. 7-16 Pictorial drawing of bolt joints 螺栓联接示意图

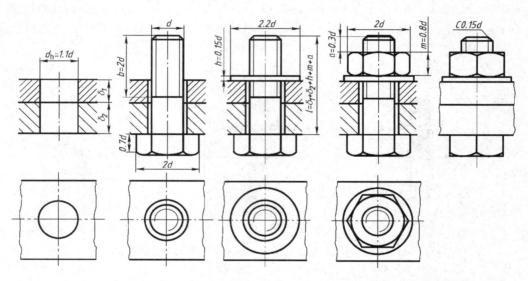

Fig. 7-17 Method and steps for drawing bolt assemblies 螺栓装配图的画法及步骤

Several points must be taken into account in bolt assembly drawings:

1) The thread major diameter and the thickness δ_1, δ_2 of the jointed parts are known, and the effective length l' of the bolt should be estimated through the following formula:

$$l' = \delta_1 + \delta_2 + 0.15d(\text{washer thickness}) + 0.8d(\text{nut thickness}) + a$$

where "a" is the length between the bolt top and the nut (the bare end), $a = 0.3d$. The immediate standard value is then chosen as the effective length l by consulting Tble 2-1 in the appendix according to the estimated value.

2) To ensure proper assembly, the hole diameter of the joint parts must be a little bit bigger than that of the thread major diameter, drawn as $d_h = 1.1d$.

3) The thread end line is specified to be lower than the top surface of the hole in order to provide sufficient length to tighten nuts firmly.

画螺栓装配图时应注意如下几点:

1) 螺纹大径和被联接件的厚度 δ_1、δ_2 已知时,螺栓的有效长度 l' 应按下式估算:

$$l' = \delta_1 + \delta_2 + 0.15d(垫圈厚) + 0.8d(螺母厚) + a$$

式中 a 是螺栓顶端伸出螺母外(露出端)的长度,且 $a = 0.3d$。然后根据估算出的数值,查附录表2-1,选取相近的标准数值,确定有效长度 l。

2) 为了保证装配工艺合理,被联接件的孔径 d_h 应比螺纹大径大些,按 $d_h = 1.1d$ 画出。

3) 螺栓上的螺纹终止线应低于通孔的顶面,以便于拧紧螺母时有足够的螺纹长度。

4) When a section plane cross the axes of bolts, nuts, screws and washer, these fastening parts are not drawn according to section drawings, but still following their outlines.

5) When two joint parts are cut, the directions of their section lines shall be reversed.

6) When the thread assembly drawing is too small, the nut and bolt head can be simply drawn as outlines as shown in Fig. 7-18.

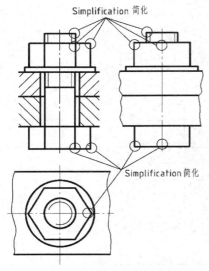
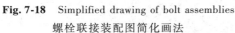

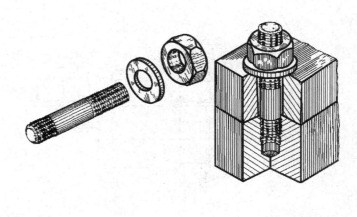

Fig. 7-18　Simplified drawing of bolt assemblies
螺栓联接装配图简化画法

Fig. 7-19　Pictorial drawing of stud joint
双头螺柱联接示意图

3. Drawing methods of a stud assembly

A stud assembly is composed of a stud, a nut and a washer. The two ends of the stud are threaded. The end that is straightway screwed in the joint parts is called a metal end, and the other end screwed down with the nut (see Fig. 7-19) is called a nut end. An illustration based on the proportional drawing method of the stud assembly is shown in Fig. 7-20a.

Several aspects must be taken into account when drawing the stud assembly:

1) The effective length l' of the stud should be estimated through the following formula:

$l' = \delta_1 + 0.15d(\text{washer thickness}) + 0.8d(\text{nut thickness}) + a$

where, $a = 0.3d$. The immediate standard value is then selected as the effective length l by consulting Table 2-2 in the appendix, according to the estimated value.

2) The length b_m is the end dimension of the stud which is screwed on to the workpiece, which is called a metal end. Length b_m is related to the parts' material: $b_m = d$ for steel, $b_m = 1.25d$ or $1.5d$ for cast iron, and $b_m = 2d$ for aluminum. In order to guarantee the reliability of the joint, the metal end must be entirely screwed into the thread hole of the workpiece, so the thread end line is leveled with its end face.

3. 双头螺柱装配图画法

双头螺柱联接由双头螺柱、螺母及垫圈组成。双头螺柱两端均有螺纹，联接时，一端直接旋入被联接零件，称为旋入端；另一端用螺母拧紧（见图7-19），称为露出端。双头螺柱装配图的比例画法如图7-20a所示。

画双头螺柱装配图时应注意如下几方面：

1) 双头螺柱的有效长度 l' 应按下式估算：

$l' = \delta_1 + 0.15d$（垫圈厚度）$+ 0.8d$（螺母厚度）$+ a$

式中 $a = 0.3d$。然后根据估算出的数值，查阅附录表2-2，选取相近的标准数值，确定其效长度 l。

2) b_m 是双头螺柱旋入机件的尺寸，对应螺柱部分称为旋入端。b_m 与机件的材料有关：材料为钢时 $b_m = d$，材料为铸铁时 $b_m = 1.25d$ 或 $1.5d$，材料为铝时 $b_m = 2d$。为保证联接可靠，旋入端应全部旋入机件螺孔内，所以螺纹终止线与机件端面应平齐。

3) The thread depth of the thread hole of the workpiece shall be bigger than the thread length b_m of the metal end and, generally, it is drawn as $b_m+0.5d$. In the assembly drawing, the depth of the drilled hole can be simply drawn according to the thread's depth.

4) Proportions between the major diameter d and the dimensions of all portions of the nut and the washer and their descriptions are the same as those of the bolt.

5) When the figure is too small, drawings of the nut and the stud can be simplified as outlines shown in Fig. 7-20b.

3) 机件螺孔的螺纹深度应大于旋入端的螺纹长度 b_m ，一般螺孔的螺纹深度按 $b_m+0.5d$ 画出。在装配图中钻孔深度可按螺纹深度简化画出。

4) 螺母和垫圈等各部分尺寸与大径 d 的比例关系及画法与螺栓中的相同。

5) 当图形较小时，螺母和双头螺柱的头部均可简化画出，如图7-20b所示。

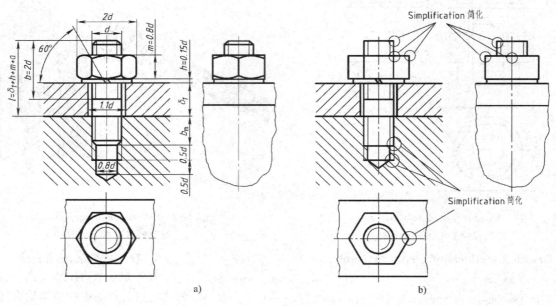

Fig. 7-20 Description of the stud assembly drawing　双头螺柱的装配图画法
a) Proportional drawing　比例画法　b) Simplified drawing　简化画法

4. Drawing methods of a screw assembly

In case of a screw joint, the nut is not used, but the screw is directly screwed into the thread hole of the workpiece. The screw joint is mostly used for the cases with small loading. According to the shape of the screw head, there are many types of screws. Fig. 7-21 illustrates several drawing methods for several commonly used ones.

Several aspects must be taken into account when drawing screw assemblies:

1) The effective length l of the screw can be estimated through the following formula, and then the immediate standard value is selected as the effective length l by consulting Table 2-3 to Table 2-6 in the appendices, according to estimated values.

$l=\delta_1+b_m$ (b_m varies according to the material of screwed parts)

2) The thread length is taken as $b_m=2d$, making the thread end line above the end face of the thread hole, to guarantee that the screw can be screwed in and tightened to the workpiece.

4. 螺钉装配图画法

螺钉联接不用螺母，而将螺钉直接拧入机件的螺孔。螺钉联接多用于受力不大的情况。螺钉根据头部形状的不同而有多种类型，图7-21所示为几种常用螺钉装配图的画法。

绘制螺钉装配图时应注意如下几点：

1) 螺钉的有效长度 l 可按下式估算，然后根据估算出的数值，查附录表2-3～表2-6，选取相近的标准数值即可。

$l=\delta_1+b_m$ （b_m 根据被旋入零件的材料而定）

2) 取螺纹长度 $b_m=2d$ ，使螺纹终止线伸出螺孔端面，以保证联接时能使螺钉旋入压紧。

3) The groove of the slotted screw head must be shown on the front view, but it's not drawn in the top view according to normal projection. It must be drawn with an inclination angle of 45° as shown in Fig. 7-21.

4) If the groove width of the slotted screw head is too small, a thick black line is adopted as shown in Fig. 7-21c and d.

3）开槽螺钉头部的沟槽槽口必须在主视图表示出来，而俯视图则可不按投影关系画，按规定应画成倾斜45°，如图7-21所示。

4）开槽螺钉头部的槽宽过小时，可采用涂黑的画法（图7-21c、d）。

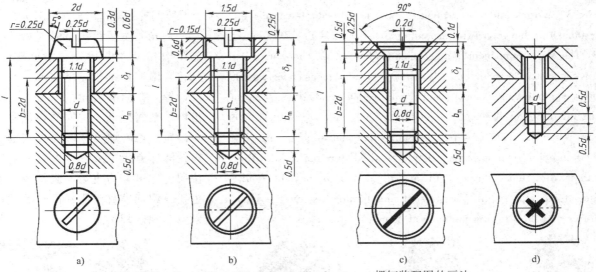

Fig. 7-21 Description of screw assembly drawings　螺钉装配图的画法
a) Slotted pan head screw　开槽盘头螺钉　b) Slotted cheese head screw　开槽圆柱头螺钉
c) Slotted countersunk flat head screw　开槽沉头螺钉　d) Philips countersunk screw　十字槽沉头螺钉

5. Drawing methods of set screws

Set screws are used to fix two parts of a device in order to prevent relative movement. Fig. 7-22a shows the drawing method of assembly with a slotted cone screw with point set. Fig. 7-22b shows the drawing method of assembly with a slotted set screw with long dog point.

5. 紧定螺钉联接画法

紧定螺钉联接用于固定两个零件，使它们不产生相对运动。图7-22a所示为开槽锥端紧定螺钉联接装配图的画法，图7-22b所示为开槽长圆柱端紧定螺钉联接装配图的画法。

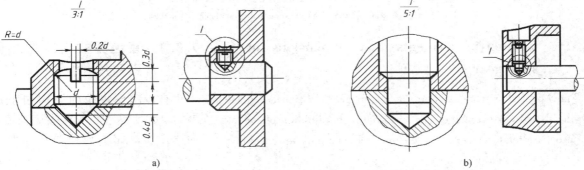

Fig. 7-22 Discription of set screw assembly drawings　紧定螺钉装配图的画法
a) Slotted set screw with cone point　开槽锥端紧定螺钉　b) Slotted set screw with long dog point　开槽长圆柱端紧定螺钉

7.3　Gears

7.3.1　Types of gears

Gears are widely known as transmission components, being used to transfer power and change the value and direction of rotation. There are different types of gears and they can be classified into three categories according to their transmission.

7.3　齿轮

7.3.1　齿轮的种类

齿轮是机械传动中广泛应用的传动零件，用于传递动力、改变转速和方向。齿轮的种类繁多，根据其传动情况可分为三类。

(1) Cylindrical gears (Spur gears, helical gears and double heilcal gears) Used to transmit the force and rotational speed between two parallel axes (see Fig. 7-23a, Fig. 7-23b and Fig. 7-23c).

(2) Bevel gears Used to transmit the force and rotational rate between two perpendicular axes (see Fig. 7-23d).

(3) Worms and wormwheels Used to transmit the force and rotational rate between two perpendicular axes (see Fig. 7-23e).

The major component of a gear is a toothed wheel. There are related national standards for both the structure and dimension of various gears. Gears following national standards are called standard gears. Those gears modified on the basis of standards are called profile shifted gears. When the teeth of a pair of cylindrical mating gears are produced on cylindrical surfaces, they are called spur gears. According to their directions, gear teeth are divided into straight teeth, helical teeth and so on. The present chapter focuses on the basic features and the drawing methods of standard spur gears.

（1）圆柱齿轮（直齿轮、斜齿轮、人字齿轮）用于传递两平行轴间的动力和转速（见图7-23a、b、c）。

（2）锥齿轮 用于传递两垂直相交轴间的动力和转速（见图7-23d）。

（3）蜗杆和蜗轮 用于传递两垂直交叉轴间的动力和转速（见图7-23e）。

齿轮的主要组成部分是轮齿，它的结构和尺寸由相关国家标准规定。凡符合国家标准规定的齿轮，称为标准齿轮。在标准齿轮的基础上，对轮齿做了某些改动，这样的齿轮称为变位齿轮。轮齿加工在圆柱外表面上的齿轮，称为圆柱齿轮。齿轮上的轮齿按照方向可分为直齿、斜齿等。本节主要介绍标准直齿圆柱齿轮的基本知识及其规定画法。

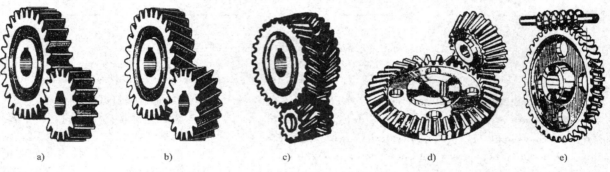

Fig. 7-23 Forms of transmission gears   传动齿轮的形式

a) A spur gear 直齿轮 b) A helical gear 斜齿轮 c) A double helical gear 人字齿轮

d) A bevel gear 锥齿轮 e) A worm and wormwheel 蜗杆和蜗轮

7.3.2 Geometric elements and dimensions of a spur gear

Fig. 7-24a shows the structure of a spur gear, which is named following its cylindrical shape. The lines between adjoining planes of gears are parallel to its axis. As the meshed curve is an involutes curve, it is also known as an involutes gear.

1. Names and dimension symbols of spur gears (see Fig. 7-24b)

(1) Tip circle Passing through the tooth's top surface, with its diameter denoted as d_a.

(2) Root circle Passing through the root of the tooth, with its diameter denoted as d_f.

(3) Reference circle Used for graduation (side set). For the reference circle of a standard gear, the arc lengths of the tooth thickness and the slot width are equal, with its diameter denoted as d.

7.3.2 直齿圆柱齿轮的几何要素和尺寸

直齿圆柱齿轮的结构如图7-24a所示。由于这种齿轮是由圆柱体加工而成，而且轮齿素线与齿轮轴线平行，故称为直齿圆柱齿轮。由于齿轮端面轮廓上参与啮合的曲线是一段渐开线，故又称为渐开线齿轮。

1. 直齿圆柱齿轮的各部分名称及尺寸代号（见图7-24b）

（1）齿顶圆 通过齿轮各齿顶轮廓的圆柱体，直径用 d_a 表示。

（2）齿根圆 通过齿轮各齿槽底部的圆柱体，直径用 d_f 表示。

（3）分度圆 用来分度（分齿）的圆，标准齿轮的分度圆位于齿厚和槽宽相等的地方，直径用 d 表示。

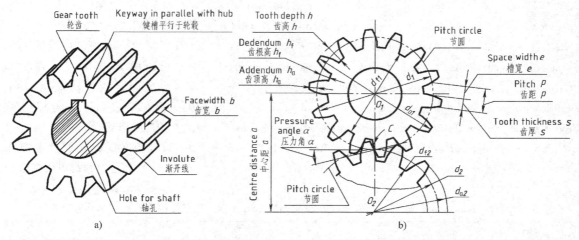

Fig. 7-24 Names and dimensions of spur gear sections   直齿圆柱齿轮的各部分名称及尺寸代号
a) Structure of a spur gear   直齿轮结构   b) Names and dimension symbol   名称及尺寸代号

(4) Addendum   The radial distance from the tip circle to the reference circle, denoted as h_a.

(5) Dedendum   The radial distance from the root circle to the reference circle, denoted as h_f.

(6) Tooth depth   The radial distance from the tip circle to the root circle, denoted as h. Consequently, $h = h_a + h_f$.

(7) Tooth thickness   The arc length between two sides of a tooth on the reference circle, denoted as s.

(8) Space width   The length of the arc on the reference circle between two teeth, denoted as e.

(9) Pitch   The length of the arc on the reference circle from one tooth end to the next, denoted as p. Consequently, $p = s + e$.

(10) Facewidth   The width of the gear's teeth (measured along the gear's axial line), denoted as b.

2. Basic parameters of spur gears

(1) Number of teeth z Total number of teeth.

(2) Module m The value p divided by π, expressed in mm. that is, $m = p/\pi$.

A module is a basic parameter used in gear design and manufacturing. Modules have been standardized to simplify the design and manufacturing processes. Table 7-4 shows their standard values.

(4) 齿顶高   齿顶圆与分度圆之间的径向距离，用 h_a 表示。

(5) 齿根高   齿根圆与分度圆之间的径向距离，用 h_f 表示。

(6) 齿高   齿顶圆与齿根圆之间的径向距离，用 h 表示。显然有 $h = h_a + h_f$。

(7) 齿厚   一个齿的两侧齿廓之间的分度圆弧长，用 s 表示。

(8) 槽宽   一个齿槽的两侧齿廓之间的分度圆弧长，用 e 表示。

(9) 齿距   相邻两齿的同侧齿廓之间的分度圆弧长，用 p 表示。显然有 $p = s + e$。

(10) 齿宽   齿轮轮齿的宽度（沿齿轮轴线方向度量），用 b 表示。

2. 直齿圆柱齿轮的基本参数

(1) 齿数 z 一个齿轮的轮齿总数。

(2) 模数 m 齿距 p 除以圆周率 π 所得的商，单位为 mm。即 $m = p/\pi$。

模数是设计和制造齿轮的基本参数。为设计和制造方便，已将模数标准化。模数的标准数值见表 7-4。

Table 7-4   Standard modules (Quoted from GB/T 1357—2008)   标准模数（摘自 GB/T 1357—2008）

No. 1 Series 第一系列	1, 1.25, 1.5, 2, 2.5, 3, 4, 5, 6, 8, 10, 12, 16, 20, 25, 32, 40, 50
No. 2 Series 第二系列	1.125, 1.375, 1.75, 2.25, 2.75, 3.5, 4.5, 5.5, (6.5), 7, 9, 11, 14, 18, 22, 28, 36, 45

Notes: No. 1 Series is recommended. Modules 6.5 in brackets should be avoided.
注：选择模数时应优先选用第一系列，其次选用第二系列，避免采用模数 6.5。

(3) Pressure angle The angle between the common normal (a direction along which the load is transferred) at the pitch point C and the tangent of the two engaging pitch circles of the gears. It is denoted as α (shown in Fig. 7-24b). Following the national standard, the pressure angle of standard gears is 20°. Only gears with equal module and pressure angle can mate.

Before preparing the design, one needs to specify the module and the number of teeth. Other parameters can be calculated by using the module and the number of teeth. Formulas are shown in Table 7-5.

（3）压力角 两相啮合轮齿齿廓在节点 C 处的公法线（力的传递方向）与两节圆的公切线的夹角，称压力角，用 α 表示（见图 7-24b）。我国标准渐开线齿轮的压力角 α 为 20°。只有模数和压力角都相同的齿轮，才能互相啮合。

设计齿轮时，先要确定模数和齿数，其他各部分尺寸都可由模数和齿数计算出来。计算公式见表 7-5。

Table 7-5 Dimension formulas for standard a spur gear 标准直齿圆柱齿轮的尺寸计算公式

Parameter 参数	Code 代号	Formulas 公式
Reference diameter 分度圆直径	d	$d = mz$
Addendum 齿顶高	h_a	$h_a = m$
Dedendum 齿根高	h_f	$h_f = 1.25m$
Tooth depth 齿高	h	$h = 2.25m$
Tip diameter 齿顶圆直径	d_a	$d_a = m(z+2)$
root diameter 齿根圆直径	d_f	$d_f = m(z-2.5)$
Pitch 齿距	p	$p = \pi m$
Tooth thickness 齿厚	s	$s = p/2 = \pi m/2$
Space width 槽宽	e	$e = p/2 = \pi m/2$
Centre distance 中心距	a	$a = (d_1+d_2)/2 = m(z_1+z_2)/2$

Notes: The module m is calculated based on the part stress and standard values are used. The number of teeth z is determined based on the motion design, and the pressure angle is usually 20°.

注：基本参数中，模数 m 由强度计算并取标准值，齿数 z 由运动设计确定，压力角为 20°。

7.3.3 Conventions in drawing gears

Methods for preparing gear drawings are specified in national standards.

1. Drawing methods of a single spur gear

A gear is generally shown through noncircular and end views (circular views), as shown in Fig. 7-25.

1) On noncircular and circular views, the tip circle is shown with continuous thick lines.

2) The reference circle is shown with thin long dashed dotted lines. On noncircular views, the thin long dashed dotted line of the reference circle should be drawn 2—3mm out of the profile.

3) On a non-sectional view, the root circle is shown with continuous thin lines, or it can be omitted.

4) On a sectional view, the root circle is shown with continuous thick line. On a circular view, the root circle is shown with continuous thin line, or omitted.

5) Section lines in teeth are not required.

6) Other gear structures should be drawn in accordance with projections.

7) With regard to helical gears and double helical gears, teeth are represented with three parallel continuous thin lines on the noncircular view.

7.3.3 齿轮的画法规定

国家标准对齿轮的画法做了一些规定

1. 单个齿轮的画法

单个齿轮的画法，一般用非圆视图和端（圆形）视图两个视图来表示，如图 7-25 所示。

1) 在非圆视图和端（圆形）视图中，齿顶圆用粗实线表示。

2) 分度圆用细点画线表示，在非圆视图中表示分度圆的细点画线应伸出轮廓线 2~3mm。

3) 不剖时，齿根线画细实线或省略不画。

4) 在剖视图中，齿根线用粗实线表示，在端（圆形）视图中齿根线仍用细实线表示或省略。

5) 轮齿部分不画剖面线。

6) 齿轮的其他结构按投影画出。

7) 对于斜齿轮和人字齿轮，可在非圆投影的视图上用三条平行的细实线表示齿线方向。

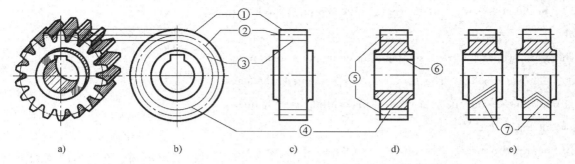

Fig. 7-25 Drawing methods of single spur gear 单个齿轮的画法
a) Axonometric drawing 轴测图 b) End view 端视图 c) Outside view 外形图 d) Full sectional view 全剖视图
e) Half sectional view of helical and double helical gears 斜齿轮和人字齿轮的半剖视图

2. Drawing methods of meshing spur gears

When two standard gears mate, the reference circle is also called pitch circle since both reference circles are cut by a tangent. It is represented as follows:

1) In noncircular sectional views, tip circles and root circles are shown with continuous thick lines. The pitch circles are tangent, and are shown with long dashed dotted lines.

The representation of both the tip circles and root circles on meshing points is as follows: Driving wheel (pinion) teeth are visible, drawn with continuous thick lines. Driven wheel teeth are invisible, and the root circle is shown with a continuous thick line while the tip circle is shown with a thin long dashed line, as shown in Fig. 7-26a and Fig. 7-27.

2. 两直齿圆柱齿轮啮合的画法

两标准齿轮相互啮合时，其分度圆应恰好处于相切位置，此时分度圆又称为节圆。其规定画法如下：

1）在非圆投影的剖视图中，齿顶圆、齿根圆画粗实线。两轮节圆相切，画细点画线。

啮合区齿顶圆、齿根圆的画法是：将主动（小齿）轮的轮齿作为可见，画成粗实线，从动轮的轮齿被遮住，其齿顶圆画成粗实线，齿根圆画成细虚线，如图 7-26a 及图 7-27 所示。

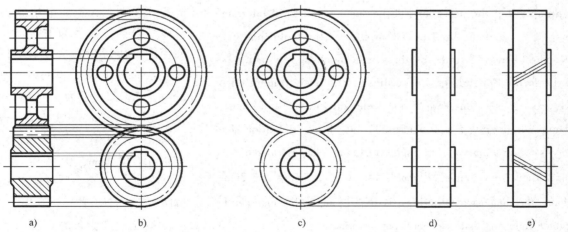

Fig. 7-26 Drawing methods of meshing gears 两圆柱齿轮的啮合画法
a) Complete sectional view 全剖视图 b) End view 端视图画法 c) Simplified end view 端视图简化画法 d) Outline of meshing gears 两直齿圆柱齿轮啮合的外形图 e) Outline of meshing helical gears 两斜齿圆柱齿轮啮合的外形图

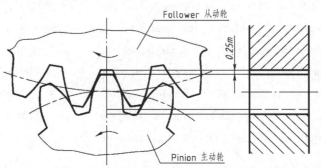

Fig. 7-27 Simplified drawing of meshing zone 啮合区的简化画法

2) On the end view (circular view), the two pitch circles are tangent.

There are two ways of representing a tip cicle and a root circle.

① Complete drawing. The tip circle is represented with continuous thick lines, and the root circle with continuous thin lines, as shown in Fig. 7-26b.

② Simplified drawing. The tip circle in the meshing zone is omitted, as well as the whole root circle, as shown in Fig. 7-26c.

3) When outlining the noncircular view, the tip circle and root circle don't necessarily have to be drawn in the meshing zone, and the pitch circle is shown with a continuous thick line, as shown in Fig. 7-26d and Fig. 7-27e.

3. An example of a spur gear detail drawing

Fig. 7-28 depicts a detail drawing of a spur gear. It is different because the top right corner shows a table, in which the module is 2.5mm and the number of teeth is 20. With these parameters, one can calculate the reference diameter $d = mz = 50$mm, the tip diameter $d_a = m(z+2) = 55$mm, and the root diameter $d_f = m(z-2.5) = 43.75$mm. As for the numbers and letters 8-7-7HK, they mean that the precision of the first tolerance group is 8, the precisions of the second and third tolerance groups are 7, and the limit deviation code of the tooth thickness is HK. The keyway's width and depth can be chosen by using Table 2-12 in the appendices, according to the basic dimension of the keyway: basic dimension 5mm, relevant width of the keyway $b = 5$mm, and JS9 (±0.015mm) as limit deviation. The keyway's width dimension is 5mm±0.015mm, keyway depth $t_2 = 2.3$mm, limit deviation is $^{+0.1}_{0}$ mm, 14mm + 2.3mm = 16.3mm, therefore, the dimension (drawing) is $16.3^{+0.1}_{0}$mm. Other data can be interpreted following that of general part drawings.

4. Drawing of a gear and a gear rack

As the diameter of a gear approaches to infinity, the tip circle, the root circle, the reference circle and the tooth curve of the gear will approach to straight lines. Therefore, the gear becomes a gear rack. When the gear and the gear rack are in mesh, as the gear rotates, the gear rack will move with linear way. The drawing of such meshed gear and gear rack is similar to the drawing of a pair of meshed cylindrical gears, namely the pitch circle of the gear should be tangent with the pitch line of the gear rack, as shown in Fig. 7-29.

2）在端（圆形）视图中，两齿轮的节圆相切。

啮合区的齿顶圆和齿根圆有两种画法：

① 完整画法。齿顶圆画粗实线，齿根圆画细实线，如图7-26b所示。

② 简化画法。啮合区中的齿顶圆省略不画，整个齿根圆均不画，如图7-26c所示。

3）非圆投影的外形图中，啮合区的齿顶线和齿根线不必画出，节圆画成粗实线，如图7-26 d、e所示。

3. 直齿圆柱齿轮的零件图举例

图7-28所示为齿轮零件图。该图与一般零件图不同之处在于：在图的右上角有一参数表。从参数表中得知，该齿轮的模数为2.5mm，齿数为20。由此可计算出齿轮的分度圆直径 $d = mz = 50$mm，齿顶圆直径 $d_a = m(z+2) = 55$mm，齿根圆直径 $d_f = m(z-2.5) = 43.75$mm，精度等级 8-7-7HK的含义是：该齿轮第一公差组精度为8级，第二、三公差组精度为7级，齿厚极限偏差代号为HK。键槽宽度和深度可通过附录表2-12选取基本尺寸：键槽基本尺寸5mm，即宽 $b = 5$mm，选极限偏差为JS9（±0.015mm），得槽宽尺寸为5mm±0.015mm，槽深 $t_2 = 2.3$mm，极限偏差为 $^{+0.1}_{0}$mm，14mm + 2.3mm = 16.3mm，由此得图中尺寸 $16.3^{+0.1}_{0}$mm。其余内容可根据零件图的读图方法进行阅读。

4. 齿轮和齿条的画法

当齿轮的直径无限大时，其齿顶圆、齿根圆、分度圆和齿廓曲线都成了直线。这时，齿轮就变成了齿条。齿轮和齿条啮合时，齿轮旋转，齿条做直线运动。齿轮和齿条啮合的画法与两圆柱齿轮啮合的画法基本相同，这时齿轮的节圆应与齿条的节线相切，如图7-29所示。

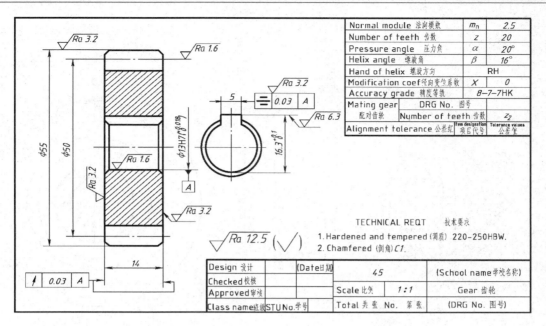

Fig. 7-28 Detail drawing of a spur gear   齿轮零件图

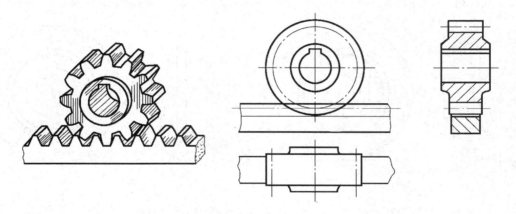

Fig. 7-29 Drawing of the meshed gear and gear rack   齿轮和齿条啮合的画法

5. Drawing of a bevel gear

Drawing of a bevel gear is similar to that of a cylindrical gear. However, it is more complicated for representation and drawing due to the characteristic feature of the bevel gear. The front view of a single bevel gear is usually shown as a sectional view. Furthermore, its tip circle of the large and small ends should be drawn with continous thick lines on the left view, and the reference circle of large end is drawn with thin long dashed dotted lines. The drawing and names of different parts of a bevel gear are shown in Fig. 7-30.

As two bevel gears are in mesh, the reference circle of one is in tangent with the other, and both tips of cone are in intersection. Therefore, the front view is usually drawn as a sectional view, as shown in Fig. 7-31a. If the external shape of a helical bevel gear is needed to be illustrated, then three parallel lines are necessary to be drawn on the external shape in order to indicate the direction of the gear's teeth, as shown in Fig. 7-31b.

5. 锥齿轮的画法

锥齿轮的画法基本上与圆柱齿轮相同，只是由于圆锥的特点，在表达和作图方法上较圆柱齿轮复杂。单个锥齿轮的主视图常画成剖视图，而在左视图上用粗实线画出齿轮大端和小端的齿顶圆，用点画线画出大端的分度圆。锥齿轮的画法及各部分名称如图7-30所示。

锥齿轮啮合时，两分度圆锥相切，它们的锥顶交于一点。画图时主视图多采用剖视表示，如图7-31a所示。当需要画斜齿锥齿轮的外形时，则在外形图上加画三条平行的细实线表示轮齿的方向，如图7-31b所示。

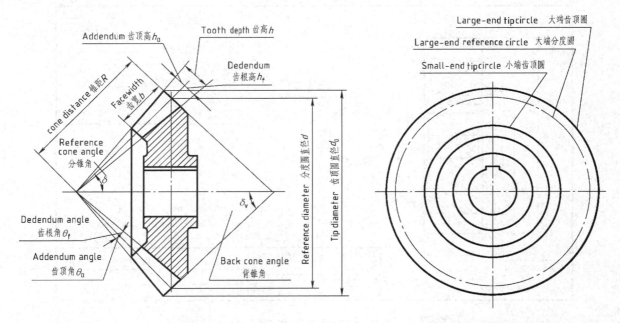

Fig. 7-30 Specifications and names of different parts of a bevel gear　锥齿轮的画法及各部分名称

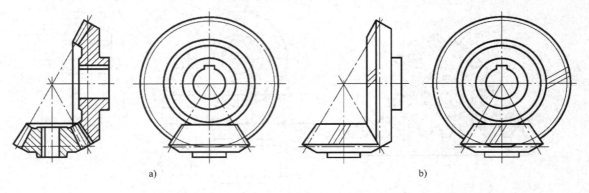

Fig. 7-31 Drawing of a pair of meshed bevel gears　锥齿轮啮合画法
a) Sectional view (straight bevel gear)　剖视图（直齿锥齿轮）　b) Outside view (helical bevel gear)　外观图（斜齿锥齿轮）

6. Drawing of worm wheels and worms

The drawing and names of different parts of worm wheels and worms are shown as Fig. 7-32. A wormwheel is actually a combination of cyclinder gear with helical gear. A worm is similar to the trapezoidal thread. As the worm rotates a full round, the worm wheel shall rotates up to one or two teeth. Therefore, the combination of the worm wheel and worm will generate a considerable ratio of deceleration in term of the transmission.

Drawing of a worm wheel is represented as follows: In a sectional view, the drawing of a tooth is the same as that of a cylindrical gear. In the view that is orthogonal to the axis of the, only the reference circle and the outer circle are illustrated on the view, while the tip circle and the root circle are omitted, as shown in Fig. 7-32a. The drawing of a worm is also the same as that of a cylindrical gear. Usually, several teeth are, however, drawn as local sectional views or partial enlargement views in order to indicate the tooth shape of the worm, as shown in Fig. 7-32b.

6. 蜗轮和蜗杆的画法

蜗轮、蜗杆的画法及各部分名称如图7-32所示，蜗轮实际上是斜齿的圆柱齿轮，蜗杆的外形和梯形螺纹相似，蜗杆转一圈，蜗轮只转过一个齿或两个齿。因此，用蜗轮、蜗杆传动，蜗杆主动时，可得到很大的降速比。

蜗轮的画法是：在剖视图上，轮齿的画法与圆柱齿轮相同。在投影为圆的视图中，只画分度圆和外圆，齿顶圆和齿根圆不必画出。如图7-32a所示。蜗杆的画法与圆柱齿轮的画法相同。为了表明蜗杆的牙型，一般都采用局部剖视画出几个牙型，或画出牙型的放大图，如图7-32b所示。

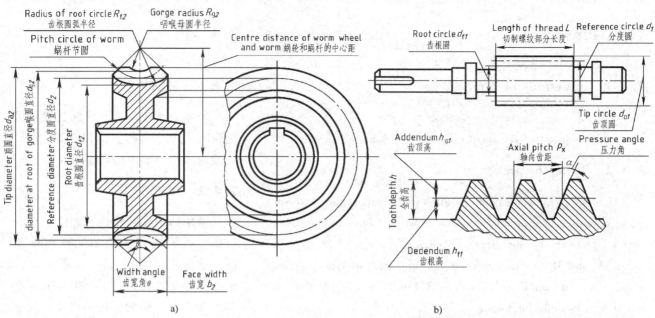

Fig. 7-32 Drawing and names of different parts of worm wheels and worms 蜗轮和蜗杆的画法及各部分名称
a) Worm gear 蜗轮 b) Worm 蜗杆

Drawing of a meshed worm wheel and worm is represented as follows: In the view that is orthogonal to the axis of the worm wheel, the reference circle of the worm wheel should be in tangent with the reference line of the worm, Furthermore in the meshing area, the root circle and the root line should be drawn with continous thick lines. In the view that is orthogonal to the worm, in the meshing area, the worm is drawn, but the worm wheel is omitted shown in Fig. 7-33.

蜗轮和蜗杆啮合的画法是：在垂直于蜗轮轴线的视图上，蜗杆的分度圆与蜗杆的分度线要画成相切，啮合区内的齿顶圆和齿顶线仍用粗实线画出；在垂直于蜗杆轴线的视图上，啮合区只画蜗杆，不画蜗轮（见图 7-33）。

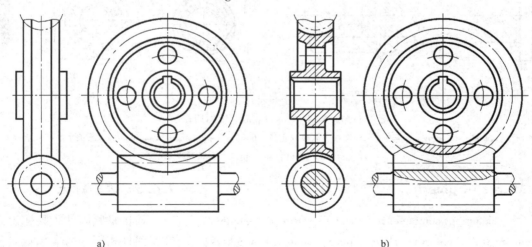

Fig. 7-33 Drawing of a meshed worm wheel and worm 蜗轮、蜗杆啮合的画法
a) Outside view 外观图 b) Sectional view 剖视图

In a sectional view, if the section plane is across the pivot of the worm wheel and is perpendicular to the pivot of the worm, the teeth of the worm should be drawn with continous thick lines, and the area which is obstructed on the worm wheel can be omitted in the drawing. If the section plane is across the pivot of the worm and is perpendicular to the pivot of the worm wheel, the tip circle and the circle at root of gorge of the worm wheel can be omitted in the meshing area, and the tip line of the worm can also be omitted (see Fig. 7-33b).

在剖视图中，当剖切平面通过蜗轮轴线并垂直于蜗杆轴线时，在啮合区内将蜗杆的轮齿用粗实线绘制，蜗轮的轮齿被遮挡的部分可省略不画；当剖切平面通过蜗杆轴线并垂直于蜗轮轴线时，在啮合区内，蜗轮的顶圆、喉圆可以省略不画（见图7-33b），蜗杆的齿顶线也可省略不画。

7.4 Pins

7.4.1 Functions of pins

Pins are used to join or position two parts (see Fig. 7-34). Pins usually only transfer small torques when joining two parts.

7.4.2 Types of pins

Commonly used pins are parallel pins, taper pins and split pins. According to different precision fittings of pins and pinholes, parallel pins are divided into two categories, i.e., A (general quench) and B (surface quench). There are two types of taper pins, i.e., A and B. See appendix Table 2-15 to Table 2-17 for pin structures and dimensions.

7.4 销

7.4.1 销的用途

销主要用于两零件之间的联接或定位（见图 7-34）。销用于联接时，一般只能传递不大的转矩。

7.4.2 销的种类

常用的销有圆柱销、圆锥销和开口销。根据销与销孔的配合精度不同，圆柱销分为 A 型（普通淬火）、B 型（表面淬火）。圆锥销也有 A 型和 B 型之分。销的结构型式及其尺寸系列见附录表 2-15～表 2-17。

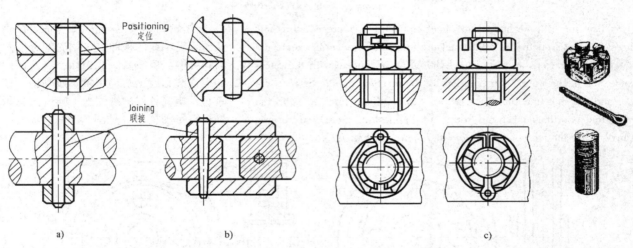

Fig. 7-34 Use of pins 销的用途
a) Parallel pin joining or positioning 圆柱销联接或定位　b) Taper pin joining or positioning 圆锥销联接或定位
c) Split pin joining and positioning 开口销联接或定位

7.4.3 Marks of pins

Pins are fastening components. Pins, like thread fasteners, are marked in the same way. Their names, standard numbers, types and dimension information should be reflected. See Table 7-6 for and marking specifications for pins.

7.4.4 Drawing methods of a pin assembly

A pin assembly drawing is shown in Fig. 7-34. If the cut plane crosses the pin's axis, the pin, being a solidpiece, will be drawn in accordance with the profile without sectioning. If the cutting plane is perpendicular to the axis, the section hatching should be added.

7.4.3 销的标记

销也是紧固件，其标记方法与螺纹紧固件相同，内容包括名称、标准编号、型式与尺寸等。销的标记规定见表 7-6。

7.4.4 销的装配图画法

销的装配图画法如图 7-34 所示。销作为实心件，当剖切平面通过销的轴线时，仍按不剖画出外形；垂直于销的轴线剖切时，应画上剖面线。

Table 7-6 Marking specifications for pins 销的标记规定

Name 名称	Parallel pin 圆柱销 GB/T 119.1—2000	Taper pin 圆锥销 GB/T 117—2000	Split pin 开口销 GB/T 91—2000
Style & dimension 结构型式及规格尺寸			
Marking example 标记示例	Pin GB/T 119.1 8m6×30 销 GB/T 119.1 8m6×30	Pin GB/T 117 A10×60 销 GB/T 117 A10×60	Pin GB/T 91 5×50 销 GB/T 91 5×50
Explanation 说明	The parallel pin, nominal diameter d = 8mm, tolerance m6, nominal length l = 30mm, material is steel, without hardening and surface treatment 公称直径 d = 8mm、公差为 m6、公称长度 l = 30mm、材料为钢、不经淬火、不经表面处理的圆柱销	The taper pin, nominal diameter d = 10mm, nominal length l = 60mm, material is 35 steel, heat treatment hardness 28—38HRC, oxidization as surface treatment, style A 公称直径 d = 10mm、公称长度 l = 60mm、材料为 35 钢、热处理硬度为 28~38HRC、表面氧化处理的 A 型圆锥销	The split pin, nominal diameter d = 5mm, nominal length l = 50mm, material is Q215 or Q235, without surface treatment 公称直径 d = 5mm、公称长度 l = 50mm、材料为 Q215 或 Q235、不经表面处理的开口销

Matching requirements for parallel pins and taper pins are precise. Usually, the pinhole is machined once the joint has been assembled. See the machining methods and dimension settings of a pinhole in Fig. 7-35.

圆柱销和圆锥销的装配要求较高，销孔一般在被联接件装配后统一加工，其加工方法与标注如图 7-35 所示。

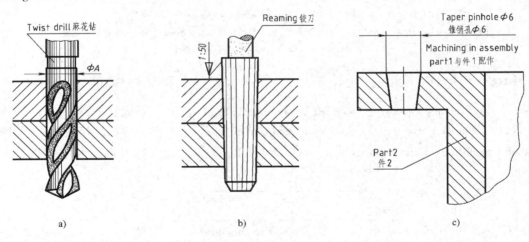

Fig. 7-35 Machining methods and dimension settings of a pinhole 销孔加工方法与标注
a) Drilling 钻孔 b) Reaming 铰孔 c) Dimension settings 尺寸注法

7.5 Keys

7.5.1 Functions of keys

A key joint is a removable joint as shown in Fig. 7-36. Such joint goes through the shaft and the hub's keyway, joining the shaft with the rotors (gears, pulleys and a clutch), and transferring torques.

7.5 键

7.5.1 键的用途

如图 7-36 所示，键联接是一种可拆联接。这种联接是将键同时嵌入轴与轮毂的键槽中，将轴及轴上的转动零件如齿轮、带轮、联轴器等联接在一起，以传递转矩。

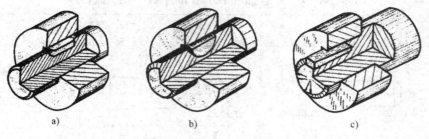

Fig. 7-36 Forms of key joints 键联接的形式
a) Square and rectangular key joint 普通型平键联接 b) Woodruff key—Normal form joint 普通型半圆键联接
c) Taper key for general use joint 普通型楔键联接

7.5.2 Types of keys

Commonly used keys are square and rectangular keys, woodruff keys—Normal form and taper keys for general use (see Fig. 7-37). One should select appropriate keys according to their features during the design process.

7.5.2 键的种类

常用的键有普通型平键、普通型半圆键和普通型楔键，如图 7-37 所示。设计时可根据其特点合理选用。

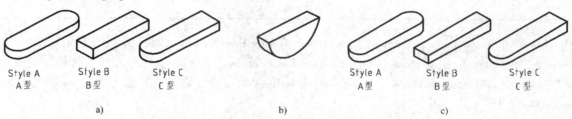

Fig. 7-37 Types of keys 键的类型
a) Square and rectangular keys 普通型平键 b) Woodruff keys—Normal form 普通型半圆键
c) Taper keys for general use 普通型楔键

7.5.3 Marks of keys

The square and rectangular keys, woodruff keys—normal form and taper keys for general use are standard components. You can purchase them according to their marks. Key marks generally include:

7.5.3 键的标记

普通型平键、普通型半圆键和普通型楔键都是标准件，使用时按其标记直接外购即可。键的完整标记通式为：

| Standard No. 标准编号 | Name 名称 | Type & specification 类型与规格 |

Table 7-7 lists the types and specified symbols of commonly used keys.

表 7-7 列出了常用键的类型和规定标记。

Table 7-7 The types and specified symbols of commonly used keys 常用键的类型和规定标记

Name 名称	Example 图例	Specified symbols 规定标记示例
Square and rectangular keys 普通型 平键 GB/T 1096—2003		Square and rectangular keys (style A), width $b = 16$mm, height $h = 10$mm, length $L = 100$mm, is marked as: GB/T 1096 Key $16 \times 10 \times 100$ 宽度 $b = 16$mm、高度 $h = 10$mm、长度 $L = 100$mm 普通 A 型平键的标记为： GB/T 1096 键 $16 \times 10 \times 100$
Woodruff keys—normal form 普通型 半圆键 GB/T 1099.1—2003		Woodruff key—Normal form, (width $b = 6$mm, height $h = 10$mm, Diameter $D = 25$mm), is marked as: GB/T 1099.1 Key $6 \times 10 \times 25$ 宽度 $b = 6$mm、高度 $h = 10$mm、直径 $D = 25$mm 普通型半圆键的标记为： GB/T 1099.1 键 $6 \times 10 \times 25$

Name 名称	Example 图例	Specified symbols 规定标记示例
Taper keys for general use 普通型 楔键 GB/T 1564—2003	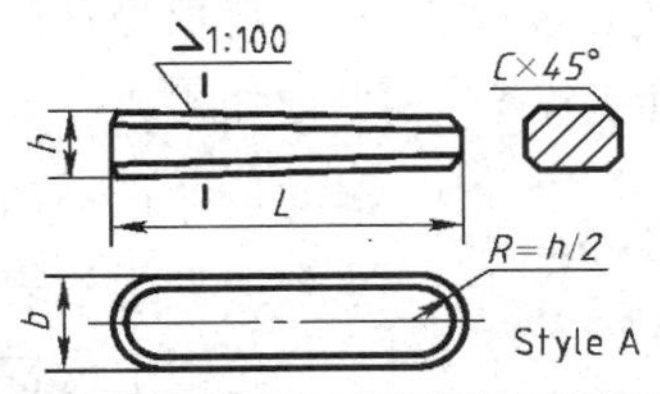 (figure: taper key, slope 1:100, C×45°, Style A)	Taper key for general use (style A), width $b = 16$mm, height $h = 10$mm, length $L = 100$mm, is marked as: GB/T 1564 Key 16×100 宽度 $b = 16$mm、高度 $h = 10$mm、长度 $L = 100$mm 普通 A 型楔键的标记为: GB/T 1564 键 16×100

7.5.4 Drawing of key joints

Drawings of square and rectangular keys, woodruff keys—normal form and taper keys for general use are shown in Fig. 7-38 and Fig. 7-39. If the cut is longitudinal, they are drawn without section lines. If the cut is cross sectional, section lines are required.

Generally, the depth of the keyway and its union with other components are conveyed by local sectional views. Both sides of the square and rectangular key and the woodruff key—normal form are working faces. The top face is not a working face. Therefore, clearance between the key and the keyway is not allowed, and one should draw a line in accordance with the interface. Clearance between the top surfaces of both the key and the hub keyway should be maintained.

7.5.4 键联接的画法

普通平键、半圆键和普通楔键联接的画法如图 7-38、图 7-39 所示。当沿着键的纵向剖切时，按不剖画；当沿着键的横向剖切时，则要画上剖面线。

通常用局部剖视图表示轴上键槽深度及与其他零件之间的联接关系。普通平键和普通半圆键的两侧面为工作面，而顶面是非工作面，故键与键槽侧面之间不留间隙，应按接触面画一条线；而键的顶面与轮毂的键槽顶面之间应留有间隙。

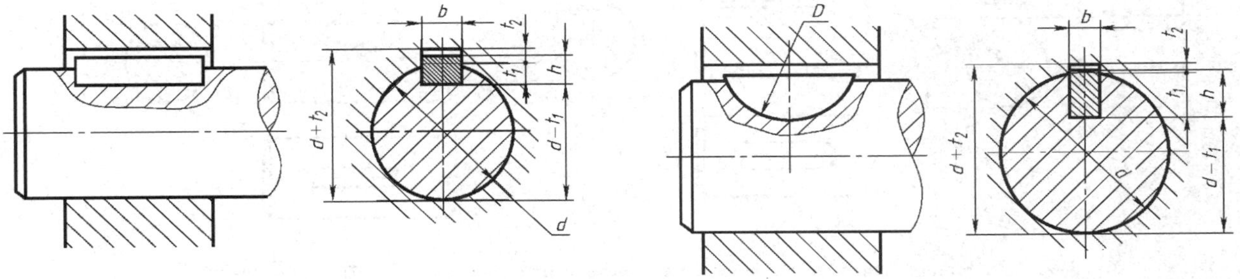

Fig. 7-38 Drawings of square and rectangular key joints and woodruff key joints　普通平键和普通半圆键联接的画法
a) Drawings of square and rectangular key joints　普通型平键联接画法　b) Drawings of woodruff key joints　普通半圆键联接画法

The top surfaces of taper keys for general use have a 1:100 slope. Drive the key into the keyway, while joining them. Since the top and bottom surfaces of the taper keys for general use are working faces, there is no clearance between them and the bottom of the keyway. The two sides of the key are not working faces. Clearance should be maintained between them and the two sides of the keyway. See Fig. 7-39.

普通楔键顶面有 1:100 的斜度。联接时将键打入键槽。普通楔键的顶面和底面同为工作面，与槽底没有间隙。而键的两侧为非工作面，与键槽两侧面应留有间隙，如图7-39所示。

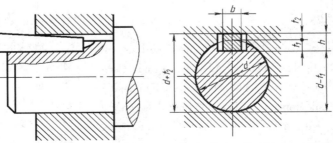

Fig. 7-39 Drawings of taper key for general use joints　普通楔键联接的画法

7.5.5 Choosing dimensions of the key and keyway

One usually first selects the type and specifications of a key according to the diameter of the design requirements. An appropriate mark symbol is then determined. The dimension of the keyway should be determined following relevant standards. Detailed information on dimensions is provided Table 2-12 to Table 2-14 in the appendices.

The keyway on the hub is generally machined on a machine with a slotting cutter. The keyway on the shaft is generally machined on a milling machine. The dimension of the keyway coincides with the dimension of the key. The depth of the keyway should be determined according to the relevant standards. The machining methods for the keyway and the relevant dimensions are shown in Fig. 7-40.

7.5.5 键的选取和键槽尺寸的确定

根据设计要求，按标准选取键的类型和规格，并给出正确的标记。键槽的尺寸也必须按标准确定。具体尺寸系列见附录表2-12~表2-14。

轮毂上的键槽一般是用插刀在插床上加工出来的，轴上的键槽一般是在铣床上加工出来的。键槽的尺寸应与键的尺寸一致，键槽的深度要按标准查表确定。键槽的加工方法和有关尺寸如图7-40所示。

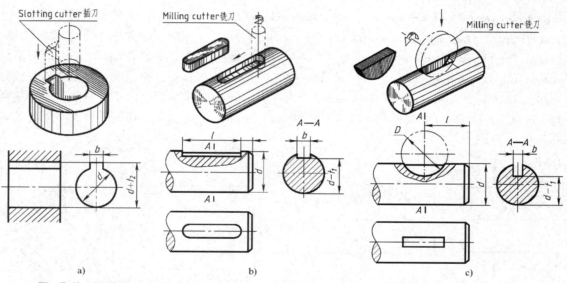

Fig. 7-40 Machining methods of the keyway and relevant dimensions　键槽的加工方法和有关尺寸
a) Keyway on a hub　轮毂孔上的键槽加工　b) Square and rectangular keyways on a shaft　轴上的平键槽加工
c) Woodruff keyway on a shaft　轴上的半圆键槽加工

7.6 Springs

7.6 弹簧

7.6.1 Functions of springs

Spring functions include shock absorption, reposition, clamping, dynamometry and accumulation of energy.

7.6.2 Types of springs

There are several types of springs. Commonly used springs are helical springs, spiral springs, leaf springs and flake springs (see Fig. 7-41). The helical spring is widely used. There are compression springs, extension springs, and torsion springs, employed according to load. In this chapter we will focus on the names and drawing methods of cylindrically helically compression springs.

7.6.1 弹簧的用途

弹簧的作用主要是减振、复位、夹紧、测力和储能等。

7.6.2 弹簧的种类

弹簧的种类很多，常用的有螺旋弹簧、涡卷弹簧、板弹簧和片弹簧等，如图7-41所示，其中螺旋弹簧应用较为广泛。根据受力情况，螺旋弹簧又分为压缩弹簧、拉伸弹簧和扭转弹簧。本节主要介绍圆柱螺旋压缩弹簧的各部分名称及画法。

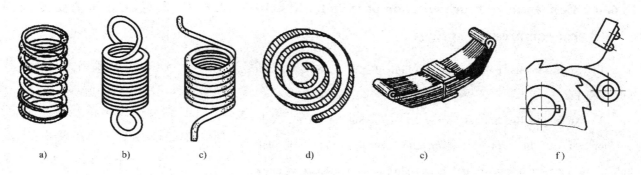

Fig. 7-41 Commonly used springs 常用的弹簧
a) Helical compression spring 螺旋压缩弹簧 b) Helical tension spring 螺旋拉伸弹簧
c) Torsion spring 扭转弹簧 d) Spiral spring 涡卷弹簧 e) Leaf spring 板弹簧 f) Flake spring 片弹簧

7.6.3 Names and parameters of a cylindrical helical compression spring (GB/T 2089—2009)

Names of a cylindrical helical compression spring are shown in Fig. 7-42.

(1) Diameter of line d The diameter of wire.

(2) Outer diametere of coil D_2 The biggest diameter of spring.

(3) Inside diameter of coil D_1 The smallest diameter of spring. $D_1 = D_2 - 2d$.

(4) Mean diameter of coil D The mean diameter is chosen according to the standard.
$D = (D_1 + D_2)/2 = D_1 + d = D_2 - d$.

(5) Pitch t The axial distance of the corresponding points on the adjacent coils except the supporting coils at the ends.

(6) Number of end coils n_2 Number of coils at the end of the spring, which are put together and assured to provide equal load and to hold other components. There are three numbers of end coils: 1.5, 2, and 2.5.

(7) Number of active coils n Number of coils engaging in elastic deformation. One should choose the number according to the standard.

(8) Total number of coils n_1 The sum of end coils number and active coils number: $n_1 = n_2 + n$.

(9) Free height H_0 Height of the spring, when unloaded:
$n_2 = 2.5$, $H_0 = nt + 2d$;
$n_2 = 2$, $H_0 = nt + 1.5d$;
$n_2 = 1.5$, $H_0 = nt + d$.

(10) Stretched length L Spring length of the wire to start:
$L = n_1 \sqrt{(\pi D)^2 + t^2} \approx n_1 \pi D$.

7.6.3 圆柱螺旋压缩弹簧各部分的名称及参数 (GB/T 2089—2009)

圆柱螺旋压缩弹簧各部分的名称如图7-42所示。

(1) 线径 d 弹簧的钢丝直径。

(2) 弹簧外径 D_2 弹簧的最大直径。

(3) 弹簧内径 D_1 弹簧的最小直径。$D_1 = D_2 - 2d$。

(4) 弹簧中径 D 弹簧的平均直径，按标准选用。
$D = (D_1 + D_2)/2 = D_1 + d = D_2 - d$。

(5) 节距 t 除两端支承圈外，弹簧上相邻两圈对应两点之间的轴向距离。

(6) 支承圈数 n_2 为使弹簧各圈受力均匀，把两端弹簧并紧磨平，工作时起支承作用的圈数。支承圈数有1.5圈、2圈和2.5圈三种。

(7) 有效圈数 n 在工作时，弹簧中起弹性变形作用的圈数，按标准选用。

(8) 总圈数 n_1 支承圈数和有效圈数之和：$n_1 = n_2 + n$。

(9) 自由高度 H_0 弹簧不受外力作用下的高度：
$n_2 = 2.5$, $H_0 = nt + 2d$;
$n_2 = 2$, $H_0 = nt + 1.5d$;
$n_2 = 1.5$, $H_0 = nt + d$。

(10) 弹簧丝展开长度 L 弹簧钢丝的展开长度：
$L = n_1 \sqrt{(\pi D)^2 + t^2} \approx n_1 \pi D$。

7.6.4 Conventional representation of cylindrical helical compression springs

1) Represent the helical line of a helical spring with a straight line. See Fig. 7-43.

2) No matter the helical spring is on the left or on the right position, it can be represented on the right. No matter the left helical spring is represented as left or right, it is necessary to mark it as "LH" (left-hand).

3) If both ends of the cylindrical helical compression spring need to be put together and to be grounded (irrespective of the number of supporting coils it may have), it should be represented as shown in Fig. 7-42, and be specified in technological requirements.

4) If the helical spring has more than four active coils, the center division can be omitted, and therefore the length of the view can be shortened accordingly (see Fig. 7-43).

7.6.4 圆柱螺旋压缩弹簧的规定画法

1) 圆柱螺旋弹簧各圈的螺旋线画成直线,如图 7-43 所示。

2) 螺旋弹簧不论左旋或右旋,均可画成右旋,但左旋弹簧不论画成左旋或右旋,一律要注出"LH"(左旋)。

3) 圆柱螺旋压缩弹簧如果要求两端并紧磨平时(不论支承圈多少),均可按图 7-42 所示的形式绘制,并且在技术要求中加以说明。

4) 有效圈数在 4 圈以上的螺旋弹簧,中间部分可以省略。中间部分省略后,允许适当缩短图形的长度,如图 7-43 所示。

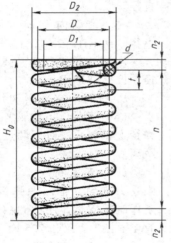

Fig. 7-42 Parameters of a cylindrical helical compression spring
圆柱螺旋压缩弹簧各部分的名称及参数

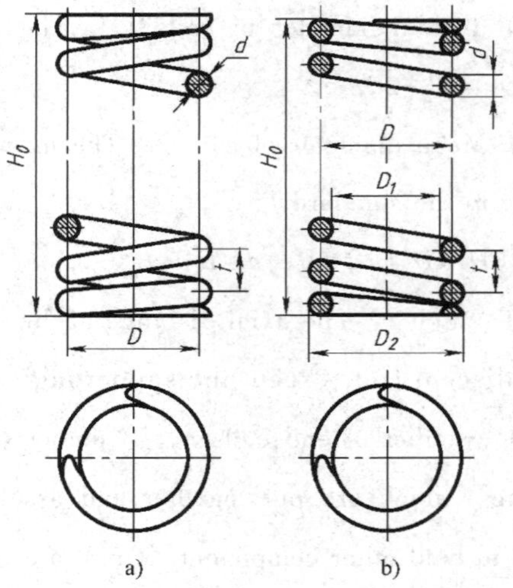

Fig. 7-43 Drawing of a cylindrical helical compression spring
圆柱螺旋压缩弹簧的画法
a) Outside view 外观图 b) Sectional view 剖视图

7.6.5 Steps for drawing cylindrical helical compression springs

Given wire diameter $d = 5$mm, mean diameter of coils $D = 40$mm, pitch $t = 10$mm, number of active coils $n = 10$, and number of end coils $n_2 = 2.5$, steps to draw the cylindrical compression spring on the right are shown in Fig. 7-44. The cylindrical helical compression spring drawing is shown in Fig. 7-45.

7.6.5 圆柱螺旋压缩弹簧的作图步骤

已知簧丝直径 $d = 5$mm,中径 $D = 40$mm,节距 $t = 10$mm,有效圈数 $n = 10$,支承圈数 $n_2 = 2.5$,右旋圆柱压缩弹簧,其画图步骤如图 7-44 所示。圆柱螺旋压缩弹簧的零件图如图 7-45 所示。

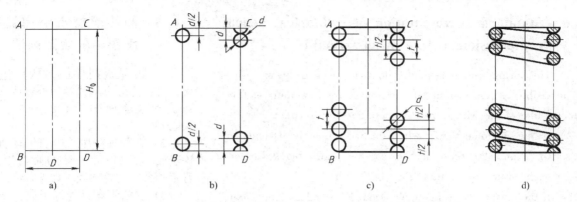

Fig. 7-44 Steps for representing a cylindrical helical compression spring 圆柱螺旋压缩弹簧的作图步骤
  a) Draw a rectangle ABCD with H_0 and D 以 H_0 和 D 画矩形 ABCD
  b) Draw the wire diameter projection of the end coils 画支承圈的簧丝投影
  c) Draw the wire diameter projection of the active coils 作有效圈的簧丝投影
  d) Draw the tangent on the right side, and section lines, and darken contour
     按右旋画簧丝切线和剖面线,加深轮廓线

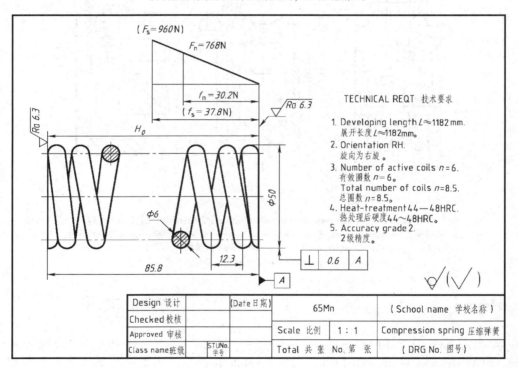

Fig. 7-45 Detail drawing of a cylindrical helical compression spring 圆柱螺旋压缩弹簧零件图

One needs to take into account the following aspects when drawing:

1) Spring parameters should be marked in the view or be included and explained in technical requirements.

2) If a note for the spring load and height ratio is required, it must be illustrated. The mechanical property curve of the helical compression spring is a straight line, in which F_n represents the maximum working load, F_s represents the experimental load, and f_n represents the greatest work deformation, f_s represents the experimental load deformation.

画图时应注意以下几点:

1) 弹簧的参数应直接注在图形上,如有困难,可在技术要求中说明。

2) 当需要说明弹簧的负荷与高度之间的变化关系时,必须用图解法表示。螺旋压缩弹簧的力学性能曲线为直线。其中,F_n 为最大工作载荷;F_s 为试验载荷;f_n 为最大工作变形量;f_s 为试验载荷下变形量。

7.6.6 Standard representation of cylindrical helical compression springs in an assembly

1) The spring's inner structure does not need to be represented. The visible section should be represented from contours or from the center line of the wire's cross section. See Fig. 7-46a.

2) In the sectional view, when the wire diameter is equal to or less than 2mm, the cross section can be drawn black. Contour lines for each circle are omitted. See Fig. 7-46b.

3) If the spring's wire diameter is equal to or less than 1mm, a drawing can be made. See Fig. 7-46c.

7.6.6 圆柱螺旋压缩弹簧在装配图中的规定画法

1) 被弹簧挡住的结构一般不画出，可见部分应从弹簧的外轮廓线或从弹簧钢丝断面的中心线画起，如图7-46a所示。

2) 在剖视图中，弹簧钢丝直径在等于或小于2mm时，其断面可以涂黑，且不画各圈的轮廓线，如图7-46b所示。

3) 弹簧钢丝直径等于或小于1mm时，允许用示意图画法，如图7-46c所示。

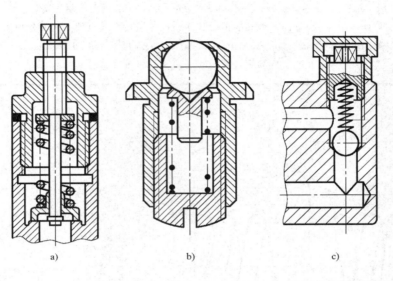

Fig. 7-46 Representation of a cylindrical helical spring in an assembly 装配图中弹簧的规定画法
a) The spring's inner structure does not need to be represented 弹簧挡住的结构一般不画
b) The wire diameter is equal to or less than 2mm 弹簧钢丝直径等于或小于2mm
c) The wire diameter is equal to or less than 1mm 弹簧钢丝直径等于或小于1mm

7.7 Bearings

Two types of bearings are used to support a shaft, i.e., sliding bearings and rolling bearings. With small friction, high rotation flexibility and easy to maintain, rolling bearings are widely used standard components of mechanical equipment. This chapter illustrates their codes, specified and simplified representation.

7.7.1 Structure of rolling bearings

The structure of rolling bearings consists of four components, shown in Fig. 7-47.

(1) Inner ring   Tightly mounted on a shaft and rotating with it.
(2) Outer ring   Mounted on the bearing's bracket hole and fastened.
(3) Rolling elements   Located between the inner rings and outer rings. They include balls, columns, and cones.
(4) Cage   Used to isolate the rolling elements.

7.7 轴承

轴承分为滑动轴承和滚动轴承，用于支承轴。滚动轴承的摩擦阻力小，转动灵活，维修方便，是机械设备中广泛应用的一种标准件。本节主要介绍滚动轴承的代号、简化画法和规定画法。

7.7.1 滚动轴承的结构

滚动轴承一般由四个元件组成，如图7-47所示。

(1) 内圈   紧密套装在轴上，随轴转动。
(2) 外圈   装在轴承座的孔内，固定不动。
(3) 滚动体   形式有圆球、圆柱、圆锥等，排列在内、外圈之间。
(4) 保持架   用来把滚动体隔开。

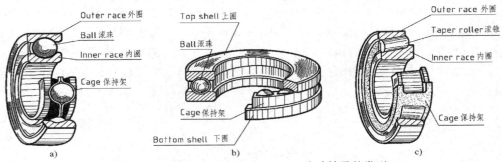

Fig. 7-47 Types of rolling bearings 滚动轴承的类型

a) Deep groove ball bearing 深沟球轴承 b) Thrust ball bearing 推力球轴承 c) Tapered roller bearing 圆锥滚子轴承

7.7.2 Types of rolling bearings

(1) Radial bearing　It is Used to bear the radial load. See the deep groove ball bearing in Fig. 7-47a.

(2) Thrust bearing　It only bears the axial load. See the thrust ball bearing in Fig. 7-47b.

(3) Radial thrust bearing　It bears both the radial load and axial load. See the tapered roller bearing in Fig. 7-47c.

7.7.3 Rolling bearing codes (GB/T 272—1993)

According to national standards, structure, dimensions, tolerance grade and technical performance are represented by the codes of the rolling bearing. Codes are composed of pre-code, base code, and post-code. The sequence is the following:

<u>Pre-code</u> <u>Base code</u> <u>Post-code</u>

The base code is essential. The pre-code and post-code are supplementary codes and added before and after the base code, respectively, when the structure, dimensions, tolerance grade and technical performance of the bearing change. If necessary, consult the relevant national standard.

Some common basic codes are listed below. The basic code consists of the bearing type code, dimension series code and inner diameter code (The dimension series code consists of width series code or height series code). Code meanings are explained with examples as follows:

Example 7-1　Bearing 6206.

7.7.2 滚动轴承的分类

(1) 向心轴承　主要用于承受径向载荷，如图7-47a中的深沟球轴承。

(2) 推力轴承　只承受轴向载荷，如图7-47b中的推力球轴承。

(3) 向心推力轴承　能同时承受径向载荷和轴向载荷，如图7-47c中的圆锥滚子轴承。

7.7.3 滚动轴承的代号（GB/T 272—1993）

按国家标准规定，滚动轴承的结构尺寸、公差等级、技术性能等特性由滚动轴承代号来表示。代号由前置代号、基本代号和后置代号组成。其排列顺序为：

<u>前置代号</u> <u>基本代号</u> <u>后置代号</u>

基本代号是轴承代号的基础，前置代号和后置代号则是轴承结构形状、尺寸、公差等级、技术性能等有改变时，在其基本代号的前、后添加的补充代号。需要时可查阅有关国家标准。

下面介绍常用基本代号。基本代号由轴承类型代号、尺寸系列代号和内径代号构成（尺寸系列代号由轴承的宽度系列代号或高度系列代号组合而成）。其所代表的意义举例如下：

例7-1　轴承6206。

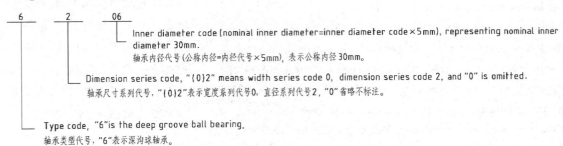

Rolling bearing marks should include the name, its code and the national standard number. The mark symbol of the above-mentioned deep groove ball bearing is the following:

Roll bearing 6206 GB/T 276—2013

Example 7-2 Bearing 51210.

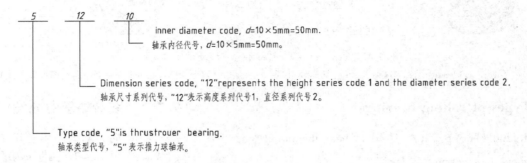

Specified symbol:

Rolling bearing 51210 GB/T 301—2015

The last two digits represent a 2-digit number of the inner diameter of the bearing. Multiply the number from 04 by 5, the result of which is the inner diameter of the bearing. Therefore, the inner diameter of the bearing is $d = 10 \times 5 \text{mm} = 50 \text{mm}$.

If the 2-digit number representing the inner diameter is below "04", the standard specifies:

00 represents $d = 10 \text{mm}$,
01 represents $d = 12 \text{mm}$,
02 represents $d = 15 \text{mm}$,
03 represents $d = 17 \text{mm}$.

7.7.4 Representation of rolling bearings (GB/T 4459.7—2017)

There is no need to draw the detail drawing for rolling bearing as it is a standard component. Therefore the specified representation and simplified representation that are specified by the national standard can be used to represent the rolling bearing in the assembly drawing. In the drawing, the main dimensions of the bearing should be referenced from the national standard according to corresponding bearing code in terms of the external diameter D, inner diameter d, and width B. And then use these main dimensions as references, draw the dimensions of rest parts with the proportion of the main dimensions. The specified and simplified representations of rolling bearings are shown in Table 7-8.

If the primary structure of the rolling bearing needs to be represented in a more detailed way, a specified representation can be adopted when drawing the assembly. If the primary structure of the rolling bearing needs to be represented in a simple way, the simplified representation, namely, the feature representation (see Table 7-8) or conventional representation (shown in Fig. 7-48) can be used.

滚动轴承的标记内容：名称、代号和国标号。所以，上述深沟球轴承的规定标记为：

滚动轴承 6206 GB/T 276—2013

例 7-2 轴承 51210。

规定标记为：

滚动轴承 51210 GB/T 301—2015

最后两位表示轴承内径的两位数字，从"04"开始用这组数字乘以 5，即为轴承内径的尺寸。故在此例中 $d = 10 \times 5 \text{mm} = 50 \text{mm}$，即为轴承内径尺寸。

表示轴承内径的两位数字，在"04"以下时，标准规定：

00 表示 $d = 10 \text{mm}$；
01 表示 $d = 12 \text{mm}$；
02 表示 $d = 15 \text{mm}$；
03 表示 $d = 17 \text{mm}$。

7.7.4 滚动轴承的画法（GB/T 4459.7—2017）

滚动轴承是标准件，不需要画零件图。在画装配图时可根据国家标准所规定的规定画法或简化画法表示。画图时，应根据轴承代号由国家标准中查出轴承的外径 D、内径 d、宽度 B 等几个主要尺寸；然后，将其他部分的尺寸按与主要尺寸的比例关系画出。常用滚动轴承的规定画法和简化画法见表 7-8。

装配图中，需要较详细地表达滚动轴承的主要结构时，可采用规定画法；在装配图中，仅需要简单地表达滚动轴承的主要结构时，可采用简化画法中的特征画法（见表 7-8）或通用画法（见图 7-48）表达。

Table 7-8 Examples for representing common rolling bearings 常用滚动轴承的画法示例

Type of bearings 轴承类型	Primary dimensions 主要尺寸	Specified representation 规定画法	Simplified representation (feature representation) 简化画法（特征画法）
Deep groove ball bearings （深沟球轴承） Type 60000 60000 型 GB/T 276—2013	D, d, B	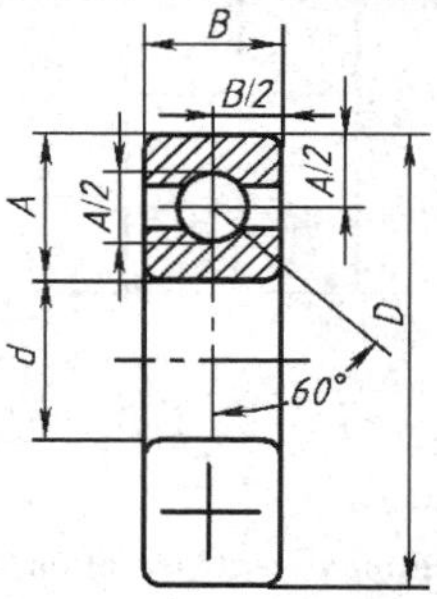	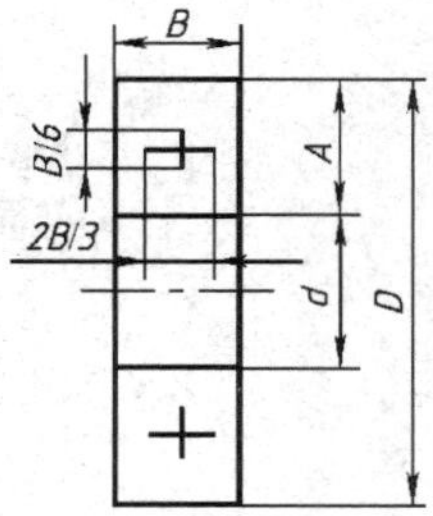
Cylindrical roller bearings （圆柱滚子轴承） Type N0000 N0000 型 GB/T 283—2007		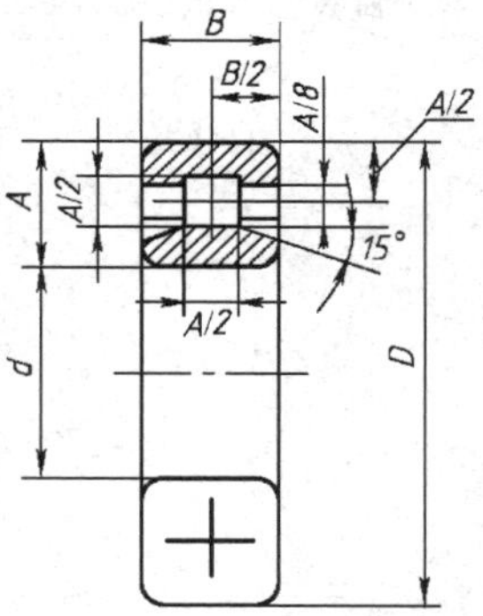	
Thrust ball bearings （推力球轴承） Type 50000 50000 型 GB/T 301—2015	D, d, H	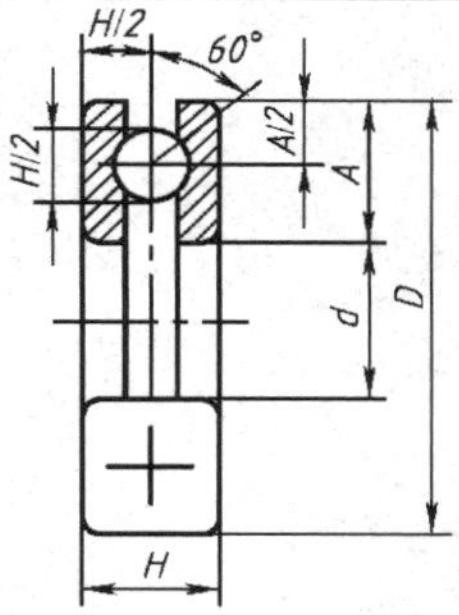	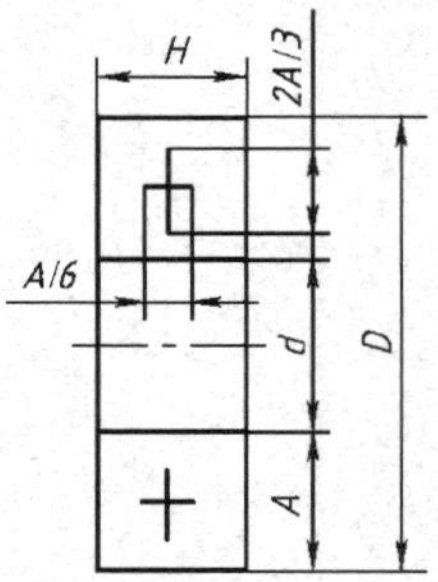
Tapered roller bearings （圆锥滚子轴承） Type 30000 30000 型 GB/T 297—2015	D, d, T, B, C	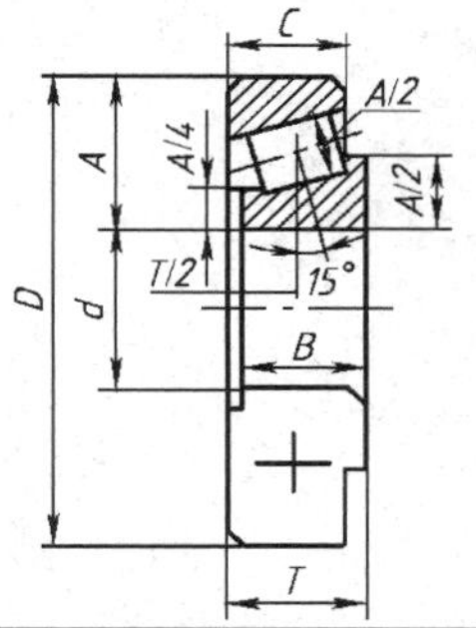	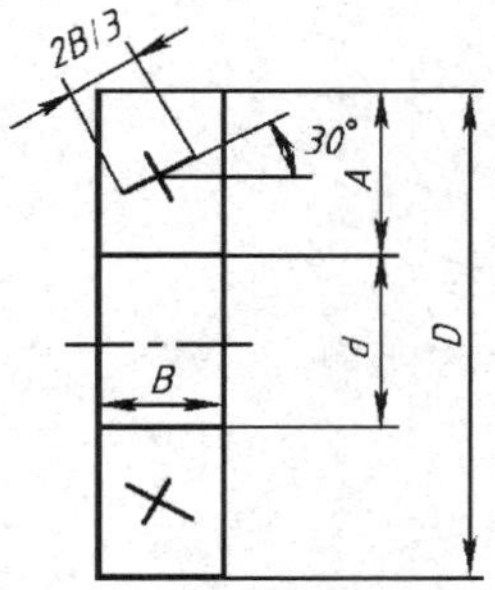
Angular contact ball bearings （角接触球轴承） Type 70000 70000 型 GB/T 292—2007		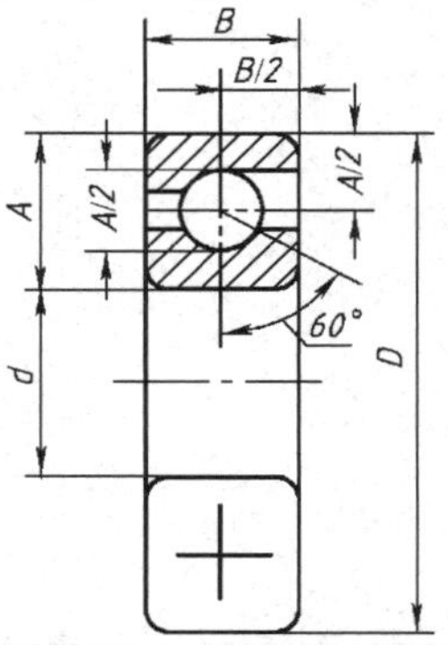	

Note 注：Dimension A in the table is calculated from the values found in standard. 表中的尺寸 A 是由标准中查得的数据计算得到的。

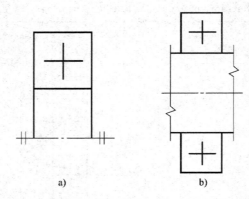

Fig. 7-48 Conventional representation of rolling bearings   滚动轴承的通用画法
a) Single bearing   单个轴承   b) Bearing and shaft assembled together   轴承与轴装配在一起

Chapter 8　Detail Drawings

A detail drawing conveys important information about the structure, shape, size and technical requirements of a single part. It is not only criteria for the part's machining and inspection, but also important technical information for the part's design and production. Fig. 8-1 shows the detail drawing of a gear pump's cover.

This chapter introduces aspects related to detail drawings, including views layout, representation, dimensioning, and technical requirements, etc.

8.1　Contents of Detail Drawings

Based on the functions and requirements, the following contents need to be included in a detail drawing as shown in Fig. 8-1.

1) A group of views (general views, sectional views, cuts, etc.) that convey the structure and shape accurately and clearly.

2) Integrate dimensions for its production, being reasonable, full and clear.

3) Technical requirements. Use designated symbols, numbers or texts to illustrate the technical requirements briefly for production and inspection.

4) Title block with name of the part, quantity, material, scale, drawing number, and the signature of the designer, checker and inspector.

第 8 章　零件图

表达单个零件结构形状、大小及技术要求的图样称为零件图。它既是加工制造和检验零件的依据，又是设计和生产过程中的重要技术资料。图 8-1 所示为齿轮泵的泵盖零件图。

本章介绍绘制零件图的有关问题，包括零件的视图布局、表达方法、尺寸标注及技术要求等。

8.1　零件图的内容

根据零件图的用途和要求，零件图一般应包括下列四方面的内容（见图 8-1）：

1) 一组视图（如视图、剖视图、断面图等）。确切、清晰地表达零件的结构形状。

2) 完整尺寸。即合理、齐全、清晰地标注出制造零件所需的全部尺寸。

3) 技术要求。用规定的符号、数字或文字简要说明制造、检验时应达到的技术要求。

4) 标题栏。包括零件名称、数量、材料、比例、图号，以及设计者、校核者和审核者的签名等。

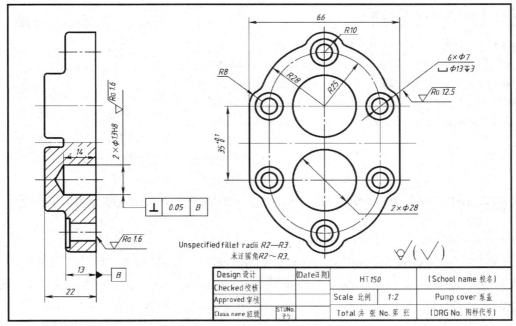

Fig. 8-1　Detail drawing of a pump cover　泵盖的零件图

8.2 Choosing Views of Detail Drawings

Before drawing a part, one should choose a suitable front view and other views, taking into account the function, installation site and machining methods of the part, as well as analyzing the shape and structure.

8.2.1 Choosing the front view

The following two factors should be considered when selecting the front view:

1. Form features

The front view should reflect the main shapes and relative positions among parts as much as possible. The front view of the pump cover in Fig. 8-1 thoroughly reflects the structures and relative positions of various holes.

2. Installation position

It refers to machining position or working position. In order to enhance comprehension for production, In the front view of the transmission shaft, hand wheel and other parts should be placed in accordance with their machining positions. In the front views of the various boxes, pumps, valves and base partsetc should be placed in accordance with their working positions (as shown in Fig. 8-1), although they are machined by using different devices and their machining positions are not the same.

8.2.2 Choosing other views

Sections not clearly conveyed in the front view should be shown in other views. Chosen views should convey key features and avoid repetition. In short, views are selected with clear objectives. They should clearly depict the part, and provide as much information as possible. Use an adequate number of views to enhance comprehension.

8.3 Representation Methods of Typical Parts

Parts have various shapes. Following their manufacturing features, they can be roughly divided into shafts and sleeves, plate wheels, fork levers, brackets, and boxes. In the following sections, we provide a representation of those parts with several typical examples.

8.2 零件图的视图选择

画零件图时,要通过对零件在机器中的作用、安装位置、加工方法的了解以及对其进行形体分析和结构分析,合理地选择主视图和其他视图。

8.2.1 选择主视图

选择主视图时,主要考虑以下两点:

1. 形状特征

主视图要较多地反映出零件各部分的形状和它们之间的相对位置。如图 8-1 所示,泵盖零件图中的主视图就较好地表达了各种不同结构的孔及其相对位置关系。

2. 安装位置

安装位置即零件的加工位置或工作位置。为使生产时便于看图,传动轴、手轮等零件在主视图按其加工时的位置摆放。各种箱体、泵体、阀体及机座等零件,需在不同的机床上加工,其加工位置亦不相同,故在其主视图中应按零件工作时的位置放置。例如图 8-1 所示泵盖零件图中的主视图,即按其工作位置投射得到。

8.2.2 选择其他视图

主视图中没有表达清楚的部分,要用其他视图表示。所选视图应有其重点表达内容,并尽量避免重复。总之,选择视图要目的明确,重点突出,使所选视图表达方案完整、清晰,数目恰当,做到看图方便,作图简便。

8.3 典型零件的表达方法

在生产中零件的形状是千变万化的,但就其结构特点来分析,大致可以分为轴套类、轮盘类、叉杆类、支架类、箱体类等。下面结合典型例子介绍这几类零件的表达方法。

8.3.1 Shafts and sleeves

Typical shafts and sleeves include rotating shafts, pins, and sleeves. Shafts are used to support driving devices (such as gears and pulleys) so as to transfer motion and dynamic force.

1. Analyzing the shape and structure

In general, shafts are composed of various cylinders of different diameters, which are called stepped shafts. Shafts usually have keyways, pinholes and tapered notches used to fix screws, in order to join gears and pulleys. The pump shaft is shown in Fig. 8-2.

8.3.1 轴套类零件

轴套类零件包括各种转轴、销轴、套筒等。其中,轴类零件用来支承传动件(如齿轮、带轮等),以传递运动和动力。

1. 形体结构分析

轴类零件通常由若干段直径不同的圆柱体组成,称为阶梯轴。为了联接齿轮、带轮等零件,在轴上常有键槽、销孔和固定螺钉的锥形凹坑等结构,如图 8-2 所示的泵轴。

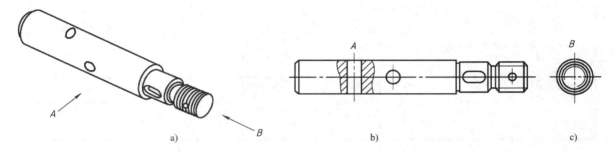

Fig. 8-2　The selection of the front view for a pump shaft　泵轴的主视图选择
a) Projection directions　投射方向　b) A direction view　A 向视图　c) B direction view　B 向视图

2. Choosing the front view

The main process of shaft machining is performed on a lathe as shown in Fig. 8-3. The axis should be positioned horizontally, to facilitate a better comprehension of the drawing during the course of production. The pump shaft in Fig.8-2a can be projected, as front views, in two directions, i.e., A and B shown in Fig. 8-2b, Fig. 8-2c. Obviously, the A direction view reflects shapes better than the B direction view。

3. Choosing other views

If the keyway structure on the shaft is not clearly shown, one can illustrate it with a cutaway view at an appropriate position in the front view as shown in Fig. 8-4. Some details of the shaft can be shown in a drawing of partial enlargenent in order to convey accurate shapes and dimensions. Therefore, shafts and sleeves are usually represented with a front view and some sectional and drawings of partial enlargement.

Fig. 8-3　Machining position of a shaft　轴加工时的位置

2. 主视图的选择

轴的主要加工工序是在车床上进行的,如图 8-3 所示。为了加工时看图方便,主视图中应将轴按水平方位放置。泵轴的主视图投射方向可以有 A 向和 B 向,如图8-2a 所示。显然,按这两个投射方向所得的主视图中,如图 8-2 b、c 所示,A 向视图比 B 向视图能够更好地反映泵轴的形状特征。

3. 其他视图的选择

轴上键槽的结构在主视图上未表达清楚时,可在主视图的适当部位用移出断面表示,如图 8-4 所示。对于一些轴上的细部结构还可采用局部放大图,以便确切地表达其形状和标注尺寸。所以,轴套类零件一般采用一个主视图和若干断面图、局部放大图来表示。

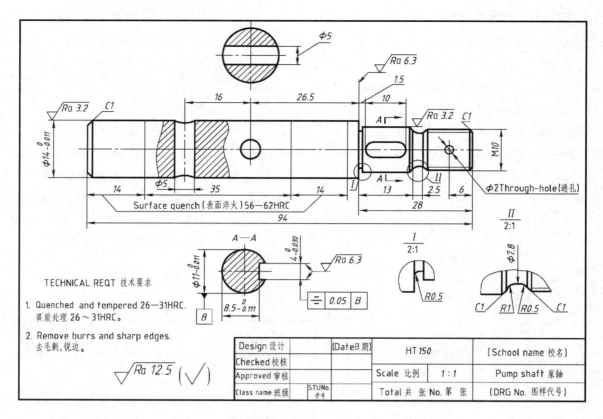

Fig. 8-4 Detail drawing of a pump shaft　泵轴零件图

8.3.2　Plate wheels

Plate wheels include various parts, such as hand wheels, gears, pulleys, flanges, glands and end covers. The end cover showing in Fig. 8-5 pertains to plate wheels.

1. Analyzing the shape and structure

These parts are represented by flat circular plates and generally have equally distributed elements such as holes, ribs, flanges, etc.

2. Choosing the front view

Front views of plate wheels can be selected to be either sectional views or external views as shown in the front view and left view of Fig. 8-6. However, with comparison, a sectional view is recommended for showing structures and relative positions clearly between each step and inner hole shape as well as conforming to the main actual machining position.

3. Choosing other views

In general, plate wheels have various kinds of flanges, equally spaced holes, grooves and ribs. Therefore only one front view may not be able to clearly show various individual features of components and additional basic views might be needed. As an example, the left view shown in Fig. 8-6 indicates four evenly arranged through-holes.

8.3.2　轮盘类零件

轮盘类零件包括各种手轮、齿轮、带轮、法兰盘、压盖和端盖等。图 8-5 所示的端盖属于轮盘类零件。

1. 形体结构分析

轮盘类零件的基本形状是扁平的盘状，上面通常有均匀分布的孔、肋及凸缘等结构。

2. 主视图的选择

轮盘类零件的主视图可选用如图 8-6 所示的剖视图，也可选用图中作为左视图的外形视图。但经过比较，选用剖视图较好，因为它层次分明，表达了各台阶与内孔的形状及其相对位置，并且也符合它的主要加工位置。

3. 其他视图的选择

一般轮盘类零件多数带有各种形状的凸缘、均布的孔、槽和肋等结构，所以仅采用一个主视图不能完整地表达零件，需要增加其他基本视图。例如，图 8-6 中的左视图用以以表达四个均匀分布的通孔。

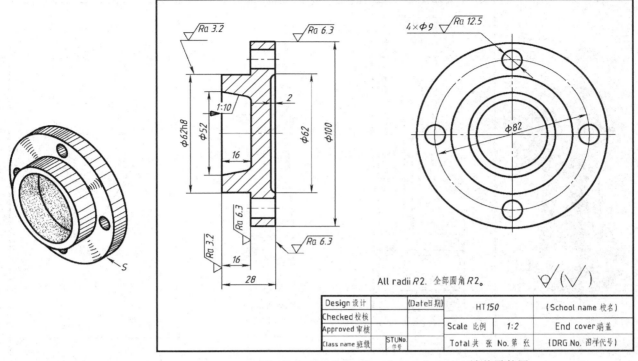

Fig. 8-5 End cover 端盖

Fig. 8-6 Detail drawing of an end cover 端盖零件图

8.3.3 Fork levers

Fork levers include transmission forks, connecting bars, valve handles, and levers. Fig. 8-7 shows the axonometric drawing of a transmission fork.

1. Analyzing the shape and structure

The characteristic of fork levers is that the workblank shope is usually complex with forks or bars.

2. Choosing the front view

While choosing front views for fork levers, one should pay attention to their working positions and shapes, due to their complex shapes and the variety of machining positions. As an example, Fig. 8-7 shows the working position of the transmission fork. The front view projected in direction S thoroughly reflects its characteristics.

3. Choosing other views

A fork lever normally needs two or more views. The details of those parts need to be represented with partial views, oblique views, sections, removed or coincident cuts. The drawing of a transmission fork is shown in Fig. 8-8.

8.3.4 Brackets

Brackets are frameworks or seats used to support other parts.

8.3.3 叉杆类零件

叉杆类零件包括各种拨叉、连杆、阀杆、杠杆等。图 8-7 所示为拨叉的轴测图。

1. 形体结构分析

叉杆类零件的结构特点是毛坯形状比较复杂,呈叉形或杆状。

2. 主视图的选择

叉杆类零件形状复杂,加工位置多变,在选择主视图时,主要考虑其工作位置和形状特征。例如图 8-7 所示为拨叉的工作位置,故以 S 向作为主视图的投射方向时完全可以反映它的形状特征。

3. 其他视图的选择

叉杆类零件常需要两个或两个以上的基本视图,并且要用局部视图、斜视图、剖视图、移出或重合断面图等表达零件的细部结构,如图 8-8 所示的拨叉零件图。

8.3.4 支架类零件

支架类零件是支承其他零件的支架或支座。

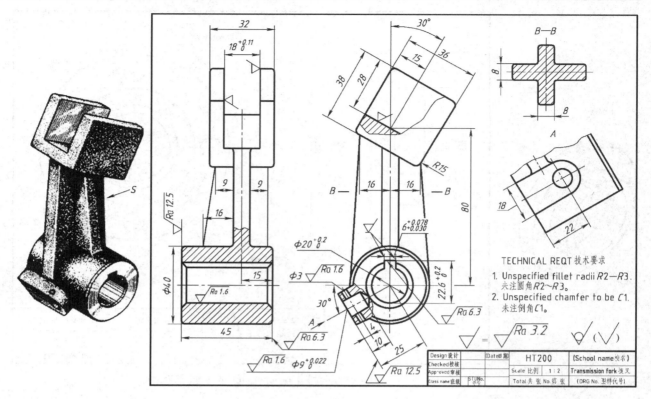

Fig. 8-7 Transmission fork 拨叉   **Fig. 8-8** Detail drawing of a transmission fork 拨叉的零件图

1. Analyzing the shape and structure

Fig. 8-9 shows the axonometric drawing of a bearing seat. The structure of the bearing seat is made up of the axle hole used to support the shaft, the base plate used to fix with other parts, and the cross brace and retaining plate for supporting.

2. Choosing the front view

Machining of the brackets requires more steps. Therefore, their positions in front views should be in accordance with the working positions of the brackets. Fig. 8-9 shows that the front view projected in direction S fully conveys the shape and structure of the bearing seat.

3. Choosing other views

The front view only conveys the fundamental shape. If some portions of the part such as the cavity of the bearing hole, the shape of the shoe plate and the cross section of the cross brace are not clearly illustrated, other views are required.

As shown in Fig. 8-10, in addition to the front view, we have used a full section as the left view to convey the inner part of the bearing hole and the shape of the cross brace. The *B—B* section represents both the shoe plate and the cross brace. *A* direction view represents the shape of the dummy club. With these views, we can represent the structure of the entire bearing seat.

1. 形体结构分析

图 8-9 所示为轴承座的轴测图。轴承座的结构由支承轴的轴孔、用来其他零件固定的底板,以及起加强、支承作用的肋板和支承板等组成。

2. 主视图的选择

支架类零件加工时其加工工序较多,故在选择主视图时,一般都按工作位置放置。如图 8-9 所示,将 S 向作为主视图的投射方向,能充分地表达该轴承座零件的形状及结构特征。

3. 其他视图的选择

主视图仅表达了零件的主要形状,但有些部分还未表达清楚,如轴承孔的内腔结构、底板的形状、肋板的断面形状等,因此还要选择一些其他视图将其表达完全。

在图 8-10 中,除主视图外,还采用全剖的左视图表达轴承孔内腔结构及肋板形状,采用 B—B 剖视图表达底板和肋板的断面形状,采用 A 向视图表达凸台形状。有了这些视图才能把整个轴承座的形状结构完整地表达出来。

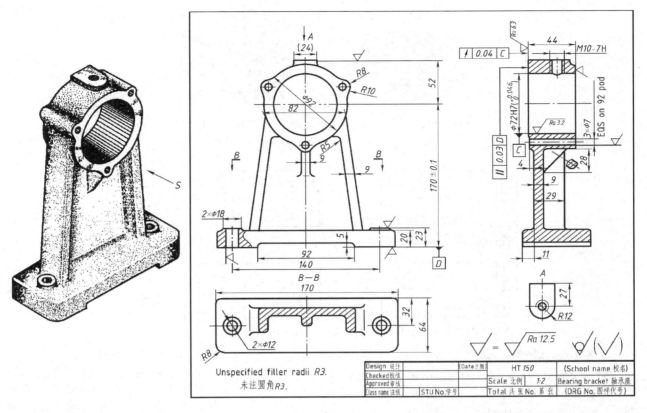

Fig. 8-9 Bearing seat 轴承座

Fig. 8-10 Detail drawing of a bearing seat 轴承座的零件图

8.3.5 Boxes

Boxes, as the name indicates, are box-shaped items, and are used to house other parts, like cabinets and machine bases of complex structures.

1. Analyzing the shape and structure

Fig. 8-11 shows the axonometric drawing of a box—an overtravel-limit switch shell. The figure tells us that the shell should have an entrance for wires, a button hole, a mounting hole to fix the shell, and the threaded hole to connect the top cover (see Fig. 8-11a).

2. Choosing the front view

The machining of a box is a complex process. The front view is in accordance with its working position and it should be projected so as to fully convey the shape and structure of the parts. Therefore, the front view of the overtravel-limit switch shell is projected in direction S (see Fig. 8-11b) and is a full section. In this way, the main structure, the inner structure ands the wall thickness of this part can be clearly represented, as shown in Fig. 8-12.

8.3.5 箱体类零件

箱体类零件的形状都具有箱体特点,用于承装其他零件,如机壳、机座,且形体结构较为复杂。

1. 形体结构分析

图 8-11 所示为箱体类零件——行程开关外壳的轴测图。从图中可以看出,该行程开关壳体上有接线的进出孔、按钮孔及固定外壳的安装孔和联接上盖的螺孔等(见图 8-11a)。

2. 主视图的选择

箱体类零件的加工工序较复杂。其主视图上一般按其工作位置放置,而其投射方向选择则以能充分显示出零件的形状、结构为原则。故行程开关外壳主视图按图 8-11b 中 S 方向投射,并取全剖视图。这样,该零件的主要结构、内部形状及各处壁厚都能表达得很清楚,如图 8-12 所示。

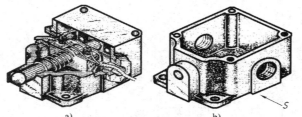

Fig. 8-11 Shell of an overtavel-limit switch 行程开关外壳
a) Inner structure 内部结构 b) Case shell 壳体

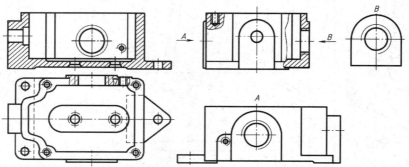

Fig. 8-12 Views of the overtravel-limit switch shell 行程开关外壳的视图

3. Choosing other views

As shown in Fig. 8-12, in addition to the front view, other views shall be adopted. In order to indicate the shapes of the shell, the shoe plate and the distribution of the holes, we have adopted the A, B direction views and the top view, and represented the dummy club with thin long dashed lines.

8.4 Manufacturing Processes of Parts

The function of a part in a machine or its assembly defines its structure. To facilitate production, one should also consider available manufacturing processes in addition to the function. Some common technological structures introduced below serve as references.

8.4.1 Casting structure

1. Pattern draft

The 1:20 (≈3°) draft along the drawing direction of the pattern facilitates to take out the pattern while molding. A casting draft is shown in Fig. 8-13a, Fig. 8-13b. If the draft is very small, it can be omitted, as shown in Fig. 8-13c. If necessary, it can be included in technical requirements.

3. 其他视图的选择

如图 8-12 所示，除主视图外还需要采用其他视图。为了表示外壳及底板的形状特征以及孔的分布情况，采用了 A 向视图、B 向视图和俯视图，并在俯视图中用细虚线表示其底面的凸台形状。

8.4 零件的结构工艺性

零件在机器或部件中的作用，决定了它的各部分结构。但在设计零件时，除考虑其作用外，还必须考虑到制造时的工艺性，以利于生产。下面介绍一些常见的工艺结构，供绘图时参考。

8.4.1 铸件结构

1. 起模斜度

铸件造型时，为便于取出模样，沿脱模方向的表面做出 1:20 (≈3°) 的起模斜度。这一斜度留在铸件上如图 8-13a、b 所示。起模斜度较小时，在图中一般不画出，如图 8-13c 所示。必要时可注在技术要求中。

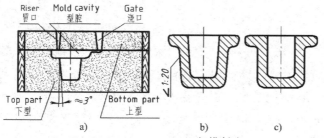

Fig. 8-13 Pattern draft 起模斜度
a) Schematic representation of sand casting 砂型铸造的示意图
b) Marks of pattern draft 起模斜度标注
c) Small pattern draft 起模斜度较小

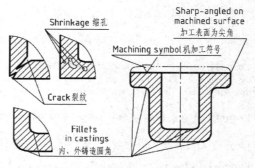

Fig. 8-14 Casting structure 铸件结构

2. Fillets in castings

There must be a fillet where two surfaces meet, as shown in Fig. 8-14, in order to avoid sand drops while removing the mould, and avoid the cracks or shrinkages while cooling. The intersection line of two surfaces is not clear due to the fillets of castings. The intersection line is needed in order to distinguish the various surfaces. The one we have provided is called arun-out line. The representation of the run-out line is similar to that of intersection line. The only difference is that the end of the run-out line is not connected with the contour line of the fillet, and the line style is a thin solid line. Drawing methods for the run-out line of commonly used castings include the following:

1) Fig. 8-15 shows the run-out line when two curved surfaces meet.

2. 铸造圆角

在铸件表面交角处应有圆角（见图 8-14），以免脱模时砂型落砂，同时防止铸件冷却时产生裂纹或缩孔。由于铸造圆角的存在，两铸造表面相交处的相贯线不很明显。为了区分不同形体的表面，仍要画出这条相贯线，这种相贯线称为过渡线。过渡线的画法和相贯线相似，只是其端点处不与圆角轮廓线接触，且线型为细实线。常见铸件的过渡线画法有：

1）两曲面相交时的过渡线画法如图 8-15 所示。

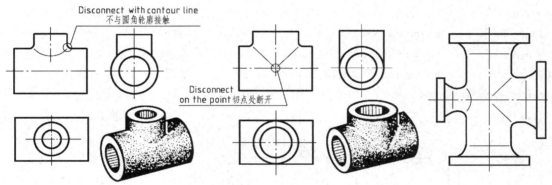

Fig. 8-15   Representation of a run-out line（Ⅰ）   过渡线画法（Ⅰ）

2) While representing the run-out line between planes or between a plane and a curved surface, one should keep both ends of the run-out line open, and draw an arc along the fillet of castings, as shown in Fig. 8-16.

2）在画平面与平面、平面与曲面相交处的过渡线时，应在交线两端断开，并按铸造圆角弯曲方向画出过渡圆弧，如图 8-16 所示。

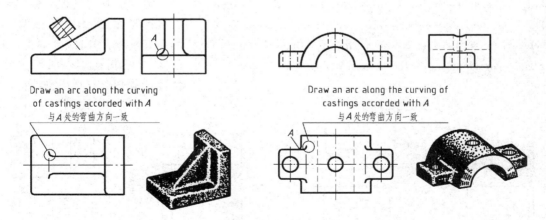

Fig. 8-16   Representation of a run-out line（Ⅱ）   过渡线画法（Ⅱ）

3) The run-out line to combine the column and the cross brace with different section shapes depends on the ratio of the tangent or intersection, as shown in Fig. 8-17.

3）不同断面形状的肋板与圆柱组合，过渡线视相切、相交的关系不同，其画法如图 8-17 所示。

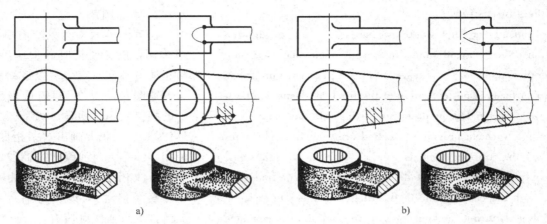

Fig. 8-17 Representation of a run-out line (Ⅲ)　过渡线画法(Ⅲ)
a) Intersection　相交　b) Tangent　相切

3. Wall thickness

To avoid shrinkages and cracks (see Fig. 8-18a) caused by different cooling speeds due to uneven thickness, during the design phase, it is a must to maintain uniform wall thickness or a smooth transition as shown in Fig. 8-18b and Fig. 8-18c. One must avoid unequal designs such as thickness abrupt varying or local thick partial areas.

3. 铸件的壁厚

为防止浇注零件时，由于壁厚不匀使冷却速度不同所产生的缩孔和裂纹（见图 8-18a），在设计铸件时，应尽量使壁厚均匀或逐渐过渡，如图 8-18 b、c 所示，以避免壁厚突变或局部粗大的不匀现象。

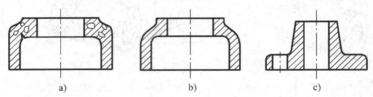

Fig. 8-18 Wall thickness　铸件的壁厚
a) Nonuniform　不均匀　b) Uniform　均匀　c) Smooth transition　逐渐过渡

8.4.2　Common technological structures

1. Boss clubs and recessed surfaces

Usually, all surfaces mounted with other parts need to be machined. The boss club or recessed surface is added to keep good contact between the surfaces of two parts and to reduce the machining area and manufacturing cost, as shown in Fig. 8-19.

8.4.2　常见工艺结构

1. 凸台与凹坑

零件上凡与其他零件接触的表面一般都要加工，为了保证两零件表面的良好接触，同时减少加工面积，以降低制造费用，在零件的需接触表面处常设计成凸台、凹坑的形式，如图 8-19 所示。

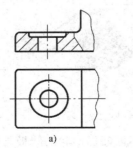 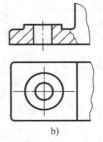 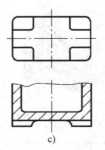 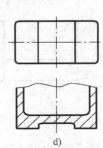

Fig. 8-19 Boss clubs and recessed surfaces　凸台与凹坑
a) Recessed surface on the top　凹坑在上部　b) Boss club on the top　凸台在上部
c) Boss club on the bottom　凸台在底部　d) Recessed surface on the bottom　凹坑在底部

2. Chamfers and roundings

To avoid cutting hands and to facilitate assembly, burrs and flashes should be eliminated, so the ends of both the shafts and the holes usually need to be chamfered, that is, the sharp edge of the shaft end is cut by a conical cutting tool with a 45° apex angle. An arc transition usually remains at the shaft shoulder, which is called a rounding, in order to avoid cracks caused by pressure concentration. Chamfer and rounding dimensions can be obtained from the mechanical design manual (See appendix Table 3-2). The representation of both the chamfer and the rounding is shown in Fig. 8-20.

3. Undercuts and grinding undercuts

The undercut, grinding or traveling limit undercut needs to be produced at the end of the surface to be machined, in order to easily withdraw the cutter or grinding wheel while tapping or grinding external diameter. The structure and dimension of the undercut are shown in Fig. 8-21. Dimension is identified as "width×depth" or "width×diameter". The structure and dimensions of the undercut and grinding undercut are shown in appendix Table 3-3 and Table 3-4.

2. 倒角与圆角

为了防止划伤人手和便于装配,应去除零件的毛刺、锐边,因此在轴和孔的端部通常需加工出倒角,即用一锥顶角为45°的圆锥刀具切除其端部锐边。在轴肩处,常常制成圆弧过渡,即倒圆,以避免因应力集中而产生裂纹。倒角和倒圆的尺寸可由机械设计手册查出(见附录表3-2),其画法如图8-20所示。

3. 退刀槽和砂轮越程槽

在车削螺纹或磨外圆时,为便于退刀或砂轮,需要在待加工面末端先切出退刀槽或砂轮越程槽。退刀槽及越程槽的结构如图8-21所示,尺寸注成"宽度×深度"或"宽度×直径"。退刀槽和越程槽的结构和尺寸见附录中表3-3和表3-4。

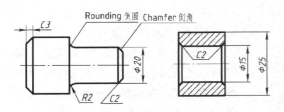

Fig. 8-20 Chamfers and roundings 倒角和倒圆

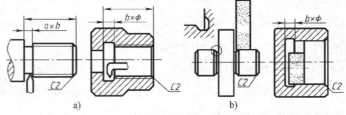

Fig. 8-21 Undercuts and grinding undercuts 退刀槽和砂轮越程槽
a) Structure and dimensions of a undercut 退刀槽结构及尺寸
b) Structure and machining of grinding undercut 砂轮越程槽结构及加工

8.4.3 Structure of drilling holes

While representing a non-through hole, since the drill head is tapered, a tool angle of 120° shall be drawn at the end of the hole. The angle dimensions are not required.

If the larger hole of stepped holes is also machined by drill, an angle of 120° at its end is required. The representation and dimension of the hole are shown in Fig. 8-22.

The axis of the drill should be perpendicular to the surface to be drilled, in order to ensure the correct drilling position and prevent the drill from breaking caused by unequal impact force. Otherwise, it is necessary to set a platform and notch structure on the oblique surface. If the angle between the axis of the drill and the oblique plane is larger than 60°, one can drill the hole directly, as shown in Fig. 8-23.

8.4.3 钻孔结构

由于钻头具有钻尖,当需要画不通孔时,其末端应画出钻头切削后留下的120°自然锥角,但不必注尺寸。

阶梯孔中的大孔用钻头钻时,在该孔的终止处也应画出120°圆锥角。孔的画法及尺寸标注如图8-22所示。

用钻头钻孔时,为保证钻孔位置准确和避免钻头因受力不均而折断,应使钻头轴线垂直于被钻孔零件的表面。否则应在与孔轴线倾斜的表面处,常需设计出平台或凹坑结构,但当钻头与倾斜面的夹角大于60°时,也可以直接钻孔,如图8-23所示。

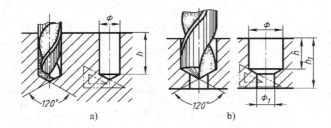 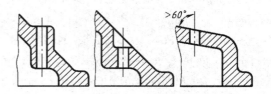

Fig. 8-22 Representation and dimension of the hole
孔的画法及尺寸标注
a) Non-through hole 不通孔 b) Stepped hole 阶梯孔

Fig. 8-23 Surface structure of the drilling hole
钻孔的端面结构

8.5 Dimensioning of Detail Drawings

Dimension is one of the major contents of a detail drawing and provides important information for machining and inspection. Therefore, it is a must to represent all dimensions correctly, completely, clearly and reasonably on a detail drawing. The previous sections have introduced the requirements for correctness, completeness and clarity. This section focuses on some general issues of dimensioning and some commonly used structures so that the dimensions will not only meet the design requirements but also other requirements to facilitate machining, measurement and inspection.

1. Selection of datums

As mentioned before, a datum provides important reference for dimensioning. In order to produce reasonable dimensions, one should appropriately select datums. When selecting a datum, one must consider part position, function, assembling, position setting and inspection. Therefore, datums should be determined according to the requirements of design, machining conditions, and inspection methods. According to the application datums can be divided into design datums and process datums, and they can be divided into primary datums and secondary datums according to their importance. The design datums and process datums are discussed below.

(1) Design datums Design datums include datum planes or datum lines which are used to locate a given part inside other parts. As shown in Fig. 8-24, select the bottom surface D as the datum for the center height 40mm±0.02mm of the bracket's axle hole.

A shaft is supported by two brackets, therefore, the centers lines of the two axle holes shall be kept in the same axial line, in order to keep the axial line horizontal. When reflecting the dimensions of the two holes on the base Board, select surface B as the lengthwise datum to keep the symmetry between the two holes and the axle hole, select the back surface C as the widthwise datum. Thus, B, D and C are the design datums.

8.5 零件图的尺寸标注

尺寸标注是零件图的主要内容之一，是制造加工和检验零件的重要依据。因此，在零件图中，必须正确、完整、清晰、合理地标注零件的全部尺寸。对于正确、完整、清晰的尺寸标注要求，前面相关章节已经做了介绍，本节着重讨论合理标注尺寸的一些基本问题和常见结构的尺寸注法，使所标注的尺寸既满足设计要求，又符合生产实际，便于加工、测量和检验。

1. 尺寸基准的选择

尺寸基准是标注尺寸的起点。要做到合理地标注尺寸，首先必须正确选择尺寸基准。在选择尺寸基准时，必须考虑零件在机器或部件中的位置、作用、零件之间的装配关系以及零件在加工过程中的定位和测量等要求。因此，基准应根据设计要求、加工情况和测量方法确定。按基准的用途可分为设计基准、工艺基准等。按基准的重要性可分为主要基准和辅助基准。下面只讨论设计基准和工艺基准。

（1）设计基准 设计基准是用于确定零件相对于其他零件工作位置的基准面或线。如图 8-24 所示，标注支架轴孔的中心高 40mm±0.02mm，应以底面 D 为基准注出。

因为一根轴要用两个支架支承，为了保证轴线的水平位置，两个轴孔的中心应在同一轴线上。标注底板两孔的定位尺寸时，长度方向以对称面 B 为基准，以保证两孔与轴孔的对称关系，宽度方向以其后表面 C 为基准，故面 B、D、C 为设计基准。

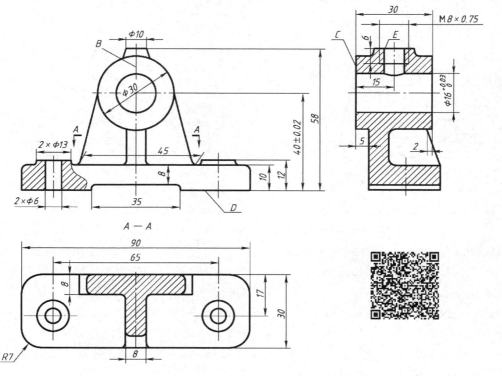

Fig. 8-24 Determining datums 确定尺寸基准

(2) Process datums Process datums include datum planes or datum lines used to machine or measure the part. As shown in Fig. 8-24, the top surface E of the dummy club on the upper position is a process datum, so it can be easily used to measure the depth of the threaded hole.

As mentioned Defore, the design datums and process datums, can be divided into primary datums, and secondary datums, but there must be a link between the two kinds of datums, like the height size of 58mm shown in Fig. 8-24. The part must have a primary datum in each direction—length, width, height, shown as B, C, and D in Fig. 8-24.

2. Direct specification of important dimensions

Important dimensions include the ones used to mount other parts, to define key relative positions and any other information related to the function of the part. All of them should be directly specified when drawing parts.

The center height of the axle hole shown in Fig. 8-25 is an important dimension. If the dimension is specified according to Fig. 8-25b, the accumulation error of dimensions b and c will make it impossible for the center height of the axle hole to meet design requirements. In addition, for mounting purposes, the distance between the centers of the two holes on the bottom board shown in Fig. 8-25a should also be directly specified. Indirectly determining l by calculating dimensions from e cannot meet matching requirements, as shown in Fig. 8-25b.

（2）工艺基准 工艺基准是零件在加工、测量时的基准面或线。在图 8-24 中，上部凸台的顶面 E 是工艺基准，以此为基准测量螺孔的深度比较方便。

如前所述基准又分为主要基准和辅助基准。两个基准之间应有联系尺寸，如图 8-24 中的高度尺寸 58mm。零件在长、宽、高三个方向都应有一个主要基准，如图 8-24 中的 B、C、D。

2. 重要的尺寸直接注出

重要尺寸是指与其他零件相配合的尺寸、重要的相对位置尺寸、影响零件使用性能的其他尺寸，这些尺寸都应在零件图上直接注出。

图 8-25 轴孔的中心高是重要尺寸。若按图 8-25b 标注，则尺寸 b 和 c 的累积误差，会使得孔中心高不能满足设计要求。另外，为装配方便，图 8-25a 中底板上两孔的中心距离也应直接注出。图 8-25b 中由标注 e 间接确定的 l 则不能满足装配要求。

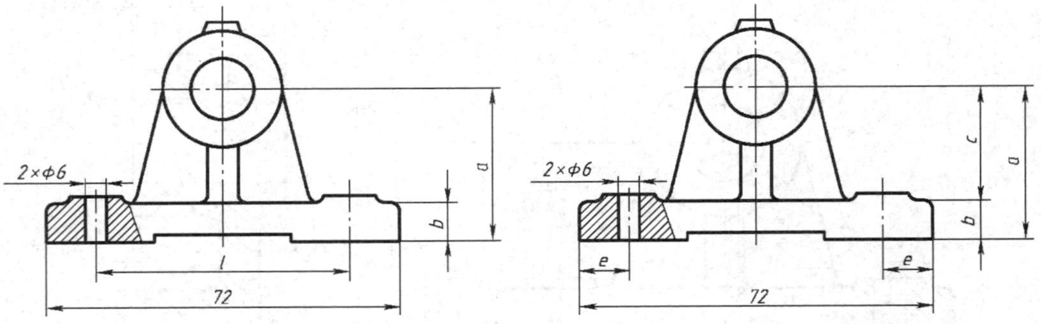

Fig. 8-25 Direct specification of important dimensions 重要尺寸直接注出
a) Mark reasonable 标注合理 b) Mark unreasonablly 标注不合理

3. Avoiding enclosing dimension chains

As shown in Fig. 8-25b, dimensions a, b and c form an enclosing dimension chain. Since $a = b + c$, if dimension a has a constant error, dimensions b and c must have very little error, which may lead to difficulties in machining. Therefore, enclosing dimension chains should be avoided and the dimension with less importance, i.e., c in this example, should be removed.

4. Producing dimensions in accordance with machining sequence

1) Dimensions should also be specified in accordance with technological requirements for machining. Fig. 8-26 shows the technological requirement for machining sequence of a shaft.

3. 避免出现封闭的尺寸链

图 8-25b 中的尺寸 a、b、c 构成一个封闭的尺寸链。由于 $a=b+c$，若尺寸 a 的误差一定，则 b、c 两个尺寸的误差就要定得很小。这样，就会使加工困难。所以应当避免封闭的尺寸链，例如在此应将一个不重要的尺寸 c 去掉。

4. 标注尺寸应符合加工顺序

1）标注尺寸应符合加工工艺要求。图 8-26 所示为轴的加工工艺要求。

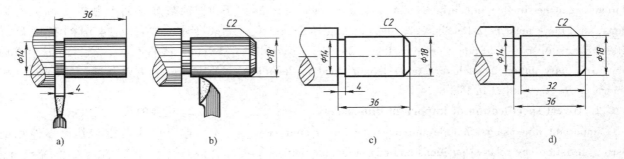

Fig. 8-26 Dimensions in accordance with technological requirements for machining
尺寸标注符合加工工艺要求
a) Turning undercut 车退刀槽 b) Turning external diameter and chamfer 车外圆、倒角
c) Mark reasonablly 标注合理 d) Mark unreasonablly 标注不合理

① Determine the location of the undercut according to the dimension 36mm, and then machine the undercut (see Fig. 8-26a).

② Turn the external diameter $\phi 18$mm and chamfer the end of the shaft (see Fig. 8-26b). Dimensioning shown in Fig. 8-26c is appropriate, but that in Fig. 8-26d is not.

2) Dimensions should be specified for easy inspection. Fig. 8-27 shows the axial dimension of the sleeve. Dimensioning A and C shown in Fig. 8-27a is convenient for measurement, while dimension C in Fig. 8-27b is not.

① 按尺寸 36mm 确定退刀槽的位置，并加工退刀槽（见图 8-26a）。

② 车 $\phi 18$mm 的外圆和轴端倒角（见图 8-26b）。图 8-26c 所示的尺寸标注合理，图 8-26d 所示的尺寸标注不合理。

2）标注尺寸应便于测量。图 8-27 所示为套筒轴向尺寸的标注。按图 8-27a 所示标注尺寸，A、C 便于测量；若按图 8-27b 所示标注尺寸 C，则不便于测量。

5. Only one link allowed along on direction between machined and non-machined surfaces

As shown in Fig. 8-28a, there are three non-machined surfaces along the height direction: B, C, and D, in which only surface B and machined surface A have a link of dimension 8mm, which is appropriate. If dimensions are specified as shown in Fig. 8-28b with three non-machined surfaces linked to A, it would be difficult to ensure the precision of linked dimensions 8mm, 34mm and 42mm while machining the surface A.

5. 同一个方向只能有一个非加工面与加工面联系

在图8-28a中，沿铸件的高度方向有三个非加工面B、C和D，其中只有B面与加工面A有尺寸8mm的联系，这是合理的。如果按图8-28b所示标注尺寸，三个非加工面B、C和D都与加工面A有联系，那么在加工A面时，就很难同时保证三个联系尺寸8mm、34mm和42mm的精度。

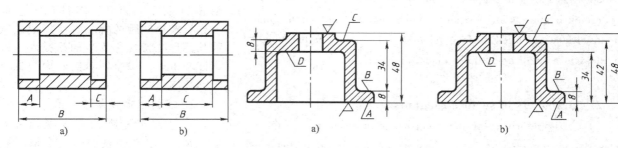

Fig. 8-27 Dimensioning for convenient measurement 标注尺寸便于测量
a) Correct 正确 b) Incorrect 不正确

Fig. 8-28 Dimensioning the rough surface 毛坯面的尺寸标注
a) Correct 正确 b) Incorrect 不正确

8.6 Technical Requirements in Detail Drawings

In order to ensure the performance for part utilization, one should specify desired technical requirements for production and inspection on the detail drawing. Technical requirements in a detail drawing include surface texture, dimension tolerances, geometrical tolerances, heat treatment and surface treatment, machining and inspection requirements, and other special requirements or remarks.

Among various technical requirements, surface texture, dimension tolerances, geometrical tolerances, heat treatment and surface treatment should be specified in accordance with related technical standards when marking dimensions with relevant code (symbol), letter and word following the standard. For those that cannot be specified by dimensioning, or that are required to be represented uniformly on the drawing, they can be addressed item by item in the blank area at the bottom of the drawing.

8.6.1 Graphical symbols for surface texture

The surface texture is a significant indicator in assessing the quality of a component. It has a direct connection with the matching, wear resistance, corrosion resistance, leak proof and appearance of a component. Therefore, surface texture requirements after machining should be specified on the detail drawing.

8.6 零件图的技术要求

为了保证零件的使用性能，必须在零件图中注明零件在制造过程中应该达到的技术要求。图样上技术要求的内容包括表面结构、尺寸公差、几何公差、热处理及表面处理要求、加工及检测要求和其他特殊要求或说明。

技术要求中的表面结构、尺寸公差、几何公差、热处理及表面处理要求，应按照有关技术标准的规定，用指定的代（符）号、字母和文字注在图形上。对无法注在图形上，或需统一说明的内容，可用简明的文字逐项写在图样下方的空白处。

8.6.1 表面结构的表示法

表面结构是衡量零件表面质量的一项重要指标。它对零件的配合、耐磨性、耐蚀性、密封性和外观都有影响。因此，应在零件图上注明零件在加工后应达到的表面结构要求。

In accordance with the national standard GB/T 3505—2009, there are three principal groups of surface texture parameters that have been standardized in connection with the complete symbol and are defined as R-parameters (roughness parameters), W-parameters (waviness parameters) and P-parameters (primary profile parameters). In this book we will only discuss R-parameters in details.

1. Definition of surface roughness

No matter how smooth and flat a surface looks like after machining, it is still rough when its micro-shape is observed under a microscope, as shown in Fig. 8-29. The microscopic geometrical property formed by the small clearance of the rise and fall on the machined surface is called surface roughness. It is also called height plainness in a small area, after machining the parts.

2. Parameters of surface roughness

(1) Arithmetical mean deviation of the assessed profile Ra

Ra is the absolute arithmetical mean of profile deviated distance (between each point of the profile surface and the base line Ox) within a sample range of the part surface, as shown in Fig. 8-30.

The formula can be expressed as follows

$$Ra = 1/l \int_0^l |y(x)| \, dx \quad \text{or approximately as}$$

GB/T 3505—2009 国家标准规定了对于表面结构有要求时，涉及的轮廓参数有三类：R 参数（粗糙度轮廓参数）；W 参数（波纹度轮廓参数）；P 参数（原始轮廓参数）。本书仅就 R 参数的标注做详细介绍。

1. 表面粗糙度概念

零件表面在加工后，无论宏观看上去多么光滑平整，在显微镜下观察其微观形状，仍然是起伏不平的，如图 8-29 所示。因此，零件加工表面上具有较小间距的峰和谷所形成的微观几何形状特性称为表面粗糙度。即零件加工后，微小区域内表面高低不平的程度。

2. 表面粗糙度参数

（1）评定轮廓的算术平均偏差 Ra　Ra 是在零件表面的一个取样长度内，轮廓偏距 y（表面轮廓上点至基准线 Ox 的距离）的绝对值的算术平均值，如图 8-30 所示。

用公式表示为

$$\text{或近似为} \quad Ra = 1/n \sum_{i=1}^{n} |y_i|$$

Fig. 8-29　Micro-shape of part surface
　　　　　显微镜下的零件表面

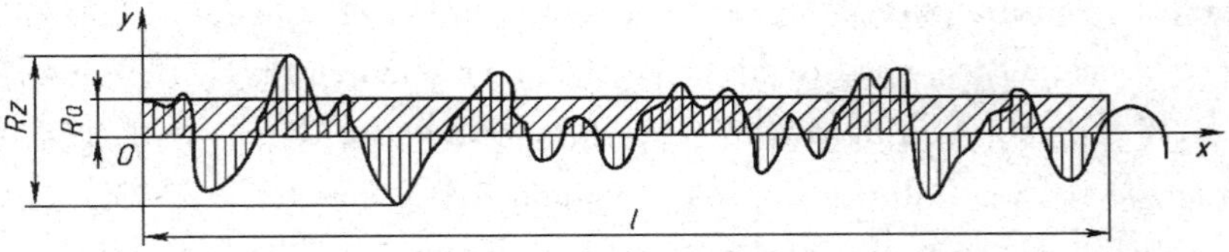

Fig. 8-30　Contour line and parameters of surface roughness
　　　　　轮廓曲线和表面粗糙度参数

(2) Maximum height of profile Rz　Rz is defined as the distance within a sample range from the peak line to the bottom line, as shown in Fig. 8-30, and the formula can be expressed as follows

$$Rz = y_{\text{峰max}} + y_{\text{谷max}}$$

When being used, the parameters of surface roughness should be selected reasonablly based on the function and application of the respective parts. In general, Ra (unit: μm) is preferably used in design and its commonly used values are given in Table 8-1.

（2）轮廓最大高度 Rz　Rz 是在一个取样长度内，轮廓峰顶线和轮廓谷底线之间的距离（见图 8-30）。用公式表示为

在使用时，应根据零件不同的作用和应用，合理选择零件的表面粗糙度参数。通常应优先选用 Ra（单位：μm），其常用数值见表 8-1。

Table 8-1　Profile arithmetic mean deviation Ra (GB/T 1031—2009)
　　　　　　　轮廓算术平均偏差 Ra 值（GB/T 1031—2009）　　　　　　　　　（μm）

Ra (Priority series 优先系列)	0.012	0.025	0.05	0.1	0.2	0.4	0.8
	1.6	3.2	6.3	12.5	25	50	100
Ra (Supplement series 补充系列)	0.008 0.160 4.0	0.010 0.25 5.0	0.016 0.32 8.0	0.020 0.50 10.0	0.032 0.63 16.0	0.040 1.00 20	0.063 1.25 32

						0.080 2.0 40	0.125 2.5 63
							80

3. Graphical symbols of surface texture

The graphical symbols and scale of surface texture are shown in Fig. 8-31. Fig. 8-32 shows mandatory positions of specifications on a graphical symbol of surface texture in cases of one single requirements and other complementary requirements. Specifications *a-e* represent the following respectively (in accordance with GB/T 131—2006/ISO 1302:2002).

3. 表面结构的图形符号

表面结构的图形符号及其数值和比例如图 8-31 所示。在完整符号中，对表面结构的单一要求和补充要求应注写在图 8-32 所示的指定位置，图中位置 *a~e* 分别注写以下内容（详细内容可查阅 GB/T 131—2006/ISO 1302:2002）。

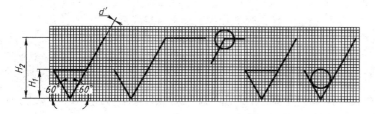

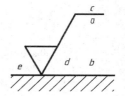

Fig. 8-31 Symbols and scale of surface texture 表面结构图形符号和比例   **Fig. 8-32** Location of specifications 注写位置

The meaning of the symbols is as following:

Position *a*—specifies one single surface texture requirement.

Positions *a* and *b*—specify two or more surface texture requirements.

Position *c*—specifies manufacturing method, surface treatment, coating, or other process.

Position *d*—specifies surface lay (such as =, ×, M, see Table 8-2);

Position *e*—specifies machining allowance (unit: mm).

表面结构符号的含义如下：

位置 *a*——注写表面结构的单一要求。

位置 *a* 和 *b*——注写两个或多个表面结构要求。

位置 *c*——注写加工方法、表面处理、涂层或其他加工工艺要求。

位置 *d*——注写表面纹理方向（如：=、×、M，见表 8-2）。

位置 *e*——注写加工余量（单位：mm）。

Table 8-2 Meaning of surface texture symbols 表面纹理符（代）号的意义

Symbol 符号	Schematic diagram 示意图	Interpretation 说明	Symbol 符号	Schematic diagram 示意图	Interpretation 说明
=		The lay is parallel to the projection plane of the view. 纹理平行于视图所在的投影面	M		The lay is multi-directional. 纹理呈多方向
⊥		The lay is perpendicular to the projection plane of the view. 纹理垂直于视图所在的投影面	R		The lay is approximately concentric. 纹理呈近似同心圆
×		The lay crosses the projection plane of the view and is in two oblique directions. 纹理呈两斜向交叉且与视图所在的投影面相交	R		The lay is approximately radial. 纹理呈近似放射状
			P		The lay is particulate, protuberant, non-directional. 纹理呈微粒、凸起，无方向

4. Specification of surface texture symbols in drawings

(1) Commonly used surface texture symbols and their meanings Commonly used surface texture symbols and their meanings are detailed in Table 8-3 and Table 8-4.

4. 表面结构符（代）号在图样中的标注

（1）常用表面结构图形符号及其含义 常用的表面结构图形符号及其含义及解释见表8-3、表8-4。

Table 8-3 Symbols of surface texture and their meanings 表面结构图形符号及其含义

Symbol 符号	Meaning 含义
∨	As a basic graphical symbol, it does not specify a manufacturing process. The symbol should not be indicated alone, without complementary. It may, however, be used for collective specification 基本图形符号，未指定工艺方法的表面。无补充符号时不能单独使用，但当通过一个注释解释时可单独使用
∨ (with bar)	As an expanded graphical symbol, it indicates that removal of material is required, such as turning, milling, drilling, grinding, shearing, polishing, eroding, electric discharge machining, or gas cutting. Only when its meaning is "treated surface", it may be used alone 扩展图形符号，用去除材料的方法获得的表面。例如：车、铣、钻、磨、剪切、抛光、腐蚀、电火花加工、气割等。仅当其含义是"被加工表面"时可单独用
∨ (with circle)	As an expanded graphical symbol, it indicates that removal of material is not permitted, such as casting, forging, forming, hot rolling, cold rolling, powder metallurgy. it is also used to keep the conditions of the previous processing step, whether this situation is removing the material or without removing the material 扩展图形符号，不去除材料的表面。例如：铸、锻、冲压变形、热轧、冷轧、粉末冶金等。也可用于表示保持上道工序形成的表面，不管这种状况是通过去除材料还是不去除材料形成的
∨ 铣	Manufacturing method: milling 加工方法：铣削
∨ M	Surface lay: the lay is multi-directional 表面纹理：纹理呈多方向
∨ ○	Indicating that the same surface texture is required on every surface around a workpiece 对投影视图上封闭的轮廓线所表示的各表面有相同的表面结构要求
3∨	Machining allowance: 3 mm 加工余量：3mm

Notes: The manufacturing methods, surface lay and machining allowance in the table are only examples.
注：这里给出的加工方法、表面纹理和加工余量仅作为示例。

Table 8-4 Meanings and interpretations of surface texture symbols 表面结构图形代号的意义及解释

Symbol 代号	Meaning and interpretation 意义及解释
∨ Rz 0.4	Indicating that material shall not be removed. An upper call out requirement. Transmission belt (default). R-parameters, maximum roughness height is 0.4μm, five sample lengths (default), 16% rule (default) 表示不允许去除材料，单向上限值，默认传输带，R参数，粗糙度的最大高度0.4μm，评定长度为5个取样长度（默认），"16%规则"（默认）
∨ Rz max 0.2	Indicating that material shall be removed. An upper call out requirement. Transmission belt (default). R-parameters, maximum roughness height is 0.2μm, five sample lengths (default), the maximum rule 表示去除材料，单向上限值，默认传输带，R参数，粗糙度最大高度的最大值0.2μm，评定长度为5个取样长度（默认），"最大规则"

(Continued 续)

Symbol 代号	Meaning and interpretation 意义及解释
√0.008-0.8/Ra 3.2	Indicating that material shall be removed. An upper call out requirement. Transmission belt is 0.008-0.8mm. R-parameters, profile arithmetical mean deviation is 3.2μm, five sample lengths (default), 16% rule (default) 表示去除材料，单向上限值，传输带0.008~0.8mm，R参数，算术平均偏差3.2μm，评定长度为5个取样长度（默认），"16%规则"（默认）
√-0.8/Ra 3 3.2	Indicating that material shall be removed. An upper call out requirement. Transmission belt is 0.8μm in accordance with GB/T 6062—2009. R-parameters, profile arithmetical mean deviation is 3.2μm, three sample lengths, 16% rule (default) 表示去除材料，单向上限值。传输带：根据GB/T 6062—2009，取样长度0.8μm。R参数，算术平均偏差3.2μm，评定长度包含3个取样长度，"16%规则"（默认）
√U Ra max 3.2 L Ra 0.8	Indicating that material shall not be removed. An upper and lower call out requirement. Transmission belt (default). R-parameters. Upper call out requirement: profile arithmetical mean deviation is 3.2μm, five sample lengths (default), the maximum rule. Lower call out requirement: profile arithmetical mean deviation is 0.8μm, five sample lengths (default), 16% rule (default) 表示不允许去除材料，双向极限值。两极限值均使用默认传输带，R参数，上限值为算术平均偏差3.2μm，评定长度为5个取样长度（默认），"最大规则"。下限值为算术平均偏差0.8μm，评定长度为5个取样长度（默认），"16%规则"（默认）

Notes: The surface texture parameters, the transmission belt / sampling length, parameter values and the selected symbols are provided in this table as examples and for references only.

注：这里给出的表面结构参数、传输带/取样长度和参数值以及所选择的符号仅作为示例。

(2) Specification of surface texture (GB/T 131—2006/ISO 1302：2002)

1) For a given surface, its roughness symbols should be marked only once on visible contours, dimension lines or the extension lines and they should be as close as possible to related dimension lines. They also can be marked by leader lines with arrows or black dots at their ends, as shown in Fig. 8-33a.

（2）表面结构要求的标注方法（GB/T 131—2006/ISO 1302：2002）

1）对一给定表面一般只标注一次表面粗糙度代号，应注在可见轮廓线、尺寸线或延长线上，并尽可能靠近有关尺寸线；也可用带箭头或黑点的指引线引出标注，如图8-33a所示。

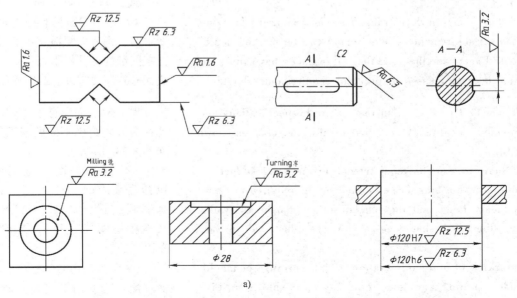

Fig. 8-33 Specification of surface texture 表面结构要求的标注方法
a) Specification of surface roughness symbol 表面粗糙度符号的注写

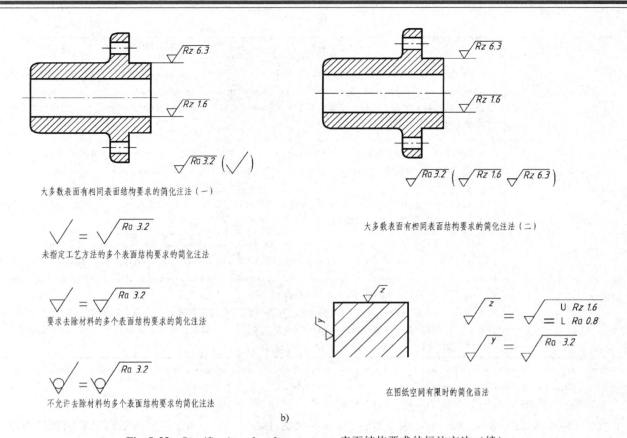

Fig. 8-33 Specification of surface texture 表面结构要求的标注方法（续）
b）Simplified specification of multiple surface textures 多个表面结构要求的简化注写

2) The tip of the surface texture symbol should point to the surface from outside and should touch the surface contours, as shown in Fig. 8-33a.

3) The surface symbol could be translated or rotated in accordance with position variation, but it should not be turned over or be distorted. The font size and direction of surface roughness value specifications should be the same as that of dimensions, as shown in Fig. 8-33a.

4) If the same surface roughness is required for some surfaces, the surface roughness requirement can be marked nearby the title block. Except for the case that all surfaces have the same requirement, individual cases of different requirements should be either marked with "（√）" in the rear with specifications collectively provided, or directly specified in the brackets, as shown in Fig. 8-33b.

5) If there are various requirements on the same surface, a thin solid line or a thin long dashed dotted line is required to represent the boundary line, and the related surface roughness symbol (code) and dimension should be shown, as illustrated in Fig. 8-34.

(3) Marking of other information The markings of machining methods, plating and machining lays are shown in Fig. 8-35.

2）表面结构图形符号的尖端应从材料外指向并接触表面，如图 8-33a 所示。

3）表面结构图形符号可随加工表面的位置不同平移或旋转，但不能翻转和变形；表面粗糙度数值中数字的大小、方向应与同一图样的尺寸数字一致，如图 8-33a 所示。

4）多个表面具有相同表面粗糙度要求时，则表面粗糙度要求可以统一标注在图纸的标题栏附近。此时，除全部表面有相同要求外，其表面结构要求的代（符）号后面应有"（√）"，或者在括号内给出不同的表面结构要求，如图 8-33b 所示。

5）同一表面上有不同的表面粗糙度要求时，需用细实线或细点画线画出其分界线，并标注相应的表面粗糙度代（符）号和尺寸（见图 8-34）。

（3）其他内容的标注 加工方法、镀（涂）覆、加工纹理方向等其他内容的标注方法如图 8-35 所示。

Fig. 8-34 Marking of multiple requirements 不同要求的标注

Fig. 8-35 Other markings 其他内容的标注

5. Selecting the surface roughness value

To reduce production cost and difficulties, the allowed surface roughness value should be as large as possible, provided that the function requirements are met. Table 8-5 shows surface features, economic machining methods and applications of typical surface roughness values.

5. 表面粗糙度数值的选用

为了降低成本，减少加工困难，在满足功能要求的情况下，表面粗糙度允许值尽可能大些。表 8-5 列出了一些典型表面粗糙度对应的表面特征、加工方法及应用举例。

Table 8-5 Ra values with their corresponding machining methods and applications

Ra 数值和与之对应的加工方法及应用举例

$Ra/\mu m$	Microcosmic surface feature 表面微观特征	Primary machining method 主要加工方法	Application 应用举例
50	Obviously visible tool marks 明显可见刀痕	Casting, forging, rough turning, rough milling, rough planing, drilling, rasping and grinding with coarse abrasion wheel 铸造、锻压、粗车、粗铣、粗刨、钻、粗纹锉刀加工和用粗砂轮加工	It is used for surfaces with the lowest roughness. It is generally used in non-working and non-contact surfaces 用于最粗糙的加工面，一般用于非工作表面和非接触表面
25	Visible tool marks 可见刀痕		
12.5	Occasional tool marks 微见刀痕	Rough turning, planing, vertical milling, plain milling, drilling 粗车、刨、立铣、平铣、钻	Non-contact and unimportant contact surfaces, such as threaded holes, chamfers and bottom surfaces of the machine base 不接触表面、不重要的接触面，如螺孔、倒角、机座底面
6.3	Visible machining marks 可见加工痕迹	Fine turning, fine milling, fine planing, reaming, boring, rough grinding 精车、精铣、精刨、铰、镗、粗磨	Contact surfaces without relative movement, such as cases, covers and sleeves. Tight contact surfaces, working surfaces of keys and keyways. Surfaces with low relative movement speed, such as brace holes, bushes and belt wheel holes 没有相对运动的零件接触面，如箱、盖、套筒；要求紧贴的表面，如键和键槽的工作表面；相对运动速度不高的接触面，如支架孔、衬套、带轮轴孔的表面
3.2	Occasional machining marks 微见加工痕迹		
1.6	Invisible machining marks 看不见加工痕迹		
0.80	Recognizable machining veins 可辨加工痕迹方向	Fine turning, fine reaming, fine broaching, fine boring, fine grinding 精车、精铰、精拉、精镗、精磨	Tight contact surfaces, such as the surfaces mounting rolling bearings and taper pin holes, contact surfaces with high-speed relative movement, such as mating surfaces of sliding bearings, and working surfaces of gear teeth 要求很好密合的接触面，如与滚动轴承配合的表面、锥销孔等；相对运动速度较高的接触面，如滑动轴承的配合表面、齿轮轮齿的工作表面
0.40	Occasional machining veins 微辨加工痕迹方向		
0.20	Invisible machining veins 看不见加工痕迹方向		
0.10	Dark varnish 暗光泽面	Grinding, polishing, super fine grinding 研磨、抛光、超精研磨	Surfaces of precision measuring tools and critical attrition surfaces, such as the inner surfaces of cylinders, the main journals of precision machine tools and jig boring machines 精密量具的表面、极重要的零件的摩擦面，如气缸的内表面、精密机床的主轴颈、坐标镗床的主轴颈
0.05	Bright varnish 亮光泽面		
0.025	Mirror shaped varnish 镜状光泽面		
0.012	Fog shaped mirror surface 雾状镜面		
0.006	Mirror 镜面		

Principles for selecting surface roughness parameters are listed as following:

1) The priority series are preferred.

2) Select the parameters according to contact conditions between the parts, matching requirements and the speed of relative movement. Generally, parameters on working surfaces are smaller than that on non-working surfaces, parameters on the moving surfaces are smaller than that on still surfaces, as shown in Table 8-6.

3) Select parameters according to production cost. Surface roughness should be as large as possible to reduce machining cost, provided design or usage requirements are met.

表面粗糙度参数值的选用原则如下:

1) 选用优先系列的数值。

2) 根据零件与零件的接触状况、配合要求、相对运动速度等来选择参数值。一般来说，工作表面比非工作表面的表面粗糙度数值小，运动表面比静止表面的表面粗糙度数值小，可参照表8-6选择。

3) 根据零件加工的经济性来选择参数值。在满足设计或使用要求的前提下，其表面粗糙度值应尽可能大，以降低加工成本。

Table 8-6　Selecting Ra for different surfaces　不同表面的 Ra 选用值

Surface condition 表面工作状况	$Ra/\mu m$	Surface condition 表面工作状况	$Ra/\mu m$
Relative movement 相对运动	0.4, 0.8, 1.6, 3.2	Non-contact 不接触	12.5, 25
Static contact 静止接触	3.2, 6.3	Non-removing material 不去除材料	Special symbol 用专用符号表示

8.6.2　Limits and fits (GB/T 1800.2—2009)

During processing, it is impossible and unnecessary to keep absolutely accurate dimensions of parts. To ensure that mounted parts maintain their mechanical properties and necessary precision, limit and fit standards have been defined. Such standards ensure the interchangeability and precision of parts. Therefore, it benefits cooperation and facilitates efficient specialized production.

1. Interchangeability of parts

In modern machine production, it is required that randomly selected parts from the same lot shall be successfully mounted to the machine and fulfill their functions without repairing. Such a property is called the interchangeability of parts. As an example, frequently used screws, nuts, bubbles and lamp holders have interchangeability. Being interchangeable benefits professional cooperation. It plays an important role in improving quality and productivity. These parts are easy to be changed and repaired.

8.6.2　极限与配合 (GB/T 1800.2—2009)

在零件加工过程中，不可能也没有必要将零件的尺寸加工得绝对准确。因此，为了保证相互接触的零件具备确定的力学性能和必要的精度，国家制定了极限与配合的标准。这些标准保证了零件的互换性和制造精度，因而既满足了生产部门广泛的协作要求，又满足了进行高效率的专业化生产要求。

1. 零件的互换性

在现代化机械生产中，要求制造出来的同一批零件，不经挑选和修配，任取一件就能顺利地装到机器上去，并满足机器性能的要求。零件的这种性能称为互换性。例如日常使用的螺钉、螺母、灯泡和灯头等都具有互换性。互换性有利于大量生产中的专业协作，对提高产品质量与生产率有着重要的作用。具有互换性的零件也便于修理和调换。

2. Basic concepts of limits and fits

(1) Nominal size   The size of the ideal part shape determined by drawing specifications, i.e., the size specified on the drawing. The nominal size of the hole and the shaft shown in Fig. 8-36 is $\phi 50$mm.

2. 极限与配合的基本概念

(1) 公称尺寸   由图样规范确定的理想形状要素的尺寸，即零件图上标注的尺寸，如图 8-36 所示的孔、轴的公称尺寸为 $\phi 50$mm。

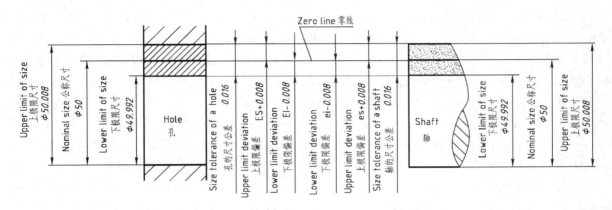

Fig. 8-36   Basic concepts of limits and fits   极限与配合的基本概念

(2) Real (integral) feature   The size obtained by measuring, such as $\phi 50.005$mm hole, and $\phi 49.995$ mm shaft.

(3) Limit size   The maximum and the minimum values of the size.

Upper limit of size: maximum size limit admissible for the feature of size, such as $\phi 50.008$ mm hole, and $\phi 50.008$mm shaft.

Lower limit of size: minimum size limit admissible for the feature of size, such as $\phi 49.992$mm hole, and $\phi 49.992$mm shaft.

(4) Deviation   It is the difference obtains by minusing its nominal size from a size. A deviation may be an upper limit deviation or a lower limit deviation. Both can be simply called limit deviations and be either positive, negative values, or zero.

Upper limit deviation code: for hole is ES, for shaft is es;
Lower limit deviation code: for hole is EI, for shaft is ei。
Upper limit deviation = Upper limit of size - Nominal size
Lower limit deviation = Lower limit of size - Nominal size

As shown in Fig. 8-36, the upper and lower limit deviations of the hole and shaft are the following

Shaft 轴 $\begin{cases} es = 50.008\text{mm} - 50\text{mm} = +0.008\text{mm} \\ ei = 49.992\text{mm} - 50\text{mm} = -0.008\text{mm} \end{cases}$

(5) Size tolerance (tolerance for short)   The allowable range for size variation is called tolerance. Tolerance is an absolute value without any sign.

(2) 实际 (组成) 要素   通过测量所得的尺寸，如孔 $\phi 50.005$mm，轴 $\phi 49.995$mm。

(3) 极限尺寸   尺寸要素允许的尺寸的两个极端。

上极限尺寸：尺寸要素允许的最大尺寸，如孔 $\phi 50.008$mm，轴 $\phi 50.008$mm。

下极限尺寸：尺寸要素允许的最小尺寸，如孔 $\phi 49.992$mm，轴 $\phi 49.992$mm。

(4) 偏差   某一尺寸减去其公称尺寸所得的代数差。尺寸偏差有上极限偏差和下极限偏差。上、下极限偏差统称为极限偏差，可以是正值、负值或零。

上极限偏差代号：孔为 ES，轴为 es；
下极限偏差代号：孔为 EI，轴为 ei。
上极限偏差 = 上极限尺寸 - 公称尺寸
下极限偏差 = 下极限尺寸 - 公称尺寸

如图 8-36 所示，轴、孔直径的上、下极限偏差为

Hole 孔 $\begin{cases} ES = 50.008\text{mm} - 50\text{mm} = +0.008\text{mm} \\ EI = 49.992\text{mm} - 50\text{mm} = -0.008\text{mm} \end{cases}$

(5) 尺寸公差 (简称公差)   它是允许尺寸的变动量。尺寸公差是一个没有符号的绝对值。

Size tolerance 尺寸公差 = |Upper limit of size 上极限尺寸 - Lower limit of size 下极限尺寸|
= |Upper limit deviation 上极限偏差 - Lower limit deviation 下极限偏差|

As shown in Fig. 8-36, the size tolerance is the following

如图 8-36 所示，其尺寸公差为

$\begin{cases} \text{Size tolerance of shaft 轴的尺寸公差} = |50.008\text{mm} - 49.992\text{mm}| = |(+0.008\text{mm}) - (-0.008\text{mm})| = 0.016\text{mm} \\ \text{Size tolerance of hole 孔的尺寸公差} = |50.008\text{mm} - 49.992\text{mm}| = |(+0.008\text{mm}) - (-0.008\text{mm})| = 0.016\text{mm} \end{cases}$

(6) Zero line Fig. 8-37 illustrates the tolerances and their matching (tolerance zone diagram) with a reference line to determine the deviation, i.e., the zero line. Generally, the zero line shows the nominal size as shown in Fig. 8-36 and Fig. 8-37.

(7) Tolerance zone In the tolerance zone diagram, the area bounded by two lines respectively representing the upper and lower limit deviations, or the upper and lower limit sizes, as shown in Fig. 8-37.

The difference between tolerance zone and tolerance is that the tolerance zone represents not only the tolerance (the size of the tolerance zone), but also the location of the tolerance relative to the zero line (location of the tolerance zone). Standards specify that the size and location of the tolerance zone are determined separately, by the standard tolerance and fundamental deviation.

3. Standard tolerance and fundamental deviation

(1) Standard tolerance standard tolerlance is any tolerance specified in the standards of limits and fits in GB/T 1800 series (See appendix Table 5-1).

Standard tolerance is represented with the symbol "IT". The grade code of the tolerance is represented with numbers and there are 20 grade codes in total: IT01, IT0, IT1, IT2, ⋯, IT18. In turn, the tolerance grade falls from IT01 to IT18. The lower the precision, the greater the tolerance will be. Tolerances of the same grade (such as IT7) with respect to a group of tolerances of nominal size are considered to have the same precision. IT01—IT11 are used for matching dimensions, while IT12—IT18 are used for non-matching dimensions. Standard tolerance values are determined by the degree of tolerance grade and nominal size, so they can be determined by consulting the appendix. For example, the tolerance grade of the $\phi 30 F8$ hole is IT8. After consulting appendix Table 5-1, one will find that the tolerance value is $33\mu m$, that is 0.033mm.

(2) Fundamental deviation Fundamental deviation is the limit deviation that determines the location of the tolerance zone relative to the zero line following the standard of limits and fits in GB/T 1800.1—2009. It can be the upper limit deviation or lower limit deviation. If the tolerance zone is under the zero line, the upper limit deviation is the fundamental deviation, as shown in Fig. 8-38.

（6）零线　在极限与配合图解（简称公差带图）中，确定偏差的一条基准直线即零线。通常零线表示公称尺寸，如图 8-36 和图 8-37 所示。

（7）公差带　在公差带图中，公差带是由代表上、下极限偏差或上、下极限尺寸的两条直线所限定的一个区域，如图 8-37 所示。

公差带与公差的区别在于，公差带既表示公差（公差带的大小），又表示公差相对于零线的位置（公差带的位置）。标准规定，公差带的大小和位置分别由标准公差和基本偏差来确定。

3. 标准公差与基本偏差

（1）标准公差　标准公差是在 GB/T 1800 系列标准极限与配合制中规定的任一公差（见附录表 5-1）。

标准公差用符号"IT"表示。数字表示公差等级，共有 20 个等级，即 IT01，IT0，IT1，IT2，⋯，IT18。从 IT01 到 IT18 公差等级依次降低，精度越低，公差值越大。一组不同公称尺寸和不同公差值对应于同一公差等级（例如 IT7）时，可以认为这些不同公称尺寸的零件具有相同精度。IT01~IT11 用于配合尺寸，IT12~IT18 用于非配合尺寸。标准公差的数值取决于公差等级和公称尺寸，可通过查附录表来确定。例如孔 $\phi 30F8$，其公差等级为 IT8，查附录表 5-1 可得公差值为 $33\mu m$，即 0.033mm。

（2）基本偏差　基本偏差是指在 GB/T 1800.1—2009 极限与配合标准中，确定公差带相对于零线位置的那个极限偏差。它可以是上极限偏差或下极限偏差；公差带在零线下方时，基本偏差为上极限偏差，如图 8-38 所示。

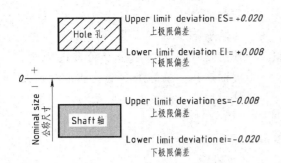

Fig. 8-37 Tolerance zone diagram 公差带图

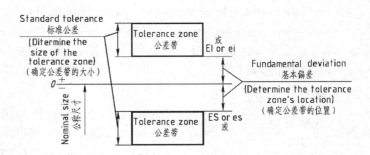

Fig. 8-38 Standard tolerance and fundamental deviation
标准公差与基本偏差

Fundamental deviation code: In the case of holes, it is represented with capital letters A, B, C, ···, ZC. In the case of shafts, it is represented with minuscule a, b, c, ···, zc. Both the upper and lower deviations have 28 grades to form fundamental deviation series, as shown in Fig. 8-39.

基本偏差代号：对孔用大写字母 A、B、C···ZC 表示，对轴用小写字母 a、b、c···zc 表示，各 28 个，形成基本偏差系列，如图 8-39 所示。

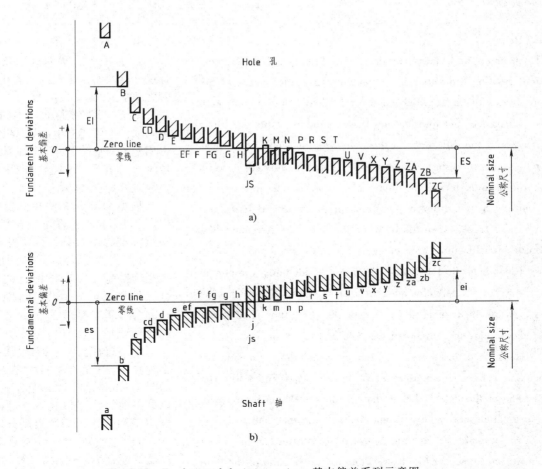

Fig. 8-39 Fundamental deviation series 基本偏差系列示意图
a) Fundamental deviations of holes 孔的基本偏差   b) Fundamental deviations of shafts 轴的基本偏差

(3) Determining the tolerance zone of the hole and shaft The sketch map of fundamental deviation series (See Fig. 8-39) shows only the location of tolerance, but not the size of the tolerance zone. One side of the tolerance zone close to the zero line is fundamental deviation. The other side is open, the deviation of which can be further determined with the fundamental deviation and standard tolerances of the hole and shaft, respectively.

(3) 孔、轴公差带的确定 在基本偏差系列示意图（见图8-39）中，只表示了公差的位置，没有表示公差带的大小。公差带中靠近零线的一端是基本偏差。另一端是开口的，其偏差取决于所选标准公差的大小，可根据孔、轴的基本偏差和标准公差算出。

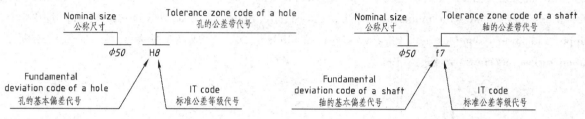

(4) Tolerance zone code of the hole and shaft It is composed of the fundamental deviation code and the stasndard tolerance grade code. Here is an example:

(4) 孔、轴的公差带代号 孔轴的公差带代号由基本偏差代号和标准公差等级代号组成。例如：

4. Fits

(1) Concept and classification of fits Fits establish the tolerance zones relationship between the matched tolerance zones of the hole and shaft with the same nominal size. When fitting a shaft with a hole, clearance or interference may appears, due to their different sizes. If the difference of the hole size and shaft size is positive (hole diameter > shaft diameter), clearance appears. If it is negative (hole diameter < shaft diameter), interference appears.

According to the different functions of two matched parts and the tightness or looseness of fits between the hole and the shaft, the state has divided fits into three categories, as follows:

1) Clearance fit. Fit with clearance (including minimum clearance zero) is called clearance fit, as shown in Fig. 8-40a. The tolerance zone of the hole is above that of the shaft. It is used where there is relative movement between the hole and the shaft.

2) Interference fit. Fit with interference (including minimum interference zero) is called interference fit, as shown in Fig. 8-40b. The tolerance zone of the hole is below that of the shaft. It is used where there is no relative movement between the hole and the shaft.

3) Transition fit. Fit having clearance or interference is called transition fit, as shown in Fig. 8-40c. The tolerance zone of the hole overlaps that of the shaft. It is used where there is a high neutral requirement.

4. 配合

(1) 配合的基本概念及种类 公称尺寸相同的、相互结合的孔与轴公差带之间的关系称为配合。孔与轴配合时，由于二者的实际尺寸不同，可能产生间隙，也可能产生过盈。孔的尺寸减去相配合的轴的尺寸所得的代数差为正数时（即孔径>轴径），产生间隙；为负数时（即孔径<轴径），产生过盈。

根据两配合零件在机器中所起的作用和孔与轴之间的配合松紧不同，国家标准将配合分为以下三类：

1) 间隙配合。具有间隙（包括最小间隙等于零）的配合称为间隙配合，如图8-40a所示。间隙配合中，孔的公差带在轴的公差带之上。间隙配合用于轴、孔有相对运动处。

2) 过盈配合。具有过盈（包括最小过盈等于零）的配合称为过盈配合，如图8-40b所示。过盈配合中，孔的公差带在轴的公差带之下。过盈配合一般用于轴、孔无相对运动处。

3) 过渡配合。可能具有间隙或过盈的配合称为过渡配合，如图8-40c所示。过渡配合中，孔的公差带与轴的公差带相互重叠。过渡配合一般用于对中性要求较高处。

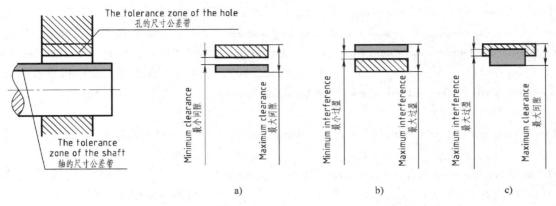

Fig. 8-40 Fit and its classification 配合及其种类示意图
a) Clearance fit 间隙配合 b) Interference fit 过盈配合 c) Transition fit 过渡配合

(2) Fit system In order to standardize the fit, reduce the quantities and specifications of fixed value cutting tools and measuring instruments, national standards proposes two systems of fits, i. e., the hole-basis system of fits and the shaft-basis system of fits.

1) Hole-basis system of fits. A system formed by the tolerance zone of the hole, with definite fundamental deviation, and that of the shaft with different fundamental deviations, as shown in Fig. 8-41. The hole of the hole-basis system is a basic hole, represented with H, the lower limit deviation of which is zero.

2) Shaft-basis system of fits. A system formed by the tolerance zone of the shaft with definite fundamental deviation, and that of the hole with different fundamental deviations, as shown in Fig. 8-42. The shaft of the shaft-basis system is a basic shaft, represented with h, the upper limit deviation of which is zero.

The type of fits can be determined by the fundamental deviation code of the shaft and hole: In the hole-basis system (or shaft-basis system) of fits, the fundamental deviations A—H (or a—h) are used in clearance fits. The fundamental deviations J—N (or j—n) are used in transition fits. The fundamental deviations P—ZC (or p—zc) are used in interference fits.

（2）配合制　为了实现配合的标准化，减少定值刀具、量具的规格和数量，方便生产，国家标准规定了基孔制配合和基轴制配合两种制度。

1）基孔制配合。基本偏差一定的孔的公差带与不同基本偏差的轴的公差带形成各种配合的一种制度，称为基孔制配合，如图 8-41 所示。基孔制的孔称为基准孔，用代号 H 表示，其下极限偏差为零。

2）基轴制配合。基本偏差一定的轴的公差带与不同基本偏差的孔的公差带形成各种配合的一种制度，称为基轴制，如图 8-42 所示。基轴制的轴称为基准轴，用代号 h 表示，其上极限偏差为零。

根据轴和孔的基本偏差代号可确定配合种类：在基孔制（或基轴制）配合中，基本偏差 A~H（或 a~h）用于间隙配合；基本偏差 J~N（或 j~n）一般用于过渡配合；基本偏差 P~ZC（或 p~zc）用于过盈配合。

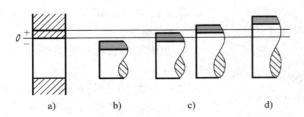

Fig. 8-41 Hole-basis system of fits 基孔制配合
a) Basic hole 基准孔 b) Clearance fit 间隙配合
c) Transition fit 过渡配合 d) Interference fit 过盈配合

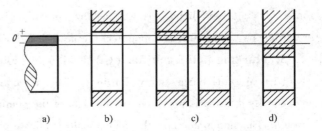

Fig. 8-42 Shaft-basis system of fits 基轴制配合
a) Basic shaft 基准轴 b) Clearance fit 间隙配合
c) Transition fit 过渡配合 d) Interference fit 过盈配合

(3) Principles of selecting a fit system

1) The hole-basis system of fit is preferred since it is usually more difficult to machine holes than machine shafts for the same degree of the tolerance. It reduces the number of fixed-value cutting tools and the cost.

2) The shaft-basis system of fits is used in the following situations:

① The shaft is made of cold drawing bar stocks without machining.

② The shafts of the same nominal size form different fits with different holes as shown in Fig. 8-43a.

③ The fit between the external ring of the rolling bearing and the bearing housing adopts shaft-basis system of fits, for a rolling bearing is a standard part, as shown in Fig. 8-43b.

（3）配合制的选择原则

1）一般情况下优先选用基孔制配合，因为加工相同公差等级的孔比轴要困难，而且减少了加工孔的定值刀具数量，从而能降低成本。

2）基轴制配合用于以下三种情况：

①用冷拉棒料做轴，不需要再进行加工。

②同一公称尺寸的轴与几个孔组成不同的配合，如图8-43a所示。

③滚动轴承的外圈与轴承座孔的配合采用基轴制配合，因为滚动轴承是标准件，如图8-43b所示。

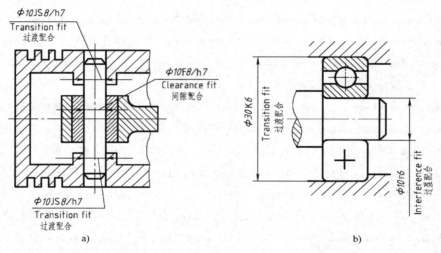

Fig. 8-43 Application of the shaft-basis system of fits　基轴制配合的应用

a) Different fits on the same shaft　同一轴上有不同配合

b) The fits between the bearing, shaft and hole　轴承与轴和孔的配合

(4) Preferable and common fits (GB/T 1801—2009) Since there are 20 standard tolerance grades and 28 fundamental deviations for shafts and holes respectively, there are a lot of possible combinations of fits. Too many tolerance zones and fits may be not economical and convenient for production. Therefore, the national standard indicates the general, common and preferred tolerance zones for shafts and holes, as well as the corresponding generally preferred fits, as shown in appendices Table 5-4 and Table 5-5.

5. Marking limits and fits (GB/T 4458.5—2003)

(1) Marking on assembly drawings The fit code marked on an assembly drawing is presented on the right of the nominal size as a formula. The numerator and the denominator indicate the tolerance zone codes of the hole and shaft, the format of which is as follows:

（4）优先、常用配合（GB/T 1801—2009）　由于标准公差等级有20个，孔、轴的基本偏差各有28个，这样可以组成大量的配合。但是公差带及配合的数量太多，既不经济也不便于生产。为此，国家标准规定了一般、常用和优先选用的孔、轴公差带以及相应的优先常用配合，见附录表5-4、表5-5。

5. 极限与配合的标注方法（GB/T 4458.5—2003）

（1）装配图中的标注　在装配图中标注的配合代号，是在公称尺寸右边以分式的形式注出的。分子和分母分别为孔和轴的公差带代号，其标注格式为：

$$\text{Nominal size}\frac{\text{Tolerance zone code of the hole}}{\text{Tolerance zone code of the shaft}} \quad \text{or} \quad \text{Nominal size}$$

$$公称尺寸\frac{孔的公差带代号}{轴的公差带代号} \quad 或 \quad 公称尺寸\frac{孔的基本偏差代号\ \ 标准公差等级代号}{轴的基本偏差代号\ \ 标准公差等级代号}$$

$$\frac{\text{Fundamental deviation code of the hole IT code}}{\text{Fundamental deviation code of the shaft IT code}}$$

A system of fits can be determined according to the fit code. If the fundamental deviation code in the numerator is H, the hole is the basic hole and the fit between the hole and the shaft follows the hole-basis system of fits, as shown in Fig. 8-44a. If the fundamental deviation code in the denominator is h, the shaft is the basic shaft and the fit between the hole and the shaft is the shaft-basis system of fits. Fig. 8-44b shows the clearance fit in which the no-minal size is $\phi12$mm, the denominator is the basic shaft code h and the tolerance grade IT7, the numerator is basic deviation code F of the hole and tolerance grade IT9, so it is expressed as $\phi12$F9/h7. The other one is $\phi12$JS8/h7 which is a transition fit.

根据配合代号的标注可以确定配合制。若分子中的基本偏差代号为 H，则孔为基准孔，轴、孔的配合一般为基孔制配合，如图 8-44a 所示。若分母中的基本偏差代号为 h，则轴为基准轴，轴、孔配合一般为基轴制配合。图 8-44b 中公称尺寸为 $\phi12$mm，分母为基准轴的代号 h 及标准公差等级 IT7，分子为孔的基本偏差代号 F 及标准公差等级 IT9，故为间隙配合，其注写形式为 $\phi12$F9/h7。另一处 $\phi12$JS8/h7 则为过渡配合。

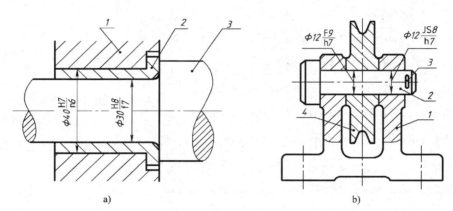

Fig. 8-44 Marking the fit code 配合代号的标注
a) 1—Base 支座 2—Sleeve 衬套 3—Shaft 轴
b) 1—Base 支座 2—Small shaft 小轴 3—Cotter pin 开口销 4—Belt wheel 带轮

（2）Marking on detail drawings　There are three ways to mark tolerances on detail drawings.

1) Marking code. Mark the nominal size and the tolerance zone code, as $\phi30$H8 and $\phi30$f7 shown in Fig. 8-45a (for massive production).

（2）在零件图上的标注　在零件图上标注尺寸公差有三种方式。

1) 代号注法。注出公称尺寸及公差带代号，如图 8-45a 中的 $\phi30$H8 和 $\phi30$f7（用于大批量生产）。

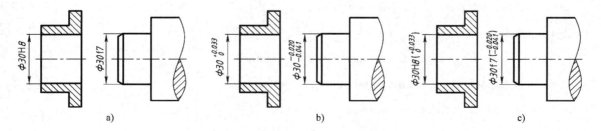

Fig. 8-45 Marking the tolerance on the detail drawing 零件图上极限与配合的标注
a) Marking tolerance zone code 标注公差带代号　b) Marking limit deviation value 标注极限偏差值
c) Marking tolerance zone code and limit deviation value 标注公差带代号及极限偏差值

2) Marking values. Mark the nominal size and the upper and lower limit deviations, for example $\phi 30^{+0.033}_{0}$ and $\phi 30^{-0.020}_{-0.041}$ shown in Fig. 8-45b (for a small lot production).

3) Comprehensive marking. Mark the nominal size with tolerance zone code and the upper and lower limit deviations of the tolerance zone. Racket shall be added to the limit deviation value as $\phi 30H8(^{+0.033}_{0})$ and $\phi 30f7(^{-0.020}_{-0.041})$ shown in Fig. 8-45c (for a periodic production).

Rules for the marking are as following:

① The text height of the limit deviation value shall be one grade smaller than that of the nominal size. Upper and lower limit deviation values should be separately marked on the top right and bottom right corners of the nominal size, expressed in mm and kept consistently with the bottom line of the nominal size value.

② Decimals in the upper and lower limit deviations should be aligned, after which digits should be the same.

③ If the upper or lower limit deviation is zero, it should still be marked and aligned to the digits on the left of the decimal in the other deviation. For positive or negative limit deviations, "+" or "-" must be marked.

④ If the upper and lower limit deviation is equal, the deviation value should be in the same height as the nominal size. And "±" should be marked between the nominal size and the deviation value, like $\phi 50\pm 0.031$.

8.6.3 Geometrical tolerances (GB/T 1182—2008)

Permitted variation between real (shape and position) feature and nominal (shape and position) feature is called geometrical tolerance. For high-precision parts, it is necessary to maintain not only their dimensional tolerances, but also their geometrical (shape and position) tolerances.

For a general part, its geometrical tolerance may be maintained by its dimension tolerance and machining precision. For parts with high precision requirements, it is necessary to specify the geometrical tolerances in the detail drawing in accordance with the design requirements.

2) 数值注法。注出公称尺寸及上、下极限偏差，如图8-45b中所示的 $\phi 30^{+0.033}_{0}$ 和 $\phi 30^{-0.020}_{-0.041}$（用于单件小批量生产）。

3) 综合注法。注出公称尺寸及公差带代号，同时注出公差带的上、下极限偏差，偏差值要加上括号，如图8-45c中的 $\phi 30H8(^{+0.033}_{0})$ 与 $\phi 30f7(^{-0.020}_{-0.041})$，（用于周期性生产）。

这种标注方法有如下规定：

① 极限偏差数值字高比公称尺寸字高小一号。上、下极限偏差数值以mm为单位，分别写在公称尺寸的右上、右下角，并与公称尺寸数字底线平齐。

② 上、下极限偏差数值中的小数点要对齐，其后面的位数也应相同。

③ 上、下极限偏差数值中若有一个为零时，仍应注出，并与另一个偏差小数点左面的个位数对齐。偏差为正或负时，"+" "-"号也必须写出。

④ 若上、下极限偏差相同，偏差值数字与公称尺寸数字的字高相同，并在公称尺寸与偏差数值之间标出"±"号，如 $\phi 50\pm 0.031$。

8.6.3 几何公差（GB/T 1182—2008）

实际（形状和位置）要素对公称（形状和位置）要素所允许的变动量称为几何公差。机器中某些精确度较高的零件，不仅需要保证其尺寸公差，而且还要保证其几何（形状和位置）公差。

对一般零件来说，它的几何公差，可由尺寸公差、加工机床的精度等来保证。对精度要求较高的零件，则需根据设计要求在零件图上注出有关的几何公差。

1. General concepts of the geometrical tolerance

There are many factors affecting the quality of a part, such as its form and structure position, not just the size.

(1) Form error Form error is the variation of the true form of a single factor to its ideal element. Make a small shaft according to Fig. 8-46a, the body of which is bent and is not cylindrical. The inaccuracy of the form is called a form error, namely, the variation between the actual axis and an ideal axis, i. e. , a kind of straightness error. While producing a cylindrical pin, it may occurs that the generatrixes of the cylinder are not straight lines so that the cylinder has a form of being thick in the middle part while thin at both ends, instead of a perfect cylinder, as shown in Fig. 8-46b. In addition to the above-mentioned situation, the straight line, plane, circle, contour line, contour plane deviating from their ideal elements, are also form errors.

(2) Location error Location error is the variation of the actual position of related factors to its ideal element. While machining a stepped shaft, the axial line of each step may not be the same, as shown in Fig. 8-46c, namely, coaxiality error. Surfaces expected to be perpendicular are not perpendicular after machined, as shown in Fig. 8-46d, this error is, perpendicularity error. In addition to the above-mentioned condition, geometrical elements' perpendicularity, parallelism, obliquity, symmetry deviating from their ideal elements, are also belong to position errors.

1. 几何公差的基本概念

评定零件质量的因素是多方面的，不仅零件的尺寸影响零件质量，零件的几何形状和结构的位置也大大影响零件的质量。

（1）形状误差 单一要素的实际形状对其理想要素的变动量称为形状误差。按图 8-46a 所示制造一个小轴，其形体弯曲，不是准确的直圆柱形。这种形状上的不准确就是形状误差，即实际轴线与理想轴线之间的变动量——直线度误差；加工圆柱销时可能出现因素线不直，而导致中间粗、两端细的情况，如图 8-46b 所示。除上述情况外，直线、平面、圆、轮廓线和轮廓面偏离理想要素的情况，都属于形状误差。

（2）位置误差 相关要素间的实际位置相对其理想要素的变动量称为位置误差。例如加工台阶轴时，可能出现各段圆柱轴线不在同一直线上的情况，如图 8-46c 所示，即属同轴度误差。要求垂直的表面加工后不垂直，如图 8-46d 所示，即为垂直度误差。除上述情况外，几何要素的相互垂直、平行、倾斜、对称等对理想要素的偏离情况，均属于位置误差。

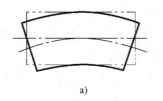 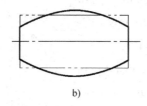 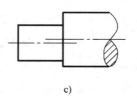 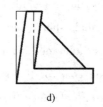

a)　　　　　　　　　　b)　　　　　　　　　　c)　　　　　　　　　　d)

Fig. 8-46 Form and location errors　形状和位置误差

a) Straightness error　直线度误差 b) Cylindricity error　圆柱度误差

c) Coaxiality error　同轴度误差 d) Perpendicularity error　垂直度误差

2. symbols of geometrical tolerances

(1) The types, geometrical tolerances and symbols Geometrical tolerance The national standard GB/T 1182—2008 specifies the types and geometric features of geometrical tolerances and their symbols, as shown in Table. 8-7.

2. 几何公差的表示

（1）类型、几何特征及符号
国家标准 GB/T 1182—2008 规定了几何公差的类型、几何特征及其符号，见表 8-7。

Table 8-7 Geometric features of geometrical tolerances and their symbols (From GB/T 1182—2008)
几何公差的几何特征和符号（摘自 GB/T 1182—2008）

Type of tolerance 公差类型	Geometric feature 几何特征	Symbol 符号	Datum needed 有无基准	Type of tolerance 公差类型	Geometric feature 几何特征	Symbol 符号	Datum needed 有无基准
Tolerance of form 形状公差	Straightness 直线度	—	No 无	Tolerance of orientation 方向公差	Parallelism 平行度	∥	Yes 有
	Flatness 平面度	▱	No 无		Perpendicularity 垂直度	⊥	Yes 有
	Roundness 圆度	○	No 无		Angularity 倾斜度	∠	Yes 有
	Cylindricity 圆柱度	⌭	No 无	Tolerance of location 位置公差	Concentricity and Coaxiality 同心度和同轴度	◎	Yes 有
					Symmetry 对称度	=	Yes 有
					Ture position 位置度	⊕	Yes or no 有或无
Tolerance of form, orientation or location 形状、方向或位置公差	Profile of a line 线轮廓度	⌒	Yes or no 有或无	Tolerance of run-out 跳动公差	Circular run-out 圆跳动	╱	Yes 有
	Profile of a surface 面轮廓度	⌓	Yes or no 有或无		Total run-out 全跳动	╱╱	Yes 有

(2) Code of geometrical tolerance Marking the geometrical tolerance in a detail drawing is a requirement for the designer, in order to control the form errors and location errors some factors. To illustrate these technical requirements, national standards specify that symbols should be reflected in the drawing. The drafting of geometrical tolerance feature symbols is shown in Fig. 8-47a.

The code of the geometrical tolerance is represented by the geometrial tolerance box. The geometrial tolerance box contains some grid, in which the characteristic symbol, tolerance value, datum symbol are marked, as shown in Fig. 8-47b. The number of small grids can be increased or decreased according to the requirements and the item's characters.

(3) Boxes of tolerance Tolerances are written in the rectangular grids which are divided into two or more parts. Each grid is filled with the following contents in order from left to right (see Fig. 8-47b):

（2）几何公差的代号 在零件图上给出几何公差，是设计人员为控制该零件上某些要素的形状或位置误差而提出的一种技术要求。为表达这些技术要求，按国家标准规定，应在图上用代号标注。几何公差特征符号的画法如图8-47a所示。

几何公差代号用几何公差框格来表示。几何公差框格由若干个小格组成，并在相应的小格中标出几何公差特征符号、公差值、基准符号等，如图8-47b所示。根据几何特征和要求，框格中的小格可以增减。

（3）公差框格 公差要求注写在划分成两格或多格的矩形框格内，各格自左至右顺序标注以下内容（见图8-47b）：

① First field—symbol of geometric feature.

② Second field—tolerance value, if the tolerance zone is a circle or a cylinder, symbol "ϕ" should be added before the tolerance value; If the tolerance zone is a ball, "$S\phi$" should be added.

③ Third field—datum, with a single letter representing a single datum while several letters representing a datum system or a public datum.

See Table 8-8 for additional symbols which are used in geometrical tolerance.

① 第一格——几何特征符号。

② 第二格——公差值，如果公差带为圆形或圆柱形，公差值前应加注符号"ϕ"；如果公差带为圆球形，公差值前应加注符号"$S\phi$"。

③ 第三格——基准，用一个字母表示单个基准或用几个字母表示基准体系或公共基准。

几何公差的附加符号见表 8-8。

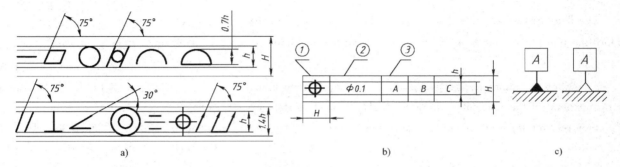

Fig. 8-47 Dimension and datum of geometric tolerance symbols and frame
几何公差符号、框格的尺寸及基准

Table 8-8 Additional symbols (From GB/T 1182—2008)
附加符号（摘自 GB/T 1182—2008）

Description 说明	Symbol 符号	Description 说明	Symbol 符号
Toleranced feature 被测要素		Datum feature 基准要素	
Datum target 基准目标	$\dfrac{\phi 2}{A1}$	Theoretically exact dimension 理论正确尺寸	50
Projected tolerance zone 延伸公差带	Ⓟ	Maximum material requirement 最大实体要求	Ⓜ
Least material requirement 最小实体要求	Ⓛ	Free status condition (Non-rigid parts) 自由状态条件（非刚性零件）	Ⓕ
All around profile 全周（轮廓）		Envelope requirement 包容要求	Ⓔ
Common tolerance zone 公共公差带	CZ	Reciprocity requirement 可逆要求	Ⓡ

(4) Datum The datum related to the measured geometric element is indicated by an uppercase letter. The letter is written in a small square which is connected to a blackened or blank triangle, indicating the datum, as shown in Fig. 8-47c. In order to avoid misunderstanding, letters such as E, I, J, M, O, P are not adopted.

(4) 基准 与被测要素相关的基准用大写字母表示，字母注在小方格内，与涂黑或空白的三角形相连以表示基准，如图 8-47c 所示。一般不采用 E、I、J、M、O、P 等字母，以免引起误解。

3. **Methods for marking geometrical tolerance**

(1) Marking toleranced feature The measured feature and the tolerance box are connected by a guide-line, which comes from either side of the tolerance box and has an arrow at the other end. The marking methods are as follows:

1) When the tolerance involves in contour lines or profile faces, the arrow should point to the contour line or its extension (staggered with the dimension lines), as shown in Fig. 8-48a and Fig. 8-48b.

2) The arrow can also point to the horizontal line of the guide-line that comes out from the measured surface, as shown in Fig. 8-48c.

3) When the tolerance involves in the centerline, center plane or center point of the element, its arrow should be on the extension of the corresponding dimension line, as shown in Fig. 8-48d and Fig. 8-48e.

(2) Marking datum feature A datum feature is marked by a datum symbol. The letter of the datum shall be marked horizontally in the tolerance box no matter how the direction of the datum on the drawing varies. The datum triangle with the datum letter should be placed according to the following rules:

1) If the datum feature is a contour line or profile surface, the datum triangle should be placed on the contour line or its extension (staggered with the dimension lines), see Fig. 8-48f.

2) When the datum is an axis, a center plane or a center point specified by a dimension line, the datum triangle should be placed on the extension of the dimension line, see Fig. 8-48g.

3) The datum triangle can also be directly placed on the horizontal line of the guide-line from a profile surface, see Fig. 8-48h.

4) If there is no enough space to place the two arrows of the dimension line of the datum element, one of them can be replaced by the datum triangle, see Fig. 8-48i.

5) If only a part of an element is set as the datum, this part should be indicated by thick long dashed dotted lines and dimensioned, see Fig. 8-48j.

6) When the datum is a single element, it is represented by a single uppercase letter, see Fig. 8-48k. When two elements work together as a public datum, it is represented by two uppercase letters connected by a hyphen, see Fig. 8-48l. If a public datum has two or more elements, the uppercase letters should be arranged in the order of priorities from left to right and placed in grids, see Fig. 8-48m.

3. 几何公差的标注方法

(1) 被测要素的标注 用指引线连接被测要素和公差框格。指引线引自框格的任意一侧，终端带一个箭头。具体标注方法如下：

1) 当公差涉及轮廓线或轮廓面时，箭头指向该要素的轮廓线或其延长线（应与尺寸线明显错开），如图8-48a、b所示。

2) 箭头也可指向引出线的水平线，引出线引自被测面，如图8-48c所示。

3) 当公差涉及要素的中心线、中心平面或中心点时，箭头应位于相应尺寸线的延长线上，如图8-48d、e所示。

(2) 基准要素的标注 基准要素由基准代号表示，表示基准的字母应在公差框格内注明。无论基准要素在图中的方向如何变化，其方格中的字母一律应水平书写。带基准字母的基准三角形应按如下规定放置：

1) 当基准要素为轮廓线或轮廓面时，基准三角形放置在要素的轮廓线上或其延长线上（与尺寸线明显错开），如图8-48f所示。

2) 当基准是尺寸要素确定的轴线、中心平面或中心点时，基准三角形应放置在该尺寸线的延长线上，如图8-48g所示。

3) 基准三角形也可直接放在该轮廓面引出线的水平线上，如图8-48h所示。

4) 如果没有足够的位置标注基准要素尺寸的两个箭头，则其中的一个箭头可用基准三角形代替，如图8-48i所示。

5) 如果只以要素的某一局部作基准，则应用粗点画线示出该部分并加注尺寸，如图8-48j所示。

6) 单个要素做基准时，用一个大写字母表示，如图8-48k所示。以两个要素建立公共基准时，用中间加连字符的两个大写字母表示，如图8-48l所示。以两个或三个要素建立公共基准体系时，表示基准的大写字母按基准的优先顺序自左至右填写在各框格内，如图8-48m所示。

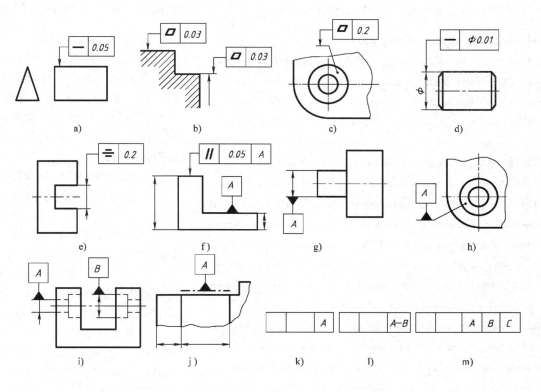

Fig. 8-48 Marking geometrical tolerances 几何公差的标注

(3) An example for marking geometrical tolerances The air valve handle shown in Fig. 8-49 is a typical example for marking geometrical tolerances. In this example, the geometrical tolerances marked are as following: cylindricity of the handle body (ϕ16mm) is 0.005mm; the coaxiality of the axis of the threaded hole (M8 × 1) to the axis (ϕ16mm) is ϕ0.1mm; and the circular run-out of the spherical surface (SR750mm) to the axis (ϕ16mm) is 0.003mm.

（3）几何公差标注示例 以图8-49所示的气门阀杆作为几何公差标注的典型示例。其中所注各几何公差的含义为：杆身ϕ16mm的圆柱度公差为0.005mm；M8×1的螺孔轴线对于杆身ϕ16mm轴线的同轴度公差为ϕ0.1mm；SR750mm的球面对于ϕ16mm轴线的圆跳动公差为0.003mm。

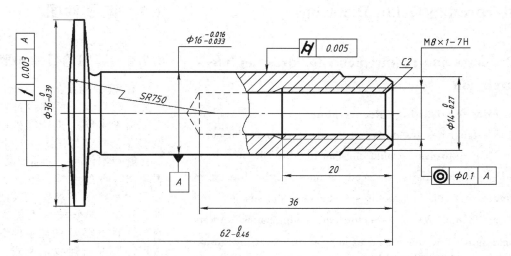

Fig. 8-49 Citing an instance of marking geometrical tolerances 几何公差标注示例

8.6.4 Marking other technical requirements

1. Marking the material of a part

Materials vary in accordance with various part functions. Therefore, in the frame with "material mark" of the title block, it is required to mark the item number of part material. See appendix 6 for descriptions, related standards, item numbers, characteristics and utilization examples of commonly used materials.

2. Marking heat treatment and surface treatment

Heat treatment is a method of altering metal characteristics. When a part is required to be heat-treated, one can briefly describe it in the technical requirement. See Fig. 8-4 as an example marked in the drawing of the pump shaft as "Quenched and tempered 26—31HRC". When partial heat treatment is required, one can specify its range with a thick long dashed dotted line, and mark necessary dimensions and descriptions, as shown in Fig. 8-50a.

Surface treatments mainly include plating, painting, chemical coating and so on. In general, one can briefly describe them in the technical requirements. For surfaces that do not require coating, one can mark them with "∇", as shown in Fig. 8-50b.

Refer to Appendix 7 for codes, descriptions and applications of commonly used heat treatment and surface treatment.

8.6.4 其他技术要求的注写

1. 零件的材料标注

零件的作用不同，所使用的材料也各不相同。因此，必须在零件图标题栏的"材料标记"栏中注明该零件所选用的材料牌号。常用材料的名称、标准、牌号、性能与应用举例，如附录6所示。

2. 热处理和表面处理标注

热处理是用来改变金属性能的一种方法。当零件全部需要进行热处理时，可在技术要求中用文字简要说明如图8-4泵轴零件图中的"调质处理26～31HRC"。当需要局部热处理时，可用粗点画线画出其范围，并标注相应的尺寸及说明，如图8-50a所示。

表面处理主要指电镀、涂漆、化学镀等，一般可在技术要求中用文字简要说明。对不需要镀涂的表面，可用符号"∇"表示，如图8-50b所示。

常用的热处理和表面处理代号、名词解释及应用举例，如附录7所示。

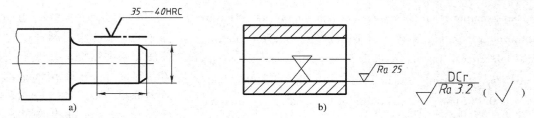

Fig. 8-50 Marking of heat treatment and surface treatment 热处理及表面处理的标注
a) Marking of heat treatment 热处理的标注 b) Marking of surface treatment 表面处理的标注

8.7 Interpreting Detail Drawings

8.7.1 Purposes and requirements for drawing interpretation

1. Purposes for drawing interpretation

When a designer designs parts, he needs to refer to the pertinent machine part drawings, so that one can interpret the drawings. When a worker produces parts, he needs to interpret the drawing in order to figure out their structures, shapes, dimensions and technical requirements. When inspectors and maintenance workers inspect or maintain parts, they need to interpret the drawings to assess whether the parts meet technical requirements. In short, all technicians should be able to interpret detail drawings.

8.7 读零件图

8.7.1 读图的目的和要求

1. 读图的目的

工程设计人员在设计零件时，经常需要参考同类机器零件的图样，这就需要会看零件图。生产制造技术人员在制造零件时，也需要看懂零件图，想象出零件的结构、形状，了解各部分尺寸及技术要求等，以便加工出零件。检验、维修技术人员在检验或维修零件时也需要查看零件图，以判断该零件是否达到技术要求。总之，从事各种工程技术专业工作的技术人员，都必须具备读零件图的能力。

2. Requirements for interpreting part drawings

1) Understanding the name, material, and functions (including functions of each section) of a part.

2) Understanding the structural shape of each section of the part.

3) Understanding the dimensioning and technical requirements. One also needs to analyze and understand the representations of various views.

8.7.2 Methods and steps to read detail drawings

1. Synoptically understanding

Read the title block to understand the name, material, scale, etc.

2. Analyzing views and figuring out the part structure of the part

Analyze the representation of the detail drawings and the projective relationship between the views, such as selected views, location of the cut plane and projection direction. Figure out the external and internal structure, and shape of the part, using the projection relation according to the shape analysis method. The basic method to analyze views is, to read the primary section first, and then the secondary section; to read the easier section first, and then the more difficult one; to read it as a whole first and then in detail.

3. Analyzing dimensions and understanding technical requirements

Determine dimension datum for each direction and understand the size dimensions, location dimensions and general dimensions. Understand the tolerance of the fitting surface, geometrical tolerance, roughness of each surface and other requirements.

4. Comprehensive imagination

Integrate the structures, shapes, dimensions, and the technical requirements to figure out the complete picture of the part. In this way, the drawing has been interpreted.

5. An example of drawing interpretation

Let's take the valve shown in Fig. 8-51 as an example to introduce the common method and steps of interpreting drawings.

2. 读零件图的要求

1) 对零件的名称、材料和功用（包括各组成形体的作用）要有所了解。

2) 读懂零件各部分的结构形状。

3) 能理解图中的尺寸注法和技术要求，并且能够分析和了解零件的视图表达方案。

8.7.2 读零件图的方法和步骤

1. 概括了解

看零件图的标题栏，了解零件的名称、材料、绘图比例等。

2. 分析视图并想象零件形状

分析零件图采用的表达方法及各视图之间的投影关系，如选用的视图、剖切面位置及投射方向等，按照形体分析等方法，利用各视图的投影对应关系，想象出零件的内外结构、形状。分析视图的基本方法是：先看主要部分，后看次要部分；先看易懂的部分，后看难的部分；先看整体，后看细节。

3. 分析尺寸并了解技术要求

确定各方向的尺寸基准，了解各部分的定形尺寸、定位尺寸及总体尺寸。了解各配合表面的尺寸公差，以及有关的几何公差、各表面的粗糙度要求及其他要达到的要求等。

4. 综合想象零件全貌

将看懂的零件的结构、形状、所注尺寸以及技术要求等内容综合起来，想象出零件的全貌，这样就能看懂一张零件图。

5. 读零件图举例

以图 8-51 所示阀体零件图为例，介绍读零件图的一般方法和步骤。

(1) Synoptically understanding Find out the name of the part in the title block. It is used as a machine component for containing, supporting and sealing of other components.

The material is cast steel (ZG230—450), after casting and aging treatment it is made by cutting.

The scale is 1∶2, that is to say, the material object is two times bigger, compared to the size shown on the detail drawing.

（1）概括了解 从标题栏中可知该零件的名称为阀体。其用途是容纳、支承、封闭其他零件等。

阀体材料是铸钢（ZG230-450），它是经铸造成形、时效处理后，再切削加工而成。

画图比例是1∶2，即采用缩小的比例，故可知实物应比图形大。

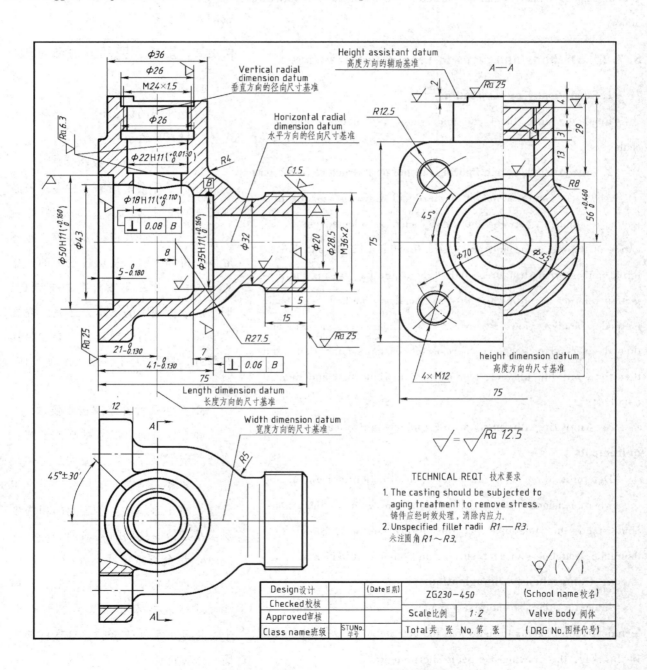

Fig. 8-51 Detail drawing of a valve 阀体零件图

(2) Analyzing views and figuring out the structure of the part

1) Analyzing the representation. The valve is represented by the front, left and top view, of which the front and left view adopt full and half section separately to convey its complex inner structure. The top view is the external view and has a local sectional view.

2) Analyzing the whole structure of the part. The top of the valve is a cylindrical pipe. The bottom of the cylindrical pipe penetrates into the internal spherical valve, where there is a horizontal gateway. There is a cylindrical pipe structure on the right of the gateway, and a square flange structure on the left. The left and right entrance of the gateway and the top cylindrical pipe have stepped holes. The whole valve is a symmetric structure (front and rear). See the top view in Fig. 8-51.

3) Analyzing detailed structure and function. The basic shape of the valve is a spherical shell. On the left square flange of the valve, there are four threaded holes to join with the valve cover using bolts. The cylindrical hole on the top has stepped, annular recess holes to mount the valve handle and for stuffing. The external thread on the right cylindrical end of the valve has been designed to access the pipe system.

Comparing the front view with the top view, there is a 45° symmetric structure convex edge to control the angle of rotation of turned parts.

(3) Analyzing dimensions and understanding technical requirements

1) Analyzing dimensions.

① Dimension datums. The dimension datums for length, width and height are shown in Fig. 8-51. In addition, there are also datums for the horizontal radial dimensions and height assistant datums. Those datums can be used to determine the overall dimensions, size dimensions and location dimensions.

② Characteristic dimensions. The external thread M36×2 on the right connecting the pipes and the internal thread M24×1.5 on the top of the valve are characteristic dimensions.

③ Fitting dimensions. The cylindrical groove $\phi50H11$ fitted to the valve cover on the left of the front view, and the valve's internal cylinder faces $\phi22H11$, $\phi18H11$ and $\phi35H11$ are hole-basis system of clearance fits.

④ Mounting dimensions. The four threaded holes 4×M12 on the square flange and its location dimension $\phi70mm$ and 45° are mounting dimensions for mounting to the pipe system.

（2）分析视图并想象零件结构

1）表达方案分析。这个阀体的零件图画出了主、俯、左三视图，其中主、左视图分别为全剖和半剖视图，以表达其复杂的内部结构，而俯视图则为外形视图并有一处局部剖视图。

2）零件整体结构形状分析。阀体上部为圆柱管状，圆柱管的下部与球形阀身相贯，球形阀身的内部有一左右方向水平管道通路，通路的右端为圆柱管形结构，通路的左端为方形凸缘结构，左、右通道口和上部圆柱管内均有切削加工的阶梯孔。整个阀体呈前后对称结构，如图 8-51 俯视图所示。

3）细部结构及功能分析。阀体的基本形状是球形壳体。左边方形凸缘上有四个螺孔，用螺柱与阀盖相连。上部的圆柱筒内孔有阶梯孔和环形槽，以便安装阀杆、密封填料等。阀体右端的外螺纹是为接入管路系统而设计的。

对照主视图和俯视图可以看出，在阀体的顶部有一个呈前后 45°对称结构的扇形限位凸块，用来控制与之相连的转动件的旋转角度。

（3）分析尺寸并了解技术要求

1）分析尺寸。

① 尺寸基准。长、宽、高三个方向的尺寸基准如图 8-51 所示。另外，对于球体还有水平方向的径向尺寸基准和高度方向的辅助基准。由这些基准出发，可确定总体尺寸、定形尺寸和定位尺寸等。

② 特性尺寸。阀体右端与管路系统相联接的外螺纹 M36×2 以及阀体上端的内螺纹 M24×1.5 均为特性尺寸。

③ 配合尺寸。主视图左端与阀盖相配合的圆柱形槽 $\phi50H11$、阀体内圆柱面的 $\phi22H11$、$\phi18H11$ 和 $\phi35H11$，这几处均为基孔制间隙配合。

④ 安装尺寸。方形凸缘上的四个螺孔尺寸 4×M12 及其定位尺寸 $\phi70mm$ 和 45°，均为安装到管路系统时的安装尺寸。

⑤ Other dimensions. Convex edge's shaping dimension 45°±30′ is to control the limit of the fittings' movement.

2) Analyzing the technical requirements.

① Surface roughness. The highest requirement is for internal cylindrical faces φ22H11 and φ18H11, As shown in Fig. 8-51, the highest requirement is 6.3μm (Ra). Besides, most of machined surfaces have requirements of 12.5μm and 25μm (Ra). Following the exception marked with "∀(√)" nearby the title block, non-marking surfaces are all casting surfaces that are not required for machining. The quality requirements on these surfaces are not high.

② Geometrical tolerance. In Fig. 8-51, there are two perpendicularity requirements in total, namely, the perpendicularity tolerance between the axis of hole φ18H11 and datum B (axis of hole φ35H11) is 0.08mm and the perpendicularity tolerance between the right end of hole φ35H11 and datum B is 0.06mm.

③ The technical requirements on the text used. They are as following: In order to release internal tension of casting part, timing treatment is required; Radius for all non-marked rounds R is 1-3mm.

(4) Comprehensive imagination Based on the above-mentioned analysis, combining with views, dimensions and technical requirements, one can figure out a general picture of the part as shown in Fig. 8-52.

⑤ 其他尺寸。限位凸块的定形尺寸 45°±30′，限定与之相配合零件的运动极限位置。

2）分析技术要求。

① 表面粗糙度。表面粗糙度要求最高的是阀体内圆柱面 φ22H11 和 φ18H11，如图 8-51 所示，其表面粗糙度值 Ra 均为 6.3μm。还有多数加工面的 Ra 值为 12.5μm、25μm。没有标注的表面均为不加工的铸造表面，这些表面的质量要求不高，由图中标题栏附近给出的符号"∀（√）"统一表示。

② 几何公差。由图 8-51 可知，共有两处垂直度要求，即：φ18H11 孔的轴线对 B 基准（φ35H11 孔的轴线）的垂直度公差为 0.08mm；φ35H11 孔的右端面对 B 基准的垂直度公差为 0.06mm。

③ 用文字说明的技术要求。共有两条：为消除铸件的内应力，需要对零件进行时效处理；所有未注圆角半径 R 均为 1~3mm。

（4）综合想象零件全貌 根据上述分析，将视图、尺寸和技术要求综合起来，即可想象出零件全貌，如图8-52所示。

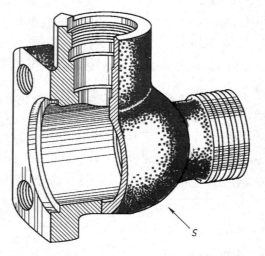

Fig. 8-52　Stereograph of a valve　阀体立体图

6. Interpreting the other detail drawings

In addition to the common typical parts introduced above, there are some sheet metal parts, studding parts and injecting parts used in telecommunication and instrumentation industry. The characteristic in the drawings for these parts would be simply stated in the following.

(1) Sheet metal parts   Most of the parts are first pressed and cut as a flat sheet, and then bent, folded or rolled into the required shape. A surface development view should be added when necessary, and marked "surface development view" in the top of the drawing, as shown in Fig. 8-53.

6. 读其他零件图

除前面介绍的常见典型零件之外，还有一些在电信、仪表工业中常见的薄板冲压零件、镶嵌零件和注塑零件等。下面简要介绍这些零件的图示特点。

（1）薄板冲压零件 这类零件大多是用板材冲裁落料、冲孔，再冲压弯曲成形。对这种零件必要时需增加一个展开图，并在该图上方注明"展开图"字样，如图8-53所示。

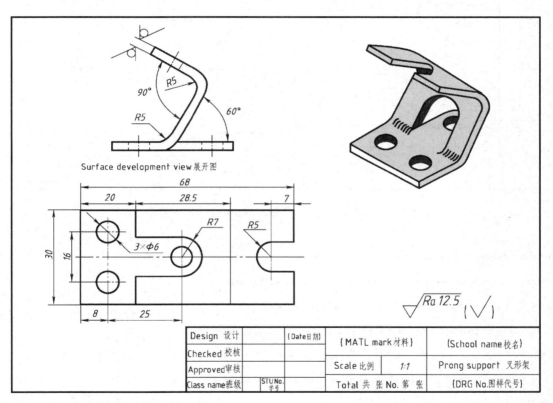

Fig. 8-53   Detail drawing of a prong bracket   叉形架零件图

There are often fillets where two surfaces meet. There are also holes and sluts on the surface for installing the electric components and installing the part on the main machine frame. These holes are often through holes. Such holes and sluts are usually drawn on a view that shows the true shape but they can be omitted on other views when the understanding of the drawing will not be impaired, as shown in Fig. 8-54.

这类零件弯折处一般有小圆角。零件的面板上有许多孔和槽口，以便安装电气元件或部件，并将该零件安装到机架上。这种孔一般都是通孔，在不致引起误解的情况下，只将反映其真实形状的那个视图画出，而在其他视图中的虚线就不必画了，如图8-54电容器架所示。

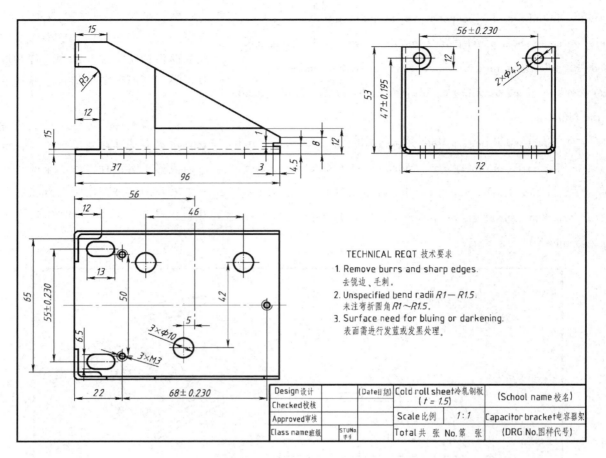

Fig. 8-54 Detail drawing of a capacitor bracket 电容器架零件图

The dimensioning principles for the sheet metal pressing parts are as follows: size dimensions are usually marked through shape-body analysis and location dimensions are usually marked between the axes of holes or marked from the hole axis to the face. Fig. 8-54 shows the location dimensions 46mm, 42mm, 56mm, 5mm for holes (3×φ10mm); and the location dimensions 50mm, 22mm, 68mm±0.230mm for screw holes (3×M3).

(2) Inlaid parts The parts are studded with metal and non-metal materials, such as the adjusting gear shaft shown in Fig. 8-55, which is studded with a metal shaft, an adjusting gear and a non-metal knob. Different sectional symbols are often used for clear representation. The studding part is considered as a single component and only one serial number is assigned on the assembly drawing.

(3) Injection part Except for the sectional symbol that should represent a non-metal material, all other representation methods are the same as that of ordinary parts.

薄板冲压零件尺寸标注的原则是：定形尺寸按形体分析方法标注，定位尺寸一般标注两孔中心或孔中心到板边的距离。例如，图8-54中孔3×φ10mm的定位尺寸46mm、42mm、56mm、5mm，螺孔3×M3的定位尺寸50mm、22mm、68mm±0.230mm等。

（2）镶嵌零件 镶嵌类零件是由金属材料与非金属材料镶嵌而成的。如图8-55所示的调节齿轴就属于镶嵌零件，它是由金属的小轴、调节齿轮与非金属的旋钮镶嵌在一起的，在表达上要以不同的剖面符号来区别。镶嵌零件作为一个组件，在装配图中只编一个序号。

（3）注塑零件 注塑零件除了剖面符号采用非金属材料的剖面符号外，其余表达方法与一般零件相同。此处不再赘述。

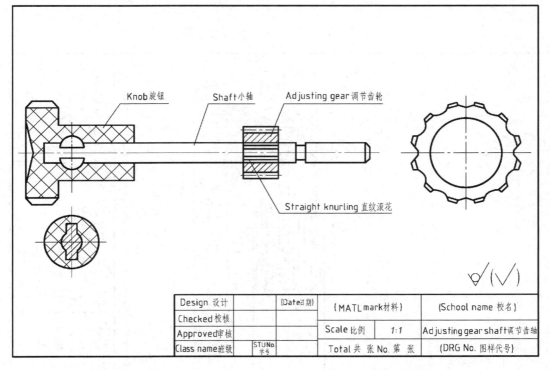

Fig. 8-55 Drawing of an adjusting gear shaft 调节齿轴组件图

7. Detail drawings of other countries

Compared with the standards of other countries, the detail drawing of ours is similar except the title block and the notation of technical requirements. See the drawing in Fig. 8-56 produced following the British standards, and Fig. 8-57 produced following the United States standards.

7. 国外零件图示例

国外零件图除标题栏和技术要求的注写形式不同外，基本上与我国的零件图形式相似。例如，图8-56所示为英国图样，图8-57所示为美国图样。

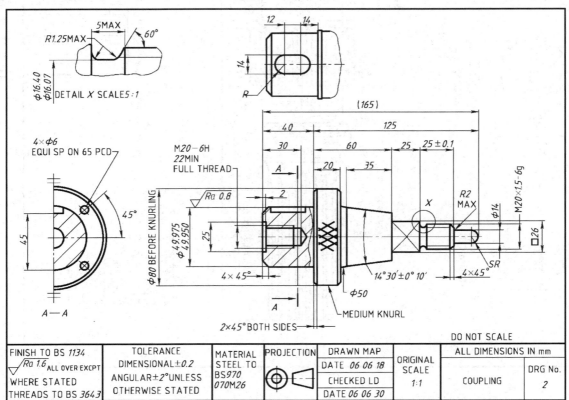

Fig. 8-56 Detail drawing of a coupling 联轴器零件图

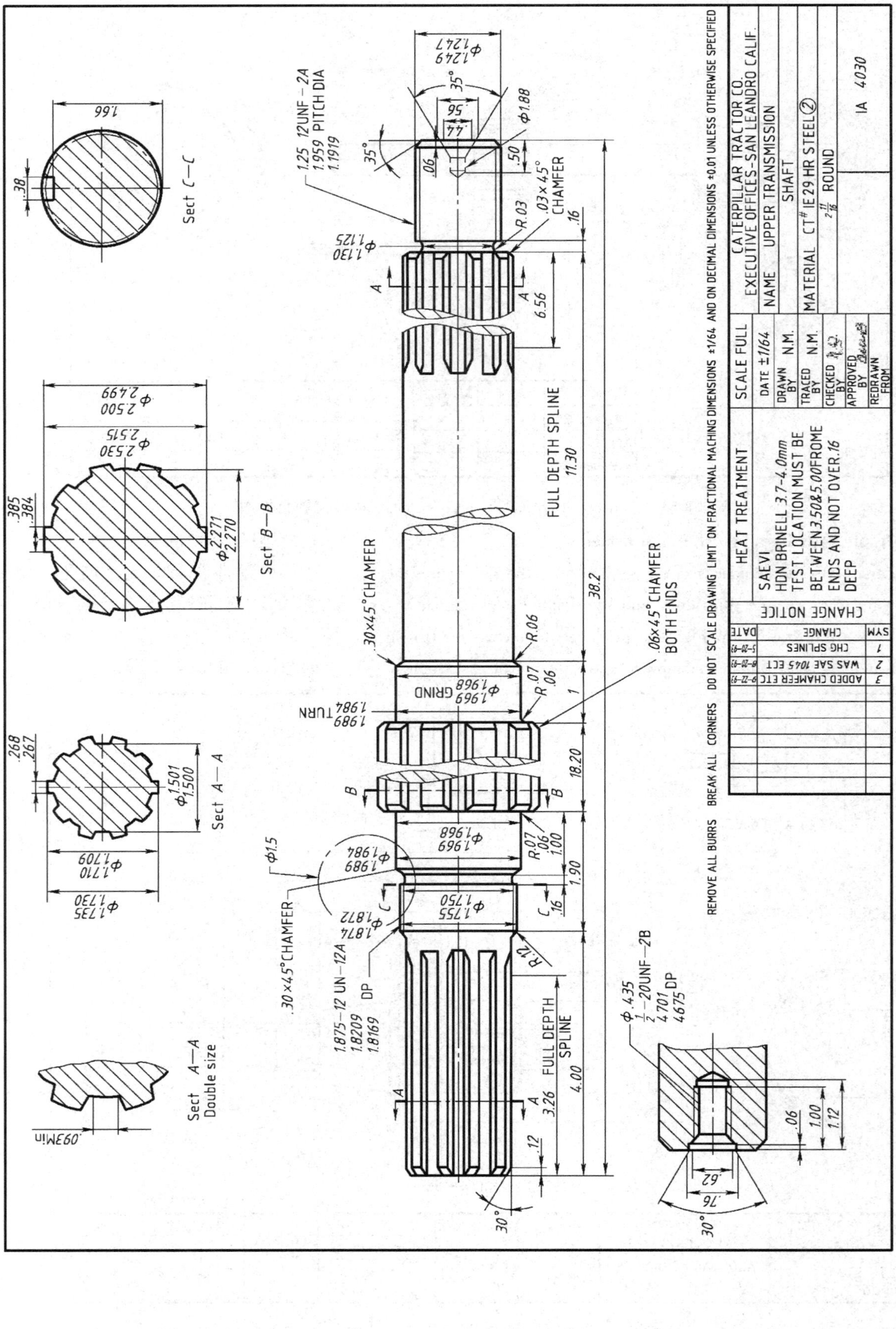

Fig. 8-57 Detail drawing of a shaft 轴零件图

8.7.3 Glossary and abbreviations for commonly used construction of parts

See Table 8-9 for glossary and abbreviations for commonly used construction of parts.

8.7.3 常见零件结构的专用术语和缩略语

常见零件结构的专用术语和缩略语列于表 8-9，以供参考。

Table 8-9　Glossary and abbreviations for commonly used construction of parts
常见零件结构的专用术语和缩略语

Construction of parts 零件结构	Terminology/Abbreviation 专用术语/缩略语	Description 说明
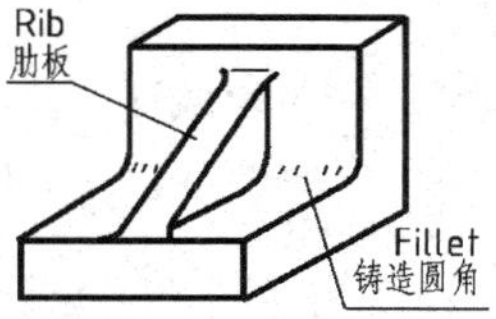 (Rib 肋板 / Fillet 铸造圆角 图示)	Rib 肋板	A reinforcement positioned to stiffen the structure usually for supporting surfaces with right angles 起加强作用，通常位于（铸件）相互垂直的二面之间
	Fillet 铸造圆角	A small round on no-machined casting surface 铸件未加工表面的小圆角
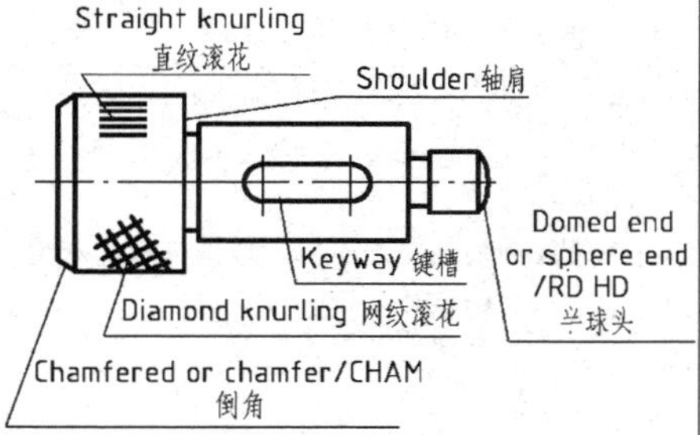 (Straight knurling 直纹滚花 / Shoulder 轴肩 / Keyway 键槽 / Diamond knurling 网纹滚花 / Domed end or sphere end /RD HD 半球头 / Chamfered or chamfer/CHAM 倒角 图示)	Straight (diamond) knurling 直纹（网纹）滚花	Pressing special rollers against the surface of the revolved component, providing a roughened surface to aid tightening or slackening of a screw by hand 在旋转机件的表面加工的条纹刻痕，称为直纹（网纹）滚花，以防打滑
	Shoulder 轴肩	A shaft end surface often used for axial positioning 轴上用作轴向定位的端面
	Chamfered or chamfer/CHAM 倒角	A small inclined plane machined at the end of a component for removing sharp edges 在机件端部加工出的小斜面
	Keyway 键槽	A slot cut on a shaft surface or hub to accommodate a key 在内孔或轴上切出的长槽，以便放入与之相配合的键
	Domed end or sphere end/RD HD 半球头	A hemisphere head machined at the end of a shaft 在轴端加工出的半球形头部
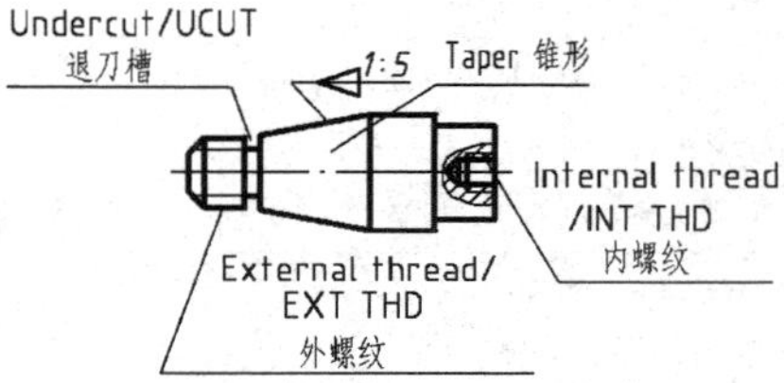 (Undercut/UCUT 退刀槽 / Taper 锥形 / Internal thread /INT THD 内螺纹 / External thread/EXT THD 外螺纹 图示)	Undercut/ UCUT 退刀槽	A circular groove at the end of a thread which permits tool escape conveniently 为方便退刀而在机件上加工的沟槽
	Taper 锥形轴	A taper surface machined on a shaft 在轴上加工出的锥形部分
	Internal thread/INT THD 内螺纹	Thread machined on the inner surface of a hole 在圆柱孔内表面加工出的螺纹
	External thread/EXT THD 外螺纹	Thread machined on the external surface of a column 在圆柱外表面加工出的螺纹
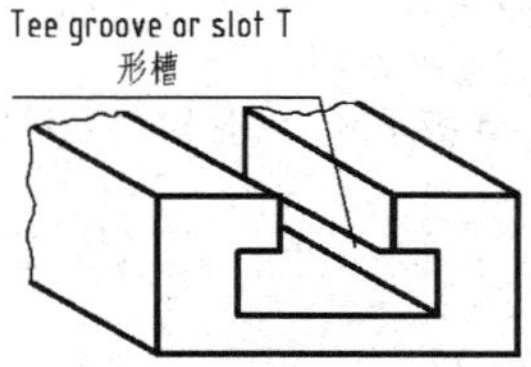 (Tee groove or slot T形槽 图示)	Tee groove or slot T形槽	Machined to "house" mating fixing bolts and prevent them from turning 可嵌入与之相配合的倒 T 形机件，以防二者相互转动

(Continued 续)

Construction of parts 零件结构	Terminology/Abbreviation 专用术语/缩略语	Description 说明
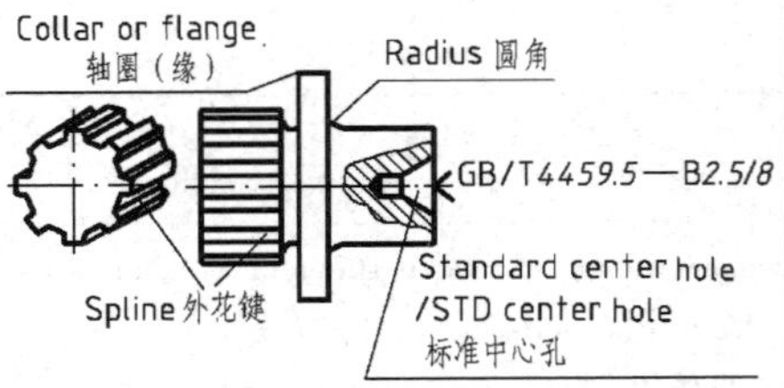	Collar or flange 轴圈（缘）	A heave ring part which is used for assembly positioning 轴上凸起的环形部分，多用于装配定位
	Radius 圆角	A small round machined at corner to prevent stress concentration 为了避免应力集中，在机件转折处加工出的小圆角
	Spline 花键	Multi-teeth machined on the surface of a shaft or a hole 在轴上或孔内加工多个齿而构成
	Standard center hole/STD center hole 标准中心孔	A small hole drilled at the end of a shaft for positioning 在轴类零件端部钻出的小孔，以便装夹定位
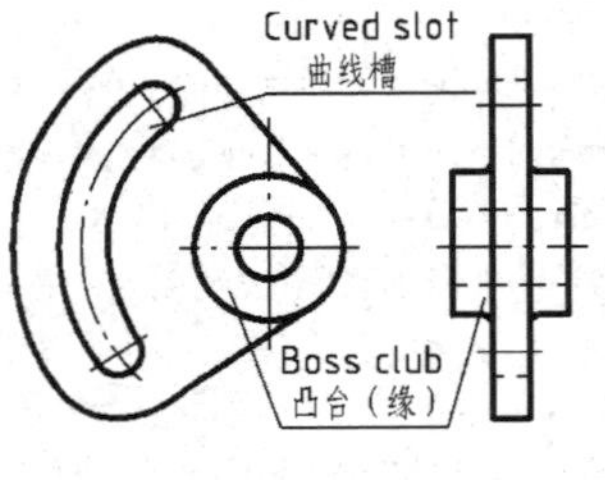	Curved slot 曲线槽	An elongated hole whose centre line lies on an arc, usually used on components whose position has to be adjusted 中心线为圆弧形的长孔，用于调节机件的位置
	Boss club 凸台（缘）、轮毂	A cylindrical projection on the surface of a component—usually a casting or a forging 在铸件或锻件表面上的圆柱形凸起部
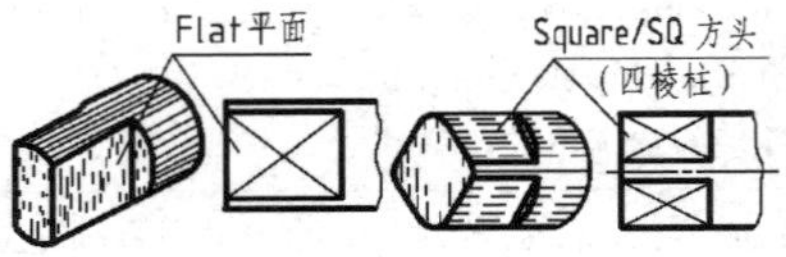	Flat 平面	A surface machined parallel to the shaft axis, in a gyration body usually used to locate and/or lock a mating component 指平行于轴线的机加工表面，一般设置在需要配合、固定或卡住的轴端处
	Square/SQ 方头（四棱柱）	Cutoff square head at shaft end 指轴端处切出的方头（四棱柱）结构
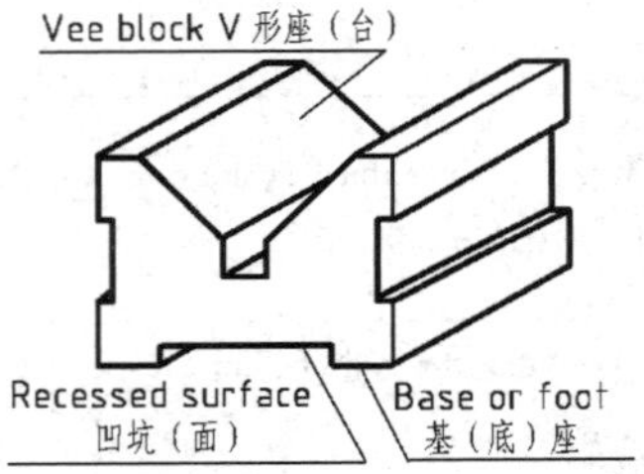	Vee block V形座（台）	Machined to support components and apply them gliding on "V" slot 支承运动件并可使其在V形槽中滑动
	Recessed surface 凹坑（面）	Ensuring better seating of the base, and minimizing machining of the base 为使机座平稳安放且加工面积最小而设计
	Base or foot 基（底）座	The part upon which the component rests 台、座、支架等的底部
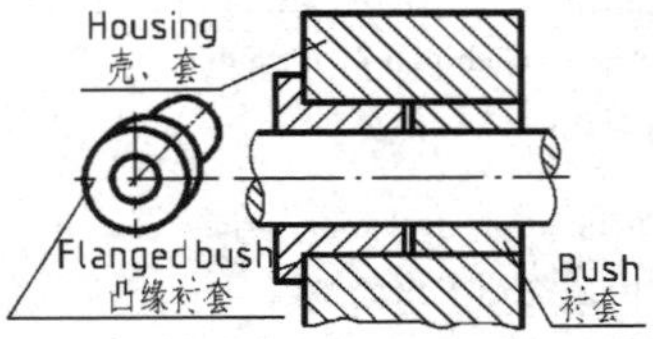	Housing 壳、套	A component in which a "male" mating part fits, sits or is "housed" 可将与之相配合的凸形件装入（落座于）其内
	Bush/Flanged bush 衬套/凸缘衬套	A removable sleeve or liner 可拆卸的套管或套筒
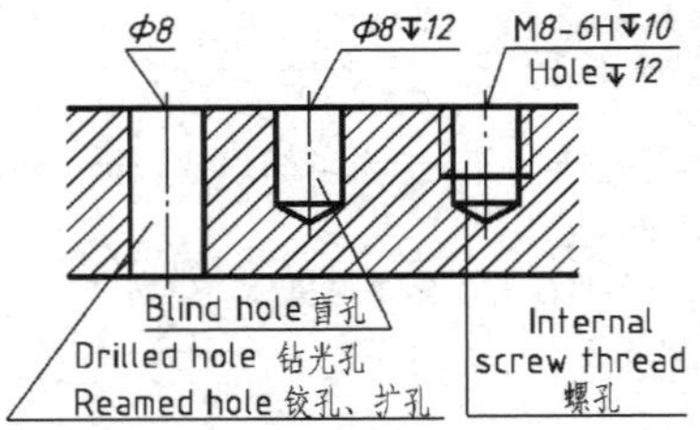	Drilled hole 钻光孔	Drill a hole in a component using a drill 用钻头在机件上钻出孔
	Reamed hole 铰孔，扩孔	If there is a high precision requirement for the hole, it is drilled first and then finished 当孔有较高精度要求时，在钻好的孔中再精加工一遍
	Blind hole 盲孔	A blind hole is drilled and then threads can be produced based on it 钻出盲孔，以便在其上再加工出螺孔

(Continued 续)

Construction of parts 零件结构	Terminology/Abbreviation 专用术语/缩略语	Description 说 明
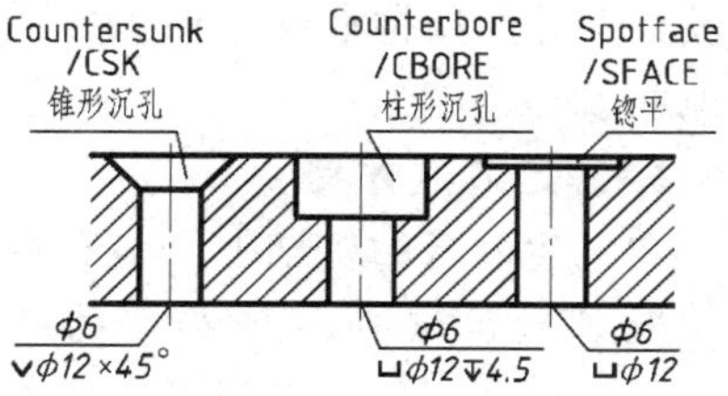	Countersunk/CSK 锥形沉孔	Provides a mating seat for a countersunk headed screw or rivet 在孔口加工出锥形以便沉入螺钉
	Counterbore/CBORE 柱形沉孔	Provides a "housing" for the heads of capscrews, bolts, ect 在孔口加工出圆台以便放入圆柱头螺钉、螺栓等
	Spot face/SFACE 锪平	A much shallower circular recess, providing a machined seat for nuts, bolt heads, washers, etc 在孔口处加工的浅凹入面，以便螺母、螺栓或垫圈等平稳安放
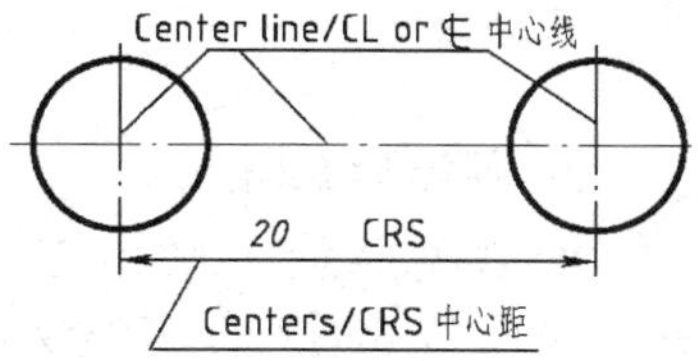	Centers/CRS 中心距	Distance between holes 孔与孔中心线之间的距离
	Center lines/CL or ℄ 中心线	Center axis of a gyration body 回转体的中心轴线
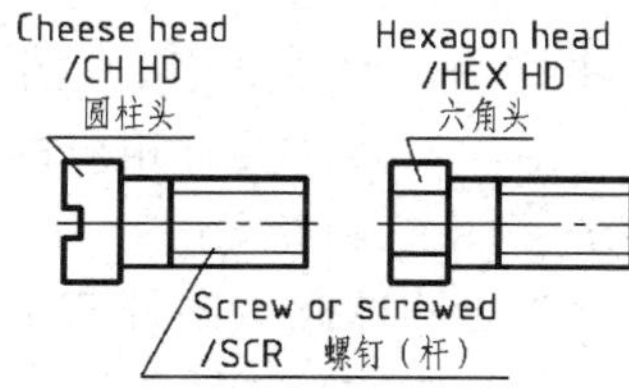	Cheese head/CH HD 圆柱头	A column head of a component 机件的圆柱形头部
	Hexagon head/HEX HD 六角头	A hexagon head of a component 机件的六角形头部
	Screw or screwed/SCR 螺钉（杆）	A component with screw threads 带有螺纹的机件
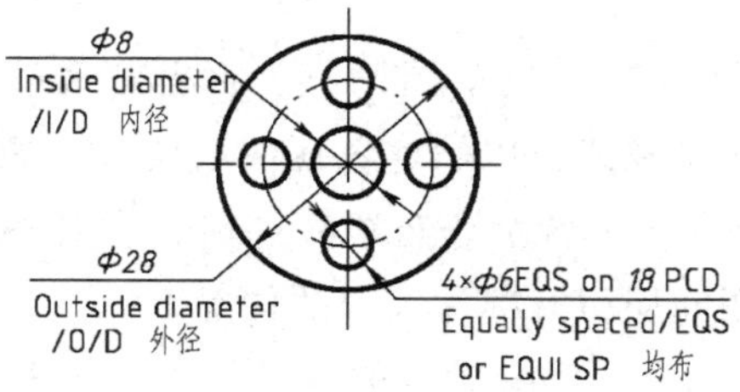	Equally spaced/EQS or EQUI SP 均布	Holes or slots machined at a wheel component, arranged in equal distance or angle 在盘类机件上加工的孔或槽等，按均匀分布的方式排列
	Pitch circle diameter/PCD 节圆直径	diameter of a pitch circle 绕圆周均布的多个孔的圆心位置的直径
	Inside diameter/I/D 内径	Inner circle diameter of a gyration body 回转体的内圆直径
	Outside diameter/O/D 外径	Outer circle diameter of a gyration body 回转体的外圆直径

8.8 Mapping Parts

To map a part drawing based on an existing physical part, one may first measure its dimensions, figure out its technical requirements, and then elaborate the working drawing according to sketches from the measurement. Part mapping is useful for duplicating an existing part or repairing a damaged part.

This chapter mainly focuses on mapping methods, steps and common measuring techniques.

8.8 零件的测绘

零件的测绘是根据实际零件画出草图，测量零件各部分的尺寸，确定技术要求，再根据草图画出零件的工作图。一般在仿制机器和修配损坏的零件时，需要进行零件测绘。

本节仅讨论测绘的方法、步骤及常用的测量方法。

8.8.1 Methods and steps to map parts

1. Analyzing the part

If one wants to map the ball valve cover shown in Fig. 8-58, first, one should know the part name, material, function, structure, shape and machining method, as well as other requirements.

2. Determining the representations

Determine the front view on the basis of the above mentioned analysis. According to the ball valve cover features shown in Fig. 8-58, one should adopt a full sectional front view to show the stepped hole and adopt a left view to convey its external structure and the distribution of the holes.

3. Producing the sketch

After visualization, draw the sketchmap, Although it is a sketch, it cannot be scratchy and it shall fully convey the part, and be clearly dimensioned. It should be drawn with clear line types, neat drawings, and upright font. Related technical requirements, border and title block are also needed. Fig. 8-59 shows the steps to map the ball valve cover. It is described as follows:

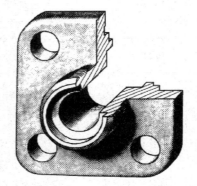

Fig. 8-58 Ball valve cover   球阀盖

1) Draw the border, title block, arrange and draw the datum lines to specify the location of each view, as shown in Fig. 8-59a. When positioning views, a place for dimensioning should be reserved.

2) Adopt appropriate representations to draw the views, so as to convey the structure and shape according to the projection relationship, as shown in Fig. 8-59b.

3) Draw the extension lines, dimension lines and arrows, and deepen the continous thick lines (see Fig. 8-59c).

4) Measure the dimensions, fill in the dimension numbers, technical requirements and title block (see Fig. 8-59d).

Compare the machined surface with the non-machined surface, determine the appropriate surface roughness values according to the function of these surfaces, and mark them in the drawing. If the dimensions are the fitting ones with respect to other parts, one should determin their tolerance values and mark the tolerance symbols correspondingly.

8.8.1 零件测绘的方法和步骤

1. 分析零件

以绘制图 8-58 所示球阀盖为例，首先了解零件的名称、材料、用途，以及各部分的结构形状、加工方法和其他要求等。

2. 确定表达方案

在上述分析的基础上，确定主视图。根据图 8-58 所示球阀盖的结构特征，采用全剖的主视图以表达阶梯通孔，并用左视图表达其外形结构及孔的分布情况。

3. 画零件草图

经目测后徒手画出零件草图。虽然是草图，但不能潦草绘画，要做到视图表达清楚，尺寸标注完整，线型分明，图面整洁，字体工整，要有相应的技术要求、图框和标题栏。图 8-59 所示为绘测球阀盖零件草图的步骤，其具体过程说明如下。

1) 画出草图图框、标题栏和作图基准线，以确定各视图的位置，如图 8-59a 所示。布置视图时，在各视图间应留有标注尺寸的位置。

2) 采用恰当的表达方法，按照投影的对应的关系画出各个视图，表达零件各部分的结构形状（见图 8-59b）。

3) 画出尺寸界线、尺寸线和箭头，并加深粗实线（见图 8-59c）。

4) 测量尺寸，填写尺寸数字、技术要求和标题栏（见图 8-59d）。

对照零件的加工面和非加工面，按这些面的作用，定出适当的粗糙度，并标注在图上。对于与其他零件有配合关系的尺寸，定出它们的公差，并标出公差代号。

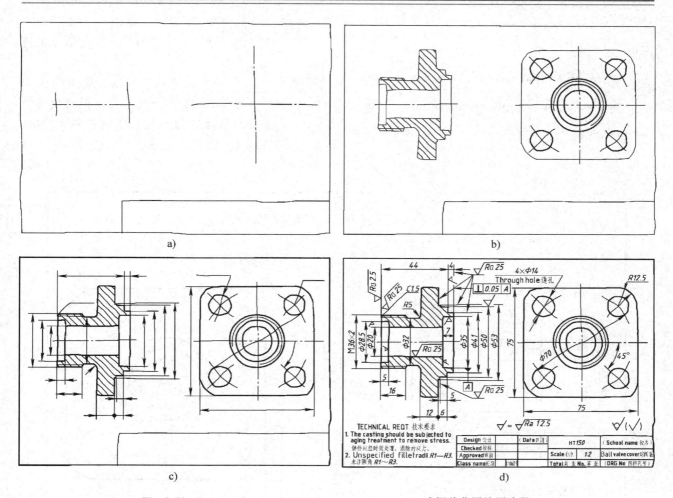

Fig. 8-59　Steps to draw a sketch for a ball valve cover　球阀盖草图绘图步骤
a) Position views　布置视图　b) Draw views　画各视图
c) Draw dimension lines and deepen lines　画尺寸线并加深图线
d) Fill in dimension values and technical requirements, etc.　注全尺寸和技术要求等

4. Producing the part drawing

After reviewing the sketch, draw the working drawing. The complete detail drawing is shown in Fig. 8-60.

8.8.2　Remarks on part mapping

1) Manufacturing defects of the part, such as sand inclusion and air holes, tool marks and wearing, should not be drawn.

2) The technological structure originated from manufacturing or mounting, such as fillet of castings, chamfer, undercut, boss club and recessed surface, should be drawn and can not be omitted.

3) Draw only basic fitting dimensions (for example, the diameters of the fitted hole and shaft). Its system of fits and corresponding tolerance should be determined by consulting the manual after the analysis.

4. 画零件图

对画好的零件草图进行复核后，再画正式的零件图。完成后的零件图如图 8-60 所示。

8.8.2　零件测绘时的注意事项

1) 零件的制造缺陷，如砂眼、气孔、刀痕以及长期使用所造成的磨损都不应画出。

2) 零件上因制造、装配的需要而形成的工艺结构，如铸造圆角、倒角、退刀槽、凸台、凹坑等，都必须画出，不能忽略。

3) 只绘出基本配合尺寸（如配合的孔和轴的直径），其配合性质和相应的公差值，应在分析考虑后，再查阅有关手册确定。

4) Non-fitting or unimportant dimensions are allowed to be adjusted to integers.

5) The results of measuring the thread, keyway, and gear teeth should be checked with corresponding standardized values and standard structures should be used to facilitate production.

4）无配合关系的尺寸或不重要的尺寸，允许将测得的尺寸适当调整为整数。

5）对螺纹、键槽、齿轮的轮齿等标准结构的尺寸，应该把测量的结果与标准值核对，采用标准结构尺寸，以利于制造。

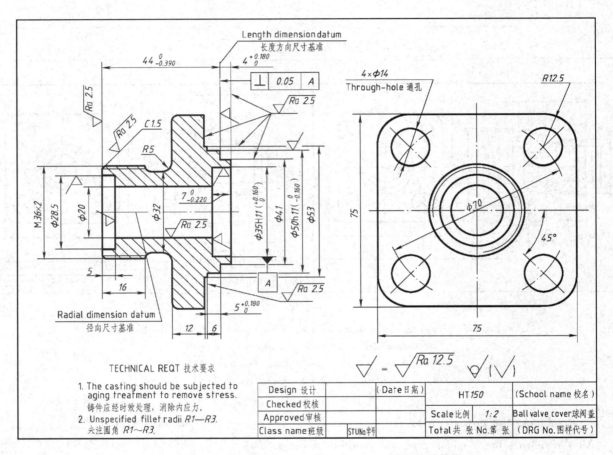

Fig. 8-60　Detail drawing of a ball valve cover　球阀盖零件图

8.8.3　Methods for measuring dimensions

Measuring is a necessary step in part mapping. Measurements should be made in a single process. In this way, it will not only improve efficiency, but also avoid errors and omissions.

While measuring the part, one should select appropriate measuring instruments, according to the dimension and precision.

1. Measuring instruments

The frequently used tools for measuring dimensions are as following: ruler, interior calipers, external calipers. Vernier calipers are used for high-precision parts.

8.8.3　零件尺寸的测量方法

测量尺寸是零件测绘过程中的一个必要的步骤。零件上全部尺寸的测量应集中进行，这样不但可以提高工作效率，还可以避免错误和遗漏。

测量零件尺寸时，应根据零件尺寸和精确程度正确选用量具。

1. 测量工具

测量尺寸常用的工具有：直尺、内卡钳、外卡钳，测量较精密的零件需用游标卡尺。

2. Common measuring methods

(1) Measuring linear dimensions The linear dimensions can be measured directly using a ruler. Sometimes one can measure them by combining a triangular rule with a regular ruler, as shown in Fig. 8-61. If precision is high, one can use vernier calipers.

2. 常用的测量方法

(1) 测量线性尺寸 一般可用直尺直接测量线性尺寸，有时也可用三角板与直尺配合进行，如图 8-61 所示。若要求精确时，则用游标卡尺测量。

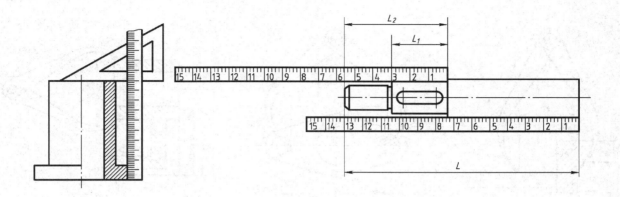

Fig. 8-61 Measuring linear dimensions 测量直线尺寸

(2) Measuring the internal (inside) and external (outside) diameters of the revolving body Measure the external diameters with external calipers and the internal diameters with internal calipers. When measuring, one should move the internal and external calipers up and down, left and right. The maximum value obtained is the internal or external diameter. The measuring method using the vernier caliper is similar, as shown in Fig. 8-62.

(2) 测量回转体的内外径 测量外径用外卡钳，测量内径用内卡钳。测量时要把内、外卡钳上下、前后移动，量得的最大值为其内径或外径。用游标卡尺测量时的方法与用内、外卡钳时相同，如图 8-62 所示。

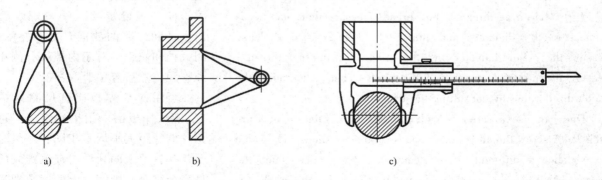

Fig. 8-62 Measuring internal and external diameters 测量内、外径
a) Measuring O/D with external calipers 用外卡钳测量外径
b) Measuring I/D with internal calipers 测量内径用内卡钳
c) Measuring I/D and O/D with a vernier caliper 用游标卡尺测量内径或外径

(3) Measuring wall thickness Combine external calipers with a ruler, as shown in Fig. 8-63 (wall thickness $X = A - B$).

(3) 测量壁厚 可用外卡钳与直尺配合使用来测量壁厚，如图 8-63 所示，壁厚为 $X = A - B$。

(4) Measuring hole pitches Measure related dimensions with external calipers, as shown in Fig. 8-64, and calculate hole pitches $A = a + d/2 + d/2$.

(5) Measuring the height of the axle hole's center As shown in Fig. 8-65, one can measure related dimensions with the external calipers and the ruler, then calculate the center height of the axle hole ($A = B + D/2$).

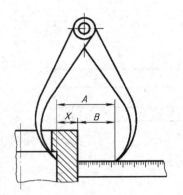

Fig. 8-63 Measuring wall thickness 测量壁厚

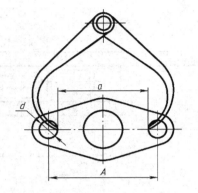

Fig. 8-64 Measuring hole pitches 测量孔间距

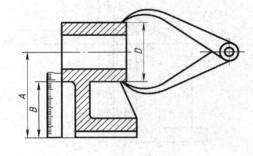

Fig. 8-65 Measuring the height of the axle hole's center 测量轴孔中心高

(6) Measuring round Fig 8-66 shows a measuring method using a radius template. Each set of the radius template has several slices, the radius of which are different. Find out the slice matching the measured section. The reading of that slice will be the radius. Casting curving can be visualized. If the technical information is available, one should select the appropriate value, instead of measuring.

(7) Measuring threads For threads measurement, one needs to measure the diameter and the pitch. For external threads, measure the major diameter and the pitch. For internal threads, measure the minor diameter and the pitch. Then, determine the standardized value by consulting the manual.

One can use a screw thread template or a ruler to measure pitch P. A screw thread template has several steel slices. The value of each slice is different. If the teeth on a given slice match the measured thread, the reading of that slice will be the pitch, as shown in Fig. 8-67a.

If a screw thread template is not available, one can press the thread on a piece of paper and measure the pitch on the paper, namely, $P = T/n$, T is n times of the pitch, and n is the number of pitches, as shown in Fig. 8-67b. Then determine the standardized pitch value by consulting the manual, according to the calculated pitch.

(4) 测量孔间距 用外卡钳测量相关尺寸，如图 8-64 所示，再进行计算，孔间距 $A = a + d/2 + d/2$。

(5) 测量轴孔中心高 如图 8-65 所示，用外卡钳及直尺测量相关尺寸，再进行计算，轴孔中心高 $A = B + D/2$。

(6) 测量圆角 图 8-66 所示为用半径样板测量圆角半径的方法。每套半径样板有很多片，其圆弧半径均不同。测量时只要在半径样板中找出与被测量部分完全吻合的一片，则片上的读数即为圆角半径。铸造圆角一般可通过目测估计其大小。若手头有工艺资料，则应选取相应的数值而不必测量。

(7) 测量螺纹 测量螺纹要测出直径和螺距的数据。对于外螺纹，测大径和螺距；对于内螺纹，测小径和螺距，然后查手册取标准值。

可用螺纹样板或直尺测量螺距 P。螺纹样板由一组钢片组成，每一钢片的螺距大小均不相同，测量时只要某一钢片上的牙型与被测螺纹的牙型完全吻合，则钢片上的读数即为螺距大小，如图 8-67a 所示。

在没有螺纹样板的情况下，可以在纸上压出螺纹的印痕，然后算出螺距的大小，即 $P = T/n$，T 为 n 个螺距的长度，n 为螺距数量，如图 8-67b 所示。根据算出的螺距再查手册取标准值。

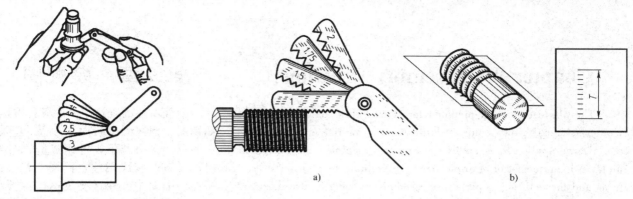

Fig. 8-66 Measuring round 测量圆角

Fig. 8-67 Measuring a thread 测量螺纹
a) Measuring the pitch with a thread gage 用螺纹规测量螺距
b) Measuring pitch with a ruler 用直尺测量螺纹压痕

(8) Measuring a gear First, measure the tip diameter and the number of teeth, and then calculate the related parameters of the gear. Finally, determine the standardized value. First, measure the tip diameter with a vernier caliper, as shown in Fig. 8-68, and then count the number of teeth and calculate the gear module. Finally, determine the standardized module and then calculate gear parameters with the standardized module.

(9) Measuring curved contours Rubbing methods can be used for measuring the curved contours, as shown in Fig. 8-69a. Then the center and radius of the approximation arc can be obtained by drawing two perpendicular bisectors using auxiliary arcs/circles, as shown in Fig. 8-69b.

（8）测量齿轮 首先测量齿轮的齿顶圆直径与齿数，然后计算齿轮的有关参数，最后归化到标准的参数数值上。先用游标卡尺测量齿轮的齿顶圆直径，如图 8-68 所示，再数齿数，然后计算齿轮模数。应与标准模数对照，归化到最靠近的一个标准模数，然后再用这个标准模数反算齿轮的各个参数。

（9）测量曲面轮廓 对于曲面轮廓的形体，可用拓印法勾出曲面轮廓，如图 8-69a 所示；然后用求圆弧上两条弦的垂直平分线的方法，作图求出圆心和半径，如图 8-69b 所示。

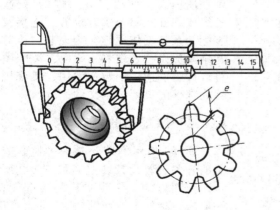

Fig. 8-68 Measuring a gear 测量齿轮

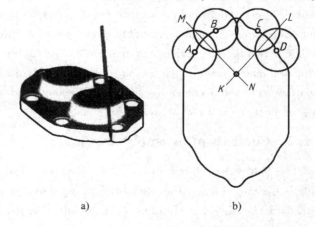

Fig. 8-69 Measuring curved contours 测量曲面轮廓

Chapter 9　Assembly Drawings

The above-discussed methods for representing parts, such as views, sectional views, cuts and drawings of partial enlargement, can also be applied to assembly drawings of machines and assembly units, which consist of multiple parts. In addition to showing structures and forms of major parts, an assembly drawing is also used to represent the operating principles and connections of individual parts of the assembly. Therefore, in comparison to the detail drawing, an assembly drawing has special and other specified representations.

This chapter introduces some special and recommended representations as well as the methods of producing and reading an assembly drawing.

9.1　Summary of Assembly Drawings

9.1.1　Purposes of assembly drawings

An assembly drawing is a drawing used to represent a machine, an assembly unit and other equipment.

To design a new product or to improve an old equipment, one often first produces an assembly drawing and then produces individual detail drawings.

In production, one should set up a machine, an assembly unit and an equipment according to an assembly drawing.

In practical use and maintenance, an assembly drawing will also help users understand the structure of the machine as well as its relative positions, joints, assembly and operating procedures, thereby providing technical information for the assembly, manufacturing and maintenance of the machine. The assembly drawing is an important technical document for the design, manufacturing, assembly, applications and maintenance of the machine.

9.1.2　Contents of assembly drawings

The ball valve shown in Fig. 9-1 is an on-off device that controls the rate of flow. The assembly drawing shown in Fig. 9-2 is a standard assembly drawing illustrating the following four aspects.

1. A group of views

Following contents are represented by views: ①relative mounting position of each part; ②transmitting motion and operating principle; ③basic structure of each part.

2. Necessary dimensions

The types of dimensions are as follows: ① external dimensions; ②assembly dimensions; ③characteristic dimensions; ④mounting dimensions; ⑤as well as, other important dimensions.

第9章　装配图

前面所述的表达零件的各种方法，如视图、剖视图、断面图以及局部放大图等，在由若干零件组成的部件或机器设备的装配图中也同样适用。但是在装配图中，除过要表达出主要零件的形状和结构外，还要表示出部件的工作原理、连接关系。因此，与零件图相比，装配图还有一些特殊的表达方法和规定画法。

本章主要介绍装配图的规定画法和特殊画法，以及绘制和阅读装配图的方法等内容。

9.1　装配图概述

9.1.1　装配图的作用

装配图是表达机器、部件或设备的图样。

在设计新产品、改进旧设备时，一般先画出装配图，再根据装配图画出全部零件图。

在生产过程中，根据装配图将零件装配成机器、部件或设备。

在使用和维修过程中，装配图可帮助使用者了解机器或部件的结构、各组成部分之间的相对位置、连接及装配关系与工作原理，为安装、检验和维修提供技术资料。所以，装配图是设计、制造、安装、使用、维修机器或部件的重要技术文件。

9.1.2　装配图的内容

图9-1所示的球阀是一种控制流量的开关装置，其装配图如图9-2所示。由图可知，一张常规的装配图，一般应包括四个方面的内容。

1. 一组视图

视图表达内容包括：①各零件间的相对位置；②传动关系及工作原理；③每个零件的基本结构。

2. 必要的尺寸

尺寸标注包括：①外形尺寸；②装配尺寸；③特性尺寸；④安装尺寸；⑤其他重要尺寸。

Chapter 9 Assembly Drawings 第 9 章 装　配　图

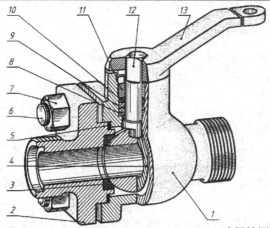

Fig. 9-1 Axonometric drawing of a ball valve 球阀轴测图
1—Valve body 阀体 2—Valve cover 阀盖 3—Valve gasket 密封垫 4—Valve core 阀芯 5—Spacer 调整垫
6—Stud 双头螺柱 7—Nut 螺母 8—Stuffing spacer 填料垫 9—Middle stuffing 中填料
10—Above stuffing 上填料 11—Packing set 压紧套 12—Valve handle 阀杆 13—Wrench 扳手

3. Technical requirements

Technical requirements mainly include: ① performance, assembly and debugging; ② operation, acceptance check; ③ decoration.

4. Title block, No., item lists

The contents in title block, part numbers and item list include: ① Name, scale, number of pieces, drawing number, specification of the machine or assembly unit; ② drawing number, name, material, quantity, standard of all the parts; ③ signature of the designer, checker, and person approving, etc.

3. 技术要求

技术要求主要有：①性能及安装调试等方面的要求；②运转及验收使用等方面的要求；③装饰等方面的要求。

4. 标题栏、序号、明细栏

标题栏、零件序号及明细栏的内容包括：①机器或部件名称、画图比例、件数、图号、规格等；②全部零件的序号、名称、材料、数量、标准等；③设计、校核、审核人员的签名等。

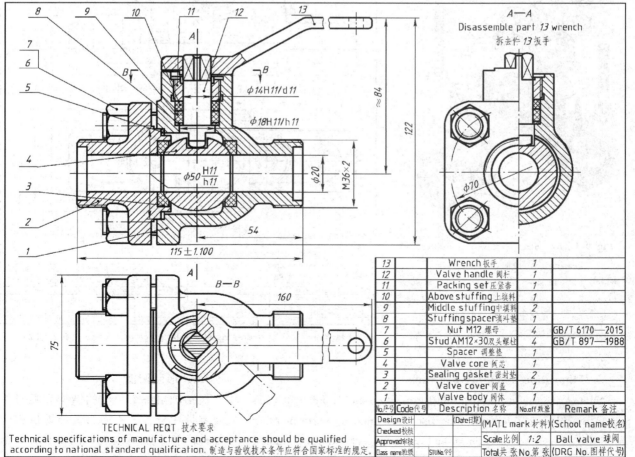

Fig. 9-2 Assembly drawing of a ball valve 球阀装配图

9.1.3 Types of assembly drawings

Assembly drawings can be divided into the following types in terms of their purposes and uses:

1. General assembly drawings

Fully and clearly represent the mounting and function of each sub-assembly, including outline drawings, sectional views, technical requirements, essential dimensions, part numbers and title blocks or item lists. These characteristics are exemplified in the assembly drawing of a ball valve in Fig. 9-2.

2. Design assembly drawings

Draw major parts together to determine their distances and dimensions, which is used to assess the feasibility of the design. Fig. 9-3 is an axonometric drawing of each part of the sliding bearing, and Fig. 9-4 shows the design assembly drawing of the sliding bearing.

9.1.3 装配图的种类

根据表达目的及用途的不同可将装配图分为以下几种：

1. 常规装配图

完整、清楚地表达各个部件的装配关系及其作用，包括外形图及剖视图、技术要求、必要的尺寸及零件序号、标题栏、明细栏等，如图9-2所示球阀装配图。

2. 设计装配图

将主要部件画在一起，以便确定其距离及尺寸关系等，常用来评定该设计的可行性。图9-3所示为滑动轴承各零件的轴测图，图9-4所示为该滑动轴承的设计装配图。

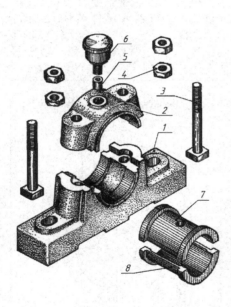

Fig. 9-3 Axonometric view of a sliding bearing 滑动轴承轴测图

1—Bearing seat 轴承座　2—Bearing cover 轴承盖　3—SQ screw 方头螺栓　4—Nut 螺母　5—Bush fastness set 轴瓦固定套　6—Oil cup 油杯　7—Above bush 上轴瓦　8—Bottom bush 下轴瓦

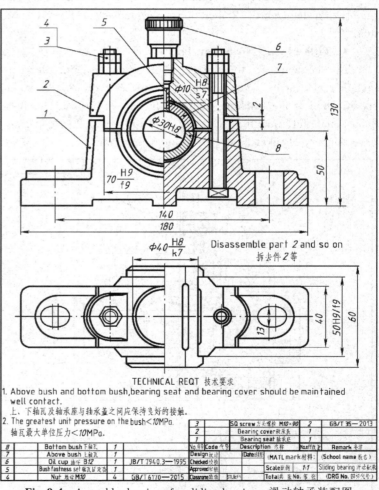

Fig. 9-4 Assembly drawing of a sliding bearing 滑动轴承装配图

3. Figuration assembly drawings

Summarize the structure, relative position and number of each part to provide a catalog and item list for selling the corresponding parts. Fig. 9-5 is a figuration assembly drawing of a multifunctional milling machine.

3. 外形装配图

概括画出各个部件的结构和相对位置及零件序号，常用来为销售提供相应零部件的目录及明细表。图9-5所示为多用途铰磨机的外形装配图。

4. Subassembly drawings

Represent the most complicated mounting parts to show how they mount onto the main structure. Fig. 9-6 is a subassembly drawing of a thread valve, while Fig. 9-7 is its general assembly drawing.

4. 局部装配图

仅将最复杂的装配部分画成局部装配图，以便于了解主要装配结构。图9-6所示为螺旋阀的局部装配图，图9-7是螺旋阀的常规装配图。

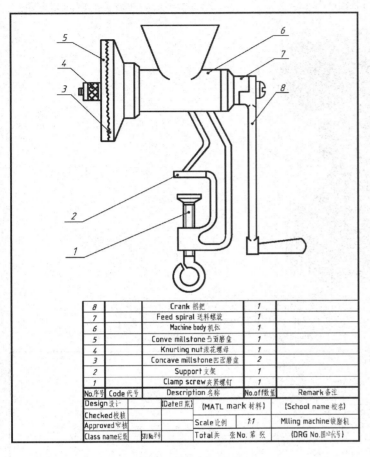

Fig. 9-5 Figuration assembly drawing of a multifunctional milling machine 多用途铰磨机外形装配图

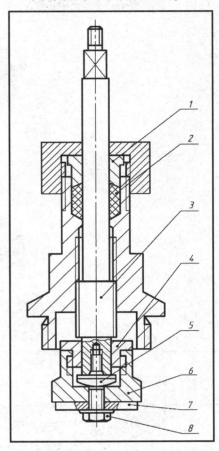

Fig. 9-6 Subassembly drawing of a thread valve 螺旋阀局部装配图

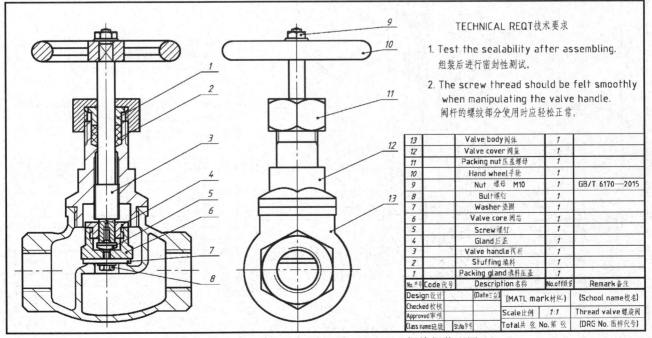

Fig. 9-7 Assembly drawing of a thread valve 螺旋阀装配图

5. Sectional assembly drawings

Using the sectional assembly drawing to represent the mounting of complex internal structures can clearly express any hidden mounting structures. Fig. 9-8 illustrates an axonometric drawing, while Fig. 9-9 shows a sectional assembly drawing of a throttle flap.

5. 剖视装配图

将内部结构复杂的装配关系画成剖视装配图，从而可将隐藏的装配结构清晰地表达出来。图 9-8 所示为节流阀的轴测图，图 9-9 所示为节流阀的剖视装配图。

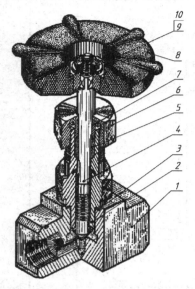

Fig. 9-8 Axonometric drawing of a throttle flap 节流阀的轴测图
1—Valve body 阀体 2—Valve cover 阀盖 3—Spacer 垫片 4—Valve handle 阀杆 5—Stuffing 填料
6—Packing nut 压盖螺母 7—Packing gland 填料压盖 8—Hand wheel 手轮 9—Nut 螺母 10—Washer 垫圈

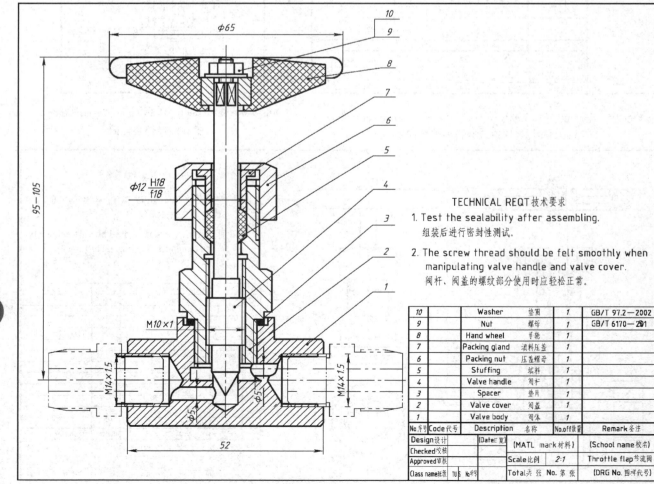

Fig. 9-9 Sectional assembly drawing of a throttle flap 节流阀的剖视装配图

9.2 Representation Methods of Assembly Drawings

9.2.1 General representation

All the above-mentioned representations are applied to the assembly drawing, including views, sectional views, cuts, and simplified representations. In addition, assembly drawing has some special and specified representations due to different objectives and purposes.

9.2.2 Conventional representation

1. Representing section lines

Section lines in two contiguous metal parts must be in opposite directions or with different section line spacing, while section lines of the same part must be consistent on different views (see Fig. 9-10).

2. Representing contact surfaces and fitting surfaces of two parts

The contact surfaces or fitting surfaces of two parts are represented as a line (see Fig. 9-10B, E), while the non-contact surfaces or non-fitting surfaces are represented as two lines even if the clearance is very small (see Fig. 9-10C, D).

9.2 装配图的表达方法

9.2.1 一般表达方法

前面所介绍的各种表达方法均适用于装配图，包括视图、剖视图、断面图、简化画法等。但由于表达的对象和目的不同，装配图还有一些规定画法和特殊画法。

9.2.2 规定画法

1. 剖面线的画法

两个相邻金属零件的剖面线应方向相反或间距不同，而同一零件的剖面线在各个视图中应保持一致（见图9-10）。

2. 零件接触面与配合面的画法

两个零件的接触面或配合表面只画一条线（见图9-10B、E），而非接触面或非配合表面处，即使间隙很小也应画成两条线（见图9-10C、D）。

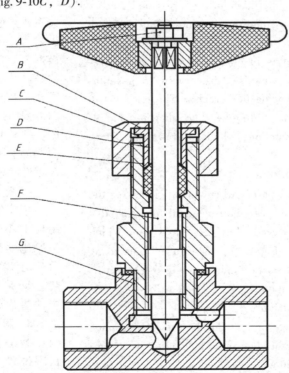

Fig. 9-10 Sketch map for throttle flap fitting　节流阀装配关系示意图

3. Representing fastening pieces and solid pieces

If the screwed fastening pieces and solid pieces (shaft, key, pin, ball, connecting rod, etc.) are longitudinally (axially) sectioned and the cut plane goes through the center of symmetry, these parts are represented as if they were not sectioned (see Fig. 9-10 A, F).

See Fig. 9-10 as an example, for a schematic representation of a throttle flap assembly.

3. 紧固件及实心件的画法

对于螺纹紧固件及实心件（轴、键、销、球、连杆等），若按纵向（沿轴线）剖切且剖切平面通过其对称中心时，这些零件均按不剖画出（见图9-10 A、F）。

以上画法参见图9-10 节流阀装配关系示意图。

9.2.3　Special representation

1. Cutting away along fitting surfaces representation

To represent the internal details of a machine or an assembly, we can represent some parts as a cut away drawing along the fitting surfaces. In this representation, the section lines of the fitting surfaces are not needed, but the section lines of parts that pass through the fitting surfaces are needed. The left view in the assembly drawing of a gear pump, shown in Fig. 9-11, is obtained by cutting away along the fitting surfaces between the pump body and the pump cover. Note that the section lines on the surface of the pump body is omitted, while the shaft and screw that pass through the cut plane are represented in the sectional view.

2. Single piece representation

If the structure of a certain part is not clearly represented, which will thus affect the understanding of its fitting relationship and function, the part must be drawn separately and marked with "×× direction" near the corresponding view indicating the projection direction with arrows and capital letters. "Part 3 C" shown in Fig. 9-11 is one example of this representing way.

3. Dismounting representation

If the fitting relationships or the parts to be represented are behind some other parts, one may assume that the interfering parts are removed, and then clearly draw the parts to be represented. Make sure to represent the removed parts in other views and mark "removed xx". In Fig. 9-2, the left view is a dismounting representation, which is drawn out by removing the wrench. The top view in the design assembly drawing of the sliding bearing, shown in Fig. 9-4, is also a dismounting representation, which is drawn out by removing the bearing cover.

4. Enlargement representation

For the tiny structures, thin parts, when the thickness or the clearance shown on the drawing is less than or equal to 2mm, this part of the drawing may be enlarged. Part 2, the washer which shown in the front view in Fig. 9-11, is drawn out using enlargement representation. The washer shown in Fig. 9-12 also illustrates an enlargement representation.

5. Imagination representation

To represent the moving range or limit location of moving parts or other parts related to the assembly but not belonging to the assembly, draw the contour using a thin long dashed double dotted line. The limit location of part 13, the wrench in the top view shown in Fig. 9-2, the assembly drawing of the ball valve, and the two pipe junctions connecting the throttle valve shown in Fig. 9-9, the sectioned assembly drawing, as well as the representation of the bolt, nut and the base line connecting the base shown in Fig. 9-11, the assembly drawing of a gear pump, are examples of imagination representation.

9.2.3　特殊画法

1. 沿结合面剖切画法

为了表达机器或部件的内部结构，可采用沿某些零件的结合面剖切的画法。其特点是在结合面上不画剖面线，但穿过该结合面被剖切到的零件应按剖视画。图9-11所示齿轮泵装配图中的左视图，即为沿泵盖和泵体结合面剖切画法，泵体表面上不画剖面线，而被剖切到的轴、螺钉需画上剖面线。

2. 单件画法

当某个零件的结构未表达清楚，且对理解配置关系或部件功能有影响时，应单独画出该零件，并在其上方注明"零件××"，在相应视图附近用箭头指明投射方向，并注明相应的大写字母。图9-11所示的"零件3 C"即属此画法。

3. 拆卸画法

当某些零件遮住了要表达的装配关系或其他零件时，可假想将这些零件拆去后，再画出欲表达部分的视图，但被拆去的零件必须在其他视图中已表达清楚。并应在其上方注明"拆去××件"字样。图9-2所示球阀装配图中的左视图即采用拆卸画法，是拆去扳手后画出的。图9-4所示滑动轴承装配图中的俯视图也采用了拆卸画法，是拆去轴承盖等零件后画出的。

4. 夸大画法

细小结构、薄件的厚度或间隙在图形上小于或等于2mm时，允许不按原比例而将其适当夸大画出。图9-11所示齿轮泵装配图主视图中的件2垫片即采用夸大画法画出。图9-12中的垫片亦采用了此画法。

5. 假想画法

为了表示运动零件的运动范围或极限位置，或与装配体有连接关系而又不属于该装配体的其他相邻部件时，可用细双点画线画出其轮廓。例如，图9-2所示球阀装配图的俯视图中零件13扳手的极限位置，图9-9所示节流阀剖视装配图中与该阀相联接的两个管接头，以及图9-11所示齿轮泵装配图中与底座相联接的螺栓和基底板的画法均为假想画法。

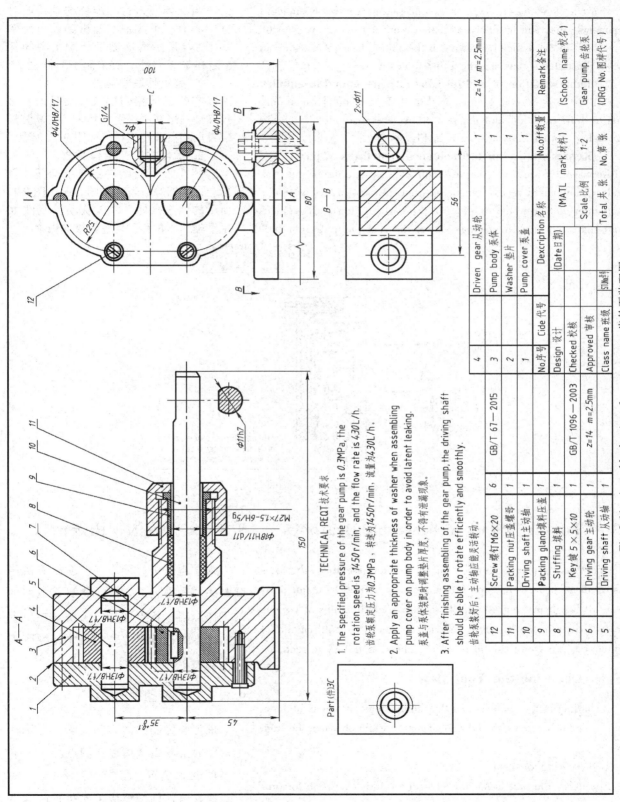

Fig. 9-11 Assembly drawing of a gear pump 齿轮泵装配图

9.2.4 Simplified representation

1. Representation of standard components

If a cut plane goes through an axis or a symmetry line of standard components, they can be drawn as if they were not cut away. Part 6, the oil cup, in the assembly drawing of the sliding bearing as shown in Fig. 9-4 illustrates a standard simplified representation.

2. Representation of multiple threaded joints and assemblies

If there are several groups of threaded part joints, one only needs to draw in detail and the rest others are positioned with thin long dashed dotted lines as shown in Fig. 9-12.

3. Representation of technological structures of parts

In assembly drawings, certain technological structures, such as round, chamfer and recess can be omitted. As shown in Fig. 9-11, the chamfer and grinding undercut of the part 3 pump body, part 5 driven-shaft and part 10 driving-shaft of the gear pump are omitted.

9.2.4 简化画法

1. 标准件的画法

当剖切面通过标准件的轴线或对称中心时，可按不剖画出其外形。例如图 9-4 所示滑动轴承装配图中的件 6 油杯是标准件，故采用了此画法。

2. 螺纹件联接画法

有若干组相同螺纹联接件时，可仅详细画出其中一组，其余只在其装配位置用细点画线表示，如图 9-12 所示。

3. 零件的工艺结构画法

零件的工艺结构，如小圆角、倒角、退刀槽等可省略不画。例如图 9-11 所示的件 3 泵体、件 5 从动轴及件 10 主动轴的倒角和砂轮越程槽均未画出。

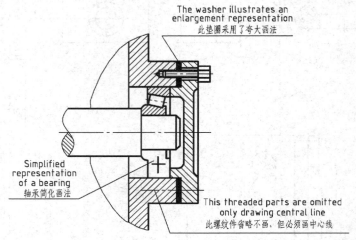

Fig. 9-12 Enlargement representation 夸大画法

9.3 Choosing Views of Assembly Drawings

The working principles of a machine or an assembly, the relationship of each part, and the fundamental structures of major parts should all be clearly represented by accurate applications of various representations in the assembly drawing to meet the requirements of production. The views should be brief, clear, and easy to read.

9.3.1 Choosing the front view

According to the contents and the requirements of assembly drawings, the following items should be considered while choosing the front view:

1. Working position

Normally, the part is positioned in the working position for producing the front view. For example, the gear pump is positioned based on its working position, shown in Fig. 9-11, the assembly drawing of a gear pump.

9.3 装配图的视图选择

为了满足生产的需要，应正确运用装配图的各种表达方法，将机器或部件的工作原理、各零件间的关系及主要零件的基本结构清晰地表达出来。视图表达方案力求简明，便于读图。

9.3.1 主视图的选择

根据装配图的内容和要求，在选择主视图时应着重考虑以下两点：

1. 工作位置

通常将部件摆放成工作位置画其主视图。例如图 9-11 所示的齿轮泵就是以其工作位置摆放的。

2. Fitting relationships

One should show as much as possible the structures and features, especially the main fitting relationships, the functions and working principles of a machine or an assembly.

As shown in Fig. 9-11, the gear pump is positioned in its working position, through the axis of a couple of engaging gears, the section plane in the front view clearly represents the main fitting relationship of the gear pump.

9.3.2 Choosing other views

One may use different views to represent various other portions not clearly represented on the front view. In such cases, one needs to consider the fitting relationship and working principle before considering the structural shape of the major part. For example, as shown in Fig. 9-11, the left view of the gear pump, the mounting position of the screw (part 12), and the external structure of the pump body (part 3) and the pump cover (part 1), are all functional representations. The shape of the oil entrance is also represented with a partial sectional view. The left view describes its working principle: If two gears rotate in accordance with the direction of the arrow shown in Fig. 9-13, the portative force generated in the left of the gear transports oil into a pump from the oil entrance. As the pump runs, it outputs oil from the exit at a certain pressure.

2. 装配关系

应考虑尽可能多地显示机器或部件的结构特征，特别是能清楚地表达机器或部件的主要装配关系、功能和工作原理。如图 9-11 所示，主视图中的剖切面通过一对啮合齿轮的轴线，清楚地反映了齿轮泵的主要装配关系。

9.3.2 其他视图的选择

针对主视图中未表达清楚的部分，还应辅以其他视图进行补充表达。此时应将装配关系及工作原理的表达放在首位，其次考虑主要零件的结构形状等。图 9-11 左视图中补充表达了齿轮泵的工作原理、螺钉（件 12）的装配位置以及泵体（件 3）和泵盖（件 1）的外形结构，并采用局部剖视反映了进油口的形状。由左视图可知其工作原理为：当两个齿轮按图 9-13 中箭头所示方向旋转时，在齿轮啮合区的左侧产生真空吸力，将油从吸油口吸入泵内，随着齿轮的转动，不断地从出油口将具有一定压力的油输送出去。

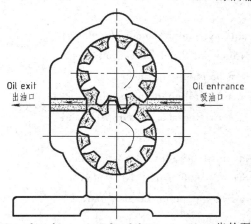

Fig. 9-13 Illustration of working principle of the gear pump 齿轮泵工作原理示意图

After determining the left view, check what other structures are not clearly represented in accordance with the needs of an assembly drawing. With this content in mind, choose other views and ensure that each view has a clear purpose. For example, as shown in Fig. 9-11, the C directional view is added to express the shape of the dummy club at the oil entrance and exit on pump body (part 3). the B—B view is added to show the structural shape at the base and the bottom cutting of pump body (part 3). The selection of views for the gear pump is now complete, but one needs to keep in mind other representation views for comparison in order to select the best representation.

左视图确定后，再根据装配图应表达的内容，检查还有哪些内容是没有表达清楚的。据此再选择其他视图，使每个视图都有明确的表达目的。图 9-11 所示的齿轮泵装配图中，增加 C 向局部视图，以表达泵体（件 3）上吸油口、出油口处凸台的形状，增加 B—B 剖视图，以表达泵体（件 3）的底座结构和下部的断面形状。至此，齿轮泵的视图选择就算完成了。但有时为了能选定一个最佳方案，最好多考虑几种表达方案，以供比较和选用。

9.4 Dimensioning and Specifications for Assembly Drawings

9.4.1 Dimensioning

On an assembly drawing, only the following dimensions are marked:

1. Characteristic and specification dimensions

The dimensions that represent machine specifications or working performance are the foundation on which users choose the right products, such as travel distances, calibers and threads. For example, the dimension G1/4 of pipe thread and the diameter $\phi 4mm$ of the oil entrance and exit in the gear pump shown in Fig. 9-11 are related to the volume of the incoming and outgoing oil, therefore these dimensions are characteristic and specification dimensions.

2. External dimensions

The dimensions that represent the general length, width, and height are useful for packing, storage and transportation. In Fig. 9-11, for example, the total length is 150mm, the total width is 80mm, and total height is 100mm.

3. Fitting dimensions

Fitting dimensions represent dimensions and related tolerances between parts or subassemblies that have the following characteristics:

1) The distance from the main axis to the mounting datum face, such as the center height of 45mm for the driving-shaft (part 10) in the assembly drawing of the gear pump shown in Fig. 9-11.

2) The distance between the main parallel shafts, such as the center distance $35^{+0.1}_{0}$mm between the driving-shaft (part 10) and driven-shaft (part 5) in Fig. 9-11.

3) The fitting relationship that must be kept after mounting, such as the fitting dimensions of driving-shaft (part 10), driven-shaft (part 5), pump body (part 3) and pump cover (part 1) in the gear pump are all $\phi 13H8/f7$.

4. Mounting dimensions

Mounting dimensions include those that are required to mount the machine or subassembly to its base or to another units. For example, the center distance 56mm of two holes and the dimension $2\times\phi 11mm$ of the hole on the base of the pump body (part3), as shown in Fig. 9-11.

5. Other important dimensions

Other important dimensions include dimensions of major parts and limit dimensions of moving parts, such as the radius $R25mm$ of the distribution of the threaded holes used to mount screws on the pump body and pump cover.

9.4 装配图的标注

9.4.1 尺寸标注

在装配图中，一般只标注下列几种尺寸：

1. 特性与规格尺寸

特性与规格尺寸表示机器规格或工作性能的尺寸，是用户选用产品的依据，如行程、口径及螺纹等。例如图9-11所示齿轮泵装配图中的吸油口、出油口管螺纹尺寸 G1/4 及口径尺寸 $\phi 4mm$，它们与泵的吸油量和出油量有关，所以是特性与规格尺寸。

2. 外形尺寸

外形尺寸即表示机器长、宽、高的总体尺寸，供装箱、存储和运输时参考。如图 9-11 中，齿轮泵的总长为 150mm，总宽为 80mm，总高为 100mm。

3. 装配尺寸

装配尺寸表示如下零部件间有配合性能要求的配合尺寸及公差等：

1) 主要轴线到安装基准面的距离。例如图 9-11 所示齿轮泵装配图中，主动轴（件 10）的中心高 45mm。

2) 主要平行轴之间的距离。例如图 9-11 所示齿轮泵装配图中主动轴（件 10）与从动轴（件 5）的中心距 $35^{+0.1}_{0}$mm。

3) 装配后两零件之间必须保证的配合关系。例如图 9-11 齿轮泵中主动轴（件 10）、从动轴（件 5）与泵体（件 3）和泵盖（件 1）的配合尺寸均为 $\phi 13H8/f7$ 等。

4. 安装尺寸

安装尺寸表示机器或部件安装在基底或整机系统上所需的尺寸。例如图 9-11 所示齿轮泵装配图中泵体（件 3）底座上两孔的中心距 56mm 及孔的尺寸 $2\times\phi 11mm$。

5. 其他重要尺寸

其他重要尺寸包括主要零件的重要尺寸及运动零件的极限尺寸等。如泵体与泵盖上安装螺钉用的螺孔分布半径尺寸 $R25mm$。

9.4.2 Numbering (GB/T 4458.2—2003)

For the convenience of reading and managing drawings and for organizing production, parts and sub assemblies need to be numbered in a specific process. The numbering should be in accordance with the national standard.

1. General specification

1) All parts and subassemblies in an assembly drawing must be numbered.

2) The same part in the same assembly drawing should use the same number and be marked only once.

3) The part number in an assembly drawing should be in accordance with that in the item block. One should first mark the number on the drawing to ensure the number sequence is in the right order and then fill in the item list.

4) All numbers should be arranged clockwise or counterclockwise in the front view along the horizontal or vertical directions. If the number can not be sequenced continously on the drawing, one can arrange them in an order along the horizontal or vertical directions.

2. Marking methods

1) Part numbers shall be marked on the horizontal line or inside the circle of the leader, or they can be directly marked at the end of the leaders. The font size shall be one or two grades larger than the dimension number on the drawing (see Fig. 9-14a).

2) A leader shall be drawn from the visible contour of the part or subassembly with a continuous thin line. The end of the leader shall have a small dot. If it is not easy to draw a dot, one can point to the contour with an arrow (see Fig. 9-14b).

3) The leader cannot be intercrossed with each other or parallel to the section lines. When necessary, a leader line may be drawn as a folding line but only can be folded once.

4) A group of fasteners or parts with a clear fitting relationship can be labeled with a shared leader (see Fig. 9-14c).

Refer to the part numbering shown in Fig. 9-14 for examples of the above-mentioned numbering.

9.4.3 Item lists (GB/T 10609.2—2009)

Item list is a table used to indicate part number, code, name, specification, quantity and material. Refer to Fig. 9-15 for the dimension specification of an item list.

9.4.2 序号标注 (GB/T 4458.2—2003)

为了便于读图、组织生产和图样管理，需要对装配图中的零部件进行编号，称为零件序号。编写序号应遵循国家标准的有关规定。

1. 一般规定

1) 装配图中的所有零部件都必须编写序号。

2) 同一装配图中相同的零件应编写同一个序号，只注一次。

3) 装配图中的零件序号应与明细栏中的序号一致。为确保零件序号顺序排列，应先在图中编写序号，然后再填写明细栏。

4) 序号应按顺时针或逆时针方向排列整齐。若在整个图上无法连续时，可只在每个水平或垂直方向顺次排列。

2. 标注方法

1) 零件序号应注写在指引线的水平线上或圆圈内，或直接注在引线末端，且字号比图中的尺寸数字大一号或两号（见图9-14a）。

2) 指引线应从零部件的可见轮廓线内用细实线引出，端部画一小圆点。若不方便画小圆点时，可用箭头指向其轮廓线（见图9-14b）。

3) 指引线不可相互交叉或与图中的剖面线平行。必要时可画成折线，但只可折一次。

4) 一组紧固件或装配关系清楚的零件组，可采用公共指引线（图9-14c）。

以上所述零件序号编写方法参见图9-14所示的零件序号标注形式。

9.4.3 明细栏 (GB/T 10609.2—2009)

明细栏是说明零件序号、代号、名称、规格、数量、材料等内容的表格。明细栏格式和尺寸如图9-15所示。

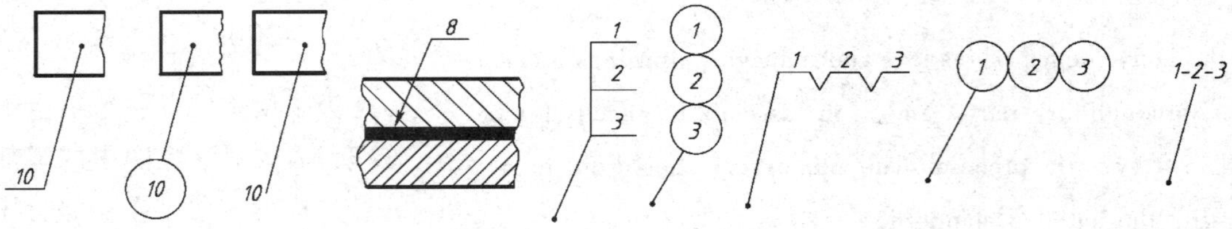

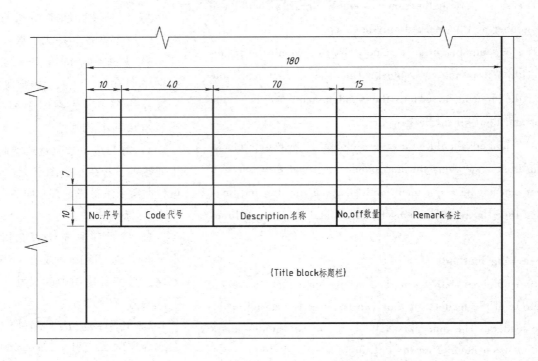

Fig. 9-14 Preparation of part serial numbers 零件序号的编写形式
a) Types of serial numbers 序号的形式 b) Serial number of a thin part 薄件的序号形式
c) Serial numbers of group parts 一组零件的序号

Fig. 9-15 Simplified format of an item list 明细栏简化格式

The item list is generally drawn on the top of the title block. Fill in the numbers from the bottom to the top. If the top space is not enough, other portions can be moved to the left of the title block and filled in numbers continously from the bottom to the top. See the item list in the assembly drawing of a gear pump shown in Fig. 9-11.

9.4.4 Technical requirements

Technical requirements are the texts or symbols indicating the requirements of the performance, fitting, mounting, inspection, debugging, use of the machine or subassembly and transportation. Technical requirements are very briefly written at the bottom of assembly drawings.

明细栏一般画在标题栏的上方，序号应自下而上填写。当上方位置不够时，可将其余部分画在标题栏的左方，继续自下而上填写，如图 9-11 所示齿轮泵装配图的明细栏。

9.4.4 技术要求

技术要求是用文字或符号说明机器或部件的性能、装配、安装、检验、调试和使用及运输等方面的要求和注意事项。技术要求一般应注写在装配图下方的空白处，力求文字简练。

9.5 Rationality of Fitting Structures

To ensure the working performance of the machine or subassembly and for the convenience of manufacturing and maintenance, one must consider the rationality of the fitting structure during the design stage. This section introduces the rationality of some commonly used fitting process structures.

9.5.1 Structures of parts contact

1. Quantity of the contact surface

1) If two parts assemble in the same direction, there should be only one pair of contact surfaces, as the bottom group shown in Fig. 9-16. Thus, reliable contact of the parts can be ensured, the processing requirements can be lowered and the cost can be reduced.

2) Two tapered parts are permitted to have only one pair of contact surfaces in two directions (axial or radial). In Fig. 9-17, the top drawings have two contact surfaces in axial and radial directions, therefore, they are inappropriate. The bottom drawings only have one contact surface in axial and radial directions, which can ensure stable contact between the surfaces.

2. Structures at contact corner positions

When two parts have contact of a pair of intersecting surfaces at the same time, it is a must to machine a chamfer, a round or an undercut at the corner position of the two intersecting surfaces for stable contact, as shown in Fig. 9-18.

9.5 装配结构的合理性

为了保证机器或部件的工作性能，并方便加工制造和拆卸维修，在设计时必须考虑装配结构的合理性。本节主要介绍一些常用的装配工艺结构的合理性。

9.5.1 接触处的结构

1. 接触面的数量

1) 两个零件在同一方向上只允许有一对接触面，如图 9-16 中下图所示。这样既保证了零件的可靠接触，又降低了加工要求和成本。

2) 两个圆锥体零件在两个方向（轴向、径向）上，只允许有一对接触面。如图 9-17 所示，上面的一组图中在轴向和径向均有两对接触面，是不合理的；而下面各图中则沿轴向和径向只有一对接触面，这样就能确保这两个表面稳定接触。

2. 接触面转角处的结构

为了保证两零件转角处接触良好，应在转角处制出倒角、圆角或退刀槽，如图 9-18 所示。

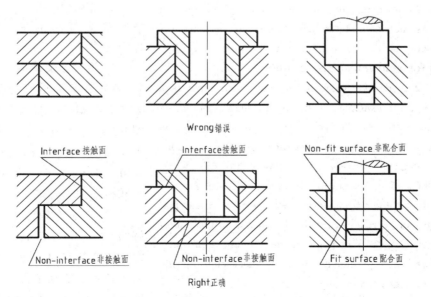

Fig. 9-16 Quantity of the contact surfaces　接触面的数量

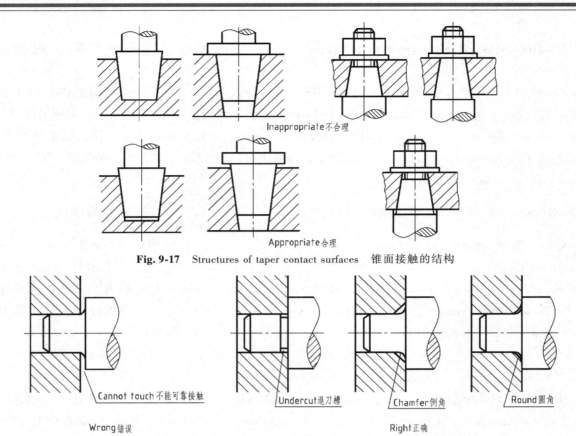

Fig. 9-17 Structures of taper contact surfaces 锥面接触的结构

Fig. 9-18 Structures at contact corner positions 接触面转角处的结构

9.5.2 Removable joint structures

1. Reliable joint structures

1) When drafting, one should ensure that the external thread fully goes into the internal thread. The external thread should have an undercut or the internal thread should have a chamfer, as shown in Fig. 9-19a.

2) If the end of a shaft is a thread joint, one should keep a segment of the thread from going into the nut to ensure that the nut can move forward, as shown in Fig. 9-19b. Fig. 9-19c does not reserve the thread length, so the nut cannot move forward.

2. Structures for convenient disassembly

1) One should consider for easy disassembly when there is a bearing or bushing by, for example, lowering the shaft shoulder or notching, as the bottom drawings in Fig. 9-20a. Alternatively one can make a small hole, as shown in the bottom drawings of Fig. 9-20b, Fig. 9-20c. The top drawings of Fig. 9-20a, Fig. 9-20b, Fig. 9-20c show incorrect structures of an irremovable bearing and bushing.

2) There should be enough space when there is a thread fastener so that the bolt can be removed. One should not only consider the space for the bolt, but also the space for the wrench when removing the bolt, as well as the convenience for disassembling the structure. For example, see the bottom drawings of Fig. 9-20d. The top drawings shown in Fig. 9-20d are inappropriate for assembly or disassembly operations.

9.5.2 可拆联接结构

1. 可靠联接结构

1）要使外螺纹全部拧入内螺纹中，外螺纹结束处应有退刀槽，或者内螺纹起始端应有倒角，如图 9-19a 所示。

2）当轴端为螺纹联接时，螺纹段应留出余量，以保证螺母可以继续旋入，如图 9-19b 所示。而图 9-19c 所示螺纹段未留出余量，导致螺母无法拧入。

2. 拆装方便结构

1）在有轴承或衬套处应考虑拆卸方便的问题，如放低轴肩、开凹槽，如图 9-20a 中下图所示；或加工一个小孔等，如图 9-20b、c 中下图所示。而图 9-20a、b、c 中的上图则是无法拆卸轴承和衬套的不合理结构，应注意避免。

2）在装有螺纹紧固件处应留有足够的空间，以便拆装。不仅应留出拧入螺栓所需的空间，同时还应考虑拆装时所用扳手的活动范围，如图 9-20d 中下图所示。而图 9-20d 中上图所示则是无法安装和拆卸的不合理情况。

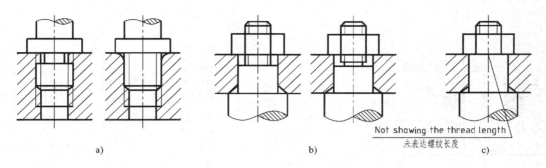

Fig. 9-19 Reliable joint structures 可靠联接结构

a) External screw thread screwed into the screw hole (right connection) 外螺纹与螺孔联接（正确）

b) Threaded shaft connection (correct) 轴端螺纹联接（正确）

c) Threaded shaft connection (wrong) 轴端螺纹联接（错误）

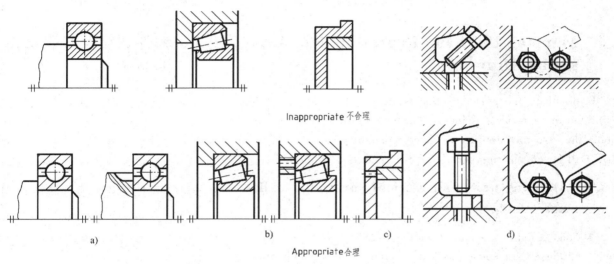

Fig. 9-20 Structures for convenient disassembly 拆装方便的结构

a) Disassembly of shaft bearings structures 轴上轴承的拆装结构 b) Disassembly of hole bearing structures 孔内轴承的拆装结构

c) Disassembly of a bush structure 衬套的拆装结构 d) Disassembly space for a bolt 螺栓的拆装空间

9.5.3 Structures of sealing devices

1. Felt ring sealing device

To use a felt ring for sealing purpose, one needs to produce an annular groove (standard structure, the dimensions of which can be found in a handbook) in the hole for mounting a shaft, insert a felt ring in the groove and fit it tightly in contact with the shaft, as shown in Fig. 9-21.

2. Stuffing box sealing device

In this case, the sealing is realized by adjusting the position of a gland using a nut, which compresses the stuffing tightly for sealing purpose. When drawing, the gland shall be positioned where it is pushed by 3—5mm, as shown in Fig. 9-22b.

3. Washer sealing device

In this case, a washer is used on the contact surfaces of two parts for sealing purpose. If the thickness of the washer shown in the drawing is less than or equal to 2mm, one should use an enlargement representation to clarify it, as shown in Fig. 9-12 and Fig. 9-21.

9.5.3 密封装置结构

1. 毡圈密封装置结构

在装轴的孔内开出一个环槽（属标准结构，尺寸可查有关手册），将毛毡圈置于槽内并使其与轴紧密接触，可起密封作用，如图9-21所示。

2. 填料函密封装置结构

通过螺母调节填料压盖的位置，将填料压紧而起到密封作用。画图时应使填料压盖处于可调节位置，一般使其调节量为3~5mm，如图9-22b所示。

3. 垫片密封装置结构

在两零件结合面常采用垫片密封。当垫片厚度在图形上小于或等于2mm时，应采用夸大画法将其涂黑，如图9-12和图9-21所示。

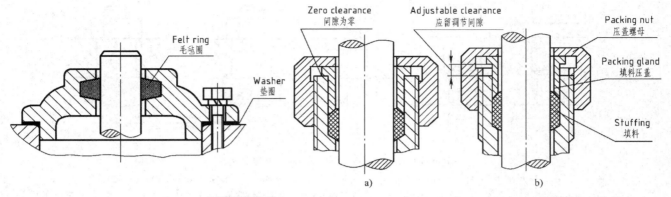

Fig. 9-21 A felt ring sealing device 毡圈密封装置

Fig. 9-22 A stuffing box sealing device 填料函密封装置
a) Wrong 错误 b) Correct 正确

9.6 Mapping Units and Representation of Assembly Drawings

In practice, the procedure that is composed of measuring the machine or equipment, producing schematic drawing and then producing the assembly drawing is called a mapping process. It is a basic skill for engineering personnel to be able to map the units.

9.6.1 Mapping units and assembly drawings of flat nose clamp

We take a flat nose clamp (see Fig. 9-23) as an example and briefly introduce the method and steps to produce an assembly drawing through mapping the units.

1. Understanding the subassembly

Produce an assembly sketch based on the understanding of the purpose, working principle, assembly relationship, transmission routes of the assembly, and the basic structures of major parts of the assembly.

(1) Understanding the purpose A flat nose clamp is used to fix the workpiece on the platen for machining.

(2) Understanding the working principle and transmission routes Turn the screw with the handle to drive the nut and to make the movable body of the clamp perform rectilinear movements along the fixed body of the pliers. This will make the mouth of the clamp open or close to clamp or unload the part.

(3) Understanding the assembly relationship, basic structures of the major parts Fig. 9-23 is an axonometric exploded drawing showing the fitting relationship of the flat nose clamp. It consists of eleven kinds of parts including a fixed clamp body, a screw, a nut and others. The nut and the movable clamp body are connected as a unit. The dummy club under the nut contacts with the fixed body of the clamp. When the screw rotates, the nut moves rectilinearly, but not rotationally. The boards are fastened to both bodies of the clamp with screws. The left side of the screw rod has a ring connected to the screw rod with a pin.

9.6 部件测绘及其装配图的画法

在生产实践中，对机器或设备进行测量，绘出草图，然后整理绘制出装配图的过程称为部件测绘。部件测绘是工程技术人员必备的一项技能。

9.6.1 平口钳部件测绘与装配图

下面以平口钳（见图9-23）为例，简述通过部件测绘画装配图的方法和步骤。

1. 了解部件

弄清装配体的用途、工作原理、装配关系、传动路线及主要零件的基本结构。

（1）了解用途 平口钳装在机床工作台上，用来夹紧工件。

（2）了解工作原理与传动路线 用手柄转动螺杆，使螺杆带动螺母，进而带动活动钳身沿着固定钳身做直线运动，从而使钳口张开或闭合，即可夹紧或卸下零件。

（3）了解装配关系及主要零件基本结构 图9-23所示为平口钳零部件装配关系轴测分解图。它由固定钳身、螺杆、螺母等11种零件组成。螺母与活动钳身用螺钉连成一体，螺母下方凸出的台阶与固定钳身接触。当螺杆转动时，螺母只做直线运动，不做旋转运动。钳口板用螺钉分别固定在活动钳身和固定钳身上。螺杆左端有一个环，用销钉与螺杆固定。

Chapter 9　Assembly Drawings　第 9 章　装　配　图

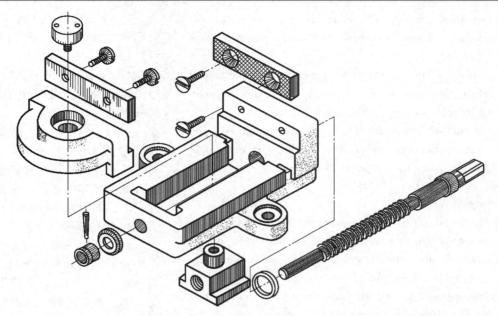

Fig. 9-23　Axonometric exploded drawing of a flat nose clamp　平口钳轴测分解图

2. Drafting the assembly schematic drawings

While understanding a new assembly or disassembling an old machine, one should draw an assembly schematic drawing in order to facilitate reassembly and as a reference. In other words, draw the contour of each part with specified symbols and lines that indicate the fitting relationship, their relative positions, transmission conditions and working principles, as shown in Fig. 9-24.

2. 画装配示意图

在了解一个新的装配体或拆卸一部旧机器的过程中，为方便装配并提供参考，应画出装配示意图，即用规定符号和简单图线画出装配体各零件的大致轮廓，用以说明各零件之间的装配关系和相对位置，以及传递情况和工作原理等，如图 9-24 所示。装配示意图的作用是为了方便将拆散的装配体复原，也可以作为画装配图时的参考。

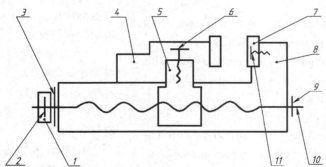

1—Ring 挡圈　2—Pin 销(GB/T 119.1—2000 A4×10)　3—Washer 垫圈(GB/T 97.1—2002 14—140HV)
4—Movable clamp body 活动钳身　5—Sheathing nut 套螺母　6—Screw 螺钉　7—Clamp plank (2 pieces) 钳口板(2个)
8—Fixed clamp body 固定钳身　9—Washer 垫圈　10—Screw rod 螺杆　11—Screw (4pieces) 螺钉(4个)(GB/T 68—2016 M6×10)

Fig. 9-24　Illustration of a flat nose clamp assembly　平口钳装配示意图

The following items should be noticed when producing an assembly schematic drawing:

1) A schematic drawing is produced as if the assembly is transparent, so both internal and external structures can be reflected.

2) Each part is represented with approximate contours or a single line.

画装配示意图应注意以下几点：

1) 示意图是将装配体假想设为透明体而画出的，因而外形轮廓和内部构造均可反映出来。

2) 每个零件只画大致轮廓，或用单线表示。

3) See specified symbols of commonly used parts in the standard *Mechanical drawings Graphical symbols for kinematic diagrams* GB/T 4460—2013.

4) Only one or two views are generally needed when producing a schematic drawing. There should be a gap between contact surfaces in order to identify the related parts.

5) Parts in the schematic drawing should be numbered. The list shows the name, quantity and material. The part name and number of the simple assembly can be written on the assembly schematic drawing, as shown in Fig. 9-24.

3. Mapping an assembly

Please note the following when mapping an assembly:

(1) Disassembly Disassembly should be performed in accordance with appropriate methods and steps to ensure the safety and integrity of the parts. One should think about the sequence of disassembly before processing and all parts should be numbered in accordance with the disassembly sequence. Small parts should be carefully stored so as to avoid losing or mixing them with other parts. Reassemble after the mapping is finished.

(2) Measurement While performing mapping, make sure to protect the machining surface and fitting surface of the part. When measuring, select the right measuring tools according to the required part accuracy. Unimportant dimensions should be changed into integers. For the important and small dimensions that require higher accuracy, decimal fractions can not be rashly changed. One should choose an appropriate value after carefully hashing. For fitting dimensions, one only needs to measure its nominal dimension, and then select the appropriate class according to its specific condition.

(3) Material of the part According to the appearance of the part, one can first determine whether the material is steel, cast iron, cast steel, or non-ferrous metal. Furthermore, compare the part with the products of the same type and choose the material brand according to the purpose and working conditions.

4. Determining the representation

Choose the projection direction of the front view and determine the drawing size and scale. Apply various representations to improve understanding.

Choose the front view for the flat nose clamp according to the above-mentioned principles, and then determine the following representations through analysis: the front view is a full sectional view cut along the front and back symmetry plane; the left view is a half sectional view or a partially sectioned view cut along the two holes of the fixed clamp body; and the top view is a partially sectioned view cut at the screw connecting to the board of the clamp mouth; use a partial direction view *B* to represent the reticulate pattern on the clamp plank; use a drawing of partial enlargement to represent the form of thread on the screw rod; use a cut to represent the cut plane structure of the right end of the screw rod, as shown in Fig. 9-25. According to scale 1∶1, one needs to use an A2 drawing sheet.

3）常用零件的规定符号参见国家标准《机械制图 机构运动简图符号》GB/T 4460—2013。

4）示意图一般只画一两个视图。两个零件接触表面应留出空隙，以便区分零件。

5）示意图中一般应编出零件序号，列表写出零件的名称、数量、材料等项目。简单装配体的装配示意图也可直接把零件名称、序号注写在示意图上，如图9-24所示。

3. 测绘装配体

测绘装配体应注意以下几点：

（1）装配体拆装问题 为了保证安全和不损坏机件，拆装要按照正确的方法、步骤进行。因此应先研究好拆装顺序，再动手拆装。拆下零件后可按拆卸顺序对其编号。要妥善保管小零件，以防丢失或发生混乱现象。测绘完成后，应将装配体复原。

（2）测量问题 在测绘中，要注意保护零件的加工面和配合面。测量尺寸时，要根据零件的精度要求选用相应的量具。对非主要尺寸，应将测得的尺寸化为整数。对精度要求较高的主要尺寸和数值比较小的尺寸，不能任意改变尺寸的小数值，而应经过认真推敲，采用合理的数值。对配合尺寸，一般只测量出公称尺寸，再根据具体条件选用合理的配合等级。

（3）零件材料 一般可先从外观判断零件材料是钢、铸铁、铸钢、有色金属等大类别，再根据零件的作用和工作条件，对比同类型产品，选定材料的牌号。

4. 确定表达方案

选择主视图投射方向，确定图幅及绘图比例，合理运用各种表达方法。

根据前面所讲的主视图选择原则，选好平口钳主视图，并经分析比较，确定以下表达方案：主视图沿前后对称面全剖；左视图沿固定钳身两孔处半剖或局部剖；俯视图在螺钉联接钳口板处局部剖；采用B向局部视图表达钳口板表面网纹形状；再用一个局部放大图表达螺杆的螺纹牙型；采用断面图表达螺杆右端部切有方头处的断面结构，如图9-25所示。按照1∶1的比例，需要采用A2号图纸

5. Preliminary drafting

First, draw the border, title block and item block. Second, position the views and draw the datum lines. Then, draw the major parts on the front view. Draw other parts according to their fitting relationships. When drafting, first map the major parts and draw the schematic drawings, then produce the assembly drawing according to the rough detail drawings and schematic drawings. General steps for preliminary drafting (Using 2H or H pencil and compass):

1) Draw the border and title block.

2) Specify the drawing's position in the border according to the ratio 3∶4∶3 (see Fig. 9-26), then draw the datum lines, symmetric lines and axes.

3) According to the three-alignment rules of different views, first draw the major parts, and then draw the neighboring parts.

4) Draw the detailed structure of each part and complete the drawing. Pay attention to the fitting relationship between contact surfaces and non-contact surfaces.

5) Draw the dimension boundaries, dimension lines, number leaders.

5. 画出底稿

先画图框、标题栏及明细栏；再布置视图，画出基准线；然后从主视图入手，画主要零件；最后根据装配关系依次画出其余零件。在画底稿时，应先将主要零件测绘后画出草图，再根据零件草图和装配示意图画装配图。

画底稿的一般步骤如下（用2H或H的硬芯铅笔及圆规）：

1）画图框及标题栏。

2）按3∶4∶3布局比例确定图形在图框中的位置（见图9-26），再画各图形的基准线、对称线、轴线等。

3）按三等规律先画其主要零件，然后根据各零件的连接关系，从相邻零件开始，依次画出其他零件。

4）画各零件的细小结构，完成全部图形。注意分清零件接触面和非接触面的装配关系。

5）画尺寸界线、尺寸线、序号指引线等。

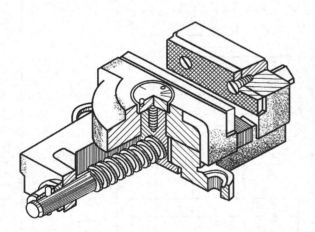

Fig. 9-25　The representation analysis of a flat nose clamp assembly　平口钳装配图表达方案分析

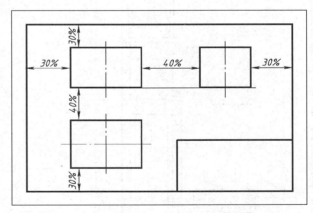

Fig. 9-26　Layout of a flat nose clamp assembly　平口钳装配图布局

6. Completing the entire drawing

Draw the section lines, mark the dimensions, set the numbering, fill in the title block and item list, specify the technical requirements, and finally complete the drawing with appropriate line styles following related standards and requirements (see Fig. 9-27).

7. Final check

Check the entire contents of the drawing. Correct any errors, then sign in the title block and complete the entire drawing.

6. 完成全图

画剖面线、标注尺寸、编排零件序号、填写标题栏、明细栏及注写技术要求，按标准加深图线，如图9-27所示。

7. 全面校核

对图中所有内容进行仔细、全面的校核，改正所有错误之处后，在标题栏内签名，完成全图。

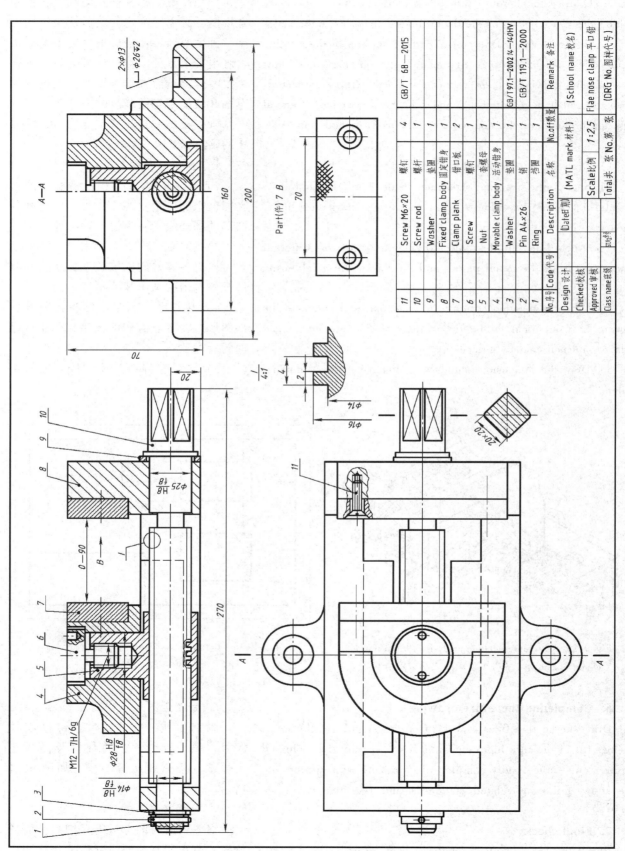

Fig. 9-27 Assembly drawing of a flat nose clamp 平口钳装配图

9.6.2 Mapping unit and assembly drawing of a single-stage cylindrical gear retarder

1. Understanding the assembly

(1) **Understanding the purpose** Retarder is used to match the speed of rotation and transmit the torque from the prime motor to the working machine of the executive device. Retarder is a kind of precision machinery, which is used to decrease the rotational speed and increase the torque.

(2) **Understanding the working principle and transmission route** Motion is firstly transmitted from the motor to the gear shaft through belt transmission, then transmitted to the spindle through gear mesh (small gear driving large gear). The aim of reducing speed is realized. Fig. 9-28 shows two transmission routes, of which the route in Fig. 9-28a is for the single-stage reducer and the route in Fig. 9-28b is for the double-stage reducer.

(3) **Understanding the assembly relationship and basic structures of the major parts** The assembly schematic drawing is helpful for understanding and showing the assembly relationship. One should draw an assembly schematic drawing in order to facilitate reassembly as a feference and illustrate the fitting relationship, relative positions, transmission conditions and working principles of each parts. Fig. 9-29 shows the assembling schematic of the single-stage cylindrical gear retarder, from which we can analysis the assembly relationship and the major parts. The retarder consists of 34 kinds of parts including the low speed shaft, gear shaft, gear, top cover, base case, through cover, end cover, adjusting ring, bearing, bolt, etc. The gear shaft and the driven spindle are supported on the case by rolliing bearings, the end cover is embedded in the case. The case is separated to the top cover and the base case, which are connected by bolt joints. In order to guarantee the installation accuracy, two taper pins are used for position fixing. The viewing port at the bottom of base case is used to monitor the oil level, and the ventilation plug on the top of the top cover is for the escaping gas in the case.

9.6.2 单级圆柱齿轮减速器部件测绘装配图

1. 了解单级圆柱齿轮减速器

(1) **了解其用途** 减速器在原动机和工作机或执行机构之间起匹配转速和传递转矩的作用。减速器是一种相对精密的机械，使用它的目的是降低转速，增加转矩。

(2) **了解其工作原理与传动路线** 动力由电动机通过带传动传送到齿轮轴，然后通过两啮合齿轮（小齿轮带动大齿轮）传送到从动轴，从而实现减速的目的，其传动路线示意图如图 9-28 所示。其中，图 9-28a 所示为单级减速器传动路线示意图，图 9-28b 所示为双级减速器传动路线示意图。

(3) **了解装配关系及主要零件基本结构** 装配示意图有助于了解装配体的装配关系，因此为了便于理解和示意装配关系，可以在着手绘制装配图之前先画出装配示意图，表示出各零件之间的装配关系和相对位置，以及传递情况和工作原理等。图 9-29 所示为单级圆柱齿轮减速器装配示意图，从图中可以分析其装配关系及主要零件。减速器是由低速轴、齿轮轴、齿轮、箱盖、箱座、透盖、端盖、调整环、轴承和螺栓等 34 种零件组成。减速器的齿轮轴和从动轴分别由滚动轴承支承在箱体上，端盖嵌入箱体内。箱体采用分离式，沿两轴线平面分为箱座和箱盖，二者采用螺栓联接。为了保证安装精度，箱体间采用两锥销定位。从箱底下部液位观察口可观察液位，箱盖上部的通气塞用于箱体内的气体排出。

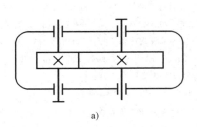
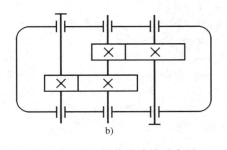

Fig. 9-28 Schematic of the retarder transmission route 减速器传动路线示意图
a) for the single-stage retarder 单级减速器传动路线示意图
b) for the double-stage retarder 双级减速器传动路线示意图

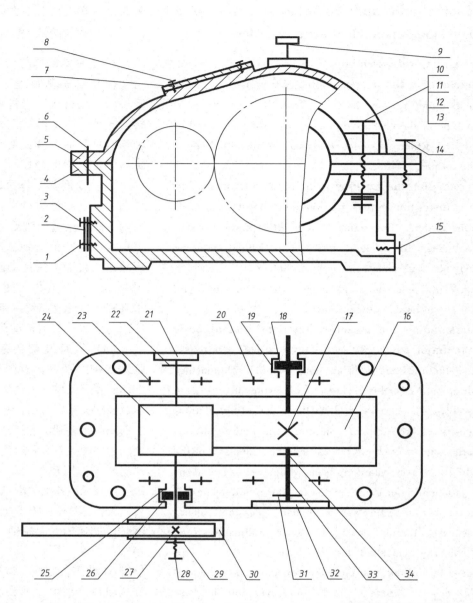

Fig. 9-29 Assembling schematic of the single-stage cylindrical gear retarder 单级圆柱齿轮减速器装配示意图

1—Chuck ring 压环 2—Observing window 液位观测口
3—Slotted countersunk flat head screws(2 pieces) 开槽沉头螺钉（2个）(GB/T 68—2016 M3×8) 4—Base case 箱座 5—Top cover 箱盖
6—Positioning pins(2 pieces)定位销(2个)(GB/T 119.2—2000) 7—Observing plate 观察板
8—Slotted cheese head screws(4 pieces)开槽圆柱头螺钉(4个)(GB/T 65—2016 M3×8) 9—Venting plug 排气塞
10—Hexagon head bolts(4 pieces) 六角头螺栓(4个)(GB/T 5782—2016 M6×40) 11—Plain washers(6 pieces)平垫圈(6个)(GB/T 97.1—2002 $d6$)
12—Spring lock washers(6 pieces) 弹簧垫圈(6个)(GB/T 93—1987 $d6$) 13—Hexagon nuts(6 pieces) 六角螺母(6个)(GB/T 41—2016 M6)
14—Hexagon head bolts(2 pieces) 六角头螺母(2个)(GB/T 5782—2016 M6×20) 15—Hexagon head bolts 六角头螺栓(GB/T 5782—2016 M6×15)
16—Big gear 大齿轮 17—Key 平键(GB/T 1096—2003) 18—Oil sealing 油封 19—Large through cover 大透盖
20—Deep groove ball bearings(2 pieces)深沟球轴承(2个)(GB/T 276—2013 6203) 21—Small end cover 小端盖
22—Small adjusting ring 小调整环 23—Deep groove ball bearings(2 pieces)深沟球轴承(2个)(GB/T 276—2913 6201)
24—Gear shaft 齿轮轴 25—Small through cover 小透盖 26—Oil sealing 油封 27—Key 平键(GB/T 1096—2003)
28—Slotted cheese head screws 开槽圆柱头螺钉(GB/T 65—2016 M4×10)
29—Plain washer 平垫圈(GB/T 97.1—2002 $d4$) 30—Belt pulley 带轮
31—Large adjusting ring 大调整环 32—Large end cover 大端盖 33—Spindle 低速轴 34—Shaft housing 轴套

2. Mapping the assembly

The content and steps for mapping a retarder assembly are following:

（1） **Disassembly and understanding of the parts** Disassembly should be performed in accordance with appropriate methods and steps to ensure the safety and integrity of the parts. One should number the parts by the sequence of disassembly. Understanding the basic structures of the parts and classifying them into standard parts and non-standard parts. Reassemble the parts when mapping is finished.

（2） **Mapping the non-standard parts** The hardcore transmission parts of the retarder are gear shaft, large gear, and spindle. The parameters of the gears should be determined interms of mapping data of the gear shaft and the large gear. Since the parts usually contain plenty of machining and fitting surfaces, pay attention to choose proper measurement accuracy and fit retationship. The boxing parts of the retarder contain the base case and the top cover, one should carefully select the front views of the boxing parts according to the working position in the assembly. The other parts on the shaft and spindle include the large and small end covers, through covers, adjusting rings, shaft housing, oil sealings, and belt pulley. The fitting dimensions to the shaft and spindle should be carefully determined. Finally, some other structures and parts like core vent, observing plate and window neet to be mapped, and for these less important dimensions, one should change the dimensions into integers.

（3） **Mapping the standard parts** For the standard parts, one can determine the nominal dimensions by measuring the parts, and choose the standard size according to the attached tables.

3. Drafting the assembly drawing

Fig. 9-30 shows the assembly drawing of a single-stage cylindrical gear retarder. The retarder has 34 kinds of parts, in which 15 kinds are standard parts. The main parts are shaft, gear, top cover, and base case, etc. In the assembly drawing, the front view and the top view are used to express the internal and external shape. The front view is determined due to the working position of the retarder, in which five local sectional views are drawn to show the internal structure of observing window, locating pin, plug screw, and bolt joining. The top view is a full sectional view, and the cutting plane lies in the fitting surface between the top cover and base case, exhibiting the transmission and assembling relationships among the parts on the two shaft systems. In addition to the two views, two removable cuts are used to illustrate joining and assembling relationships between the pulley and the gear shaft, the gear and the spindle.

2. 测绘装配体

减速器装配体测绘内容及步骤如下：

（1）拆卸装配体并了解各零部件 为了保证安全和不损坏机件，应按照正确的拆卸方法进行拆卸。零件拆下后可按拆卸顺序把零件编上号码。了解各零件的基本结构并将其分为标准件和非标件，测绘零件，测绘完成后，应将装配体复原。

（2）测绘非标件 减速器的核心传动零件包括齿轮轴、大齿轮和低速轴。根据齿轮轴和大齿轮的测绘数据，确定齿轮的参数。由于这部分零件通常具有较多的加工面和配合面，因此应注意选择合适的测绘精度以及配合关系。减速器的外壳部分，包括箱座和箱盖，对于这种箱体类零件，应结合减速器的工作位置来选择其主视图。减速器轴上配套零件部分，包括大小端盖、大小透盖、大小调整环、轴套、油封和带轮。测绘时要特别注意轴的装配尺寸。此外，对于其他零件结构，如减速器外侧排气塞、观察窗和液位观察口的测绘，其非主要尺寸可尽量选取整数。

（3）测绘标准件 对于标准件，仅需根据实际测量选择合理的公称尺寸，通过查附录来确定标准件的规格。

3. 画出装配图

单级圆柱齿轮减速器装配图如图9-30所示。减速器有34个零件。其中标准件15种，主要零件是轴、齿轮、箱盖和箱座等。减速器装配图采用主、俯视图表达其内外结构形状。主视图按照减速器的工作位置确定，通过六处局部剖视图表达出观察窗、定位销、油标。螺塞和螺栓联接处的内部结构。俯视图采用沿箱座与箱盖的结合面剖切的全剖视图，重点表达两个轴系上的各零件及其传动与配合关系。此外，为了表达带轮与齿轮轴之间、大齿轮与从动轴之间的键联接以及配合关系，还采用两个移出断面图表达其断面结构。

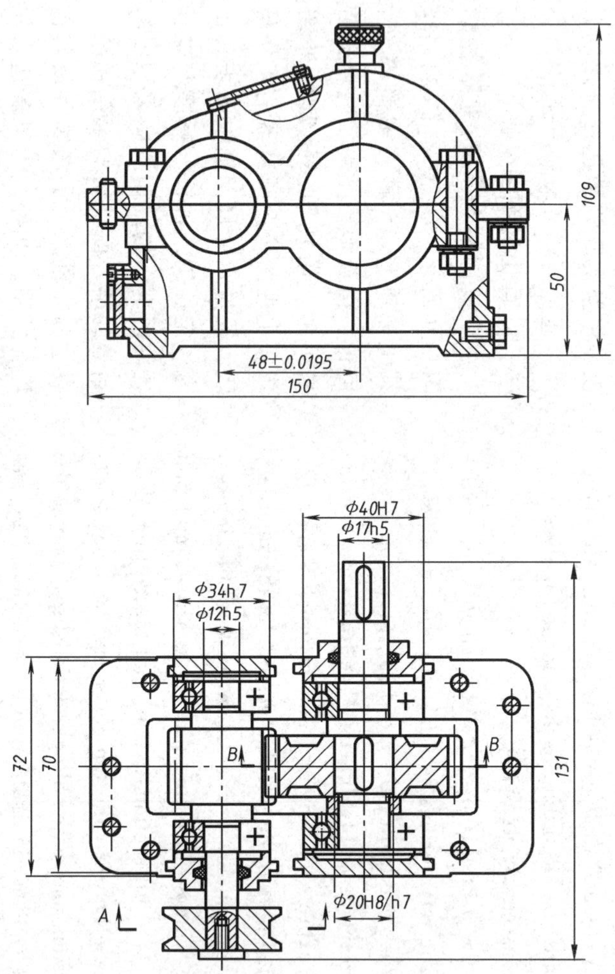
109
50
48±0.0195
150
φ40H7
φ17h5
φ34h7
φ12h5
72
70
B
B
131
φ20H8/h7
A

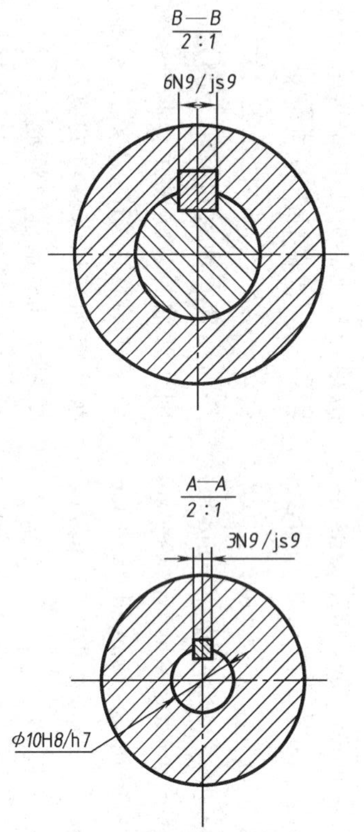
B—B
2:1
6N9/js9
A—A
2:1
3N9/js9
φ10H8/h7

TECHNICAL REQT 技术要求:

1. 装配前按照明细栏准备好所有零件，并清理毛刺。
2. 装配前用柴油清洗所有零件。
3. 装配前清洁箱体内的杂物，箱体内壁涂耐油油漆。
4. 装配前按照图样检查配合零件尺寸，合格零件才能装配。
5. 齿轮装配后用涂色法检查接触斑点，沿齿高不小于30%，沿齿宽不小于50%。
6. 减速器各密封面可涂黄油或密封胶。
7. 减速器外表涂浅灰色光亮油漆。
8. 减速器内装150号中负荷工业齿轮油，空载48h运行，平稳无冲击。

No.序号	Code代号	Description名称	No.of数量	Remark备注
34		Shaft bushing轴套	1	
33		Spindle低速轴	1	
32		Large end cover大端盖	1	
31		Large adjusting ring大调整环	1	
30		Pulley带轮	1	
29	GB/T 65—2016	Plain washer平垫圈	1	M4×10
28		Slotted cheese head screw开槽圆柱头螺钉	1	
27	GB/T 1096—2003	Key平键	1	4×4×14
26		Oil sealing油封	1	
25		Small through cover小透盖	1	
24		Gear shaft齿轮轴	1	
23	GB/T 276—2013	Deep groove ball bearings轴承	2	6201
22		Small adjusting ring小调整环	1	
21		Small end cover小端盖	1	
20	GB/T 276—2013	Deep groove ball bearings轴承	2	6203
19		Large through cover大透盖	1	
18		Oil sealing油封	1	
17	GB/T 1096—2003	Key平键	1	6×6×18
16		Big gear大齿轮	1	m=2mm,z=32
15	GB/T 5782—2016	Hexagon head bolt六角头螺栓	1	M6×15
14	GB/T 5782—2016	Hexagon head bolts六角头螺栓	2	M6×20
13	GB/T 41—2016	Hexagon nuts六角螺母	4	M6
12	GB/T 93—1987	Spring lock washers弹簧垫圈	4	d=6
11	GB/T 97.1—2002	Plain washers平垫圈	4	d=6
10	GB/T 5782—2016	Hexagon head bolts六角头螺栓	4	M6×40
9		Venting plug排气塞	1	
8	GB/T 65—2016	Slotted cheese head screws开槽圆柱头螺钉	4	M3×8
7		Observing plate观察板	1	
6	GB/T 119.2—2000	Positioning pins定位销	2	φ5
5		Top cover箱盖	1	
4		Base case箱座	1	
3	GB/T 68—2016	Slotted countersunk flat screws开槽沉头螺钉	2	M3×8
2		Observing window液位观察口	1	
1		Chunk ring压环	1	

Design设计			MATL mark(材料)	(School name校名)
Checked校核			Scale比例 1:1	Retarder减速器
Approved审核			Total共 张 No.第 张	(DRG No.图样代号)
Class name班级		STU No.学号		

Fig. 9-30 Assembly drawing of a sing-stage cyclindrical gear retarder 单级圆柱齿轮减速器装配图

9.7 Interpreting Assembly Drawings and Extracting Detail Drawings

In the design, manufacturing, use, maintenance and technical communication of the machine or subassembly, the assembly drawing is needed. Therefore, it is a basic skill for engineering personnel to understand assembly drawings and to be able to produce detail drawings according to the assembly drawings.

9.7.1 Objectives and requirements

There are three objectives or requirements when reading assembly drawings:

1) Understand the performance, function and working principles of the machine or subassembly.

2) Clarify the purpose, relative position, fitting relationship of each part, connection and fixing methods, and sequence for disassembly.

3) Read the structural shape of each part.

9.7.2 Methods and steps

Let us take the ball valve in Fig. 9-2 as an example to explain the methods and steps for reading and interpreting an assembly drawing.

1. General understanding

Get the general information about the name, function, principle, purpose, specification, structural shape, scale, quantity, material, standard, etc.

1) The name, purpose and specification can be acquired by consulting the title block, specification and investigation.

For example, known from the title block in Fig. 9-2, the name of that assembly is ball valve. The valve is a subassembly to open or shut the pipeline and adjust the rate of flow. Thus, the valve is a switch gear that controls the rate of flow. The valve core is a ball, so it is called ball valve. The scale is 1∶2.

2) In order to understand the composition of the machine or subassembly, look up the name, quantity, position of the standard part or subassembly and the non-standard part or subassembly. Locate these parts or subassemblies on the assembly drawing by comparing the number of the part or subassembly.

For example, known from the item list in Fig. 9-2, the ball vale has 13 types of parts, in which there are two types of standard parts, including four studs (part 6) and four nuts (part 7) used to connect the valve cover and valve body.

9.7 读装配图和拆画零件图

在机器或部件的设计、制造、使用、维修和技术交流中，都要用到装配图。因此，读懂装配图和由装配图拆画零件图，是每个工程技术人员必备的基本技能之一。

9.7.1 目的和要求

读装配图的目的和要求，主要有以下三点：

1) 了解机器或部件的性能、功用和工作原理。

2) 弄清各个零部件的作用和它们之间的相对位置、装配关系、联接和固定方式，以及拆装顺序等。

3) 看懂各零件的结构形状。

9.7.2 方法和步骤

现以图 9-2 所示球阀装配图为例，说明读装配图的方法和步骤。

1. 概括了解

对机器或部件的名称、功能、原理、用途、规格及其形状结构和画图比例、数量、材料、标准等做大致的了解。

1) 了解机器或部件的名称、用途和规格。可通过查阅标题栏、说明书或调查研究获知。

例如，由图 9-2 标题栏可知，该装配体的名称为球阀。它是用于管路系统中启闭和调节流体流量的部件，即控制流体流量的开关装置。由于其阀芯是球形的，故取名为球阀。绘图比例为 1∶2。

2) 了解机器或部件的组成，查阅标准零部件和非标准零部件的名称、数量及位置。通过对照零部件序号，在装配图上查找这些零部件的位置。

例如，由图 9-2 中的明细栏可知，球阀共有 13 种零件，其中有 2 种标准件，双头螺柱（件 6）和螺母（件 7）各 4 个，用于联接阀盖和阀体。

One can have a basic impression on the profile of the machine or subassembly by understanding the above-mentioned contents and consulting the related dimensions.

2. Analyzing views

The aim is to understand the relative positions, joints, fitting requirements, transmission relations (driving—driven), and mounting relation (movable—fixed), with which one can analyze the working principles and understand the dismounting sequences, inspection conditions, usage and maintenance considerations.

1) Analyze each view. According to the representation of the assembly drawing, find out the location and projection direction of each view, sectional view and cut view. Note whether there is any special representation, conventional representation or simplified representation. Make sure that the contents of each view are clear.

The cut plane in the front view in Fig. 9-2 goes through the front and back symmetry plane of the ball valve, the front view clearly reflects the fitting relationships between the major parts and the working principles of the ball valve. The left view adopts the dismounting representation, which represents the external structure and fitting relationships not clearly represented in the front view. The top view adopts the $B-B$ partial sectional view to reflect the relationships between the wrench (part 13) and the valve body top (part 1), the local dummy club and the valve handle (part 12).

2) Start from the view showing the clear movement relationship. Make sure that the relative movement between the moving part and nonmoving part is clear. Then analyze the working principles and transmission routes.

The working principles of the ball valve: the square hole of wrench (part 13) holds the quadrangular prism of the valve handle (part 12). When the wrench is in the position as shown in the front view of Fig. 9-2, the valve is open; when the wrench is in the thin long dashed double dotted line position as shown in the top view of Fig. 9-2, the valve is closed by clockwise rotating 90 degrees.

3) Start from the view with a clear picture of the main assembling relations. Make sure the fitting relationships of each part including joint, fixing, and sealing positions as well as the relative positions of each part.

For example, the fitting relationships of the ball valve is: both the valve body (part 1) and the valve cover (part 2) have square flanges and they are connected with 4 studs. The tightness between the valve core (part 4) and the sealing gasket (part 3) can be adjusted by using an spacer (part 5). A dummy club under the valve handle (part 12) connects to the notch on the valve core (part 4). The seal between the valve body (part 1) and valve handle (part 12) connects stuffing spacer and stuffing (part8, 9 and 10), as well as the packing set (part 11).

通过对以上内容的初步了解，并参阅有关尺寸，便可对该部件的大体轮廓与内容有一个概略的印象。

2. 分析视图

由各视图对应起来看懂各零件的相互位置、联接形式、配合要求及传动关系（主—从）和装配关系（动—静），由此分析其工作原理，并了解部件的装拆顺序、验收条件和使用、维修注意事项等。

1）对各视图进行分析。根据装配图上的表达情况，找出各个视图、剖视图、断面图等配置的位置及投射方向，注意观察是否采用了装配图的特殊画法、规定画法及简化画法等，从而搞清各视图的重点表达内容。

如图9-2所示，主视图中的剖切平面通过球阀的前后对称平面，清楚地反映了球阀主要零件之间的装配关系和工作原理。左视图采用拆卸画法，并补充了主视图还未表达清楚的外形结构和装配关系。俯视图采用$B-B$局部剖视，反映扳手（件13）与阀体（件1）、上部限位凸块和阀杆（件12）的关系。

2）从清楚表达运动关系的视图入手。搞清运动零件与非运动零件的相对运动关系，从而分析部件的工作原理和传动路线。

球阀的工作原理是：扳手（件13）的方孔套住阀杆（件12）上部的四棱柱，当转动扳手至图9-2主视图所示的位置时，则阀门全部开启，管道畅通；当扳手处于图9-2俯视图所示的细双点画线位置时，即顺时针旋转90°时，则阀门关闭，管道断流。

3）从反映主要装配关系最清楚的视图入手。弄清各零件间的装配关系。包括联接、固定和密封位置，以及各零件的相对位置。

例如，球阀的装配关系包括：阀体（件1）和阀盖（件2）均带有方形的凸缘，采用4个双头螺柱联接，并用合适的调整垫（件5）调节阀芯（件4）与密封垫（件3）之间的松紧程度。在阀杆（件12）的下部有凸块，榫接阀芯（件4）上面的凹槽。为了密封，在阀体（件1）与阀杆（件12）之间加进填料垫和填料（件8、9和10），再旋入填料压紧套（件11）。

There are three fits: The cylindrical surface fit of $\phi 50$ H11/h11 between the valve body (part 1) and the valve cover (part 2) is a clearance fit; the cylindrical surface fit of $\phi 18$H11/h11 between the valve body (part 1) and valve handle (part 12) is a clearance fit; the cylindrical surface fit of $\phi 14$H11/d11 between the valve handle (part 12) and the handle packing set (part 11) is a clearance fit, too.

4) Analyze the structure and dimension of each part or subassembly. Get the general information by analyzing the projection and referring to the corresponding dimensions according to the relation of lines and surfaces. Generally main parts are analyzed first, then secondary ones. And begin with the view that is easier to distinguish the projection contours, and then proceed to other views.

For example, the main shape of the part in the ball valve can be viewed in Fig. 9-2, but the complex structural shapes of the valve body and valve cover will not be described here as they have been introduced in part drawing.

3. Imaging the structural shape

By combining dimensioning and technical requirements, through further analyzing the general structures, transmission relations and working principles, one can create an image of the assembly structure, as shown in Fig. 9-1.

9.7.3 Detail drawings from assembly drawings

When designing the subassembly, one needs to produce detail drawings based on an assembly drawing, which is called assembly drawing separation or individual part drawing extraction. It is an important step in design.

1. Methods and steps for extracting component drawings from the assembly drawing

In order to extract component drawings, one must first understand the assembly drawing. Based on the understanding, the methods and steps for extracting component drawings are as follows:

(1) Separating the key parts

1) Locate the parts according to their names and the item list.

2) Separate the part from others according to the consistence of the section line in each view.

3) According to the three-alignment rule and analysis of lines and faces, determine the structural shape of the part.

其配合有 3 处：阀体（件 1）与阀盖（件 2）联接处的 $\phi 50$H11/h11 圆柱面，配合为间隙配合；阀体（件 1）与阀杆（件 12）的 $\phi 18$H11/h11 圆柱面配合为间隙配合；阀杆（件 12）与填料压紧套（件 11）的 $\phi 14$H11/d11 圆柱面配合也是间隙配合。

4）分析零部件的结构及尺寸。根据线面关系仔细分析其投影特点，参阅有关尺寸，可以对零件的内、外结构形状有大致的了解。一般先分析主要零件，后分析次要零件。先从容易区分零件投影轮廓的视图开始，再看其他视图。

例如，球阀中各零件的主要形状，大多可以从图 9-2 中看出，而其中形状结构最复杂的主要零件阀体和阀盖，已在零件图的例图中介绍过，在此不做赘述。

3. 想象结构形状

在以上各步骤的基础上，结合尺寸标注及技术要求等有关内容，进一步综合分析总体结构、传动关系和工作原理，通过归纳总结，想象出装配体的整体结构形状，如图 9-1 所示。

9.7.3 由装配图拆画零件图

在设计部件时，需要根据装配图拆画零件图，简称拆图。这是设计过程中的一个重要环节。

1. 拆画零件图的方法和步骤

拆图是在看懂装配图的基础上进行的。拆画零件图的方法和步骤如下：

（1）分离出关键零件

1）根据零件名称和明细栏找出其在图中的位置。

2）根据同一零件的剖面线在各视图中一致的特点，将该零件与其他零件区分开来。

3）根据三等规律和线面分析法，研究确定该零件的结构形状。

(2) Determining the representation　Analyze what type the part belongs to ⇨ Determine how many views are needed (Where to use sectional or cut view) ⇨ Determine the projection direction of the front view.

Note: When choosing the views, one should choose the best representation according to its structural features instead of copying from the assembly drawing, due to different starting points and requirements.

(3) Refining the structures of the part　Refine the structures of the part with what is omitted in the assembly drawing, such as chamfer, undercut and round.

(4) Dimensioning and tolerance

1) Dimensions not marked in the assembly drawing should be measured for calculation (round off to integers). Choose the dimension datum for the convenience of machining and measuring.

2) Consult related standards before dimensioning standard structures of an individual part, such as the keyway or thread.

3) Mark the tolerance for dimensions having fitting relationships, according to the appropriate dimensions given in the drawing.

(5) Indication of surface texture　From the fitting relationships shown in the assembly drawing, one can find out machined surfaces, non-machined surfaces and technological and structural requirements, and mark the corresponding surface roughness values. The surface roughness values should be small for the surfaces that have relative movement or fitting requirements, usually $Ra0.8\mu m$, $Ra1.6\mu m$, $Ra3.2\mu m$, $Ra6.3\mu m$, etc. The surface roughness values for the surfaces having sealing, antirust and decoration requirements should also be small. The surface roughness values for non-fitting surfaces or static surfaces can be a little larger, such as $Ra12.5\mu m$, $Ra25\mu m$, etc.

(6) Filling in the technical requirements　Determine the technical requirements according to the utilization, structure and related information.

(7) Checking and completing the drawing　Check for errors, then complete the drawing with appropriate line styles following related standards.

2. Examples for producing part drawings from an assembly drawing

Let us take the frame (part 1) shown in Fig. 9-31 as an example. The methods and steps to separate the drawing are described as follows:

(1) Brief summary　From the title block of the assembly drawing, we know that it is a projector lens frame. By consulting related information, we also know that the projector lens frame is used to hold the lens on the film projector and that it is important to adjust the focal length in order to appear a clear image. As known from the item list and the part number, the frame consists of seven types of parts, including one type of standard part and six types of nonstandard parts, in which the adjusting gear (part 6) is a sub-assembly part.

（2）确定其表达方案　分析该零件属于哪类零件⇨确定需用几个视图表达（何处用剖视或断面图等）⇨确定主视图的观察方向。

注意：由于表达的出发点和要求不同，所以在选择视图表达方案时，不能盲目抄袭装配图，而应根据零件本身的结构特点选择最佳表达方案。

（3）补全零件的结构　补齐该零件在装配图中缺少和省略的结构，如倒角、退刀槽、圆角等。

（4）标注尺寸及公差

1）装配图中未注的尺寸，应量取后根据图上的比例换算求得（取整数）。应注意选择合理的尺寸基准，以便加工和测量。

2）零件上的标准结构，如键槽、螺纹等，应查表后再标注。

3）对有配合关系的尺寸，应根据图中给出的相关尺寸，标注其尺寸公差。

（5）表面结构的标注　由装配图所示该零件与其他零件的装配关系，判断出非加工面与加工面及其工艺结构要求，并注明相应的表面粗糙度。对有相对运动或配合要求的表面，表面粗糙度数值应取小些，一般取 $Ra0.8\mu m$、$Ra1.6\mu m$、$Ra3.2\mu m$、$Ra6.3\mu m$ 等。对有密封、耐蚀、装饰要求的表面，表面粗糙度数值也应小些。对非配合表面或静止表面，则表面粗糙度数值应取大些，如 $Ra12.5\mu m$、$Ra25\mu m$ 等。

（6）填写技术要求　根据零件的作用、结构，参考有关资料，拟订技术要求。

（7）校核后加深图线　检查有无错漏，按标准线型加深全图。

2. 拆画零件图举例

现以拆画图 9-31 所示镜头架装配图中的件 1 架体零件图为例，对拆图方法和步骤说明如下：

（1）概括了解　从图 9-31 镜头架装配图的标题栏可以看出，该部件名称为镜头架。通过查阅有关资料可知，镜头架是电影放映机上用来夹持放映镜头，调节焦距，使图像清晰的一个重要部件。从明细栏及零件编号可知，它由 7 种零件组成，其中标准件 1 种，非标准件 6 种。调节齿轮（件 6）为组合件。

The external dimensions of the projector lens frame are length 112mm, width over 60mm and height 100mm. Therefore, the total volume of the frame is relatively small.

(2) Analyzing views The assembly drawing of a projector lens frame consists of two basic views. The front view A—A is a full sectional view with two parallel cutting planes that represents the main fitting relationship and working principles of the projector lens frame. In the front view, imaging representation is implemented. The projector lens frame is located with pins and fastened to the film projector with screws. In the left view, B—B partial sectional view is used to mainly represent the contour of the frame (part 1) and convey the engagement of the adjusting gear (part 6) and rack on the eye ring (part 2). At the same time, it represents the actual shape of the handle knob on the adjusting gear (part 6).

(3) Fitting relationships and working principles

1) Fitting relationships. The front view on the assembly drawing of a projector lens frame fully represents its fitting relationships. Known from Fig. 9-31, all the parts are mounted to the major part frame (part 1) and it is located and mounted with two pins and two screws to the film projector. The big hole (ϕ70mm) of the frame (part 1) has an eye ring (part 2) which can move forward or backward. The axis of the horizontal hole (ϕ22mm) of the frame (part 1) is a main fitting line, on which there is a lock sleeve (part 7), and it gives a clearance fit of ϕ22H7/g6. In the lock sleeve (part 7), there is an adjusting gear (part 6), the fits between them are ϕ15H11/c11, ϕ6H8/f7, which are also clearance fits. When the adjusting gear (part 6) and the eye ring (part 2) take their places, the adjusting gear (part 6) can only rotate in the lock sleeve (part 7) due to the axial locating of the screw (part 5). The external thread on the right of the lock sleeve is mounted to the lock nut (part 4). When the nut rotates, it pulls the lock sleeve (part 7) to the right. The groove on the lock sleeve (part 7) forces the eye ring to contract so that the lens is locked.

The assembly process of the projector lens frame is shown as follows: Put the washer (part 3) on the lock sleeve (part 7); screw on the lock nut (part 4); insert the adjusting gear (part 6) into the lock sleeve (part 7); put them together to the ϕ22mm hole of the projector lens frame; then make sure the groove of the lock sleeve (part 7) is up. Insert the eye ring (part 2) into ϕ70mm hole of frame (part 1); make sure its rack is down and engaged with the adjusting gear (part 6); tighten the screw (part 5). One can create a satisfactory projector lens frame if it is adjusted according to the technical requirements. Finally, fasten the projector lens frame to the film projector with two pins and two screws.

经过初步观察，镜头架的外形尺寸是长112mm、宽比60mm稍大、高100mm，可知其体积并不大。

（2）分析视图 镜头架装配图采用了两个基本视图。主视图A—A是用两个平行剖切面剖切的全剖视图，表达了镜头架的主要装配关系和工作原理。在主视图中，还采用了假想画法，可看出镜头架是被销钉定位，并用螺钉夹紧固定在电影放映机上的；左视图采用B—B局部剖视，主要反映了架体（件1）的外形轮廓，并表达了调节齿轮（件6）与内衬圈（件2）上的齿条相啮合的情况，同时反映了调节齿轮（件6）上捏手的形状。

（3）装配关系和工作原理

1）装配关系。镜头架装配图的主视图完整地表达了它的装配关系。由图9-31可知，所有零件都装在主要零件架体（件1）上。架体（件1）的大孔（ϕ70mm）中，套有能前后移动的内衬圈（件2）。架体（件1）的水平圆孔（ϕ22mm）的轴线是一条主要装配干线。在该装配干线上装有锁紧套（件7），它们采用ϕ22H7/g6的间隙配合。锁紧套（件7）内装有调节齿轮（件6），它们的配合分别为ϕ15H11/c11、ϕ6H8/f7，也都是间隙配合。因有螺钉（件5）的轴向定位，当调节齿轮（件6）与内衬圈（件2）就位后，调节齿轮（件6）只能在锁紧套（件7）中做旋转运动。锁紧套右端的外螺纹处装有锁紧螺母（件4），旋紧螺母，则锁紧套（件7）向右移动，锁紧套（件7）上的圆柱面槽就迫使内衬圈收缩而锁紧镜头。

镜头架的装配过程如下：将锁紧套（件7）套上垫圈（件3），旋上锁紧螺母（件4），将调节齿轮（件6）装入锁紧套（件7），并将它们一起装入架体（件1）下部ϕ22mm的圆孔中，使锁紧套（件7）圆槽向上。将内衬圈（件2）装入架体（件1）上部的ϕ70mm圆孔中，使其齿条向下并与调节齿轮（件6）啮合，旋紧螺钉（件5）。按技术要求内容调整，即可得到满意的镜头架。最后将镜头架用两个销和两个螺钉固定在电影放映机上。

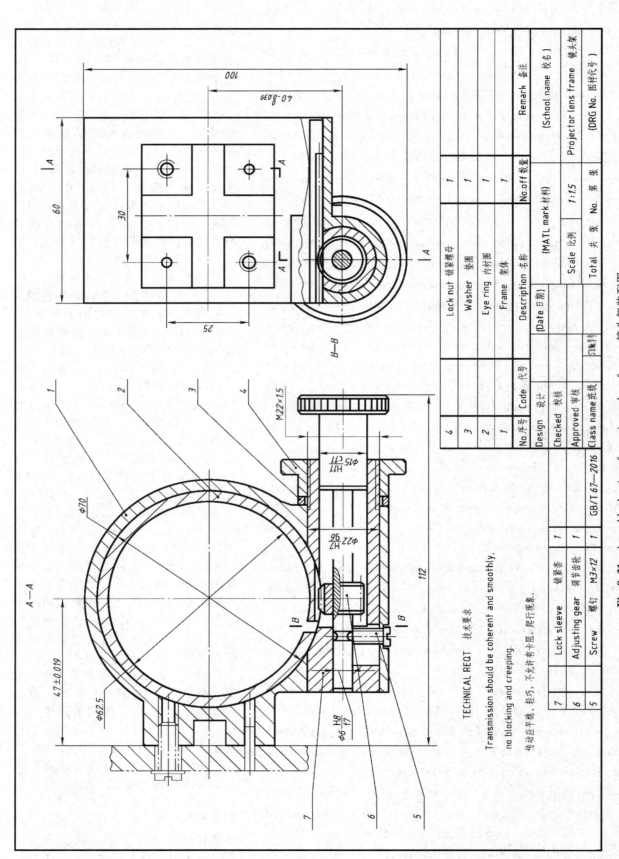

Fig. 9-31 Assembly drawing of a projector lens frame 镜头架装配图

2) Working principle. Loosen the lock nut (part 4), and put the lens into the ϕ62.5mm hole of the eye ring (part 2). When rotating the handle knob of the adjusting gear (part 6), the engagement of the gear and the rack will drive the eye ring (part 2) forward and backward to adjust the focal length. Tighten the lock nut (part 4) and the lock sleeve (part 7) will move to the right. The joint on the top of the groove and cylindrical surface of the eye ring (part 2) makes the eye ring (part 2) deforme and contract radially so that the lens is clamped.

(4) Analyzing the parts and extracting the detail drawings Here we only analyze the structure of the frame (part1) and extract its detail drawing. Readers can analyze other parts in a similar way.

1) Analyzing the structural shape of the frame (part 1). The frame (part1) is the major part of the projector lens frame. We can see its approximate structure from the assembly drawing. The frame mainly consists of two cylinders, a big one and a small one. The two cylinders are perpendicular to each another, and the internal walls of the cylindrical holes intersect. Insert the eye ring (part 2) with a rack into the big one and insert the lock sleeve (part 7) into the small one. To locate and mount the frame on the film projector, a quadrangular prism protrudes out from the left part of the outer surface of the big cylinder. On its left end face there are another four small quadrangular prisms, each having a thread throughhole or a straight pin through-hole. The bottom of the small cylinder is a half cylinder. And the top of the small one is a quadrangular prism whose front and back walls are tangent to the half cylinder. On the lower side wall of the small cylinder, there is a thread through-hole with a spotface, which matches the screw (part 5) for axial positioning the adjusting gear (part 6), as shown in Fig. 9-32.

2) Extracting the frame (part 1) from the assembly drawing. The contours extracted from the front view and left view on the assembly drawing do not form a complete drawing as shown in Fig. 9-33. One should add missing lines based on the above-mentioned structural analysis.

3) Determining the representation for the frame (part 1). Based on the two views of the frame after being separated and missing lines added, one can then infer the following: The front view not only represents the contour in one direction, but also clearly reflects the internal details for adopting stepped section; the left view is a partial sectional view, the contour is not fully represented and the hole on the partial sectional view has already been shown in the front view. Therefore, according to the working position of the frame (part 1), the projection direction resumes its original direction, but the stepped section has changed, which from the threaded hole to the smooth hole and to the big hole on the bottom. The left view should focus on representing the part contour instead of the partial section. By drawing the contour lines of the big cylinder with thin long dashed lines, the external structure of the part should be clearly and fully represented. Thus, the left view is changed with such an adjustment.

2) 工作原理。松开锁紧螺母（件4），将镜头放入内衬圈（件2）的ϕ62.5mm圆孔中。当旋转调节齿轮（件6）的捏手时，通过齿轮齿条啮合带动内衬圈（件2）做前后方向的直线移动，从而达到调整焦距的目的。当旋紧锁紧螺母（件4）时，锁紧套（件7）向右移，通过其上部圆槽与内衬圈（件2）外圆柱表面的贴合作用，使内衬圈（件2）变形，径向收缩，从而夹紧镜头。

(4) 分析零件，拆画零件图 在此只分析架体（件1）的结构形状，并拆画其零件图。其余零件请读者通过阅读，自行分析。

1) 架体（件1）结构形状分析。架体（件1）是镜头架上的主体零件。从装配图中可以看出它的大致结构形状。这个架体主要是由一大一小相互垂直偏交的两个圆筒组成，它们的圆柱孔内壁相交贯通。大圆柱筒中装入带齿条的内衬圈（件2），小圆柱筒内装入锁紧套（件7）。为了将架体定位、安装在电影放映机上，在大圆柱筒外壁的左侧伸出一个四棱柱，并在这个四棱柱的左端面上，分别设置有螺纹通孔和圆柱销孔的四个方形凸起的台阶面。小圆柱筒的下部是半个圆柱体，上部是前、后壁与半圆柱面相切的四棱柱。在小圆柱筒的下部半圆柱壁上，有一个经过锪平的螺纹通孔，旋刀螺钉（件5），可对调节齿轮（件6）进行轴向定位，如图9-32所示。

2) 从装配图中分离架体（件1）。从装配图的主、左视图中区分出架体的视图轮廓，它是一幅不完整的图形，如图9-33所示。应结合上述结构分析，补全所缺的图线。

3) 确定架体（件1）表达方案。观察分离补全后的架体两视图可以看出：主视图不仅表示出一个方向的外部轮廓形状，而且由于采用了两个平行的剖切面剖切，因而清晰地反映了内部的结构形状；左视图采用局部剖，则使其外形的表达有不完整感，而且局部剖处的孔已经在主视图中表达清楚了。所以，根据架体（件1）的工作位置，主视图仍选择原观察方向，并采用由螺孔到光孔，再到下部大孔中心剖切的全剖视图。左视图应以表达外形结构轮廓为主，而不必用局部剖，只需用细虚线画出大圆筒内壁的上、下两条转向轮廓线，就可以清晰、完整地表达内外形状。通过这样的调整，改画了左视图。

4) Fully marking the dimensions of the frame (part.1) and the technical requirements. According to the requirements of the detail drawing, fully and clearly mark the dimensions of the parts and the technical requirements. Note that the tolerance in the drawing shall be in accordance with the tolerance in the assembly drawing. Complete the detail drawing after checking for accuracy with appropriate line styles, then fill in the title block. The separation of the frame (part 1) detail drawing is now completed as shown in Fig. 9-34.

4）注全架体（件1）的全部尺寸和技术要求。按零件图的要求，完整、清晰地标注出全部尺寸和技术要求。注意图中的尺寸公差应与装配图中已经注明的尺寸公差相同。最后经检查核实无误后，描深图线，并填写标题栏等。至此，架体（件1）零件图拆画完成，如图9-34所示。

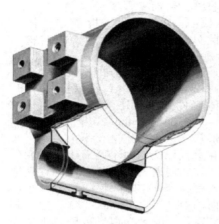

Fig. 9-32　Part 1 frame　件1架体

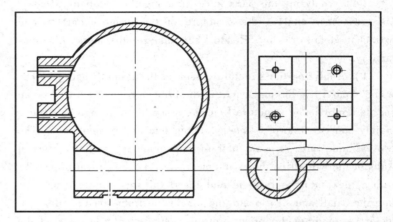

Fig. 9-33　Extracting the part 1 frame from the assembly　从装配图分离件1架体

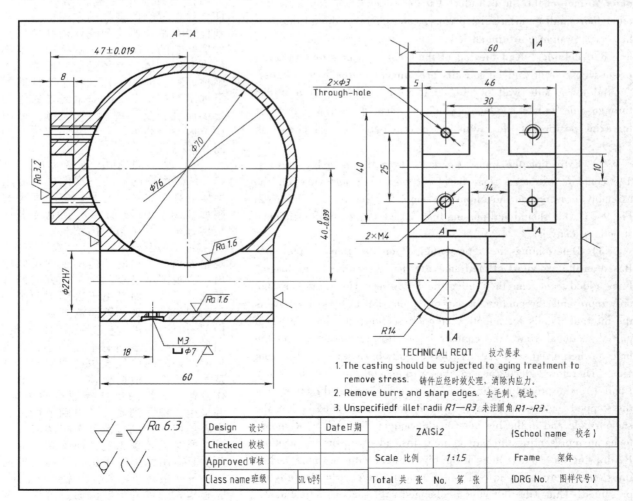

Fig. 9-34　Details drawing of the frame　架体零件图

3. Examples of foreign assembly

Comparing with the standards of other countries, the assembly drawing of ours is similar except the title block, the item block and the notation of technique requirements, as shown in Fig. 9-35 and the Fig. 9-36 produced following the British standard.

3. 国外装配图示例

国外装配图的标题栏、明细栏、技术要求的注写形式及零件序号编写与国内的要求不同。以英国图样为例，如图 9-35、图 9-36 所示。其余内容基本上与我国的装配图形式相似。

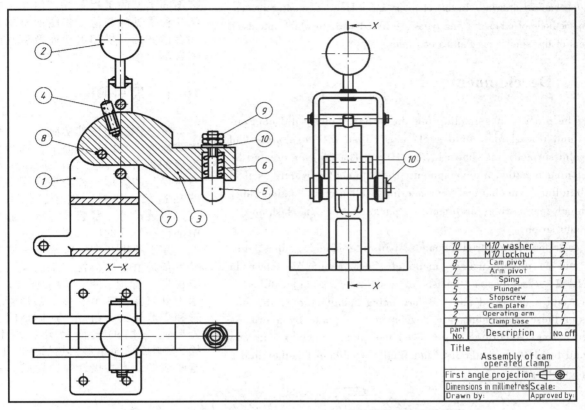

Fig. 9-35 Assembly of cam operated clamp 凸轮操纵式夹具装配图

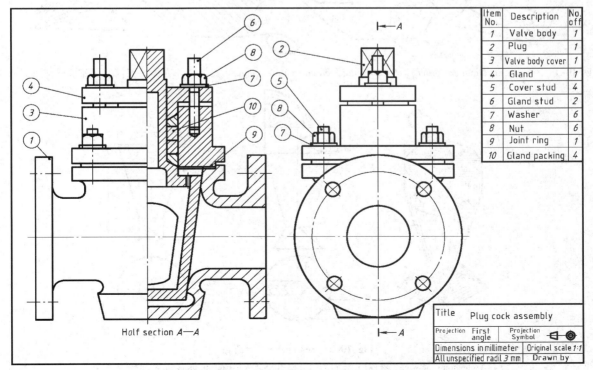

Fig. 9-36 Plug cock assembly 旋塞装配图

Chapter 10　Other Drawings

In practical manufacture, there are also many drawing types other than mechanical drawing. In this chapter, basic knowledge and representation methods of other drawing types are introduced, to which interested readers of the similar specialties can refer.

10.1　Development

The process of spreading out the surfaces of a solid continuously and true-shapely onto a plane is called the development of the solid surfaces, as shown in Fig. 10-1. The drawing created by developing is called a development. Development is widely used in shipbuilding, mechanical engineering, electronic engineering, chemical engineering, architecture, packaging design, fashion design and so on.

In terms of industrial demands, parts like shields, pipes, funnels and so on are usually made of sheet metal. As shown in Fig. 10-2, a segregator consists of a cylinder, a taper tube, a siphon and a deformed joint. Before being manufactured, the surfaces of these components are developed to a plane by a process called lofting. According to the developed plane drawings, one can cut and bend sheet materials, and finally weldthem together into a complete shape.

第 10 章　其他工程图

在实际生产中，根据需要还有许多不同类型的工程图样。本章主要介绍几种常见的其他工程图的基本知识和表达方法，供相关专业及有兴趣的读者学习参考。

10.1　展开图

将立体表面按其实际大小，依次连续地摊平在同一平面上，称为立体表面的展开，如图 10-1 所示。展开后所得到的图形，称为展开图。展开图在造船、机械、电子、化工、建筑、包装设计、服装设计等工业部门中都得到广泛的应用。

在工业生产中，常常有一些零件或设备是由金属板材加工制成的，如防护罩、管道、漏斗等。图 10-2 所示为分离器，它由圆柱、锥形管、弯管和变形接头所组成。在制造时，首先要把它们的表面展开（俗称放样），画成平面图形，然后按图下料，弯曲成形，再将接缝焊接而成。

Fig. 10-1　Surface development process of a pyramid, cuboid, tube, cone　棱锥、长方体、圆筒、圆锥立体表面的展开

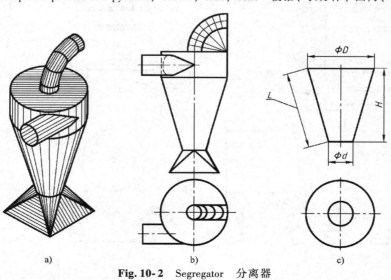

Fig. 10-2　Segregator　分离器
a) Axonometric drawing　轴测图　b) View　视图　c) View of the taper tube　圆锥管视图

There are two methods to create a development, i.e., graphic method and calculation method. This section focuses on the graphic method as it is simple as well as intuitive.

10.1.1 Development of polyhedral solid surfaces

1. Development of a prism pipe

Because the side edges of a prism are parallel to each other, once the dimensions of all the side edges and bottom lines are given, the development of the prism can be obtained, in which all developed side edges remain parallel to each other. This development method is called a parallel-line development method.

Fig. 10-3a and Fig. 10-3b are respectively the axonometric drawing and the orthogonal drawing of an erective quadrangular prism pipe with an oblique mouth. As the bottom cap ABCD is parallel to H-plane, its horizontal projection reflects the real shape. Edges EA, FB, GC and HD are all plumb lines and their frontal projections reflect the real lengths.

Based on the fact above, the real shapes of four side faces of the prism can be easily obtained. The construction process of the development is illustrated in Fig. 10-3c.

Outer border of a development shall be rendered with continuous thick lines, but the inner folding edges shall be rendered with continuous thin lines. The welding seam is better to be placed over the shortest edge in order to optimize the welding budget.

画展开图的方法有两种，即图解法和计算法。由于图解法具有作图简捷、直观等优点，故本章重点介绍这种方法。

10.1.1 平面立体表面的展开图

1. 棱柱管的展开图

棱柱管的各条棱线互相平行，因此只要求出各棱线和底边的实长，就可画出其表面的展开图，且展开后的各棱线仍然保持互相平行的关系。这种绘制展开图的方法称为平行线展开法。

图 10-3a、b 所示为斜口直立四棱柱管的轴测图和正投影图。由于四棱柱底面 ABCD 平行于 H 投影面，其 H 面投影反映实形。各棱线 EA、FB、GC 和 HD 均为铅垂线，其正面投影反映实长。

根据这个关系就可以画出四个侧面的实形。展开图的作图过程如图 10-3c 所示。

展开图的外边界线应画为粗实线，其内部的折叠线画为细实线。为减少焊接费用，最好将接缝置于最短的边线处。

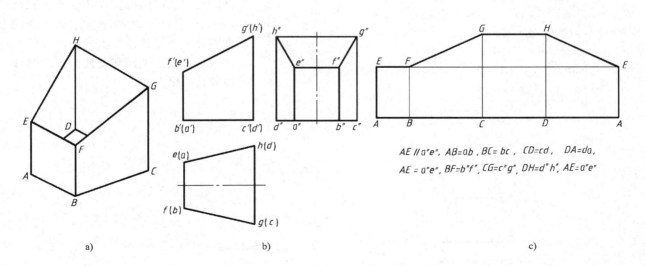

Fig. 10-3 Surface development process and development of a perpendicular quadrangular prism pipe with an oblique mouth
斜口直立四棱柱管的表面展开过程及展开图
a) Axonometric drawing 轴测图　b) Three views 三视图　c) Development 展开图

2. Development of a hexagonal prism with an oblique mouth

Fig. 10-4 shows the development of a hexagonal prism with an oblique mouth, in which the length of the bottom line is the summation of the lengths of all the edges on the bottom cap.

The height at each corner point on the top cap can be obtained by the front view, as shown in Fig. 10-4. View D is the direction view of the top cap along direction D.

2. 斜口六棱柱的展开图

图 10-4 所示为一个六棱柱展开图，其底边长度为六棱柱底面各边长度之和。

每一个拐点对应的高度可由主视图投影获得，如图 10-4 所示。棱柱顶面图形（向视图 D）可由 D 方向观察顶部实形而获得。

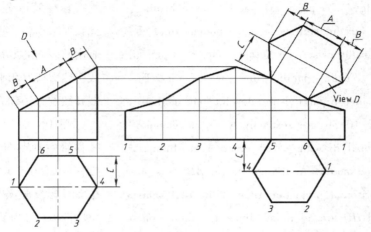

Fig. 10-4 Development of a hexagonal prism with an oblique mouth   斜口六棱柱展开图

3. Development of a truncated pyramid

First, all the side edge lengths should be determined. Then, according to the lengths of the bottom lines, the development can be created by drawing out real shapes of all the faces. Because all the side edges of a pyramid intersect at the vertex, looking like radial lines eradiating from it, this development method is called a radial-line development method.

Fig. 10-5 shows the surface development process and development of a truncated pyramid. The length of edge SA is obtained by the right triangle S_1OA_1, where $S_1A_1 = SA$. Thus triangles SAB, SBC, SCD and SDA can be created according to the lengths of the side edges and the bottom edges. By cutting the part from the top cap upward from the drawing and adding the real shapes of the top and the bottom cap on the drawing, the development of the truncated pyramid is obtained.

3. 棱锥台的展开图

首先应确定各棱线的实长，即得各棱面的实形，便可画出其展开图。由于棱锥的所有棱线汇交于锥顶，这种求作展开图的方法称为放射线展开法。

图 10-5 所示为一个棱锥台表面展开过程及展开图。其棱线的实长 SA 由直角三角形 S_1OA_1 求得为 S_1A_1。再以棱线和底边的实长依次作出三角形 SAB、SBC、SCD、SDA，得到四棱锥展开图。将各棱线上延长的棱线实长截去，加上顶面和底面图形即为该棱锥台的展开图。

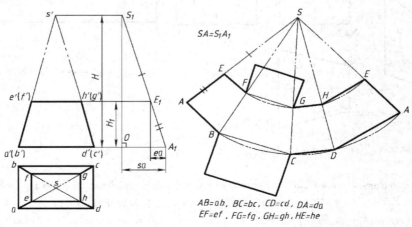

Fig. 10-5 Surface development process and development of a truncated pyramid   棱锥台表面展开过程及展开图

10.1.2 Development of curved solid surfaces

1. Development of a right angle elbow

In Fig. 10-6a, a right angle elbow consists of two parts. To create the development of the pipe, first evenly partition the cylinder surface of part Ⅰ into 12 pieces and get 12 point-pairs, 1-a', 2-b'... for example. The planar drawing of Part Ⅰ consists of 12 parts separated by lines like 1-A, 2-B and so on, which are created by projecting the dividing lines to the plane first and then connecting the end points smoothly. Generally the shortest dividing line is selected as the joint seam of the development of the pipe. However, the form of the development of Part Ⅱ, which is not the case, can reduce the material cost, see Fig. 10-6a.

Multi-piece cylindrical bends are widely used as ventilation pipes and heat pipes. Fig. 10-6b shows the development of a right-angle elbow consisting of 5-piece mitre pipes of which the middle three, parts Ⅱ, Ⅲ and Ⅳ, are mitred at both ends while two end ones, parts Ⅰ and Ⅴ have only one beveled end. All the five pieces can be developed by the method mentioned shown in Fig. 10-6b. The plane on which the intersection line between two adjacent pipe pieces are located is called the intersection plane. The angles between each axis of two pipe pieces and the corresponding intersection plane are equal to each other. So if rotate parts Ⅱ and Ⅳ 180° around their axes and then connect all the pieces, the bend will become a cylinder as shown in Fig. 10-6b. Based on this fact, the development of the right-angle elbow can be obtained by putting all the development of its pieces together, with parts Ⅱ and Ⅳ rotated, as the right figure shown in Fig. 10-6b, which can simplify the drawing process and reduce the material cost.

10.1.2 曲面立体表面的展开图

1. 直角弯管的展开图

如图 10-6a 所示，直角弯管由两部分组成。首先将件Ⅰ管的圆周长度等分为 12 份，得到 12 对点，即 1-a'、2-b'……相应地，件Ⅰ管的平面图也可认为是由 12 个部分组成，它们被 1-A、2-B 等分隔线隔开，将分割线投影到平面上，然后光滑连接各端点，得到展开图。将展开图接缝置于最短的边缘，这是通常的做法。然而，根据件Ⅱ管的展开图可以更经济地进行切割下料，如图 10-6a 所示。

多节圆柱弯管常用于通风管道和热力管道中。图 10-6b 所示为一个五节直角弯管展开图。弯头由五节斜口圆管组成，中间的三节是两端倾斜的全节，两头的两节是一面倾斜的半节，这些截头圆柱面都可以用前述圆柱面展开的方法展开。由于弯管各节圆柱管的斜口与轴线的倾角 α 相同，如果将第Ⅱ、Ⅳ节圆柱管旋转 180°后依次叠合各管，恰好构成一个完整的圆筒，如图 10-6b 中间图形所示。因此，可按图 10-6b 右面图形那样，将展开图拼合画出，这样可使划线简单，节省材料。

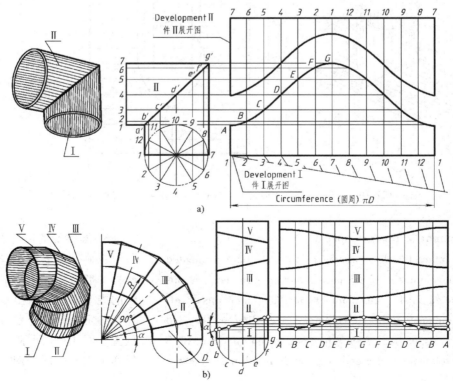

Fig. 10-6 Development of a right angle elbow 直角弯管展开图

2. Development of a straight tee

There are various kinds of bifurcated pipes in pipe construction. To create the development of such a bifurcated pipe, specify firstly the intersection lines on the pipe, then cut the pipe into several basic bodies with its intersection lines being borders between them. Develop each of these basic bodies respectively by the methods mentioned above and thus obtain the development of the whole pipe. Fig. 10-7 shows the development of a straight tee. To create the development of the intersection hole on part Ⅱ, specify the points A, B, C, D, E, F and G on the intersection line by the projection method shown in Fig. 10-7 first. Then the development of the hole is obtained by smoothly connecting these points and drawing the symmetric one of the curve ABCDEFG.

2. 正交等径三通管的展开图

在管道施工中常遇到各种各样的叉管，这类叉管表面展开，应首先确定相贯线，然后以相贯线为界限，将叉管分割为若干个基本形体，再按基本体的展开法画出各自的展开图。图10-7 所示为正交等径三通管展开图，其中画圆柱管件Ⅱ上相贯孔的展开图时，应先求出相贯线上各点 A、B、C、D、E、F、G，再光滑连接各点，最后画出曲线 ABCDEFG 的对称部分即可。

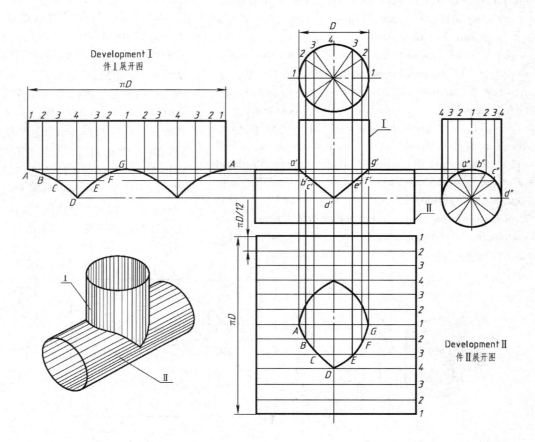

Fig. 10-7 Development of a straight tee　正交等径三通管展开图

3. Development of a cone pipe

When one creates the development of a cone pipe or truncated cone pipe, the development of the corresponding right cone is obtained first by the radial-line development method, then the development of the intersection line is obtained according to the real length, which is specified from the points on it. Fig. 10-8 illustrates the developments of a cone pipe and an oblique truncated cone pipe respectively.

3. 圆锥管的展开图

在画圆锥管或切口圆锥管展开图时，一般应首先用放射线法求出正圆锥的展开图，再求出被截切部分的实长，确定截交线上各点在展开图上的位置。图 10-8 所示分别为圆锥管和斜切圆锥管的展开图。

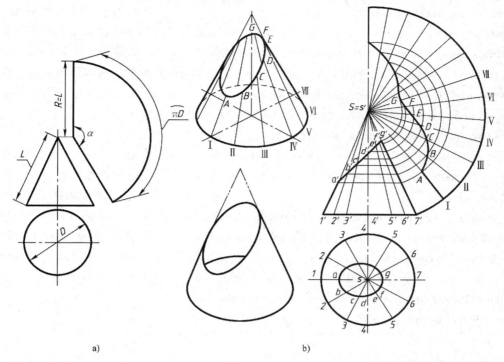

Fig. 10-8 Developments of a cone pipe and an oblique truncated cone pipe　圆锥管和斜切圆锥管的展开图

a) Development of a cone pipe　圆锥管展开图

b) Development of an oblique truncated cone pipe　斜截口圆锥管展开图

4. Development of a deformed joint

Deformed joints are commonly used as connecting parts between circular pipes and square tubes. as shown in Fig. 10-9, a deformed joint with a round top open and a square bottom open is used as a transitive part in a gas hood pipeline of a smith forging furnace. It is composed of four isosceles triangles and four partial oblique cones. Due to the joint is symmetrical, only one group of element lines need to be determined necessarily. First the real lengths of the element lines AⅠ、AⅡ、AⅢ、AⅣ are obtained by real-shape projection. By the real lengths of the element lines and the lengths of the top chords and the bottom edges, a set of triangles can be created successively, which form the development of the deformed joint, as shown in Fig. 10-9.

4. 变形接头的展开图

工程上在圆形管道与方形管道之间，常用变形接头作为过渡连接，如图 10-9 所示的"上圆下方"过渡连接件，其用于手锻炉烟罩管道连接。它是由四个等腰三角形和四个部分斜圆锥组成的。由于该变形接头是对称的，只需要确定其中一组圆锥素线的实长即可。作实长图以确定四条素线 AⅠ、AⅡ、AⅢ、AⅣ 的实长，用各素线的实长、弦长和下底的边长依次画出各个连续的三角形表面，即为所求变形接头的展开图，如图 10-9 所示。

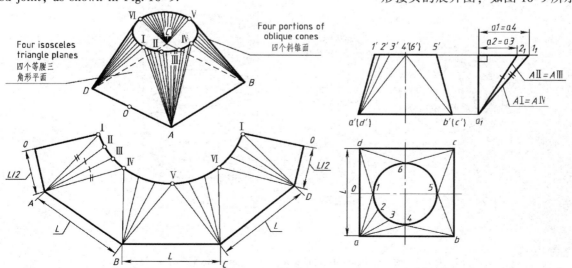

Fig. 10-9 Development of a deformed joint　变形接头的展开图

5. Approximate development of a sphere

Some surfaces can not be developed, such as spherical and spiral face, ring surface, etc. For the development of this kind of surfaces, approximate method is adopted.

As shown in Fig. 10-10a, a sphere can be divided into several horizontal layers (here is 7 layers). Each layer can be regarded as a frustum of a cone whose conic node S_i is the intersection point of two extended chords, except the middle one that can be represented by a cylinder. This method is known as stratified spherical development method.

Another method is slice method (willow leaf method), in which a set of longitudes are drawn on the sphere and the spherical surface is evenly divided into pieces, here is in 12 parts (see Fig. 10-10b). Each piece can be replaced by a partial cylinder tangent. Develop all these partial cylinders and put them together to acquire the approximate development of the sphere.

5. 球面的近似展开图

一些表面属于不可展表面，如球面、螺旋面、圆环面等。这类表面无法展平，只能采用近似的方法展平。

如图 10-10a 所示，可将球面割成若干水平层（图中为 7 层）。每层可视为一锥台，其顶点为二弦延长线之交点 S_i，中间一层可用圆柱面代替。此法即为球面分层展开法。

另一种方法是分条法（又称柳叶法），即在球面上画出若干条经线，将球面分割成 12 等份，如图 10-10b 所示。每一等份近似地用部分外切圆柱面代替。展开这些外切圆柱面，得到该球面的近似展开图。

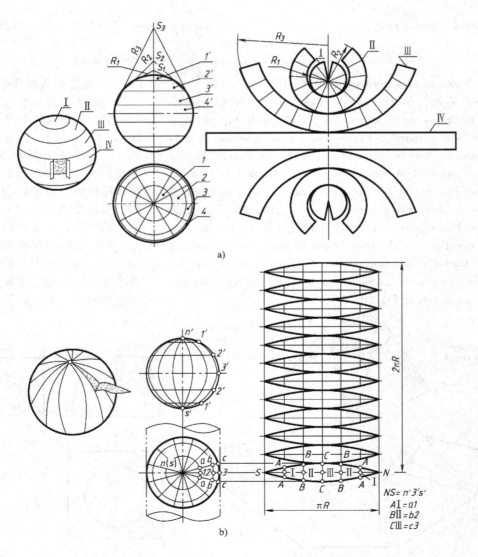

Fig. 10-10 Approximate development of a sphere　球面的近似展开图

10.2 Welding Drawings

Welding is a kind of dismountable connection and widely used in mechanical engineering, electronic engineering, shipbuilding, chemical engineering, architecture and so on.

Common welding joints include butt joints, lap joints, T-joints, corner joints and so on. The joined area of the weldment is called weld. As shown in Fig. 10-11, weld types include butt weld, spot weld, fillet weld and so on.

10.2 焊接图

焊接是一种不可拆连接，在机械、电子、造船、化工、建筑等行业中都得到了广泛的应用。

常见的焊接接头有对接接头、搭接接头、T形接头和角接接头等。焊件经焊接后所形成的结合部分称为焊缝。焊缝形式主要有对接焊缝、点焊缝和角焊缝等，如图10-11所示。

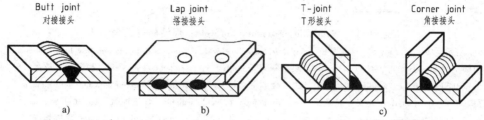

Fig. 10-11 Common forms of welding joints and welds　常见的焊接接头和焊缝形式
a) Butt weld　对接焊缝　b) Spot weld　点焊缝　c) Fillet weld　角焊缝

10.2.1 Representation of welding symbols on drawing (GB/T 324—2008)

1. Representation of welds

1) Generally a visible weld is drawn with a thick continuous line, as shown in Fig. 10-12a.

2) In sectional and cut views that perpendicular to the welding joints, a darken area is used to represent the section of a weld, as shown in Fig. 10-12.

3) A weld can also be represented using either gate lines, as shown in Fig. 10-12b and Fig. 10-12c, or bold lines, as shown in Fig. 10-12d and Fig. 10-12e. However. However, only one line type can be used in one drawing.

4) Weld sizes and welding symbols can be represented using views of partial enlargement, as shown in Fig. 10-12f.

10.2.1 焊缝符号表示法（GB/T 324—2008）

1. 焊缝的图示法

1) 一般情况下，用粗实线表示可见焊缝，如图10-12a所示。

2) 在垂直于焊缝的剖视图和断面图中，通常用涂黑表示焊缝的截面形状，如图10-12所示。

3) 在视图中，可用栅线表示焊缝，如图10-12b、c所示；也可用加粗线表示焊缝，如图10-12d、e所示。但在同一图样中，只允许采用一种画法。

4) 焊缝尺寸和焊缝符号可用局部放大图表示，如图10-12f所示。

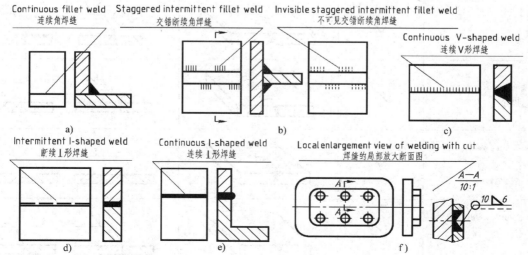

Fig. 10-12 Weld's representation examples　焊缝的画法示例

2. Welding symbols

Welding symbols generally consist of elementary symbols and leader lines, as shown in Fig. 10-13. If necessary, supplementary symbols can be used, see 10 and 13 in Fig. 10-13.

2. 焊缝符号

焊缝符号一般由基本符号和指引线组成,如图 10-13 所示。必要时还可加注补充符号,如图 10-13 中的序 10 和 13 所指符号。

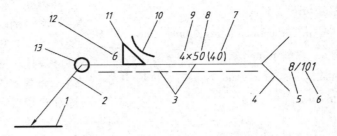

Fig. 10-13 Symbolic presentation of weld   焊缝的符号表示法

1— Weld 焊缝 2—An arrow line 箭头线 3—Reference line (with continuous thin line and thin long dashed line) 基准线(用细实线和细虚线) 4—Tail symbol 尾部符号 5—The number of the same weld 相同焊缝数量 6—Welding method (Metal arc welding) 焊接方法(金属电弧焊) 7—Distance between adjacent weld elements 断续焊缝间隔 8—Length of weld 焊缝长 9—The number of weld elements 焊缝段数 10—Concave surface 凹面 11—Elementary symbol 基本符号 12—Leg length 焊脚高 13—Weld all around 周围焊缝

(1) Elementary symbols Elementary symbols are used to represent basic forms and features of weld cross-sections, as shown in Table 10-1.

When marking double-sided welds or joints, elementary symbols can be used in combination, as shown in Table 10-2.

(1) 基本符号 基本符号表示焊缝横截面的基本形式或特征,见表 10-1。

标注双面焊焊缝或接头时,基本符号可以组合使用,见表 10-2。

Table 10-1 Elementary symbols 基本符号

Designation 名称	Symbol 符号	Diagrammatic sketch 示意图	Example for indication 标注示例
Butt weld between flanged plates 卷边焊缝	八		
I-shaped weld I 形焊缝	‖		
V-shaped weld V 形焊缝	V		
Single-V butt weld 单边 V 形焊缝	V		
V-shaped butt weld with a root face 带钝边 V 形焊缝	Y		
Single-V butt weld with a root face 带钝边单边 V 形焊缝	Y		

(Continued 续)

Designation 名称	Symbol 符号	Diagrammatic sketch 示意图	Example for indication 标注示例
U-shaped butt weld with a root face 带钝边 U 形焊缝	 Y		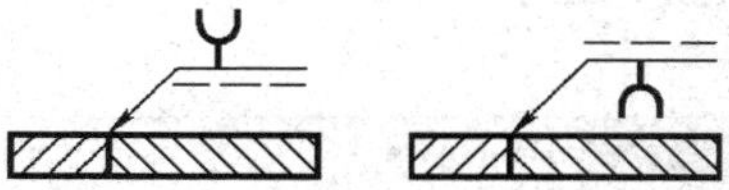
J-shaped butt weld with a root face 带钝边 J 形焊缝	P	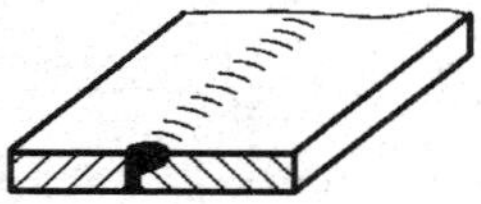	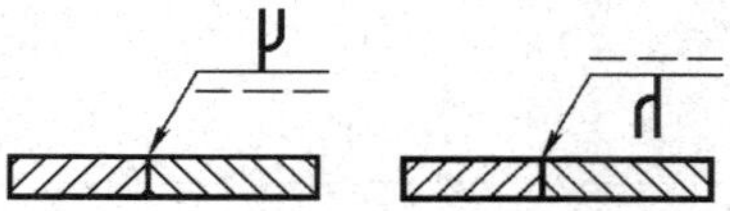
Back weld 封底焊缝	 ⌣	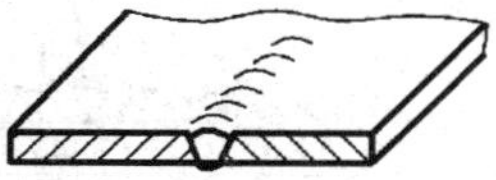	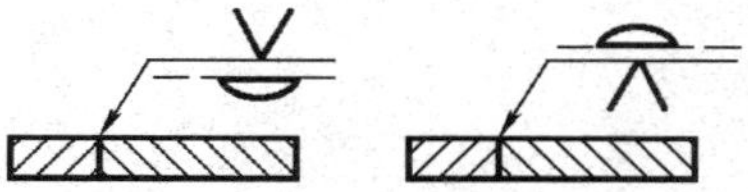
Fillet weld 角焊缝	△	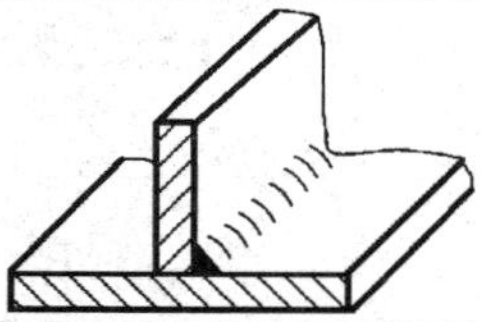	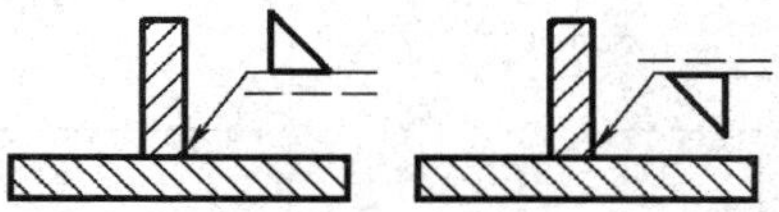
Plug weld or slot weld 塞焊缝或槽焊缝	⊓	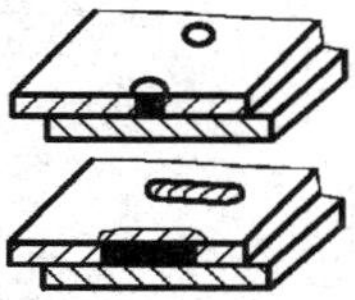	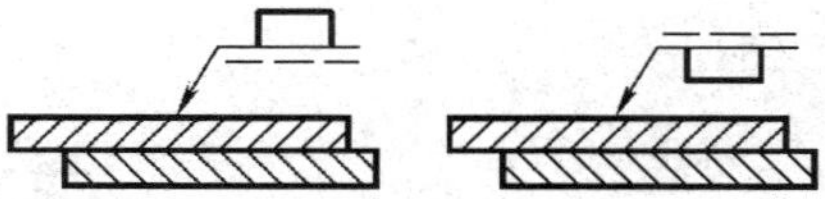
Spot weld 点焊缝	○	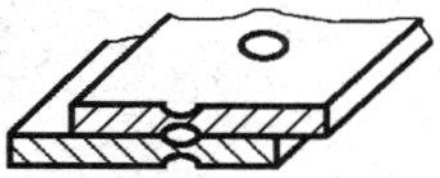	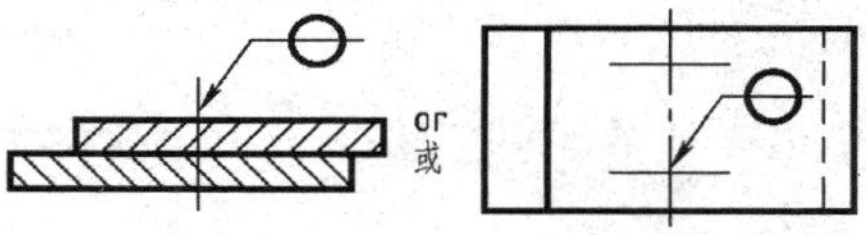
Seam weld 缝焊缝	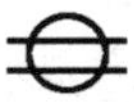 ⊖	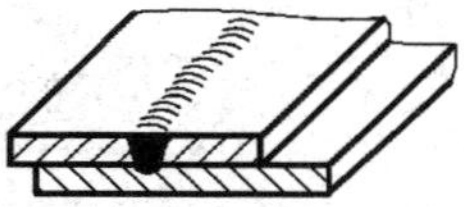	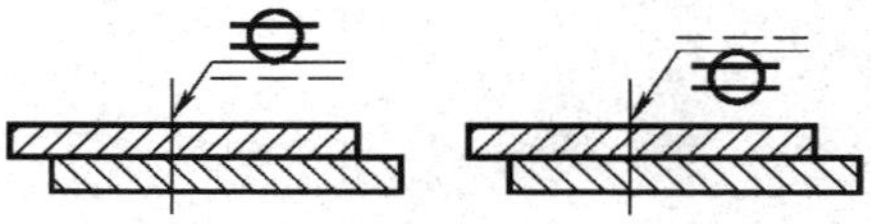

Table 10-2　Combinations of elementary symbols　基本符号的组合

Designation 名称	Symbol 符号	Diagrammatic sketch 示意图	Example for indication 标注示例
Double-sided V-shaped weld (X weld) 双面 V 形焊缝（X 焊缝）	X		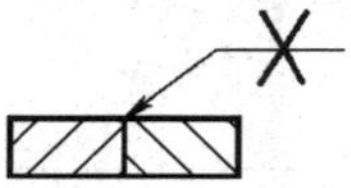
Double-sided unilateral V-shaped weld (K weld) 双面单 V 形焊缝（K 焊缝）	 K	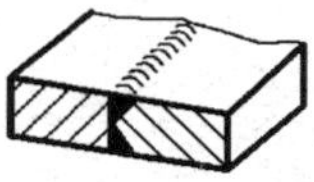	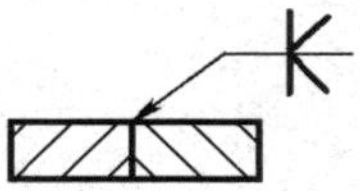
Double-sided V-shaped weld with a root face 带钝边的双面 V 形焊缝	X		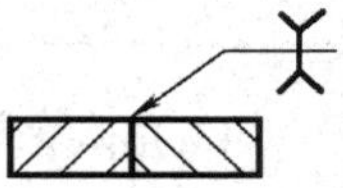
Double-sided unilateral V-shaped weld with a root face 带钝边的双面单 V 形焊缝	 K	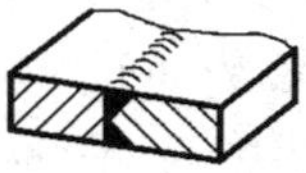	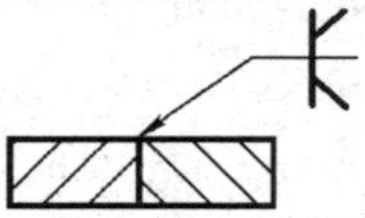
Double-sided U-shaped weld 双面 U 形焊缝	X		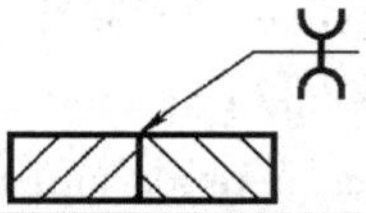

247

(2) Supplementary symbols Supplementary symbols are used to mark additionally on some weld or joint features, as shown in Table 10-3.

（2）补充符号 补充符号用来补充说明有关焊缝或接头的某些特征，见表10-3。

Table 10-3 Supplementary symbols 补充符号

Designation 名称	Symbol 符号	Diagrammatic sketch 示意图	Example for indication 标注示例
Flat 平面	—		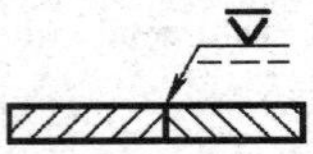
Concave 凹面	⌣	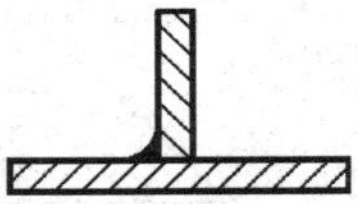	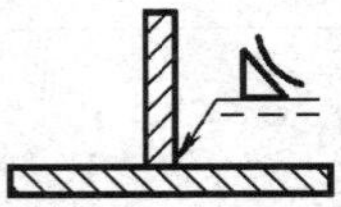
Convex 凸面	⌢		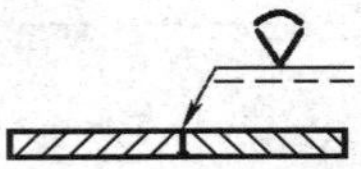
Smooth welding toe 圆滑过渡	⌊	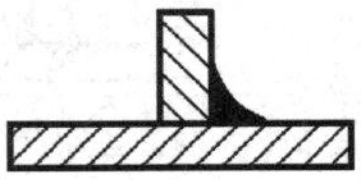	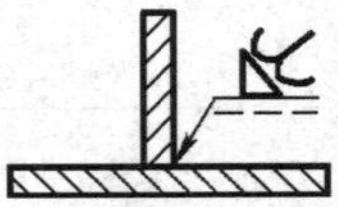
Permanent liner 永久衬垫	M		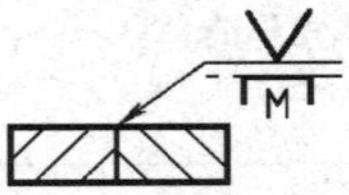
Temporary liner 临时衬垫	MR		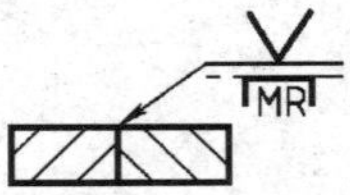
Weld on three sides 三面焊缝	⊐	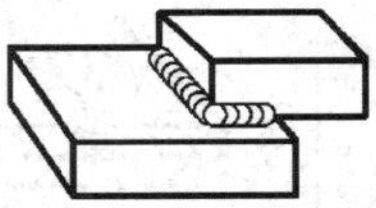	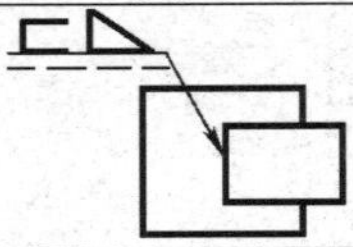
Weld all around 周围焊缝	○	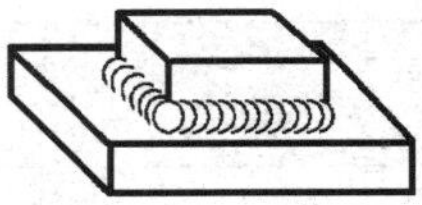	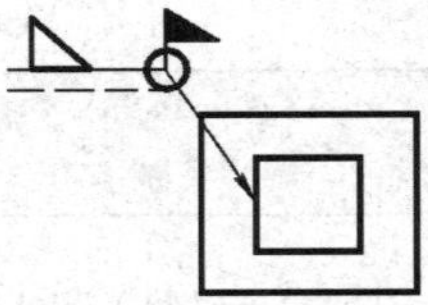
Site weld 现场焊缝	▶		
Tail 尾部	<	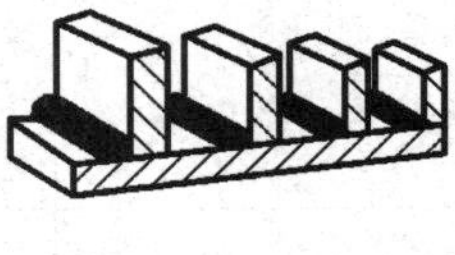	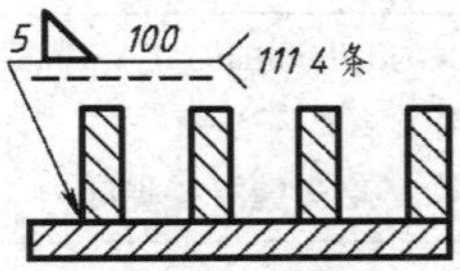

(3) Codes of welding methods There are many welding methods, such as arc welding, resistance welding, electro slag welding, spot welding and soldering welding and so on, of which arc welding is the most widely used one. Welding methods can be specified either in the technical requirements of the drawing or in the rear part of each weld symbol. Table 10-4 shows some common welding methods and their codes.

（3）焊接方法代号 焊接的方法有很多，常用的有电弧焊、电阻焊、电渣焊、点焊和钎焊等，其中以电弧焊最为广泛。焊接方法可用文字在技术要求中注明，也可用代号直接注写在尾部符号中。常用焊接方法及代号见表10-4。

Table 10-4 Common welding and allied processes—Nomenclature of processes and reference numbers (GB/T 5185—2005)
常用焊接及相关工艺方法代号 (GB/T 5185—2005)

Methods of welding 焊接方法	Code 代号	Methods of welding 焊接方法	Code 代号
Arc welding 电弧焊	1	Pressure welding 压力焊	4
Submerged arc welding 埋弧焊	12	Ultrasonic welding 超声波焊	41
Plasma arc welding 等离子弧焊	15	Friction welding 摩擦焊	42
Metal arc welding 金属电弧焊	101	Diffusion welding 扩散焊	45
welding rod arc welding 焊条电弧焊	111	Gas pressure welding 气压焊	47
Gravity station welding 重力焊	112	High energy beam welding 高能束焊	5
Atomic-hydrogen welding 原子氢焊	149	Electron beam welding 电子束焊	51
Carbon arc welding 碳弧焊	181	Laser welding 激光焊	52
Resistance welding 电阻焊	2	Gas laser welding 气体激光焊	522
Spot welding 点焊	21	Aluminium heat welding 铝热焊	71
Seam welding 缝焊	22	Electro-slag welding 电渣焊	72
Projection welding 凸焊	23	Induction welding 感应焊	74
Butt resistance welding 电阻对焊	25	Ray irradiation welding	75
Double-sided spot welding 双面点焊	212	Resistance stud welding 电阻螺柱焊	782
Gas welding 气焊	3	Flame cutting 火焰切割	81
Oxygen gas welding 氧燃气焊	31	Arc cutting 电弧切割	82
Oxygen acetylene welding 氧乙炔焊	311	Laser cutting 激光切割	84
Oxygen propane welding 氧丙烷焊	312	Hard soldering 硬钎焊	91
Hydrogen oxygen welding 氢氧氧焊	313	Soft soldering 软钎焊	94

3. Mark of welding symbols

In order to precisely indicate the position of a weld, the relative positions between the baselines and the elementary symbols should satisfy the following specifications:

1) If the elementary symbol is marked on the solid line side, it means that the weld is on the same side of the mark arrow, see Fig. 10-14a.

2) If the elementary symbol is marked on the thin long dashed line side, it means that the weld is on the opposite side of the mark arrow, see Fig. 10-14b.

3) The thin long dashed line can be omitted for symmetrical weld, see Fig. 10-14c.

4) The thin long dashed line can also be omitted for double-sided weld if the location of the weld is indicated clearly already, see Fig. 10-14d.

3. 焊缝符号的标注

为了在图样上确切地表示焊缝的位置，对基本符号与基准线的相对位置有下列规定：

1) 基本符号在实线侧时，表示焊缝在箭头侧，如图 10-14a 所示。

2) 基本符号在细虚线一侧时，表示焊缝在非箭头侧，如图 10-14b 所示。

3) 对称焊缝允许省略虚线，如图 10-14c 所示。

4) 在明确双面焊缝分布位置的情况下，有些双面焊缝也可省略虚线，如图 10-14d 所示。

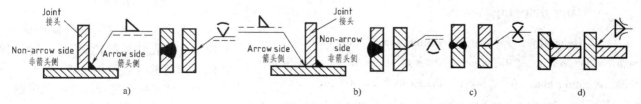

Fig. 10-14 Marks of welding symbols 焊缝符号的标注

4. Dimensioning welds

When there are strict size requirements for welds, weld dimensions must be marked. For related terms please refer to Fig. 10-15.

4. 焊缝尺寸的标注

对于有严格尺寸要求的焊缝，必须标注焊缝尺寸。与焊缝尺寸规范有关的焊接名词如图 10-15 所示。

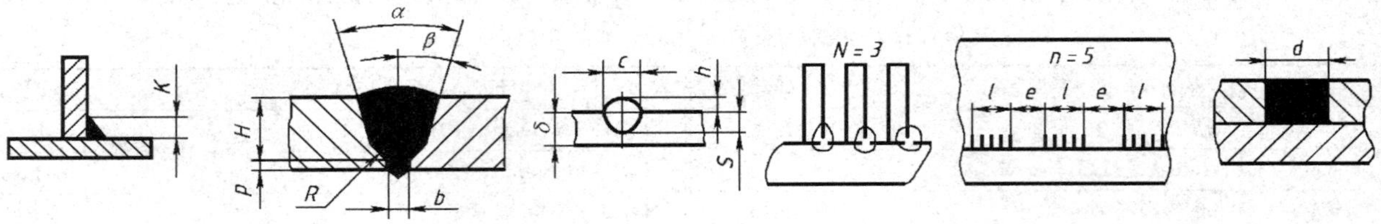

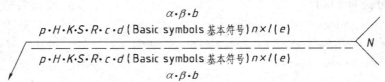

Fig. 10-15　Dimensioning welds　焊缝尺寸标注
K—Leg length　焊脚高度　P—Root face　钝边　H—Groove depth　坡口深度　R—Root radius　根部半径
α—Groove angle　坡口角度　β—Bevel angle　坡口面角度　b—Root opening　根部间隙　δ—Workpiece thickness　工件厚度
c—Weld width　焊缝宽度　h—Weld reinforcement　余高　S—Actual throat　焊缝厚度　N—The number of the same kind of welds　相同焊缝数量
n—The number of weld elements　焊缝段数　l—Weld length　焊缝长度　e—The distance between adjacent weld elements　焊缝间距
d—Diameter of spot welding nugget or plug welding hole　点焊熔核直径或塞焊孔径

For locations of weld dimensions please refer to Fig. 10-16. Rules of weld dimensioning are as follows:

1) Horizontal dimensions are placed on the left side of the elementary symbol.

2) Vertical dimensions are placed on the right side of the elementary symbol.

3) The groove angle α, the bevel angle β and the root opening b are placed above or below the elementary symbol.

4) The number of the same type of welds is placed at the tail of the dimension.

各焊缝尺寸在焊接符号中的位置如图10-16所示。焊缝尺寸的标注规则为：

1) 横向尺寸标注在基本符号的左侧。

2) 纵向尺寸标注在基本符号的右侧。

3) 坡口角度α、坡口面角度β、根部间隙b，标在基本符号的上侧或下侧。

4) 相同焊缝数量符号标在尾部。

Fig. 10-16　Principles of weld dimensioning　焊缝尺寸符号标注规则

5. The order of the Mark tail

When there are many items in the mark tail, they should be separated by the symbol "/" and sorted in the following order: number of the same kind of welds/code of welding method/grade of defect and quality/welding position/welding material/other items, see Fig. 10-13.

5. 尾部标注内容的次序

尾部需要标注的内容较多时，每个款项用斜线"/"分开，可参照如下次序排列：相同焊缝数量/焊接方法代号/缺陷质量等级/焊接位置/焊接材料/其他，如图10-13所示。

10.2.2　Examples of welds dimensioning and welding drawings

Generally a weld is represented simply with welding symbols on the contour of the view, instead of being drawn up with details. Welding mark examples are shown in Table 10-5.

Welding drawings are assembly drawings of all welded parts. A welding drawing includes all the contents that an assembly drawing should have. Furthermore, it contains weld symbols, and weld dimensions if necessary. If the welded parts are simple, their detail drawings are not compulsory and they can be represented by simply marking the sizes in the welding drawing, see Fig. 10-17.

10.2.2　焊缝尺寸标注示例及焊接图示例

在图样上，焊缝一般只用焊接符号简单标注在视图的轮廓上，不画细节图。焊缝的标注示例见表10-5。

焊接图实际上是焊接件的装配图，它应包括装配图的所有内容。此外，还应标注焊缝符号，必要时还应标注焊缝尺寸。如果焊接件比较简单，应将各组成构件的全部尺寸直接标注在焊接图中，从而不必画各组成构件的零件图，如图10-17所示。

Table 10-5 Examples of weld dimensioning 焊缝的标注示例

Type of joint 接头形式	Type of weld 焊缝型式	Example for indication 标注示例	Interpretation 说明
Butt joint 对接接头			V-shaped weld of groove angle is α, root opening is b, the number of weld elements is n, and the length of weld is l 表示V形焊缝的坡口角度为α，根部间隙为b，有n段长度为l的焊缝
T-joint T形接头			Single-fillet weld, leg length is K 表示单面角焊缝，焊脚高度为K
			Double-face intermittent fillet weld, the number of weld elements is n, the length of weld is l, the distance between adjacent weld elements is e, leg length is K 表示有n段长度为l的双面断续角焊缝，间隔为e，焊脚高度为K
Corner joint 角接接头			Double-face weld, above for the single V butt weld, the fillet weld at below 双面焊接，上面为单边V形焊缝，下面为角焊缝
			Weld on three sides, single-fillet weld, leg length is K 表示三面焊接，焊脚高度为K的单面角焊缝
Lap joint 搭接接头			That is the point of n spot weld. The solder nuclear diameter is d, the distance between adjacent weld elements is e 表示有n个焊点的点焊缝，焊核直径为d，焊点的间距为e

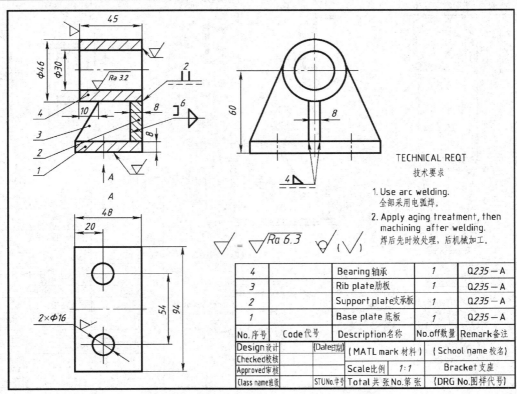

Fig. 10-17 Welding drawing of a bracket 支座焊接图

Fig. 10-17 shows the welding drawing of a bracket which consists of a bearing, a rib plate, a support plate and a base plate. These components are connected by welding. The weld between the bearing and the support plate is a I-shaped weld with a 2mm root opening. The weld between the support plate and the base plate is a single-fillet weld with a 4mm leg length. The welds between the rib plate and the other components are twin fillet welds on three sides with 6mm welding heights. From the technical requirements it can be seen that arc welding is used for all the welds.

图 10-17 是支座的焊接图。支座由轴承、肋板、支承板和底板组成。这些组成构件均焊接在一起，轴承和支承板之间采用 I 形焊缝，根部间隙为 2mm；支承板和底板之间采用单面角焊缝，焊脚高度为 4mm；肋板与其他构件之间均采用双面角焊缝三面施焊，焊脚高度均为 6mm。从技术要求可看出，焊接方法全部采用电弧焊。

The cylindrical shell shown of a dust catcher in Fig. 10-18 is used in a ventilation device for dust separation and collection. The device is made of several steel sheets which are processed to various shapes and then welded together.

Fig. 10-19 illustrates the general drawing of the $\phi300$mm cylinder shell and Fig. 10-20 shows the detail drawings of a circular tube, a taper pipe, an discharge pipe and a screw cap in the shell. Readers can refer to the drawings for skills and specifications of welding drawing.

图 10-18 所示的除尘器筒体用于通风设备中，其作用是分离和收集灰尘。它由钢板制成所需要的各种形状后焊接而成。

图 10-19 和图 10-20 所示分别为 $\phi300$mm 筒体的总图以及各组成部分的零件图，如圆管零件图、圆锥管零件图、排出管零件图和螺旋盖零件图。读者可参考学习焊接图的绘制技巧与规定。

Fig. 10-18 Cylindrical shell of a dust catcher 除尘器筒体

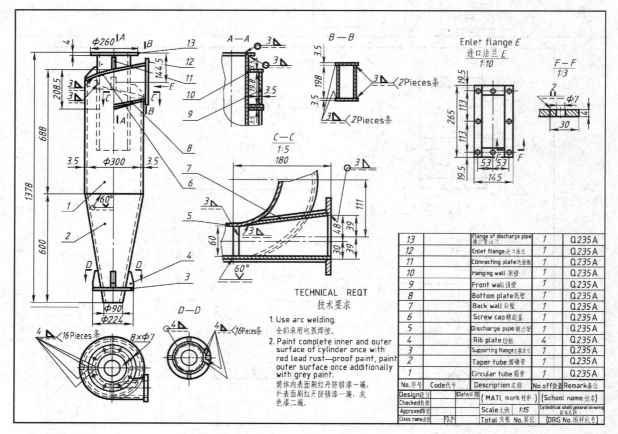

Fig. 10-19 General drawing of the cylindrical shell of a dust catcher 除尘器筒体总图

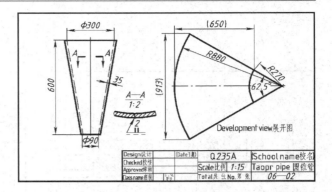
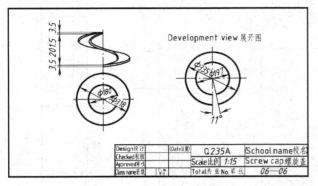
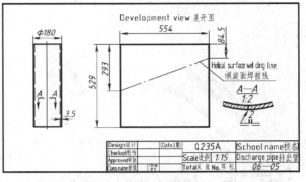

Fig. 10-20 Cylindrical shell detail drawings of a dust catcher 除尘器筒体零件图

10.3 Chemical Drawings

Chemical drawings mainly include chemical equipment drawings, chemical process diagrams and equipment layouts. This section focuses on the basic knowledge and representation methods of these four types of drawings.

10.3.1 Chemical equipment drawings

1. Representation methods

(1) View selection and configuration As chemical equipments are usually rotating bodies, generally only two basic views are needed to represent the major structures of an equipment. Usually the top view or the side view is generally placed in an empty space near the front view in the equipment drawing, with the view name being marked, see Fig. 10-21. Alternatively it can be provided with one additional drawing. If so, remarks like "top view (or side view) is shown in the other drawings" are needed on the major drawing. Meanwhile, remarks of "top view (or side view) of the ×× equipment" should be represented in the additional drawing. All symbols and numbers used on the drawings should be consistent and be labeled orderly on the major equipment drawing without repetition.

For non-standard components in an equipment of simple structures, their detail drawings can be placed in the same drawing sheet with the basic views of the equipment. In this case, the basic views of the equipment should be arranged in the major positions of the sheet. For those simple components consisting of plates, rods, blocks or pipes, it is enough to descript them by only offering their feature parameters like materials, diameters, thicknesses, lengths and so on in the item block of the assembly drawing, instead of providing their detail drawings.

10.3 化工制图

化工制图主要包括化工设备图、化工工艺图、化工设备布置图和管道布置图四类。本节主要简介这几种图的基本知识和表达方法。

10.3.1 化工设备图

1. 表达方法

(1) 视图的选择及其配置 由于化工设备多为回转体，因此一般采用两个基本视图来表达设备的主体结构即可。为了布图合理，可将俯（左或右）视图放在主视图旁边的图幅空白处，并注明视图名称，如图10-21所示。也可将俯（左或右）视图画在另一张图纸上，但必须注明"俯（左或右）视图在第二张图纸上"。同时，在第二张图纸的视图上方或标题栏中注明"××设备之俯（左或右）视图"。此时，在各张图样上采用的符号及编号等应在第一张图样中顺次标注，不得重复，以免混乱。

对于结构简单的设备所采用的少量非标准零部件，其零件图可与设备基本视图配置在同一张图纸上，但应将基本视图布置在图幅的主要部位上。对于用板、棒、块以及管材或型材组成的简单的零件，均可不画零件图，而在装配图的明细表中注明其材料、直径、厚度、长度等特征尺寸。

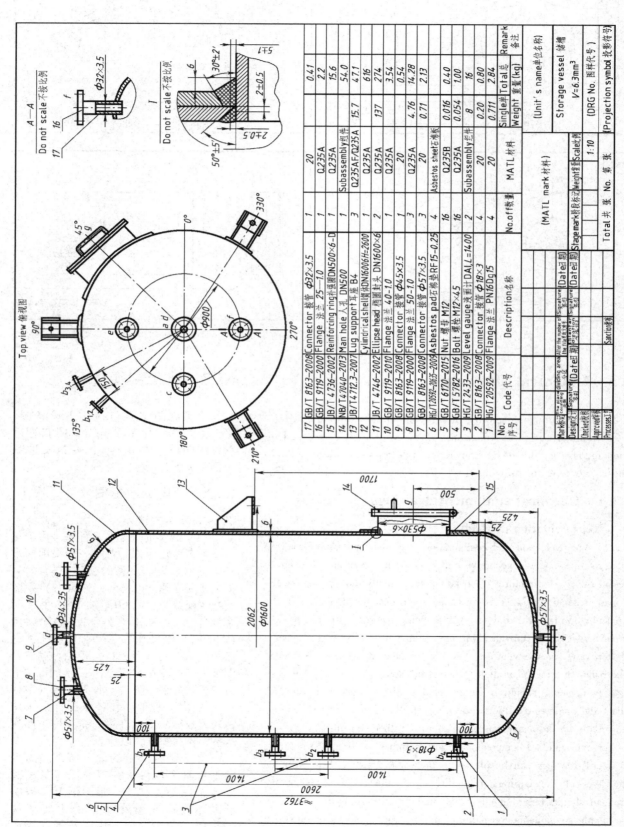

Fig. 10-21 Assembly drawing of a storage tank 储槽装配图

(2) Representation method of multi-rotation　In order to clearly represent the structures and positions of joints and openings on a shell equipment, multi-rotation method is adopted. In the method the structures are imaginarily rotated to the positions where their axes are parallel to the projection planes. In this way their views are obtained easily. No mark is needed in the drawings created by multi-rotation method and there must be no overlaps between views, as shown in Fig. 10-21.

(3) Drawing of partial enlargement and exaggerated representation　There are usually significant differences in component sizes for a chemical equipment. If only one scale is used in its assembly drawing, many components would be too tiny to be drawn up clearly. To overcome this problem, partial enlargement drawing and exaggerated representation are adopted, which accord with the national standards of mechanical drawings and are also called partial detail drawing or joint drawing in chemical engineering.

(4) Representation method of cut and segmentation (layer)　Usually diameters of chemical equipments are much less than their heights or lengths. In order to reduce sheet area and simplify drawing process, cut method is often adopted for the representation of the equipments, which has the same or regularly-changed structures along the height or length direction, as shown in Fig. 10-22. Also the equipments can be expressed segment-by-segment by segmentation representation method as shown in Fig. 10-23, or layer-by-layer by layer representation method.

（2）多次旋转的表达方法　为了将化工设备壳体上的接口管和开孔等结构形状和位置表达清楚，通常采用多次旋转的表达方式，即将设备各个方位上的接口管和开孔等结构分别采用假想的旋转画法，将其轴旋转到与投影面平行的位置，画出其视图或剖视图。这种多次旋转画法一般不作标注，但必须注意不可使图形重叠，如图10-21所示。

（3）局部放大图和夸大的表达方法　化工设备零部件的尺寸大小相差较大，在装配图中只用一种比例绘制时，会导致尺寸较小的复杂零件的图线过于密集而表达不清。为了清晰地表达这些细部结构，可在图纸空白处用局部放大图和夸大画法予以补充表达，这与机械制图标准相吻合。这种化工图亦称为局部详图或节点图。

（4）断开和分段（层）的表达方法　一般情况下，化工设备的直径与高（长）度尺寸大小相差很大。为了节约图幅、简化作图，对于沿高（长）度方向的形状或结构相同或按规律变化的过高（或过长）的设备，可采用断开画法，如图10-22所示。也可将其图形分段（层）画出，如图10-23所示。

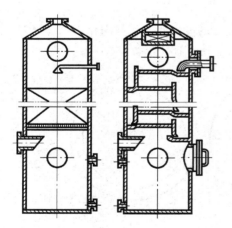
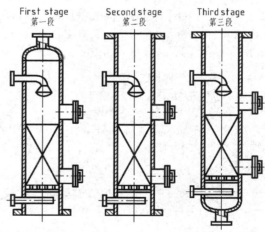

Fig. 10-22　Cut representation　断开表达方法　　**Fig. 10-23**　Segmentation (layer) representation　分段（层）表达方法

(5) Simplified representation　Besides the simplified representation methods specified by the national standards of mechanical drawing, following ones are also widely adopted in chemical engineering:

1) Simplified representations of parts and components. For parts and components by outsourcing, only their outlines need to be provided by thick continuous lines according to their main dimensions and scales, as shown in Fig. 10-24.

A glass-conduit liquid level meter can be represented with long dashed dotted line, with liquid level indicated by symbol "+" using thick line. A flange can be drawn by the method shown in Fig. 10-24. Its characteristics such as size and form of sealing surface can be provided in item block and nozzle table.

（5）简化画法　在化工设备图中，除采用机械制图国家标准中规定的简化画法外，还广泛采用了下面几种简化画法：

1）零部件的简化画法。对外购的零部件，只需根据主要尺寸，按比例用粗实线画出其外形轮廓，如图10-24所示。

对液面计的玻璃管可用细点画线表示，液面用粗"+"字线画出。法兰均可按图10-24所示的画法绘制。其规格、密封面形式等内容可在明细栏和管口表中说明。

2) Simplified representations of repeated structures

① A bolt-hole can be omitted and represented by its center lines or axis instead. A bolt connection can be simplified using symbol "+" or "×" with continous thick line, as shown in Fig. 10-24.

2) 重复结构的简化画法

① 螺栓孔可省略圆孔的投影，只画中心线或轴线；螺栓联接可用"+"或"×"粗实线符号简化画出，如图10-24所示。

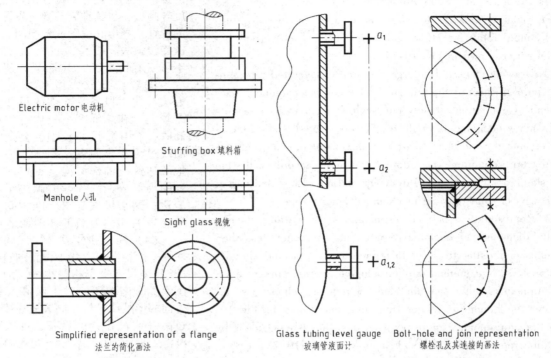

Fig. 10-24　Simplified representations of parts and components　零部件的简化画法

② When holes on tube sheets, baffle plates and column plates in a heat exchanger are aligned in a designed sequence, the drawing can be simplified, as shown in Fig. 10-25.

② 换热器中的管板、折流板或塔板上的孔按规则排列时，其简化画法如图10-25所示。

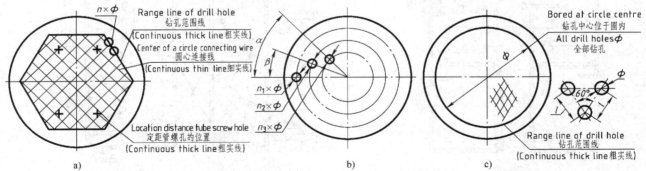

Fig. 10-25　The simplified representation of holes on perforated plate　多孔板上孔的简化画法
a) Holes in the usual array 孔按一般顺序排列　b) Holes in accordance with concentric circles 孔按同心圆排列　c) No requirement for the number of holes 对孔数要求较低

③ When there are a dense set of tubes built in an equipment and the tubes are aligned in a designed sequence, one tube needs to be drawn clearly at least, whereas the remaining ones can be simplified with thin long dashed dotted line, as shown in Fig. 10-26.

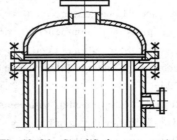

Fig. 10-26　Simplified representation of a dense set of tubes
密集管子的简化画法

③ 当设备上密集的管子按一定的规律排列成管束时，图中至少画出一根管子，其余管子用细点画线表示，如图10-26所示。

④ When there are fillers built in equipments, like porcelain ring and grit, which have identical model, material, and settings, we can use intersected thin continuous lines together with dimension information and literal explanation to describe the fillers on section views, as shown in Fig. 10-27a and Fig. 10-27b. Simplified representation of stuffing in a stuffing box is shown in Fig. 10-27c.

④ 当设备中装有同一规格、同一材料和同一堆放方式的填充物（如瓷环、砂砾）时，在剖视图中，可用相交的细实线表达，并注写有关的尺寸和文字说明，如图10-27a、b所示。填料箱中填料的简化画法如图10-27c所示。

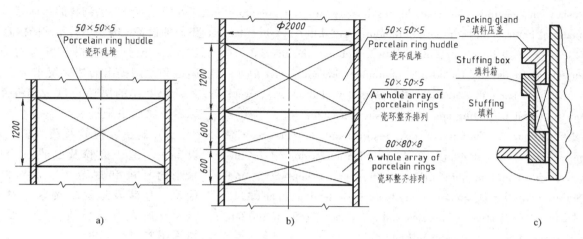

Fig. 10-27　Simplified representations of fillers　填料的简化画法
a) Unified fillers　单一填料　b) Multi-layered fillers　多层填料　c) Sealed fillers　密封填料

3) Single line schematic representation method. When a structure has been expressed clearly in other views, it can be simplified using a single thick continuous line in the equipment drawing, as shown in Fig. 10-28.

3）单线示意简化画法。设备上的某些结构，在其他视图中已经表达清楚时，设备图上允许用单线（粗实线）表示，如图10-28所示。

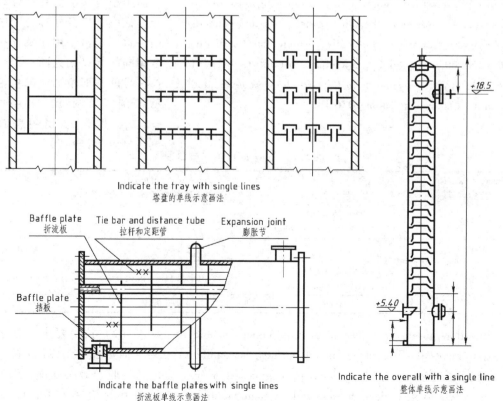

Fig. 10-28　Single line schematic representations　单线示意简化画法

2. Technical data and requirements in chemical equipment drawings

(1) Technical data sheet Different equipments have different technical data. There are mainly five basic columns in the technical data sheet.

1) Basic data column. This column contains general characteristics as well as the standards, specifications and laws with which the design must comply, for example working pressure, temperature and material name, etc. Different equipments have different contents.

2) Design data column. This column includes the special requirements and measures for design, manufacturing and usage of the equipment, for example corrosion allowance of wall thickness, weld coefficient, heat insulating and anticorrosion measure.

3) Column of manufacturing, testing and acceptance. The column contains the standards and specifications adopted in the course of equipment manufacturing, testing and acceptance, and the requirements for equipment manufacturing, testing, packaging and delivery. Orientations of nozzles and supports, "refer to the left view" for instance, must be specified.

4) Welding rod table. It is filled with welding methods, weld rod brands, standards of welding groove forms and welding seam sizes.

5) Nozzle table. It includes the data of all nozzles and openings. Nozzles must be numbered with nozzle symbols using lowercase Chinese pinyin letters, which should be consistent with those in the equipment drawing. All nozzles of the same model, use and connection surface form are numbered identically, which can be distinguished from each other by adding numeral subscript to each of them, for example b_1, b_2, b_3 and b_4 in Fig. 10-21. A recommended format of the table is shown in Table 10-6.

2. 化工设备图的技术数据和技术要求

（1）技术数据表 根据设备种类的不同，其内容也有所不同，主要有五个基本栏目。

1）基本数据栏。填写通用特性以及设计所依据的标准、规范和法规，如工作压力、温度、物料的名称等。对于不同的设备，需要填写不同的内容。

2）设计数据栏。填写对设计、制造和使用提出的特殊要求或专用措施，如壁厚的腐蚀裕度、焊缝系数、保温及防腐措施等。

3）制造、检验及验收栏。填写设备在制造、检验及验收过程中所采用的标准和规范，以及对制造、检验、包装及运输等要求。对于管口及支座方位，必须填写，如"按左视图"。

4）焊条表栏。填写焊接的方法、焊条的牌号、焊缝的坡口形式和尺寸所遵循的标准等内容。

5）管口表栏。填写各管口和开孔的有关数据等内容。设备上的管口一律用小写汉语拼音字母编写管口符号，栏中管口符号应与图中管口符号相一致。规格、用途及连接面形式完全相同的管口编为同一编号，但必须在符号的右下角加注阿拉伯数字，以示区别，如图10-21中的b_1、b_2、b_3、b_4，其推荐格式见表10-6。

Table 10-6 Nozzle table 管口表 (mm)

Symbol 符号	Nominal diameter 公称直径	Connection size and standard 连接尺寸和标准	Connection surface form 连接面形式	Usage or name 用途或名称
a	250	PN0.25 DN25 HG/T 20592~20653—2009	Raised face 凸面	Air inlet××气进口
b_{1-2}	400×300			Manhole 人孔
c	150	PN0.25 DN25 HG/T 20592~20653—2009	Raised face 凸面	Water inlet 水进口
d	20	G3/4	Screw 螺纹	Outlet 放净口

(2) Technical requirements Technical requirements specify the special requirements for materials, manufacturing, assembly, acceptance, surface treatment, lubrication, packaging, delivery etc., and standards and specifications adopted in manufacturing and testing of the equipment.

（2）技术要求 技术要求是用文字简明地表示对材料、制造、装配、验收、表面处理、润滑、包装和运输等的特殊要求，以及设备在制造、检验时应遵循的标准和规定。

3. Numbering of parts and components in chemical equipment drawings

All parts and components in an equipment drawing must be numbered. Each standard component such as man hole, stuffing box, sight glass and level gauge can be regarded as one single part and has only one sequence number, though they may consist of several parts. While for a non-standard component, if it has an additional component assembly drawing, it can be assigned a single sequence number as well in the equipment drawing, as shown in Fig. 10-21.

10.3.2 Chemical process diagrams

1. Scheme flow diagram (SFD)

(1) Functions and contents of SFD SFD is also called process flow diagram, representing process in a factory or a workshop. SFD helps to determine process scheme in the stage of preliminary design and provides a basis for the drawing of construction flow diagram in next step.

A SFD (see Fig. 10-29) generally includes the contents as follows:

1) Machines and equipments engaged in the production process. See equipment classification codes in Table 10-7 (HG/T20519.35—1992).

2) Process flow by which raw materials are made into products or semi-finished products.

3. 化工设备图的零部件编号

化工设备图上的所有零部件均需编号。对标准化的部件，如人孔、填料箱、视镜、液面计等，虽然由若干个零件组成，但可视为一体，只编写一个序号。而那些非标准化的部件，如果另有部件装配图，可在设备装配图上只编一个序号，如图10-21所示。

10.3.2 化工工艺图

1. 方案流程图

（1）方案流程图的作用及内容 方案流程图又称流程示意图或流程简图，表达工厂或车间流程，用于初步设计阶段工艺方案的确定，为下一步设计施工流程图提供依据。

方案流程图（见图10-29）一般包括以下内容：

1) 生产过程中所采用的各种机器、设备。设备类别代号见表10-7（HG/T 20519.35—1992）。

2) 物料由原料转变为半成品或成品的运行程序——工艺流程线。

Table 10-7 Equipment classification codes 设备类别代号

Equipment class ification 设备类别	Tower 塔	Pump 泵	Compressor/Fan 压缩机、风机	Heat exchanger 换热器	Reactor 反应器	Vessel 容器	Industrial furnace 工业炉	Chimney 烟囱	Other machinery 其他机械
Code 代号	T	P	C	E	R	V	F	S	M

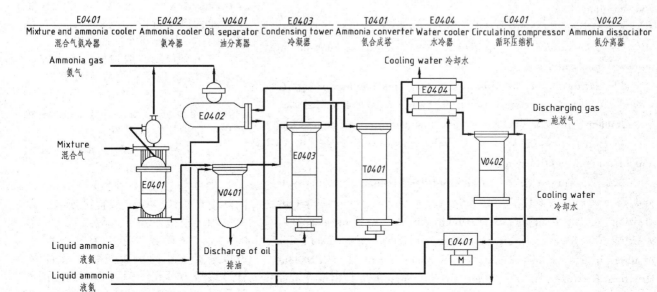

Fig. 10-29 Process flow diagram of synthetic ammonia 合成氨的工艺流程图

(2) Drawing of SFD SFD is a kind of schematic development drawing of equipments, with necessary marks and illustrations. Equipment item numbers and names are located in the upper part of the diagram and near the figures of the corresponding equipments, as shown in Fig. 10-29.

The drawing method of a SFD is as follows:

1) Draw up the sketchy outlines of all the equipments using thin continuous lines according to their approximate relative sizes.

2) Keep enough spaces between the equipments. The heights of the equipments and the positions of important nozzles should be approximately consistent with the real conditions.

3) Only one delegate equipment needs to be drawn for a set of equipments of the same type. Standby equipments can be omitted.

4) Draw process flow lines of major materials using thick continuous lines, with arrows indicating the flow directions of the materials. The names, sources and destinations of the materials are labeled at the beginnings and ends of the process flow lines.

5) If there are overlaps in the diagram between process flow lines and equipments while actually there is no connection between them, some process flow lines should be broken or pass by the overlapped elements so that the process flow lines look clear and tidy.

6) All main process flow lines need to be drawn, while subsidiary ones can be omitted or simplified.

2. Construction flow diagram (CFD)

(1) Functions and contents of CFD CFD is a more detailed process flow diagram based on SFD. It is also called process piping and instrument flow diagram or pipe installation flow diagram with control points. CFD illustrates the equipments, pipes, valves, pipe attachments and instrument control points involved in the production process. It is the basis of equipment and pipe arrangement and the guidance to equipment construction, operation, running and maintenance.

A CFD usually contains following contents:

1) Illustrations of equipment item numbers, equipment names, spools and nozzles.

2) Various kinds of piping flow diagrams containing pipes, valves and instrument control points including points of temperature measurement, pressure measurement, flow measurement and so on.

3) Illustrations of symbols of valve isosceles triangle fittings and instrument control points.

(2) Drawing of CFD In different design stages, the contents and requirements of CFD are different. A CFD is constructed by drawing up all diagram units which can be working sections, working procedures, equipment item numbers, names or pipe numbers in a workplace. Generally A1 sheet is used for CFD construction. However, if a diagram is very simple, A2 sheet also works well.

（2）方案流程图的绘制 方案流程图是一种示意性的设备展开图，并加上必要的标注和说明。在流程图上方和靠近设备图形的位置列出设备的位号及名称，如图10-29所示。

绘制方法如下：

1）用细实线画出设备的大致轮廓线，一般不按比例绘制，但应保持它们的相对大小。

2）各设备之间应保留适当的距离，其高度及设备上重要接管口的位置应大致符合实际情况。

3）同样的设备可只画一套。备用设备可以省略不画。

4）用粗实线画出主要物料的工艺流程线，用箭头标明物料流向，并在流程线的起始和终了位置注明物料的名称、来源及去向。

5）当流程线之间、流程线与设备之间交错或重叠，而实际并不相连时，应将其中一条断开或曲折绕过，使各设备间流程线的表达清晰明了、排列整齐。

6）一般只画出主要工艺流程线，其他辅助流程线不必一一画出。

2. 施工流程图

（1）施工流程图的作用及内容 施工流程图又称工艺管道及仪表流程图，或称带控制点的管道安装流程图。它是在方案流程图的基础上设计绘制的内容较详细的工艺流程图，是用图示法把化工生产的工艺流程和所需的设备、管道、阀门、管道附件及仪表控制点表示出来的一种图样。它是设备布置和管道布置设计的依据，也是施工、操作、运行及检修的指南。

施工流程图一般包括以下内容：

1）设备位号、名称和接管口。

2）含有管道、阀门以及仪表控制点（测温点、测压点、测流量点等）的各种管道流程图。

3）对阀门等腰三角形管件和仪表控制点图例符号的说明。

（2）施工流程图的绘制 按不同的设计阶段，施工流程图所表达的内容和要求也不同。一般以工艺装置的主项（工段或工序）为单元绘制，或以装置（车间）设备位号及名称、管道编号为单元绘制。施工流程图通常采用A1图幅，对特别简单的施工流程图可采用A2图幅。

The drawing method of a CFD is as follows:

1) Draw the legends of all equipments and machines, which are specified in HG/T 20519.34—1992, using thin continuous lines from left to right according to the process flow, as shown in Fig. 10-30.

施工流程图的绘制方法如下：

1）根据流程，自左向右将设备、机器的图形（按照 HG/T 20519.34—1992 标准中的图例）用细实线绘制，如图 10-30 所示。

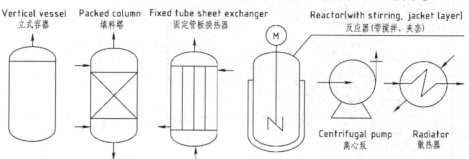

Fig. 10-30 Illustrations of equipments　设备示意画法

2) All nozzles including manhole, hand hole, discharge opening, etc., should be drawn if possible, among which the pipes connecting to external environment such as drainage valves, air outlets, discharging openings, meter interfaces and so on must be drawn, as shown in Fig. 10-31.

3) All equipments and machines should be arranged reasonably in the drawing so that their marks and the connection pipes between them are easy to be drawn. For equipments and machines among which there are close material relations, high tank and holding tank, pump and cooling tower for instance, their relative positions in the drawings must be decided according to their real layout in the workshop.

（3）Equipment annotation　Each device should be marked up with device serial number and device name; the format of usage is shown in Fig. 10-32.

（4）Drawing and dimensioning of pipes

1) All piping flow lines are drawn using thick continuous lines. Auxiliary pipes and pipes of synergic system need not to be drawn completely, instead only short segments connecting the equipments need to be drawn, on which material codes and the numbers of the flow diagrams to which they belong are marked. For connection pipes between flow diagrams, the numbers of the connected diagrams, equipments and spool pieces should be marked in an area of 30mm×6mm above their ends, as shown in Fig. 10-33a.

2) Marking of pipe codes. A pipe code consists of three sections, i.e., material code, principal term of code and sub-order of pipes, as shown in Fig. 10-33b. Common material codes are listed in Table 10-8.

3) Marking of pipe diameters. Only nominal diameter is marked. When process is simple and there are only few pipe types, piping classes can be omitted while pipe sizes, marked as "outer diameter × wall thickness", and piping material codes need to be marked, as shown in Table 10-9.

4) Drawing of pipes. Generally all process pipes, as well as short segments of auxiliary pipes related to the process flow, are required to be drawn. For legends, line types and line widths used on pipe drawing, please refer to Table 10-10.

2）设备、机器的全部接口（包括人孔、手孔、卸料口等）尽可能画出。其中与配管有关以及与外界有关的管口（如直连阀门的排放口、排气口、放空口及仪表接口等）则必须画出，如图 10-31 所示。

3）各设备、机器在图上的布局位置要便于连接管线和标注，相互间物料关系密切者（如高位槽与储槽，泵与冷却塔等）的高低相对位置要与设备在车间的实际布置相吻合。

（3）设备的标注　每个设备都应该编写设备位号，注写设备名称，其组成如图 10-32 所示。

（4）管道的绘制与标注

1）工艺管道流程线均用粗实线绘制。辅助管道、公用系统管道，只绘出与设备（或工艺管道）相联接的一小段，并在此管道上标注物料代号及该辅助管道或公用系统管道所在流程图的图号。各流程图间相衔接的管道，应在始末端注明其连续图的图号（写在 30mm×6mm 的框内）及来去的设备位号或管段号，如图 10-33a 所示，一般标注在管道上方。

2）管道的标注。管道号由物料代号、主项代号和管道分段顺序号组成，如图 10-33b 所示。常见物料代号见表 10-8。

3）管径的标注。一律标注公称直径。对于工艺流程简单、管道品种不多时，管道等级可省略标注，管子尺寸可直接填写"外径×壁厚"，并标注工程规定的管道材料代号，见表 10-9。

4）管道的绘制。一般应画出全部工艺管道及与工艺有关的一小段辅助管道。管道的图例、线型及线宽见表 10-10。

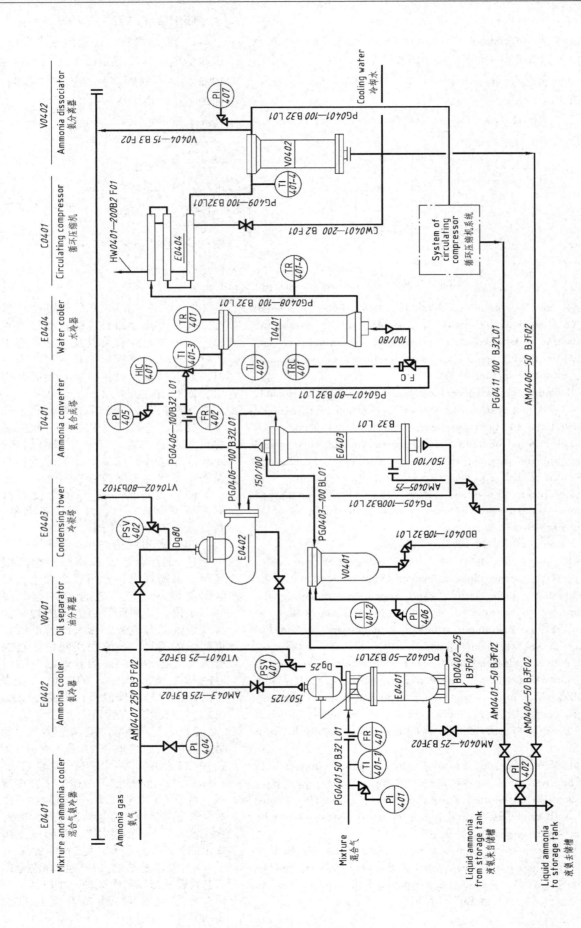

Fig. 10-31 Piping and instrument flow diagram of ammonia synthesis work section 氨合成工段管道及仪表流程图

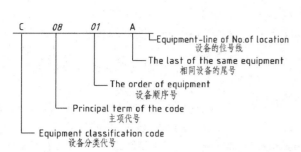

Fig. 10-32　The composition of device serial number
设备位号的组成

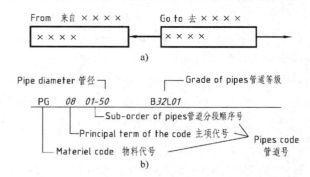

Fig. 10-33　Marking of a pipe　管道的标注

Table 10-8　Common material codes　常见物料代号

Name 名称	Code 代号	Name 名称	Code 代号
Process air 工艺空气	PA	High-pressure steam 高压蒸汽	HS
Process gas 工艺气体	PG	Medium-pressure steam 中压蒸汽	MS
Process liquid 工艺液体	PL	Low-pressure steam 低压蒸汽	LS
Process solid 工艺固体	PS	Steam condensate 蒸汽冷凝水	SC
Process water 工艺水	PW	Recirculating cooling water return 循环冷却水回水	CWR
Compressed air 压缩空气	CA	Recirculating cooling water supply 循环冷却水上水	CWS
Instrument air 仪表空气	IA	Hot water return 热水回水	HWR
Fuel gas 燃料气	FG	Hot water supply 热水上水	HWS
Fuel oil 燃料油	FO	Liquid ammonia 液氨	AL
Natural gas 天然气	NG	Drain 排液	DR
Refrigerating brine water return 冷冻盐水回水	RWR	Vacuum exhaust 真空排放气	VE
Refrigerating brine water supply 冷冻盐水上水	RWS	Vent 放空	VT
Inert gas 惰性气体	IG	Flare vent 火炬排放气	FV
Raw water 原水	RW		

Table 10-9　Piping material codes　管道材料代号

Material 材料	Cast iron 铸铁	Carbon steel 碳钢	Ordinary low alloy steel 普通低合金钢	Alloy steel 合金钢	Stainless steel 不锈钢	Nonferrous metal 有色金属	Nonmetal 非金属	Lining and inner anticorrosion 衬里及内防腐
Code 代号	A	B	C	D	E	F	G	H

Table 10-10　Pipe legends, line types and widths　管道图例、线型及线宽

Name 名称	Legend 图例	Line type and width 线型及线宽/mm
Principal material pipeline 主物料管道	——	Thick continuous line $b=0.9—1.2$ 粗实线 $b=0.9\sim1.2$
Auxiliary material pipeline 辅助物料管道	——	Medium continuous line $b=0.5—0.7$ 中粗线 $b=0.5\sim0.7$
Instrument pipe 仪表管	——	Thin continuous line $b=0.15—0.3$ 细实线 $b=0.15\sim0.3$
Tracing pipe 伴热（冷）管道	– – –	Medium dashed line $b=0.5—0.7$ 中粗虚线 $b=0.5\sim0.7$
Electrical tracing pipe 电伴热管道	–·–·–	Thin dashed-dotted line $b=0.15—0.3$ 细点画线 $b=0.15\sim0.3$
Heat insulation layer 管道隔热层	▨	Thin continuous lines except pipe lines 除管道外其他线为细实线
Jacketed piping 夹套管	⊨⊨⊨	

(5) Drawing of valves, pipe fittings and pipe attachments

1) Represent valves, pipe fittings and attachments using specified standard symbols with thin continuous lines, and label their specification codes, see Table 10-11.

2) Mark valves, pipe fittings and attachments. Reducers can be marked as "nominal diameter at large end × nominal diameter at small end".

（5）阀门、管件和管道附件的绘制

1）用细实线按标准所规定的符号（见表10-11）绘制出管道上的阀门、管件和管道附件，并标注其规格代号。

2）注出管道上的阀门、管件和管道附件。对异径管道应标注"大端公称直径×小端公称直径"。

Table 10-11　Symbols of common valves, pipe fittings and attachments　常用阀门、管件和管道附件的图形符号示例

Name 名称	Gate valve 闸阀	Stop valve 截止阀	Throttle valve 节流阀	Ball valve 球阀	Hydrophobic valve 疏水阀	Spring safety valve of angle type 角式弹簧安全阀
Symbol 符号						

Name 名称	Eccentric reducer 偏心异径管		Elbow 弯头	Tee 三通	Four-way 四通	Flame arrester 阻火器
	Flat bottom 平底	Flat top 平顶				
Symbol 符号						

Name 名称	Stub end 管端法兰	Cap 管帽	Discharging cap 放空帽	Discharging pipe 放空管	Concentric reducer 同心异径管	Sight glass 视镜
Symbol 符号						

(6) Drawing of instrument control points Instrument control points are drawn on their corresponding pipes using specified symbols with thin continuous lines. The symbols consist of graphic symbols and letter codes, which jointly express measured variables, functions and names of instruments, equipments, components and pipelines.

1) Graphic symbol. A graphic symbol is a circle of 12mm or 10mm diameter drawn with thin continuous line, in which an instrument item number is labeled.

2) Letter code. A letter code includes two letters. The first letter is used to represent the measured variable and the second one stands for the instrument function. Common measured variables and instrument functions are formulated in HG/T 20505—2014, as shown in Table 10-12.

（6）仪表控制点的绘制　仪表控制点应用细实线在相应的管道上用规定的符号画出。符号包括图形符号和字母代号，它们组合起来表达被测变量和功能，或表示仪表、设备、元件和管线的名称。

1）图形符号。仪表的图形符号是直径为12mm（或10mm）的细实线圆圈，圆圈中注上仪表位号。

2）字母代号。字母代号由两个字母组成，第一个字母表示被测变量，第二个字母表示仪表的功能。常用被测变量和仪表功能代号按HG/T 20505—2014标准，见表10-12。

Table 10-12　Codes of common measured variables and instrument functions　常用被测变量和仪表功能代号

Measured variable 被测变量		Instrument function 仪表功能	
T	Temperature 温度	I	Indicate 指示
P	Pressure 压力	R	Record 记录
F	Flow 流量	C	Control 控制
L	Location 物位	A	Alert 报警

3) Instrument item number. An instrument item number consists of letter code and loop number. The loop number, generally a 3—4 digit number, includes a workshop section code and a serial number. The letter code is filled in the upper part, the circle, the graphic symbol of the instrument, and the loop number is placed in the lower part, as shown in Fig. 10-34.

3) 仪表位号。仪表位号由字母代号和回路编号两部分组成。回路编号由工段号和顺序号组成，一般用3~4位阿拉伯数字表示。字母代号填写在表示仪表的图形符号——圆圈的上半部分中，回路编号填写在圆圈的下半部分中，如图10-34所示。

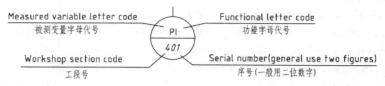

Fig. 10-34 Instrument item number 仪表位号标注

4) Symbols of instrument installation. They are shown in Table 10-13. In the table, "main position" means the installations of all instruments are centralized in control room, while "auxiliary position" means the installations are centralized on site.

4) 仪表安装位置的图形符号。表10-13列出了仪表安装位置的图形符号，表中"主要位置"是指仪表引向控制室集中安装，"辅助位置"是指仪表现场集中安装。

Table 10-13 Symbols of instrument installation 仪表安装位置的图形符号

Installation 安装位置 Function 仪表功能	Main position (for monitoring) 主要位置（操作员监视用）	Installation on site (normally no monitoring) 现场安装（正常情况下操作员不监视）	Auxiliary position (for monitoring) 辅助位置（操作员监视用）
Discrete instrument 离散仪表	⊖	○	⊖
Shared display, shared control 共用显示、共用控制	⊡	▢	⊡

10.3.3 Equipment layout

(1) Functions and contents of equipment layout Equipment layout represents the relative positions between equipments and buildings. It contains following contents:

1) A group of views. They display the basic structures of buildings and the arrangements of equipments in the factory.

2) Necessary dimensions. They display sizes of equipment arrangements, and position axis numbers of buildings and equipment item numbers and names.

3) Orientation mark of installations. It indicates the orientation basis of equipment installations.

4) Equipment list. It illustrates important equipment parameters such as item numbers, names, technical specifications and so on with a table.

5) Title block. It contains drawing information like drawing name, drawing number, scale and name of designer.

10.3.3 设备布置图

(1) 设备布置图的作用及内容 设备布置图是表示车间或工段内外的设备之间、设备与建筑物之间的相对位置的图样。一般包括以下内容：

1) 一组视图。用于表达厂房建筑的基本结构及设备在其内外的布置情况。

2) 必要的尺寸。用于表达设备布置的相关尺寸、建筑定位轴线编号、设备位号及名称等。

3) 安装方位标。用于表示设备安装方位基准的图标。

4) 设备一览表。用于将设备的位号、名称、技术规格以及有关参数做列表说明。

5) 标题栏。用于填写图名、图号、比例、设计者等内容。

(2) Representation methods of equipment layout

1) Division. Equipment layout is drawn according to the process flow. When equipment sizes are large or there are too many equipments, the equipment layout can be divided into several small areas and each area is drawn individually. The relative positions between the areas are indicated in the general drawing of the equipments. Area limits are drawn using thin long dashed double dotted lines.

2) Drawing sheet and scale. Generally A1 sheet with 1∶100, 1∶200 or 1∶500 scale is adopted for equipment layout. For a large-sized equipment, segmentation method can be used for layout. In this case, all segments should have the same scale.

3) Representation methods and configuration of views. A group of plane graphs and sectional views are commonly used to represent an equipment layout.

① Plane graphs. A plane graph is the main representation method in equipment layout. For example, the equipment layout in a building can be obtained by imaginatively removing the roof of the building and then creating its top view. If the building is multistoried, the equipment layout has to be finished story by story. If all these plane graphs are placed in one sheet, they should be arranged from top to bottom or from left to right, starting from the first story. Under each plane graph its elevation is labeled using letters and numbers, "EL 100.00, 4.00 plane" for instance, where "EL" stands for elevation.

② Sectional views. When plane graphs are not enough to express clearly the equipment layout, sectional or partial views can be used as an auxiliary means to illustrate the equipment arrangements along height direction. In sectional views, equipments are represented as if they are not sectioned. However, the cutting positions and the projective directions should be labeled clearly in the drawing, and an uppercase name should be placed under each sectional view, as shown in Fig. 10-35.

③ Representation of equipments. Equipments are represented by drawing their outlines and installation bases using thick continuous lines. Equipments of complicated shapes are simplified by ignoring the details of their outlines. If there are more than 3 equipments of the same item number, only the first one and the last one need to be drawn, while the others are simplified and represented by their bases or rectangles drawn with thin long dashed double dotted lines.

④ Representation of buildings and their components. Buildings and their components are drawn using thin continuous lines. Structures, like walls, pillars, grounds, floors, platforms, rails, mounting holes, pits, crane beams, equipment bases, etc., should be represented with specified proportional drawings. In sectional views, components that have little relations to equipment arrangement, doors and windows for instance, can be omitted. However, in plane graphs their positions and opening directions must be indicated. For bearing walls and pillars, their architecture position axes should be drawn using thin dashed dotted lines, being used as location datum of buildings and the components and equipments on them.

（2）设备布置图的表达方法

1）分区。设备布置图是按工艺流程绘制的。当装置尺寸较大或设备较多时，可将设备布置图分成若干个小区绘制。各区的相对位置在装置总图中标明，分区范围用细双点画线表示。

2）图幅与比例。设备布置图一般采用 A1 图幅。常采用的比例为 1∶100，也可采用 1∶200 或 1∶500。但对于大的装置分段绘制设备布置图时，必须采用同一比例。

3）视图的表达方法与配置。设备布置图常用一组平面图和立面剖视图来表达。

① 平面图。设备布置图一般只画平面图，即假想掀去厂房的屋顶或上层楼板的俯视图，以表达某层厂房设备布置情况。当厂房为多层建筑时，应依次分层绘制各层的设备布置图。若在同一张图纸上绘制几层平面图时，应从最底层平面开始，在图纸上由上向下或由左向右按层次顺序排列，并在图形的下方用字母和数字注明相应的标高，如"EL 100.00，4.00 平面"等。"EL"（Elevation）表示标高。

② 立面剖视图。当平面图不能表达清楚设备布置状况时，可绘制立面方向的剖视图或局部视图，来表达设备沿高度方向的布置安装情况。规定设备按不剖画，其剖切位置及投射方向应在图样上标注清楚，并在剖视图的下方标注相应的大写字母，如图 10-35 所示。

③ 设备的表达方法。用粗实线绘制设备的外形轮廓及其安装基础。外形复杂的设备只画其基本外形。同一位号的设备多于三台时，只画首末两台设备的外形，其余的仅画出基础，或用细双点画线的方框表示。

④ 建筑物及其构件的表达方法。用细实线绘制建筑物及其构件的轮廓。墙、柱、地面、楼板、平台、栏杆、安装孔、地坑、吊车梁及设备基础等，应采用规定的比例画法。与设备布置关系不大的门、窗等构件，在剖视图上一般不画，只在平面图上画出它们的位置及门的开启方向即可。对于承重墙、柱等结构，用细点画线绘制出其建筑定位轴线，用作建筑物、构件和设备的定位基准。

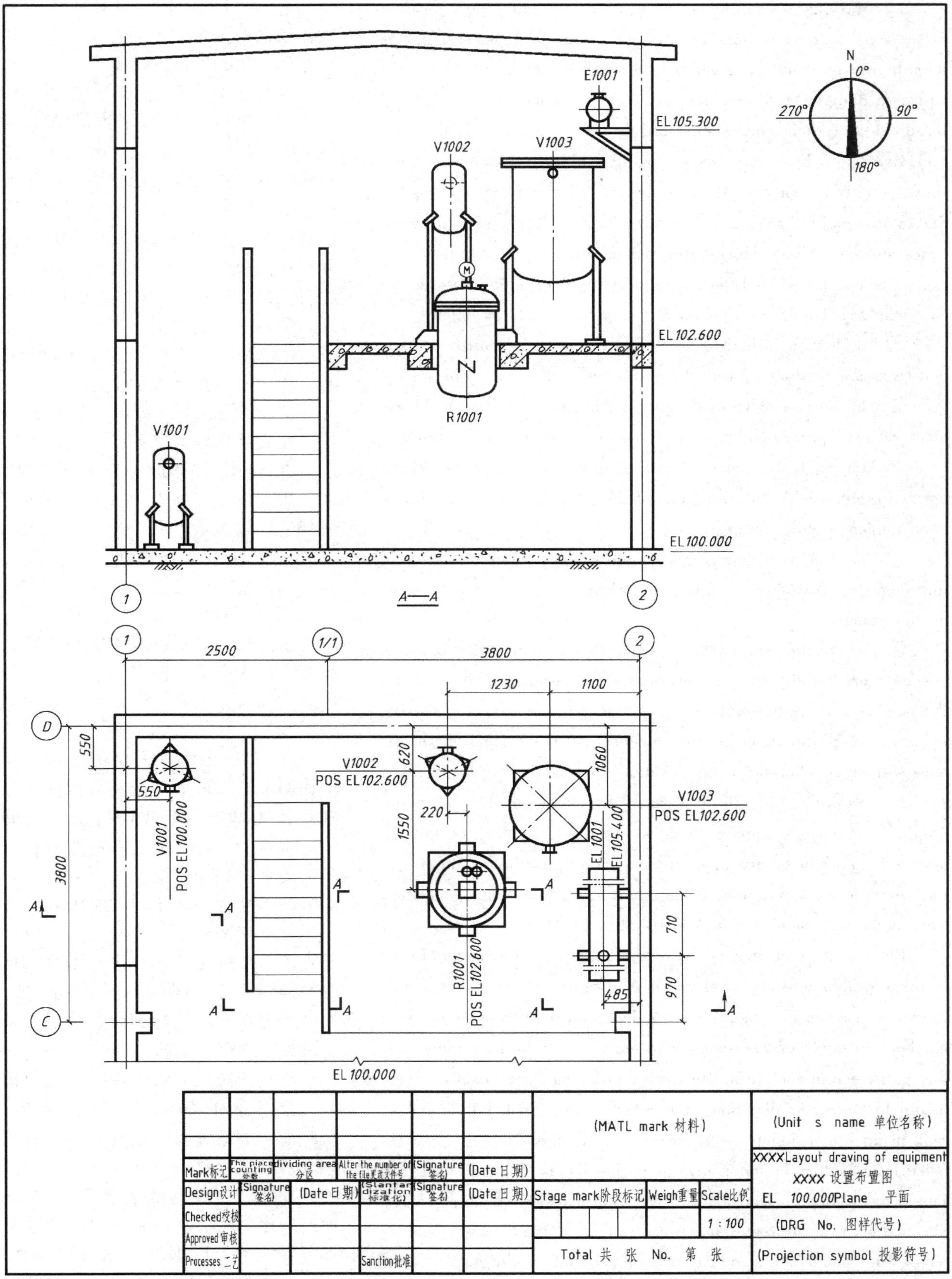

Chapter 10　Other Drawings　第 10 章　其他工程图

Fig. 10-35　Layout drawing of equipment　设备布置图

(3) Marking of equipment layout

1) Marking of equipments. In equipment layout, only location dimensions need to be marked. No unit needs to be specified. The default unit of length and width is mm while the height unit is m (reserve 3 digits after the decimal point). Generally, the axes of buildings and their components are selected as dimensioning datum. Extension lines are drawn using thin continuous oblique lines. Usually indoor ground of the factory building is selected as the datum in height direction, determining the heights of equipment bases and centerlines. Height dimensions are expressed with elevations, which are labeled together with equipment item numbers, as shown in Fig. 10-35. Equipment elevation is marked as follows:

① The heights of horizontal head exchangers and horizontal slots and tanks are represented with the elevations of their centerlines.

② The heights of vertical head exchangers and plate head exchanges are represented with the elevations of their bearing points.

③ The heights of reactors, towers and vertical slots and tanks are represented with the elevations of their bearing points or bottom head tangent welding seams.

④ The heights of pumps and compressors are represented with the heights of their floors, i.e., the elevations of the top surfaces of their bases.

⑤ For special equipments, the heights of those with brackets can be represented with the elevations of the bearing points, while the heights of those horizontal ones without brackets can be represented with the elevations of their centerlines and the heights of vertical ones without brackets with the elevations of specified pipe centers.

2) Marking of buildings and their components. In plane graphs, dimensions such as lengths, widths of factory buildings, distances between pillars and walls, location positions of holes and pits reserved for equipment installations are marked using position axes as datum, as shown in Fig. 10-35.

Position axes of bearing walls and pillars should be numbered in plane graphs and sectional views. And the numbers must be consistent with those in related construction drawings. To mark the axis numbers, a circle of 8—10mm diameter is drawn by thin continuous line at the end of each axis. The circles are aligned horizontally or vertically. In horizontal direction, the numbers are labeled from left to right using Arabic numbers, while in vertical direction, they are labeled from bottom to top using uppercase letters (letters I, O, Z should be avoided for fear that they are confused with Arabic numbers). If auxiliary axes are required between two axes, fractions are used for numbering, numerator standing for the number of an auxiliary axis. For example, "1/A" means the first auxiliary axis after axis A.

(3) 设备布置图的标注

1) 设备的标注。设备布置图中只标注设备的定位尺寸，长、宽定位尺寸的单位为 mm，高度尺寸的单位为 m（取小数点后 3 位），在图上均不注尺寸单位。一般以建筑物及其构件的轴线作为基准，尺寸界线可用细斜线绘制。高度方向的定位基准一般选择厂房室内地面，以此确定设备基础或设备中心线的高度尺寸。高度尺寸用标高表示，注写时与设备位号的注写相结合，如图 10-35 所示。设备标高的表示方法如下：

① 对于卧式换热器和卧式槽、罐，用中心线标高表示。

② 对于立式换热器、板式换热器，用支承点标高表示。

③ 对于反应器、塔和立式槽、罐，用支承点标高表示，或下封头切线焊缝标高表示。

④ 对于泵和压缩机，用其底板高，即基础顶面标高表示。

⑤ 对于特殊设备，若有支耳，用支承点标高表示；若无支耳的卧式设备，用中心线标高表示；若无支耳的立式设备，用某一管口的中心标高表示。

2) 建筑物及其构件的标注。在平面图上，以厂房建筑及其构件定位轴线作为基准，标注出厂房建筑的长、宽尺寸，柱、墙之间的距离，以及为安装设备预留的孔洞以及沟、坑等的定位尺寸，如图 10-35 所示。

在平面图或立面剖视图上，应对承重墙、柱的定位轴线进行编号，且其与建筑图上的轴线编号必须一致。在图形与尺寸线之外的明显位置，于各轴线的端部画直径为 8~10mm 的一个细实线圆圈，水平或垂直方向排列。水平方向编号采用阿拉伯数字，从左向右顺序编写；垂直方向编号采用大写拉丁字母（I、O、Z 三个字母不宜使用，以免与数字混淆），自下而上顺序编写。在两轴线之间如需附加分轴线时，则编号可用分数表示，分子表示附加轴线的编号（用阿拉伯数字顺序编写），如"1/A"表示 A 号轴线后附加的第一根分轴线。

3) Orientation mark of installations. It is also called direction mark, a symbol used to indicate the equipment installation orientations. The orientation mark of installations is usually placed in the upper-right part of the equipment layout, as shown in Fig. 10-35. It is represented by a circle of 20mm diameter drawn using thick continuous line. The arrow has a 3mm bottom width. The north direction of the direction mark (N) should be consistent with that in the general drawings of the equipments.

4) Equipment list. It is a table located above the title block and illustrates equipment information like item numbers, names, specifications and so on. When plane graphs are drawn on different sheets, the equipment list can be either placed on the one with EL 0.00, or provided in a separate design file.

(4) Drawing of nozzle orientations and detail drawing of equipment installations

1) Drawing of nozzle orientations. It indicates nozzle orientations and relative positions between nozzles and other components such as supports and anchor bolts, as shown in Fig. 10-36.

3）安装方位标。安装方位标又称方向标，是确定设备安装方位的符号，一般画在设备布置图的右上方，如图 10-35 所示。图中直径为 20mm 的圆圈用粗实线绘制，其箭头底部宽度为 3mm。方向标的北向（N）应与总图的设计北向相一致。

4）设备一览表。将设备的位号、名称、规格等在标题栏的上方列表说明，即为设备一览表。当平面图绘制在几张图纸上时，该表应附在标高为"0.00"的平面图上，也可单独制表在设计文件中附出。

（4）管口方位图及设备安装详图

1）管口方位图。管口方位图是确定各管口方位、管口与支座及地脚螺栓等相对位置的图样，如图 10-36 所示。

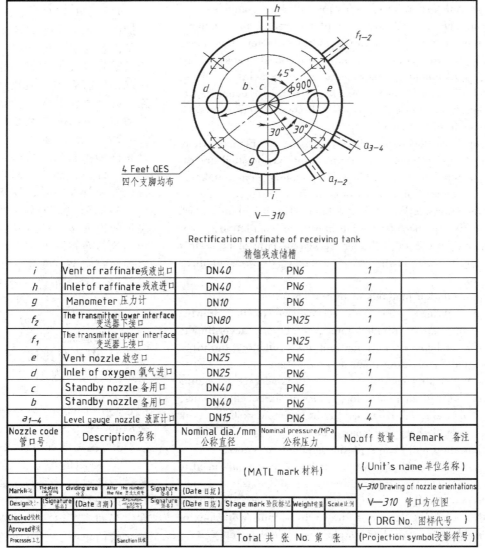

Fig. 10-36 Drawing of nozzle orientations 管口方位图

In a drawing of nozzle orientations only one view that can reflect the nozzle orientations needs to be drawn. This view is represented by simplification method and on it all nozzles are labeled alphabetically (a, b, c, d……) using lowercase letters. Above the title block there is a nozzle list, providing nozzle information including nozzle symbols, names, nominal diameters, working pressures, etc.

Generally a group of equipments sharing one item number are represented only with one drawing of nozzle orientations. For multi-layer equipments of too many nozzles, the drawing of nozzle orientations should be drawn in layers.

2) Detail drawing of equipment installations. Its drawing method is similar to that of mechanical drawing except that insignificant outlines can be drawn using thin continuous lines or thin long dashed double dotted lines.

10.3.4　Piping layout

(1) Functions and contents of piping layout As an important basis of pipe installation, piping layout (also called "piping diagram") expresses pipe connections between equipments and machines, and installation positions of valves, pipe fittings, pipe attachments and instrument control points.

A piping layout generally includes the following contents:

1) A group of views. A group of plane graphs and sectional views are adopted to represent basic structures of factory buildings, equipment shapes, layouts and installations of pipes, valves, pipe fittings, pipe attachments, instrument control points and so on. Positions of all vessels, heat exchangers and pipes connected to equipments, platforms, vessel supports and steel structures are indicated to the scale of the views.

2) Dimensions. Positions and elevations of planes where pipes, a part of pipe fittings and control points are located are dimensioned. Axis numbers of buildings, equipment item numbers, spool numbers, control point codes need to be marked as well.

3) A orientation mark and a title block. The orientation mark is a mark indicating the orientation datum of pipe installation, located at the upper-right corner of the piping layout. Title block mainly contains drawing information like drawing name, drawing number, scale and so on.

(2) Representation methods of piping layout

1) Division. When layout area is oversize of the drawing sheet, the drawing can be divided into several parts according to the layouts of workshops or working sections. Multistory buildings can be represented by dividing them into single stories.

2) Sheet and scale. Generally scale 1∶25 is adopted for piping layout. However, only if there are long pipes or large equipments, scale 1∶50 is also allowed. Drawings of the same workshop, work section or layer should share the same scale.

3) Configuration and representation methods of views

① Configuration of views. A piping layout is commonly used to represent the layout of pipes and equipments, and it is a top view drawn by hypothetically takeing off the ceiling or upper floor. Generally, a piping layout is drawn by taking workshop or work section as drawing unit. For simple cases, the layout can be exploited as equipment layout.

管口方位图中只简化画出一个能反映设备管口方位的视图，并注明各管口的符号，用小写拉丁字母按顺序编写（a、b、c、d……）。在标题栏的上方列出管口表，用来注明各管口的符号、名称、公称直径、压力等内容。

对具有相同位号的设备一般只绘制一张管口方位图。当多层设备且管口较多时，则应分层画出管口的方位图。

2) 设备安装详图。设备安装详图与机械制图相近，但次要的轮廓线可用细实线或细双点画线绘制。

10.3.4　管道布置图

（1）管道布置图的作用及内容

管道布置图又称配管图，是表达设备、机器间的管道连接及阀门、管件、管道附件、仪表控制点等安装位置的图样，是管道安装施工的重要依据。

管道布置图一般包括以下内容：

1) 一组视图。用一组平面图、立面图表示整个车间建筑物的基本结构、设备形状以及管道、阀门、管件、管道附件、仪表控制点的布置及安装情况。所有容器、换热器、与设备相连接的管道、平台、容器支座以及钢结构的位置都按比例注明。

2) 尺寸和标注。标注管道、部分管件和控制点的平面位置尺寸和标高，并标注出建筑物的轴线编号、设备位号、管段序号、控制点代号等。

3) 方位标与标题栏。方位标是表示管道安装方位基准的图标，一般画在管道布置图的右上角。标题栏中主要填写图名、图号、比例等内容。

（2）管道布置图的表达方法

1) 分区。当绘图区域较大而图幅有限时，可采用分车间或工段画图，多层建筑一律采用分层绘制。

2) 图幅与比例。管道布置图一般采用的比例为1∶25，仅当有大管道或大设备时，可采用1∶50的比例。同一车间、工段或分层的图样应采用同一比例。

3) 视图的配置与表达方法

① 视图的配置。管道布置图一般只画管道和设备的平面布置图，是假想掀去屋顶或上层楼板的俯视图，常以车间或工段为单元绘制。管道布置比较简单时，管道布置图可兼作设备布置图。

If there are local parts that are incomprehensible, sectional views and vertical views can be applied as complements, however, the dimensions labels of which can be omitted but only mark the elevations. The sectional views or vertical views shall be drawn either in appropriate positions on the layout drawing or on other separate sheets. The cutting positions and the projection directions of the sectional views shall be labeled clearly. The corresponding uppercase name shall be given on the bottom of each sectional view, as shown in Fig. 10-37.

② Marking and representation methods of equipments and buildings. Outlines of equipments and buildings are drawn to scale using thin continuous lines according to the equipment layout. The centerlines and axes of equipments and buildings are compulsory to draw, however, contents irrelevant with piping can be omitted.

Label the item numbers and location dimensions of equipments according to the layout. The axes numbers of buildings, distances between axes and elevations shall be in accordance with equipment layout.

③ Representation and dimensioning of pipes. Pipes with different nominal diameters (DN) are drawn using different line types. A pipe with a DN not less than 400mm or 16 feet is represented by double-line method, while a pipe with a DN not larger than 350mm or 14 feet is represented by single-line method. If there are few pipes of large diameters, a pipe with a DN not less than 250mm or 10 feet is represented by double-line method, while a pipe with a DN not larger than 200mm or 8 feet is represented by single-line method. For single line width please refer to Table10-10. As for the lengths of pipes, they should be drawn to the drawing scale.

In piping layout pipe combination numbers shall be marked with the same values as those in the process pipe and instrument flow diagrams, and location dimensions need to be marked, too. In plane graphs, axes of buildings, equipment centerlines, nozzle centerlines and end faces of flanges can be selected as datum of location dimensions. Elevations are used as location dimensions in height direction. An elevation is marked as "EL×××.×××" if its datum is a pipe centerline, or "BOP EL×××.×××" if its datum is a bottom of a pipe, where BOP is the code of the bottom of the pipe. A single pipe can be dimensioned using leader line, as shown in Fig. 10-37.

④ Representation and marking of piping supports Pipes are installed on various kinds of piping supports. Piping supports are represented using legends, beside which serial numbers of the piping supports are labeled. The types of piping supports include anchor (code A), guide (code G), sliding support (code R), etc. Generally piping supports are planted in such structures as concrete structure (code C), ground foundation (code F), steel structure (code S)、equipment (code V), wall (code W) and so on.

A 2-digit number is employed to represent the serial number of a piping support of general type. For legends, serial numbers and marking methods of piping supports, please refer to Fig. 10-38, in which all circles have a 5mm diameter.

当平面图中局部表示不够清楚时，可绘制其剖视图或立面图，不需标注尺寸，只需标注标高。应将剖视图或立面图置于图纸的适当位置或在单独的图纸上画出，其剖切位置及投射方向应在图上标注清楚，并在剖视图下方标注相应的大写字母，如图10-37所示。

② 设备及建筑物的表达方法及标注。根据设备布置图，按比例用细实线绘制出设备和建筑物的简略外形轮廓。必须画出设备和建筑物的中心线或轴线。与管道无关的内容可略去。

按设备布置图标注设备的位号和定位尺寸。建筑物的轴线号、轴线间的尺寸和标高的标注与设备布置图相同。

③ 管道的表达方法及标注。管道布置图中，不同的公称通径（DN）的管道采用不同的线型绘制：公称通径（DN）大于或等于400mm或16feet的管道用双线表示；公称通径（DN）小于或等于350mm或14feet的管道用单线表示。如果管道布置图中大口径的管道不多时，公称通径（DN）大于或等于250mm或10feet的管道用双线表示；公称通径（DN）小于或等于200mm或8feet的管道用单线表示。单线线宽按表10-10绘制。管道的长度应按比例画出。

管道布置图上除标注与工艺管道及仪表流程图相同的管道组合号外，还应标注定位尺寸。在平面图上，其定位尺寸常以建筑物的轴线、设备中心线、设备管口中心线、法兰的一端面等作为基准进行标注。管子高度方向的定位尺寸以标高来表示。标高以管道中心线为基准时，标注"EL×××.×××"；以管底为基准时，加注管底代号BOP，如"BOP EL×××.×××"。单根管道也可用指引线引出标注，如图10-37所示。

④ 管架的表达方法及标注。管道安装在各种型式的管架上。管架用图例表示在管道布置图中，并在图旁标注管架序号。管架的类型有：固定架（代号A）、导向架（代号G）、滑动架（代号R）等几种。管架装入下列结构中：混凝土结构（代号C）、地面基础（代号F）、钢结构（代号S）、设备（代号V）、墙（代号W）等。

对于通用型管架，其管架编号中用两位数表示管架序号。图例、管架序号和标注如图10-38所示，图中圆圈直径均为5mm。

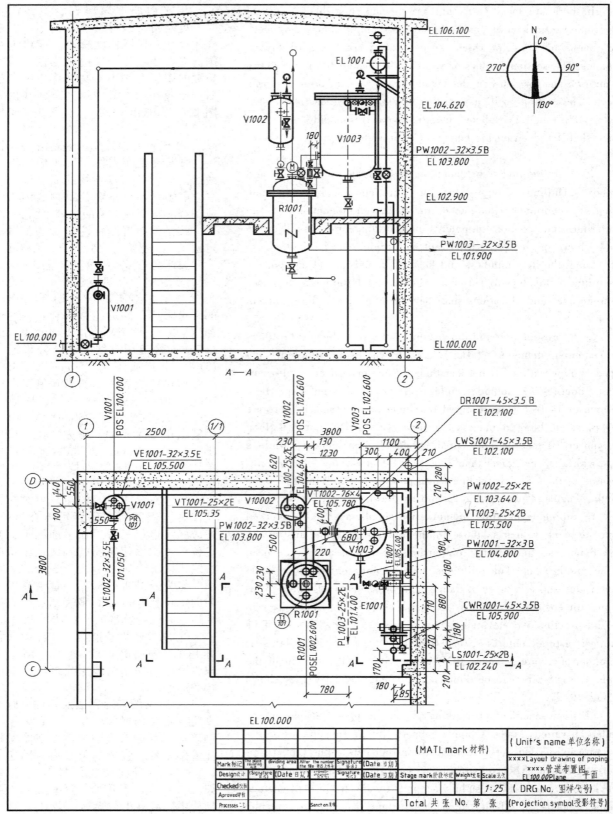

Fig. 10-37 Layout drawing of piping 管道布置图

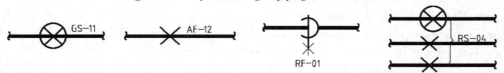

Fig. 10-38 Legends and marking methods of piping supports 管架图例及其标注方法

⑤ Representation and marking of valves, pipe fittings, pipe attachments and instrument control points. All these components are represented by specified graphic symbols drawn using thin continuous lines. Common symbols are shown in Table 10-11. For instance, a valve is represented by the symbol of the control hand wheel and its installation orientation. Fig. 10-39 shows stop valves connected with flanges.

⑤ 阀门、管件、管道附件、仪表控制点的表达方法及标注。按规定的图形符号用细实线绘制管道上的阀门、管件、管道附件，常用图形符号见表10-11。一般绘制出阀门的控制手轮及安装方位，如图10-39所示。图中的阀门是与法兰连接的截止阀。

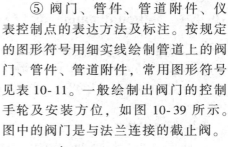

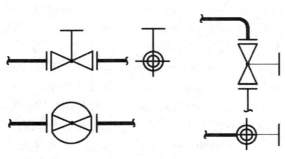

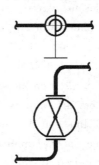

Fig. 10-39 Valves and their installation orientations　阀门及安装方位

For connection forms between pipes, valves, pipe fittings and pipe attachments please refer to Fig. 10-40.

管道、阀门、管件、管道附件间的连接形式可按图10-40所示绘制。

Fig. 10-40 Piping connection forms　管道连接形式

In piping layout, all instrument control points should be drawn. Each control point appears in only one view, where its installation position can be clearly indicated. The symbols of control points are the same with those in the process pipe and instrument layout. Instruments on pipelines, for instance, pressure meters and thermometers, are represented by $\phi 10mm$ circles. As shown in Fig. 10-31, one end of the leader line of an instrument symbol points to the nozzle, while the other end is connected with the circle inside which the instrument code is located.

Generally spool lengths need not be marked on piping layout. Location dimensions of valves, pipe fittings and pipe attachments can be marked or omitted according to practical requirements.

⑥ Representation and marking of orientation. The mark methods are the same as above.

管道布置图上还应画出所有仪表控制点的符号。每个控制点一般仅在能清楚地表达其安装位置的一个视图上画出，其图示符号与工艺管道及仪表流程图相同。管道上的检测仪表（如压力表、温度表等）在平面图上用$\phi 10mm$的圆圈表示。将指引线一端指向接管口，另一端同圆圈相接触，圆圈内所注写的代号如图10-31所示。

管道布置图上，一般不标注管段的长度尺寸，而阀门、管件、管道附件的定位尺寸可按需注出。

⑥ 方向标的表达方法及标注同前，略。

10.4 Electric Diagrams

Electric diagram is used to express electrical work principles, describe structures and functions of electric productions and provide illustrations of production installation and use. In this section, drawing methods and basic knowledge of electric diagram are introduced.

10.4 电气制图

电气图是用来表达电气工作原理，描述电气产品的构成和功能，并提供产品安装和使用方法的图样。本节将简要介绍电气图的画法和相关基本知识。

10.4.1 National standards of electric diagram

1. Size and layout of drawing sheets

Please refer to *Technical drawings—Size and layout of drawing sheets* GB/T 14689—2008 for the size and layout of drawing sheets.

2. Line styles

Common line styles for electric diagrams and their usages are listed in Table 10-14. Line width can be selected from a recommended width series (unit, mm): 0.13, 0.18, 0.25, 0.35, 0.5, 0.7, 1, 1.4 and 2. The width d of thick lines is selected in the range of 0.5—1.4mm, based on the size and the complexity of the drawing. Generally, there are only two line width types used on a drawing, namely thick line and thin line between which the width scale is 2∶1. In some special applications, more than two line width types may be required. In these cases, line widths adopted increase progressively by a factor of 2.

10.4.1 电气制图的有关制图标准

1. 图纸幅面和格式

图纸幅面和格式可参照《技术制图 图纸幅面和格式》（GB/T 14689—2008）选用。

2. 线型

绘制电气图时常用的线型及其用途见表10-14。图线宽度的推荐系列为：0.13、0.18、0.25、0.35、0.5、0.7、1、1.4、2（单位：mm）。应根据图形大小和复杂程度，在0.5~1.4mm的范围内选择粗实线的宽度 d。通常同一图样中只选用两种宽度的图线，即粗线和细线，其宽度比为2∶1。在某些特殊情况下，可能需要两种以上宽度的图线，此时线的宽度应以2的倍数依次递增。

Table 10-14 Common Line styles for electric diagrams and their usages 电气路图常用线型及其用途

Description 图线名称	Line styles 线型	Usage 用途
Continuos line 实线	————	To indicate basic lines, graphic symbols and connecting wires, visible outlines and conductors 基本线、简图主要内容用线（图形符号及连线）、可见轮廓线、可见导线
Long dashed line 虚线	— — — —	For auxiliary lines, shield lines, mechanical connecting lines, invisible outlines and conductors 辅助线、屏蔽线、机械连接线（液压、气动等）、不可见轮廓线、不可见导线
Dashed dotted line 点画线	—·—·—	Used as dividing lines for structure and function grouping, box lines, control and signal lines (for power and lighting) 分界线（结构、功能分组用）、围框线、控制及信号线路（电力及照明用）
Long dashed double dotted line 双点画线	—··—··—	For auxiliary box lines, power and lighting lines with voltage lower than 50V. 辅助围框线、50V以下电力及照明线路

3. Lettering

For lettering of characters and numbers please refer to *Technical drawings—Lettering* GB/T 14691—1993. Heights of all characters and numbers can not be shorter than 3.5mm.

4. Scale

For drawing scales please refer to "*Technical drawings—Scales*" GB/T 14690—1993. Following scales are recommended for layout drawing: 1∶10, 1∶20, 1∶50, 1∶200, 1∶500.

5. Arrow and leader line (excerpted from GB/T 6988.1—2008)

Arrows on signal lines and conductors can be solid, hollow, bottom opened or short oblique lines, as shown in the upper part of Fig. 10-41.

Leader line is a thin line pointing to the element it marks. It ends with a darken point if the end is inside outlines, or an arrow if on an outline, or short oblique lines if on dimension lines or wires. The line tip can be directly led out from the dimension line, as shown in the lower part of Fig. 10-41.

3. 字体

字体按照GB/T 14691—1993《技术制图 字体》的规定书写。字体高度不应小于3.5mm。

4. 比例

绘图比例，可参考GB/T 14690—1993《技术制图 比例》规定在画布置图时，可以从下列比例系列中选取：1∶10，1∶20，1∶50，1∶200，1∶500。

5. 箭头与指引线（摘自GB/T 6988.1—2008）

信号线和连接线上的箭头可以是实心的、空心的，也可以是开口的或短斜线，如图10-41中上图所示。

指引线应是细实线，指向被注释处，并在其末端加注如下标记：若末端在轮廓线内，指引线端部为黑点；若末端在轮廓线上，指引线端部为箭头；若末端在尺寸线或电路线上，指引线端部为短斜线；指引线端部可直接由尺寸线引出，如图10-41中下图所示。

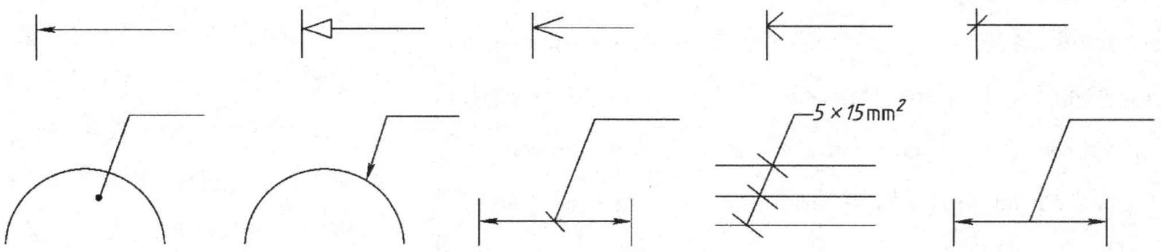

Fig. 10-41 Forms of arrows and leader lines　箭头与指引线的形式

6. Connecting wire（excerpted from GB/T 6988.1—2008）

Existing connecting wires are represented by continuous lines while those to be extended by dashed lines. A connecting wire can neither change its direction when it intersects with another wire, nor cross an existing intersection point of other wires. In order to emphasize or differentiate some circuits or functions, conductor symbols, signal pathway and connecting wires can be represented by lines of different widths. For example in Fig. 10-42a, a three-phase power transformer's switches and control devices are shown, where the power supply circuit is displayed outstandingly by continous thick lines.

If there are many parallel connecting wires, in order to improve the readability of the drawing, they should be grouped according to their functions, otherwise be grouped arbitrarily. Each group contains no more than three wires and distances between groups are larger than those between wires, as shown in Fig. 10-42b.

When a single wire joins a group of connecting wires simplified by a single trunk line, the single wire and the trunk line are drawn using continous thin line and continous thick line respectively so that they can be easily differentiated. At the joining point, the single wire is replaced by a short oblique line whose direction should help to judge the direction in which the single wire enters or leaves the trunk wire. Generally, marking symbols are required to be labeled at the end of each wire, as shown in Fig. 10-42c.

6. 连接线（摘自 GB/T 6988.1—2008）

确实存在的连接线用实线表示，计划扩展的内容用虚线表示。一条连接线不应在与另一条线交叉处改变方向，也不应穿过其他连接线的连接点。为了突出或区分某些电路、功能等，导线符号、信号通路、连接线等可采用不同粗细的图线来表示。例如图 10-42a 所示，一个三相电力变压器以及与之有关的开关装置和控制装置的一部分，其中电源电路用粗实线表示。

如果有多条平行连接线，为了便于看图，应按功能进行分组。不能按功能分组时，可以任意分组。每组不多于三条。组间距离应大于线间距离，如图 10-42b 所示。

如果有单根导线汇入用单线表示的一组连线（干线），为便于看图，单根导线用细实线绘制，干线用粗实线绘制。汇入处要用斜线表示，其方向应便于识别导线进入或离开干线的方向。通常，还需要在每根导线的末端注上标记符号，如图 10-42c 所示。

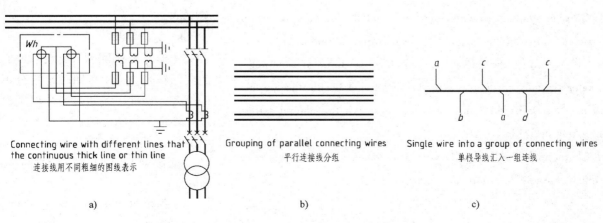

a)　　　　　　　　　　　b)　　　　　　　　　　　c)

Fig. 10-42 Forms of connecting wires　连接线的形式

7. Graphic symbols

In electric engineering there are always massive electric elements, equipments and connecting wires whose structures and installation methods are very different. It is unnecessary and impossible to truly reconstruct their shapes in the drawing. Therefore a series of simple graphic symbols are used in electric diagram to illustrate and differentiate their properties and the relations between these electric elements and equipments. A graphic symbol is so defined as a graph, a marker or a character standing for an equipment, an element or a concept.

Drawing methods of graphical symbols for electrical diagrams are specified in *Graphical symbols for electrical diagrams* GB/T 4728.1—13—2005/2008. There are mainly two kinds of symbols: block symbols and general symbols.

Block symbols are used to represent combinations of elements and equipments and their functions. A block symbol can be a square, a rectangle or a circle and is generally applied in electric connection diagram, providing neither details of elements and equipments nor connection relations in it.

General symbols are used to represent a kind of equipments and their characteristics, consisting of symbol elements and limit symbols. Symbol elements are a group of simple graphs, each one bearing certain meanings. A single symbol element can not represent any equipment or concept unless it works together with some specified symbols. Limit symbols are such symbols that are put on other symbols and provide additional information. Examples of graphic symbols are shown in Table 10-15. Graphic symbols of common electric elements and components are listed in Table 10-16. For more details please refer to related national standards.

8. Character symbols

Graphic symbols provide common symbols for equipments or electric elements of the same kind. To further differentiate them, especially those of the same kind but owning different functions, character symbols must be provided and labeled beside the graphic symbols, providing additional information, like the capital letters in brackets in Table 10-16.

Character symbols usually consist of basic character symbols, auxiliary character symbols and digits.

7. 图形符号

构成电气工程的元件、设备和连接线很多，结构类型千差万别，安装方式多种多样，没有必要也不可能一一在电气图中画出其外形。因此，在电气工程图中，为了描述和区分这些电气工程元件和设备的特征及其相互关系等，一般用一些简单的符号表示，这些符号就是图形符号。图形符号是用于电气图中表示一个设备、元器件或一个概念的图形、标记或字符。

GB/T 4728.1~13—2005/2008《电气简图用图形符号》规定了电气图中图形符号的画法，主要有方框符号和一般符号两种。

方框符号是用以表示元件、设备等的组合及其功能。一种简单的图形符号，如正方形、长方形、圆形图形符号。它既不给出元件、设备的细节，也不考虑所有的连接关系，通常用在电气接线图中。

一般符号是表示一类设备或此类设备特性的一种通常很简单的符号。一般符号由各种符号要素和限定符号组成。符号要素是一种具有确定意义的简单图形，必须同其他图形组合，以构成一个设备或概念的完整符号。限定符号是用以提供附加信号的一种加在其他符号上的符号。表10-15是各限定符号和符号要素组合成一般符号的例子。常用元器件的图形符号见表10-16。要了解更详细，可查阅相关标准。

8. 文字符号

图形符号提供了一类设备或元件的共同符号。为了更明确地区分不同的设备、元件，尤其是区分同类设备或元件中不同功能的设备或元件，还必须在图形符号旁标注相应的文字符号，见表10-16中括号内的大写字母。

文字符号通常由基本文字符号、辅助文字符号和数字组成。

Table 10-15 General symbols consisting of limit symbols and symbol elements 限定符号和符号要素组合成一般符号

Limit symbol 限定符号		Symbol element 符号要素		General symbol 一般符号	
Graphic symbol 图形符号	Description 说明	Graphic symbol 图形符号	Description 说明	Graphic symbol 图形符号	Description 说明
	Contactor function 接触器功能		Front (normal open) contact 动合（常开）触点		Position switch, front contact; limit switch, front contact 位置开关，动合触点；限制开关，动合触点
	Limit switch function and position switch function 限制开关功能、位置开关功能				Contactor (off contact on non-action position) 接触器（在非动作位置触点断开）
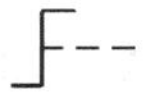	Rotation operation 旋转操作				Twist switch (locked) 旋钮开关（闭锁）
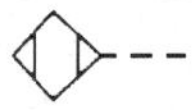	Proximity effect operation 接近效应操作			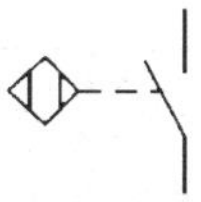	Proximity switch, front contact 接近开关，动合触点
	Isolating switch function 隔离开关功能				Isolating switch 隔离开关

9. Alphanumeric symbols of connecting terminals

Connecting terminals are marked by alphanumeric symbols. As specified in *Basic and safety principles for man-machine interface, marking and identification—Identification of equipment terminals and of conductor terminations* (GB/T 4026—2010), L1, L2, L3, N and PE stand for the first, the second, the third phase, the neutral line and the protective earthing of a three-phase AC supply. L^+, L^- and M stand for the anode, the cathode and the midline of a DC supply. The lead cables of a three-phase power equipment are marked orderly by U, V, W. The head ends of the winding of a three-phase asynchronous motor are marked by U_1, V_1, W_1, the tail ends by U_2, V_2, W_2, and the center taps by U_3, V_3, W_3. If there is more than one motor, their symbols are differentiated by adding numbers before the alphanumeric symbols. For example, 1U, 1V and 1W are used for motor M1 and 2U, 2V and 2W for motor M2. If there are middle units between conductors of a three-phase power supply and a three-phase load, the symbols of the connecting wires between them are represented by U, V and W plus numbers after them, in an ascending ordered from top to bottom.

Control circuit wires are numbered with Arabic numerals, being marked using "equipotential" principle from left to right and from top to bottom. All connecting terminals isolated by other electric elements such as coils, contact points, resistors and capacitors should be marked with different wire numbers.

9. 接线端子标记

电气图中各电器接线端子用字母数字符号标记。GB/T 4026—2010《人机界面标志标识的基本和安全规则 设备端子和导体终端的标识》规定：三相交流电源引入用 L1、L2、L3、N、PE 标记；直流系统的电源正、负、中间线分别用 L^+、L^- 与 M 标记；三相动力电器引出线分别按 U、V、W 顺序标记。三相异步电动机的绕组首端分别用 U_1、V_1、W_1 标记，绕组尾端分别用 U_2、V_2、W_2 标记，电动机绕组中间抽头分别用 U_3、V_3、W_3 标记。对于数台电动机，在字母前冠以数字加以区别。如：对 M1 电动机其三相绕组接线端标以 1U、1V、1W；对 M2 电动机其三相绕组接线端标以 2U、2V、2W。两三相供电系统的导线与三相负载之间有中间单元时，其相互连接线用字母 U、V、W 后面加数字来表示，且从上至下、从小至大的数字来表示。

控制电路各线号采用数字编号，标注方法按"等电位"原则进行。其顺序一般为从左到右、从上至下。凡是被线圈、触点、电阻、电容等元件所隔离的接线端子，都应标以不同的线号。

Table 10-16 Graphic symbols of common electric elements and components 常用元、器件图形符号

No.	Symbol	Name	No.	Symbol	Name	No.	Symbol	Name
1		Resistor 电阻	14		Voltmeter 电压表	27		Three phase asynchronous motor 三相异步电动机
2	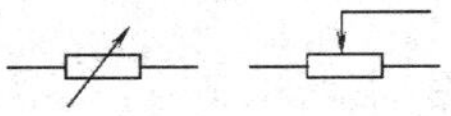 (RP)	Resistance potentiometer 电位器	15	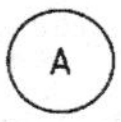	Ampere meter 电流表	28	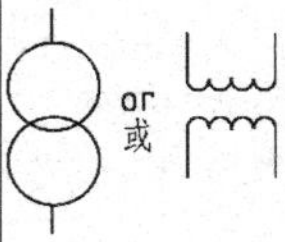 or 或	Double wound transformer 双绕组变压器
3	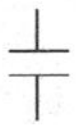	Capacitor 电容	16		Wattmeter 功率表	29		Permanent magnet stepper motor 永磁步进电动机
4		Electrolytic capacitor 电解电容	17		Antenna 天线	30	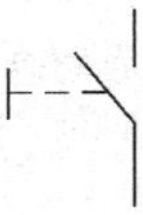	Manual switch 手动开关
5	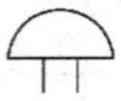	Electric bell 电铃	18		Ground connection 接地	31	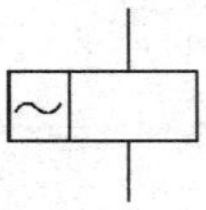	AC relay 交流继电器
6	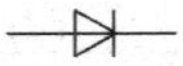	Diode 晶体二极管	19		Shell connection 接机壳	32	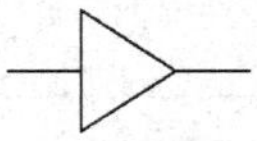	Amplifier 放大器
7		PNP Transistor PNP 型晶体管	20		Fuse 熔断器 (FU)	33		Filter 滤波器
8	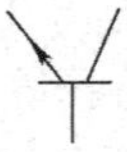	NPN Transistor NPN 型晶体管	21		Lightning protector 避雷器	34	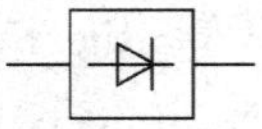	Wave detector 检波器
9		Alternating current 交流电	22		Optical fiber 光纤（缆）	35	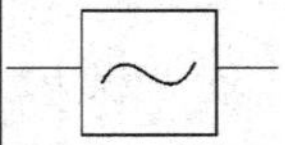	Oscillator 振荡器
10	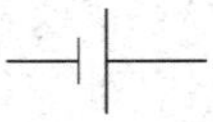	Battery 电池	23		Light 信号灯	36		Commutator 整流器
11	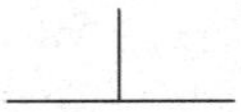	Wire joint 导线连接	24	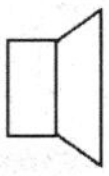	Buzzer 扬声器	37	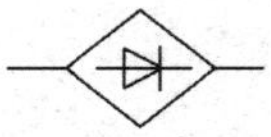	Full wave bridge rectifier 桥式全波整流器
12		Inductor 电感器	25	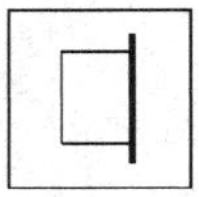	Monitor 监听器	38		N Semiconductor FET N 型场效应半导体管
13		Beeper 蜂鸣器	26	 (X)	Plug and socket 插头和插座	39		P Semiconductor FET P 型场效应半导体管

10.4.2 Common electric diagrams

Any drawing in the field of electric engineering can be called an electric diagram. Common electric diagrams include block diagrams, circuit diagrams, electric location plans, wiring diagrams and printed circuit board diagrams. This section focuses on the basic knowledge and representation methods of the first three types of drawings.

1. Block diagrams

Block diagram is widely used in design, manufacturing, installation, use and maintenance of electric equipments. In a bolock diagram, symbols and boxes together with remarks are adopted to represent basic components, relationships and main characteristics of equipments, as shown in Fig. 10-43.

Drawing methods of block diagram are specified in *Preparation of documents used in electrotechnology—Part*1: *Rules* GB/T 6988.1—2008. It includes following requirements:

1) Each unit circuit is represented by a square or a rectangle block in which the circuit name is labeled.

2) Blocks are connected by continous thin lines and arrows, indicating relationships between components.

10.4.2 几种常用的电气图

电气图是电气工程领域中各种图的总称，常用的有框图、电路图、电器位置图、接线图和印制电路板图。本节主要简单介绍这几种图的基本知识和表达方法。

1. 框图

框图是电气设备设计、生产、安装、使用和维修的过程中经常使用的电气图。它是用符号或带注释的框概略表示设备的基本组成、相互关系及其主要特征的一种简图，如图10-43所示。

绘制框图时可参照 GB/T 6988.1—2008《电气技术用文件的编制 第1部分：规则》中的规定，具体要求是：

1）每个单元电路用正方形或长方形方框表示，并在方框内注写出该单元电路的名称。

2）方框之间用细实线和箭头进行连接，以说明各组成部分之间的相互关系。

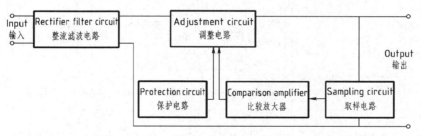

Fig. 10-43 Block diagram of a regulated power supply   某稳压电源框图

2. Circuit diagram

Circuit diagram consists of graphic symbols and character symbols arranged by their working sequence, illustrating in detail the basic compositions and connecting relations of the circuit and equipment control system. In circuit diagram, real positions of the circuit components are not concerned, as shown in Fig. 10-44 and Fig. 10-45. Since circuit diagram is mainly used to illustrate the working principles of the electric control system, it is also called electric schematic diagram. Circuit diagram is an important basis for wiring diagram generation and provides necessary technical information for equipment installation and maintenance.

Please refer to GB/T 6988.1—2008 for drawing methods of circuit diagram. Following are some concrete drawing requirements:

2. 电路图

电路图是用图形符号、文字符号按其工作顺序排列，详细表示电路、设备控制系统的基本组成和连接关系，而不考虑其实际位置的一种简图，如图10-44、图10-45所示。

电路图主要表示电气控制系统的工作原理，故也被称为电气原理图。电路图是编制接线图的重要依据，也为安装和维修提供必需的技术信息。

绘制电路图可参照 GB/T 6988.1—2008 中的规定，具体要求如下：

1) Circuit diagram is layout by function modules, that is to say, main circuit, control circuit, lamp circuit and signal circuit are drawn separately and arranged from left to right or from top to bottom according to the causal relationships between them. At the same time, these sub-diagrams should be located based on their working sequence as much as possible. Circuits in the diagram can be drawn in either horizontal or vertical direction. On a circuit diagram with horizontal arrangement, all circuits are horizontal except power supply lines which are vertical, and dissipative elements in control circuit are drawn in the most right part of the diagram, as shown in Fig. 10-44. On a circuit diagram with vertical arrangement, all circuits are vertical except power supply lines which are horizontal, and dissipative elements in control circuit are drawn on the bottom part of the diagram, as shown on Fig. 10-45.

1) 电路图在布局上按功能分开画出，即按主电路、控制电路、照明电路及信号电路分开绘制。应按因果关系从左到右或从上到下布置，并尽可能按工作顺序排列。电路图的电路可水平或垂直布置。水平布置时，电源线垂直画，其他电路水平画，控制电路中的耗能元件画在电路的最右端，如图 10-44 所示；垂直布置时，电源线水平画，其他电路垂直画，控制电路中的耗能元件画在电路的最下端，如图 10-45 所示。

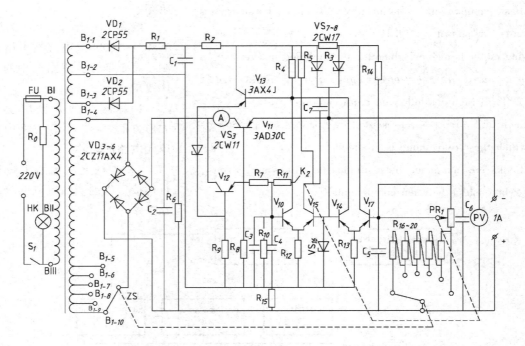

Fig. 10-44 Circuit diagram of a regulated power supply 某稳压电源电路图

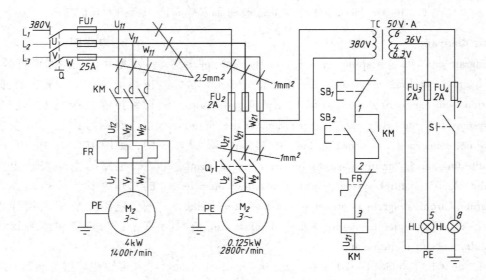

Fig. 10-45 Circuit diagram of CW6132 lathe CW6132 型车床的电路图

2) All electric elements and components on the diagram are drawn according to *Graphical symbols for electrical diagrams* GB/T 4728.1—13—2005/2008 and marked with character symbols Components belong to the same electrical apparatus must be marked using identical character symbols according to their functions in the circuit, and their graphic symbols can be located separately.

3) Main electric elements should be arranged horizontally or vertically. Electric elements and components of the same type should be aligned in horizontal or vertical direction.

4) Generally, movable parts of electric elements are drawn as if they are in non-excited states, or off-position. For example, coils of relays, contactors, brakes should be in non-excited states, mechanical control travel switches and buttons should be in the unpressed positions and hand control switches with zero operation should be in the zero positions.

5) Lines representing conductors, signal pathways and connecting wires should have as few intersections and bends as possible, and be arranged horizontally or vertically. All bends should be right angles and intersection points should be solid circular points.

3. Electric location plan

In an electric location plan, relative positions of electric apparatuses are drawn strictly to scale. As an example, Fig. 10-46 shows the electric location plan of CW6132 lathe.

Before drawing an electric location plan, relative positions of electric apparatuses should be obtained according to related standards of mechanical drawing. Character symbols of the electric elements in the electric location plan should be consistent with those on corresponding circuit diagrams.

2) 电路图中的各电气元器件，一律采用 GB/T 4728.1~13—2005/2008《电气简图用图形符号》中规定的图形符号绘出，并用规定的文字符号标记。同一电器的各个部件按其在电路中所起的作用必须用相同的文字符号标注，其图形符号可以不画在一起。

3) 各主要元件尽量排列在同一水平线或垂直线上；同类元器件尽量在横向上或纵向上对齐。

4) 电路图中的所有电气元器件的可动部分通常表示在电器非激励或不工作的状态和位置。例如，继电器、接触器、制动器等的线圈处在非激励状态；机械控制行程开关和按钮在其未受机械压合的状态；零位操作的手动控制开关在零位状态。

5) 表示导线、信号通路、连线等的图线都应是交叉和拐弯最少的直线，并应水平（或垂直）地布置。拐弯处应为直角，线路相交处应为实心圆点。

3. 电器位置图

在电器位置图中，需按比例详细绘制出电气设备中各电器的相对位置。图 10-46 所示为 CW6132 型车床的电器位置图。

绘制电器位置图时，应按机械制图有关标准表示清楚各电器之间的相对位置，并且图中各电器元件的文字符号应与有关电路图中电器元件的文字符号相同。

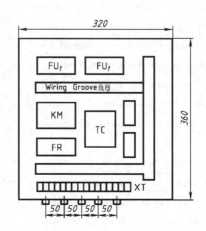

Fig. 10-46 Electric location plan of CW6132 lathe　CW6132 型车床的电器位置图

4. Wiring diagram

Wiring diagram is used to represent the relative positions and electrical connecting relationships between electric elements and components of a product. Based on circuit diagrams, it is created according to assembly and construction requirements and relative positions of electric elements and equipments. Wiring diagram is widely used on site for wiring and equipment inspection and maintenance, as shown in Fig. 10-47.

4. 接线图

接线图是表示产品内部各元器件相对位置关系以及它们之间的电气连接关系的略图,即实体布线图。它是在电路图基础,根据装配和施工要求,按照各个电器元件和设备的相对安装敷设位置绘制而成的。接线图主要用于配线、检查和维修之中,故在生产现场得到广泛的应用,如图10-47所示。

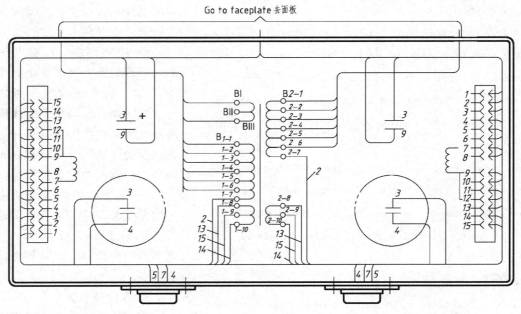

Fig. 10-47 Wiring diagram of a regulated power supply　某稳压电源接线图

Wiring diagrams can be classified into unit connection diagrams, interconnection diagrams and terminal connection diagrams, according to the objects they represent and their functions. Unit connection diagram reflects the connection relations inside the unit. Interconnection diagram expresses the connection relations between the external terminal boards of the unit, as shown in Fig. 10-48. Terminal connection diagram provides the connection relations between terminals and outer conductors.

Drawing methods of wiring diagram are specified in GB/T 6988.1—2008. It mainly includes following requirements:

1) Outlines of all electric elements and components should be drawn to scale, though they can be simplified or replaced by graphic symbols. Graphic symbols, character symbols and connection numbers between them should be consistent with those on corresponding circuit diagrams. Fixed parts or elements independent of wiring can be omitted.

2) In wiring diagram, relative positions between items, item numbers, electric connecting relations between terminals, terminal numbers, conductor numbers and conductor types and section areas need to be marked. All these contents can be listed in an additional table attached to the wiring diagram.

根据表达对象和用途的不同,接线图可分为单元接线图、互连接线图和端子接线图等。单元接线图仅反映单元内部的连接关系;互连接线图仅反映单元的外接端子板之间的连接关系,如图10-48所示。端子接线图表示单元和设备的端子及其与外部导线的连接关系。

绘制接线图,可参照 GB/T 6988.1—2008 中规定的规则,主要有以下要求:

1) 元器件的外形尺寸应按比例绘制,元器件的图形可以简化或用图形符号代替。图中各元器件的图形符号、文字符号以及它们之间的连接编号均应以电路图为准,并保持一致。与接线无关的固定元器件不予画出。

2) 接线图一般应标出项目的相对位置和项目代号,以及端子间的电气连接关系、端子号、导线号、导线类型、截面积等。或者可以将这些内容列表并附在接线图后。

3) Wires are generally drawn using contionus thin lines. However, a trunk line constituted by a group of wires should be drawn using a continous thick line. At the joining point of the wires and the trunk line, the wires are replaced by short oblique lines of 45° or short arcs to indicate the directions of the wires.

4) In interconnection diagram, boxes of all unit items are drawn using long dashed dotted lines.

5) All inner electric elements of a control box or a control panel can be connected directly, while inner elements of a control box or a panel can not be connected to outer elements unless there are terminal blocks between them.

Electric connections in a wiring diagram are not the real routes of the wires. In practical wiring operations, workers may select optimal routes by their experiences.

3）导线通常均用细实线绘制，但由若干根导线组成的线束部分（干线）用粗实线绘制。在导线与干线交接处应画成小圆弧或45°斜线，以表示导线走向。

4）互连接线图中，各个单元项目的外形轮廓围框用点画线表示。

5）同一控制箱或控制屏的各电器元件可直接相连，而箱（或屏）内与外部电器元件相连时，必须经接线端子排。

接线图上所有表示的电气连接，一般并不表示实际走线的路径。在配线时，由电工师傅根据经验选择最佳路径。

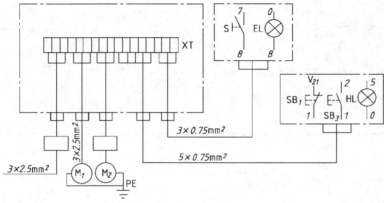

Fig. 10-48 Electric interconnection diagram of CW6132 lathe　　CW6132型车床的电气互连接线图

5. Printed circuit board (PCB) diagram

PCB is made of laminated epoxy plastic substrate covered by copper foils. By PCB diagram, the electric connections between graphic symbols in circuit diagram are converted and instantiated to real ones between electric elements and components. At the same time, PCB diagram works as a structure support in electric engineering. Its drawing methods combine orthogonal projection method and symbol representation method. For details of PCB diagram methods please refer to *Mechanical drawings—General principles of presentation—Views* GB/T 4458.1—2002 and *Graphical symbols for electrical diagrams* GB/T 4728.1—13—2005/2008. Fig. 10-49 shows a PCB diagram of the high frequency board of a radio brand.

Specifications of drawing PCB diagram are provided in *Printed board drawing* GB/T 5489—1985. The points are as follows:

1) When dimensioning by coordinate grid method in cartesian coordinate system, grid dimension for printed circuitsis is selected according to *Grid systems for printed circuits* GB/T 1360—1998 and 2.5mm is recommended. Designers can decide the number of grid lines according to the density and the scale of the graph, as shown in Fig. 10-50a.

5. 印制电路板图

印制板是由覆有铜箔的层压环氧塑料基板制成。它将电路图中各有关图形符号之间的电气连接转变成所对应的实际元器件之间的电气连接，同时在电气工程中也起着结构支撑的作用。其画法也是采用正投影法结合符号表示，具体可参考GB/T 4458.1—2002《机械制图 图样画法 视图》和GB/T 4728.1~13—2005/2008《电气简图用图形符号》。图10-49所示为某收录机高频板的印制板图。

GB/T 5489—1985《印制板制图》对绘制印制电路板做了具体规定，主要有：

1）采用直角坐标系的坐标网格法标注尺寸时，直角坐标网格的间距按GB/T 1360—1998《印制电路网格体系》选取，一般间距常取2.5mm。网格线间距由设计者根据图形的密度和比例确定，如图10-50a所示。

2) Conductive patterns are generally drawn using double line contours, as shown in Fig. 10-50a. The double contours can be filled with colors or section lines as shown in Fig. 10-50b and Fig. 10-50c. If the width of a conductor is less than 1mm, or the conductor has a uniform width, it can be drawn using a single line. In this case, conductor width, minimal space and land size should be dimensioned.

2)导电图形一般用双线轮廓绘制,如图10-50a所示;也可在双线轮廓内涂色,如图10-50b所示;或在双线轮廓内画剖面线,如图10-50c所示。当印制导线宽度小于1mm或宽度基本一致时,导电图形可用单线绘制,此时应注明导线宽度、最小间距和连接盘的尺寸数值。

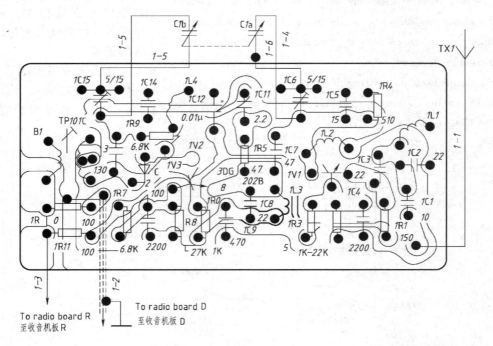

Fig. 10-49 PCB diagram of the high frequency board of a radio brand　收录机高频板的印制电路板图

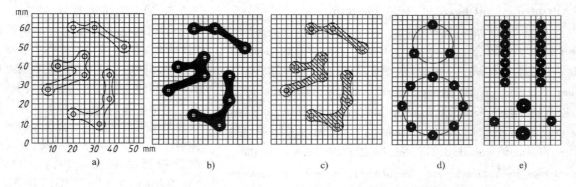

Fig. 10-50 Drawing methods of PCB diagram　印制电路板图的画法

3) When drawing a single aperture, the center of the aperture must align to an intersection point of grid lines. When drawing an aperture group, common center point of a circular aperture group must be located on an intersection point of grid lines, also there is at least one aperture in the group that has center positioned on a grid line passed the given intersection point, as shown in Fig. 10-50d. For a non-circular distributed group of apertures, there is at least one aperture centered on an intersection point of grid lines and there is at least one of other apertures has centre positioned on a grid line passed the given intersection point, as shown in Fig. 10-50e.

3)绘制单孔时,孔的中心必须位于坐标网格线的交点上;绘制组孔时,呈圆形排列的孔的公共中心点,必须位于坐标网格线的交点上,并且至少有一个孔的中心位于上述交点的同一坐标网格线上,如图10-50d所示;呈非圆形排列的一组组孔中,至少有一个孔的中心必须位于坐标网格线的交点上,其他孔至少有一个孔的中心位于上述交点的同一坐标网格线上,如图10-50e所示。

4) Marking symbol diagram. Marking symbol diagram is drawn according to the positions of components. All contents are represented by their graphic symbols, simplified outlines and item numbers that are the same with those in corresponding principle diagrams and logic diagrams. Generally, item numbers are marked near above or left to the graphic symbols of the electric elements or components, as shown in Fig. 10-51.

4）标记符号图。标记符号图按元器件装接位置绘制。一般用元器件的图形符号、简化外形和它在原理图、逻辑图中的位号表示。位号一般标注在靠近该元器件图形符号或外形图的左方或上方，如图10-51所示。

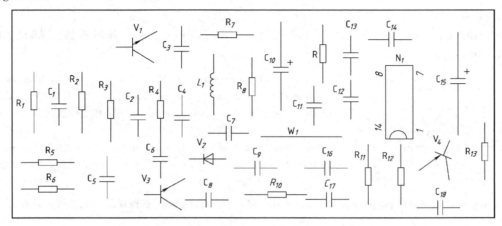

Fig. 10-51 Marking symbol diagram 标记符号图

5) PCB parts drawing. Generally, a single-side PCB part drawing is represented with one view and a double-side PCB parts drawing with two views (front view and rear view). The rear view of double-side PCB parts drawing can be omitted if the conductive patterns are clear enough in the front view. For multiplayer PCB, a view shall be drawn for each conductor layer and then label the layer number in sequence. Starting from the component surface, every wire layer in the views should be sequentially numbered. If a view is a rear view, it shall be marked with "rear view" on top.

6) PCB assembly. In the condition of clearly representing the assembly connections, electric elements and components in PCB assembly can be represented in simplified formats or shall be in accordance with *Graphical symbols for electrical diagrams* GB/T 4728.1—13—2005/2008. If there are direction requirements for the elements and components on the assembly drawing, the direction marks must be labeled. If complete and detailed assembly relations are required, the elements and components in the PCB assembly shall be drawn with mechanical drawing methods in accordance with "*Mechanical drawings—General principles of presentation—Views*" GB/T 4458.1—2002.

5）印制电路板零件图。单面印制板的图样一般用一个视图表示；双面印制板的图样一般用两个视图（主视图、后视图）表示；当后视图上的导电图形能在主视图中表示清楚时，也可以绘一个视图；多层印制电路板的每一导线层应绘制一个视图，视图上应标出层次序号。其编号方法为：从元件表面开始，依次序对每一导线层进行编号。当视图为后视图时，应在视图上方标注"后视图"字样。

6）印制电路板装配图。在清楚地表示装配关系的前提下，印制电路板装配图中的元器件一般采用简化外形或按 GB/T 4728.1~13—2005/2008《电气简图用图形符号》绘制。当元器件在装配图中有方向要求时，必须标出定位特征标志。在需要完整、详细地表示装配关系时，印制电路板装配图中的结构件和元器件按 GB/T 4458.1—2002《机械制图　图样画法》规定绘制。

Appendices 附 录

Appendix 1 Screw Threads 附录1 螺纹

1. Metric thread（Quoted from GB/T 193—2003，GB/T 196—2003）普通螺纹（摘自 GB/T 193—2003，GB/T 196—2003）

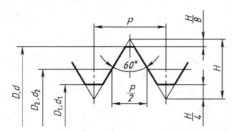

Mark examples

Coarse plain thread, nominal diameter 24mm, pitch 3mm, right-handed, is marked as:

M24

Fine plain thread, nominal diameter 24mm, pitch 1.5mm, left-handed, is marked as:

M24×1.5-LH

标记示例

公称直径为24mm、螺距为3mm的粗牙右旋普通螺纹的标记为：

M24

公称直径为24mm、螺距为1.5mm的细牙左旋普通螺纹的标记为：

M24×1.5-LH

Table 1-1 General purpose metric screw threads—General plan 普通螺纹 直径与螺距系列

（单位：mm）

Nominal diameter 公称直径 D, d		Pitch 螺距 P		Coarse threads minor diameter 粗牙小径 D_1, d_1	Nominal diameter 公称直径 D, d		Pitch 螺距 P		Coarse threads minor diameter 粗牙小径 D_1, d_1
1ST series 第一系列	2ND series 第二系列	Coarse threads 粗牙	Fine threads 细牙		1ST series 第一系列	2ND series 第二系列	Coarse threads 粗牙	Fine threads 细牙	
3		0.5	0.35	2.459		22	2.5	2, 1.5, 1	19.294
	3.5	0.6		2.850	24		3	2, 1.5, 1	20.752
4		0.7		3.242		27	3	2, 1.5, 1	23.752
	4.5	0.75	0.5	3.688	30		3.5	(3), 2, 1.5, 1	26.211
5		0.8		4.134		33	3.5	(3), 2, 1.5	29.211
6		1	0.75	4.917	36		4	3, 2, 1.5	31.670
8		1.25	1, 0.75	6.647		39	4		34.670
10		1.5	1.25, 1, 0.75	8.376	42		4.5		37.129
12		1.75	1.25, 1,	10.106		45	4.5	4, 3, 2, 1.5	40.129
	14	2	1.5, 1.25*, 1,	11.835	48		5		42.587
16		2	1.5, 1	13.835		52	5		46.587
	18	2.5	2, 1.5, 1	15.294	56		5.5	4, 3, 2, 1.5	50.046
20		2.5		17.294					

Notes 注：1. The 1ST series are priority choice, better to use other values instead of using the value in brackets. 优先选用第一系列，尽可能不用括号内尺寸。

2. Pitchs D_2, d_2 are excluded, the 3RD series are excluded. 中径 D_2、d_2 未列入，第三系列未列入。

3. The clata marked with symbol "*" on the right-upper part is used only in spark plug. 右上角"*"号标记的数据仅用于火花塞。

Table 1-2 The relation between the pitch and minor diameter of fine plain threads

细牙普通螺纹螺距与小径的关系

（单位：mm）

Pitch 螺距 P	Pitch diameter 中径 D_2, d_2	Minor diameter 小径 D_1, d_1	Pitch 螺距 P	Pitch diameter 中径 D_2, d_2	Minor diameter 小径 D_1, d_1
0.35	$D(d)-1+0.773$	$D(d)-1+0.621$	1.5	$D(d)-1+0.026$	$D(d)-2+0.376$ $D(d)-3+0.835$
0.5	$D(d)-1+0.675$	$D(d)-1+0.459$	2	$D(d)-2+0.701$	$D(d)-4+0.725$
0.75	$D(d)-1+0.513$	$D(d)-1+0.188$	3	$D(d)-2+0.051$	$D(d)-5+0.670$
1	$D(d)-1+0.350$	$D(d)-2+0.917$	4	$D(d)-3+0.402$	$D(d)-7+0.505$
1.25	$D(d)-1+0.188$	$D(d)-2+0.647$	6	$D(d)-4+0.103$	

Notes 注：Minor diameter in the table can be calculated using the formulas: $D_1 = d_1 = d - 2 \times 5/8 H$, $H = \sqrt{3}/2 P$.

表中的小径按 $D_1 = d_1 = d - 2 \times 5/8H$, $H = \sqrt{3}/2 P$ 计算得出。

2. Metric trapezoidal screw threads (Quoted from GB/T 5796.3—2005) 梯形螺纹（摘自 GB/T 5796.3—2005）

Mark examples

Trapezoidal screw thread, nominal diameter 40mm, pitch 7mm, single, right-handed, TH tolerance zone for pitch diameter, is marked as:

Tr 40×7

Trapezoidal screw thread, nominal diameter 40mm, lead distance 14mm, pitch 7mm, double, left-handed, Te tolerance zone for pitch diameter, is marked as:

Tr 40×14（P7）LH-Te

标记示例

公称直径为 40mm、螺距为 7mm、单线、右旋、中径公差带为 7H 的梯形螺纹的标记为：

Tr 40×7

公称直径为 40mm、导程为 14mm、螺距为 7mm 双线、左旋中径公差带为 7e 的梯形螺纹的标记为：

Tr 40×14（P7）LH-7e

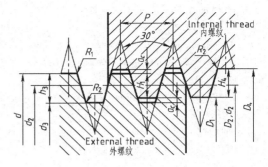

Table 1-3 Basic dimensions of metric trapezoidal screw threads 梯形螺纹的基本尺寸

（单位：mm）

| Nominal diameter 公称直径 d | | Pitch 螺距 P | Pitch diameter 中径 $d_2 = D_2$ | Major diameter 大径 D_4 | Minor diameter 小径 | | Nominal diameter 公称直径 d | | Pitch 螺距 P | Pitch diameter 中径 $d_2 = D_2$ | Major diameter 大径 D_4 | Minor diameter 小径 | |
1ST series 第一系列	2ND series 第二系列				d_3	D_1	1ST series 第一系列	2ND series 第二系列				d_3	D_1
8		1.5	7.25	8.30	6.20	6.50			3	24.50	26.50	22.50	23.00
	9	1.5	8.25	9.30	7.20	7.50		26	5	23.50	26.50	20.50	21.00
		2	8.00	9.50	6.50	7.00			8	22.00	27.00	17.00	18.00
10		1.5	9.25	10.30	8.20	8.50			3	26.50	28.50	24.50	25.00
		2	9.00	10.50	7.50	8.00	28		5	25.50	28.50	22.50	23.00
	11	2	10.00	11.50	8.50	9.00			8	24.00	29.00	19.00	20.00
		3	9.50	11.50	7.50	8.00			3	28.50	30.50	26.50	27.00
12		2	11.00	12.50	9.50	10.00		30	6	27.00	31.00	23.00	24.00
		3	10.50	12.50	8.50	9.00			10	25.00	31.00	19.00	20.00
	14	2	13.00	14.50	11.50	12.00			3	30.50	32.50	28.50	29.00
		3	12.50	14.50	10.50	11.00	32		6	29.00	33.00	25.00	26.00
16		2	15.00	16.50	13.50	14.00			10	27.00	33.00	21.00	22.00
		4	14.00	16.50	11.50	12.00			3	32.50	34.50	30.50	31.00
	18	2	17.00	18.50	15.50	16.00		34	6	31.00	35.00	27.00	28.00
		4	16.00	18.50	13.50	14.00			10	29.00	35.00	23.00	24.00
20		2	19.00	20.50	17.50	18.00			3	34.5	36.50	32.50	33.00
		4	18.00	20.50	15.50	16.00	36		6	33.00	37.00	29.00	30.00
	22	3	20.50	22.50	18.50	19.00			10	31.00	37.00	25.00	26.00
		5	19.50	22.50	16.50	17.00			3	36.50	38.50	34.50	35.00
		8	18.00	23.00	13.00	14.00		38	7	34.50	39.00	30.00	31.00
24		3	22.50	24.50	20.50	21.00			10	33.00	39.00	27.00	28.00
		5	21.50	24.50	18.50	19.00			3	38.50	40.50	36.50	37.00
		8	20.00	25.00	15.00	16.00	40		7	36.50	41.00	32.00	33.00
									10	35.00	41.00	29.00	30.00

Notes 注：1. The 1ST series of diameter are priority choice. 优先选用第一系列的直径。

2. Underlined pitch data are priority choice. 每一直径的螺距系列中带下划线的数据为优先选用的螺距。

3. Pipe threads with 55 degree thread angle where pressure-tight joints are not made on the threads (Quoted from GB/T 7307—2001) 55°非密封管螺纹（摘自 GB/T 7307—2001）

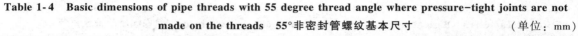

Mark examples	标记示例
1 1/2 inner left pipetap is marked as:	尺寸代号为 1 1/2 的左旋内螺
G1 1/2 LH	纹的标记为:
Rightscrew threads are not marked	G1 1/2 LH
1 1/2A grade external pipetap is marked as:	右旋不标
G1 1/2A	1 1/2A 级外螺纹的标记为:
1 1/2B grade external pipetap is marked as:	G1 1/2A
G1 1/2B	1 1/2B 级外螺纹的标记为:
Inner and external pipetap fit is marked as:	G1 1/2B
G1 1/2A	内外螺纹配合的标记为:
	G1 1/2A

Table 1-4 Basic dimensions of pipe threads with 55 degree thread angle where pressure-tight joints are not made on the threads 55°非密封管螺纹基本尺寸   （单位：mm）

Size code 尺寸代号	Tooth number per 25.4mm 每 25.4mm 内的牙数 n	Pitch 螺距 P	Tooth height 牙高 h	Basic diameter 基本直径		
				Major diameter 大径 $d=D$	Pitch diameter 中径 $d_2=D_2$	Minor diameter 小径 $d_1=D_1$
1/16	28	0.907	0.581	7.723	7.142	6.561
1/8	28	0.907	0.581	9.728	9.147	8.566
1/4	19	1.337	0.856	13.157	12.301	11.445
3/8	19	1.337	0.856	16.662	15.806	14.950
1/2	14	1.814	1.162	20.955	19.793	18.631
5/8	14	1.814	1.162	22.911	21.749	20.587
3/4	14	1.814	1.162	26.441	25.279	24.117
7/8	14	1.814	1.162	30.201	29.039	27.877
1	11	2.309	1.479	33.249	31.770	30.291
1 1/8	11	2.309	1.479	37.897	36.418	34.939
1 1/4	11	2.309	1.479	41.910	40.431	38.952
1 1/2	11	2.309	1.479	47.803	46.324	44.845
1 3/4	11	2.309	1.479	53.746	52.267	50.788
2	11	2.309	1.479	59.614	58.135	56.656
2 1/4	11	2.309	1.479	65.710	64.231	62.752
2 1/2	11	2.309	1.479	75.184	73.705	72.226
2 3/4	11	2.309	1.479	81.534	80.055	78.576
3	11	2.309	1.479	87.884	86.405	84.926
3 1/2	11	2.309	1.479	100.330	98.851	97.372
4	11	2.309	1.479	113.030	111.551	110.072
4 1/2	11	2.309	1.479	125.730	124.251	122.722
5	11	2.309	1.479	138.430	136.951	135.472
5 1/2	11	2.309	1.479	151.130	149.651	148.172
6	11	2.309	1.479	163.830	162.351	160.872

Notes 注：This standard fits for pipes, valves, pipe joints, faucets and other pipeline accessories in terms of thread joints.
本标准适用于管子、阀门、管接头、旋塞及其他管路附件的螺纹联接。

Appendix 2　Commonly Used Standard Parts　附录2　常用标准件

1. Bolt 螺栓

（1）Hexagon head bolts—Product grade C（Quoted from GB/T 5780—2016）六角头螺栓—C 级（摘自 GB/T 5780—2016）

（2）Hexagon head bolts—Product grades A and B（Quoted from GB/T 5782—2016）六角头螺栓—A 和 B 级（摘自 GB/T 5782—2016）

GB/T 5780—2016

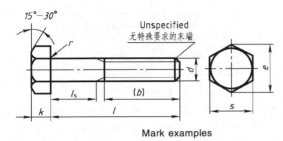

Mark examples

The hexagon head bolt（thread specification $d=$ M12；nominal length $l=$ 80mm；level of performance is 8.8, oxidization as surface treatment, grade A）is marked as：

Bolt　GB/T 5782　M12×80

GB/T 5782—2016

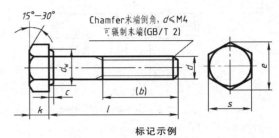

标记示例

螺纹规格 $d=$ M12、公称长度 $l=$ 80mm、性能等级为 8.8 级、表面氧化的 A 级六角头螺栓，其标记为：

螺栓　GB/T 5782　M12×80

Table 2-1　Sizes of hexagon head bolts　六角头螺栓尺寸　　　　　　　　（单位：mm）

Thread specification 螺纹规格 d		M3	M4	M5	M6	M8	M10	M12	M16	M20	M24	M30
b(Reference 参考)	$l\leqslant 125$	12	14	16	18	22	26	30	38	46	54	66
	$125<l\leqslant 200$	18	20	22	24	28	32	36	44	52	60	72
	$l>200$	31	33	35	37	41	45	49	57	65	73	85
c　max		0.4	0.4	0.5	0.5	0.6	0.6	0.6	0.8	0.8	0.8	0.8
d_w min	Production class 产品等级 A	4.57	5.88	6.88	8.88	11.63	14.63	16.63	22.49	28.19	33.61	—
	B，C	4.45①	5.74②	6.74	8.74	11.47	14.47	16.47	22	27.7	33.25	42.75
e min	Production class 产品等级 A	6.01	7.66	8.79	11.05	14.38	17.77	20.03	26.75	33.53	39.98	—
	B，C	5.88①	7.50②	8.63	10.89	14.20	17.59	19.85	26.17	32.95	39.55	50.85
k（Nominal 公称）		2	2.8	3.5	4	5.3	6.4	7.5	10	12.5	15	18.7
r　min		0.1	0.2	0.2	0.25	0.4	0.4	0.6	0.6	0.8	0.8	1
s（Nominal 公称）		5.5	7	8	10	13	16	18	24	30	36	46
l（Commodities specification ranges 商品规格范围）		20—30	25—40	25—50	30—60	40—80	45—100	50—120	65—160	80—200	90—240	110—300
l Series 系列		20，25，30，35，40，45，50，55，60，65，70，80，90，100，120，130，140，150，160，180，200，220，240，260，280，300										

Notes 注：1. A class bolts are used for $d\leqslant$ M24 and $l\leqslant 10d$ or $l\leqslant 150$, B class bolts are used for $d>$ M24 and $l>10d$ or >150.

A 级用于 $d\leqslant$ M24 和 $l\leqslant 10d$ 或 $l\leqslant 150$ 的螺栓；B 级用于 $d>$ M24 和 $l>10d$ 或 >150 的螺栓。

2. Thread specification d ranges: for GB/T 5780 is M5—M64; for GB/T 5782 is M1.6—M64.

螺纹规格 d 范围：对于 GB/T 5780 为 M5~M64；对于 GB/T 5782 为 M1.6~M64。

3. Nominal length l ranges: for GB/T 5780 is 25—500; for GB/T 5782 is 12—500.

公称长度 l 范围：对于 GB/T 5780 为 25~500；对于 GB/T 5782 为 12~500。

①、②Only be suitable for the bolt with B grade。

只适用于 B 级螺栓。

2. Stud 螺柱

(1) Double end studs—$b_m = 1d$ (Quoted from GB 897—1988)　双头螺柱—$b_m = 1d$（摘自 GB 897—1988）

(2) Double end studs—$b_m = 1.25d$ (Quoted from GB 898—1988)　双头螺柱—$b_m = 1.25d$（摘自 GB 898—1988）

(3) Double end studs—$b_m = 1.5d$ (Quoted from GB 899—1988)　双头螺柱—$b_m = 1.5d$（摘自 GB 899—1988）

(4) Double end studs—$b_m = 2d$ (Quoted from GB 900—1988)　双头螺柱—$b_m = 2d$（摘自 GB 900—1988）

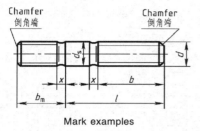

Mark examples

Double end stud (Two ends are coarse, d = M10, l = 50mm, level of performance is 4.8, no surface treatment, style B, $b_m = 1d$) is marked as:

Stud　GB 897　M10×50

Double end stud (Coarse thread for metal end; fine threads with P = 1mm, d = M10, l = 50mm for nut end, level of performance is 4.8, no surface treatment, style A, $b_m = 1.25d$) is marked as:

Stud GB 898 AM10—M10×1×50

标记示例

两端均为粗牙普通螺纹、d = M10、l = 50mm、性能等级为 4.8 级、不经表面处理、B 型、$b_m = 1d$ 的双头螺柱，其标记为：

螺柱　GB 897　M10×50

旋入端为粗牙普通螺纹、紧固端为螺距 P = 1mm 的细牙普通螺纹、d = M10、l = 50mm、性能等级为 4.8 级、不经表面处理、A 型、$b_m = 1.25d$ 的双头螺柱，标记为

螺柱　GB 898 AM10—M10×1×50

Table 2-2　Sizes of studs　螺柱尺寸　　　　　　　　　　　　　　（单位：mm）

Thread specification 螺纹规格 d	b_m (Nominal 公称) GB/T 897—1988	GB/T 898—1988	d_s max	d_s min	x max	b	l (Nominal 公称)
M5	5	6	5	4.7	1.5P	10	16—(22)
						16	25—50
M6	6	8	6	5.7		10	20, (22)
						14	25, (28), 10
						18	(32)—(75)
M8	8	10	8	7.64		12	20, (22)
						16	25, (28), 10
						22	(32)—90
M10	10	12	10	9.64		14	24, (28)
						16	30—(38)
						26	40—120
						32	130
M12	12	15	12	11.57		16	25—30
						20	(32)—40
						30	45—120
						36	130—180
M16	16	20	16	15.57		20	30—(38)
						30	40—(55)
						38	60—120
						44	130—200
M20	20	25	20	19.48		25	35—40
						35	45—(65)
						46	70—120
						52	130—200

Notes 注：1. GB 899—88, GB 900—88 are excluded in this table. 本表未列入 GB 899—88、GB 900—88 两种规格。
2. "P" represents pitch of coarse threads. P 表示粗牙螺纹的螺距。
3. l (Series 长度系列): 16, (18), 20, (22), 25, (28), 30, (32), 35, (38), 40, 45, 50, (55), 60, 65, 70, (75), 80, 90, (95), 100—200 (decade carry 十进位). Better to use other values instead of the values in brackets. 长度系列中，尽可能不用括号内数值。

3. Screw 螺钉

（1）Slotted cheese head screws（Quoted from GB/T 65—2016）开槽圆柱头螺钉（摘自 GB/T 65—2016）

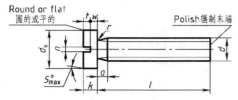

Mark examples

The slotted cheese head screw（d = M5, l = 20mm, level of performance is 4.8, and no surface treatment, grade A）, is marked as:

Screw GB/T 65 M5×20

标记示例

螺纹规格 d = M5、公称长度 l = 20mm、性能等级为 4.8 级、不经表面处理的 A 级开槽圆柱头螺钉，其标记为：

螺钉 GB/T 65 M5×20

Table 2-3　Sizes of slotted cheese head screws　开槽圆柱头螺钉尺寸　　（单位：mm）

Thread specification 螺纹规格 d	M4	M5	M6	M8	M10
P（Pitch 螺距）	0.7	0.8	1	1.25	1.5
b　min	38	38	38	38	38
d_k　nom	7	8.5	10	13	16
k　nom	2.6	3.3	3.9	5	6
n　nom	1.2	1.2	1.6	2	2.5
r　min	0.2	0.2	0.25	0.4	0.4
t　min	1.1	1.3	1.6	2	2.4
Nominal length 公称长度 l	5—40	6—50	8—60	10—80	12—80
l（Series 系列）	5, 6, 8, 10, 12, (14), 16, 20, 25, 30, 35, 40, 45, 50, (55), 60, (65), 70, (75), 80				

Notes 注：1. Nominal length l ≤ 40 socket screws, full threads. 公称长度 l ≤ 40 的螺钉，制出全螺纹。

2. Thread specification d = M1.6-M10, nominal length l = 2-80mm. 螺纹规格 d = M1.6-M10, 公称长度 l = 2~80mm。

3. Better to use other values instead of the values in brackets. 尽可能不采用括号内的规格。

（2）Slotted pan head screws（Quoted from GB/T 67—2016）开槽盘头螺钉（摘自 GB/T 67—2016）

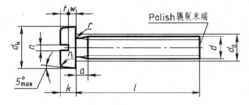

Mark examples

The slotted pan head screw（d = M5, l = 20mm, level of performance is 4.8, and no surface treatment, grade A）, is marked as:

Screw GB/T 67　M5×20

标记示例

螺纹规格 d = M5、公称长度 l = 20mm、性能等级为 4.8 级、不经表面处理的 A 级开槽盘头螺钉，其标记为：

螺钉 GB/T 67　M5×20

Table 2-4　Sizes of slotted pan head screws　开槽盘头螺钉尺寸　　（单位：mm）

Thread specification 螺纹规格 d	M1.6	M2	M2.5	M3	M4	M5	M6	M8	M10
P（Pitch 螺距）	0.35	0.4	0.45	0.5	0.7	0.8	1	1.25	1.5
b min	25	25	25	25	38	38	38	38	38
d_k　nom	3.2	4	5	5.6	8	9.5	12	16	20
k　nom	1	1.3	1.5	1.8	2.4	3	3.6	4.8	6
n　nom	0.4	0.5	0.6	0.8	1.2	1.2	1.6	2	2.5
r　min	0.1	0.1	0.1	0.1	0.2	0.2	0.25	0.4	0.4
t　min	0.35	0.5	0.6	0.7	1	1.2	1.4	1.9	2.4
Nominal length 公称长度 l	2—16	2.5—20	3—25	4—30	5—40	6—50	8—60	10—80	12—80
l（Series 系列）	2, 2.5, 3, 4, 5, 6, 8, 10, 12, (14), 16, 20, 25, 30, 35, 40, 45, 50, (55), 60, (65), 70, (75), 80								

Notes 注：1. Socket screws M1.6—M3, nominal length l≤30mm, full threads; socket screws M4—M10, nominal length l≤40mm, full threads.

M1.6~M3 的螺钉，公称长度 l≤30mm 的，可制出全螺纹；M4~M10 的螺钉，公称长度 l≤40mm 的，制出全螺纹。

2. Better to use other values instead of the values in brackets. 尽可能不采用括号内的规格。

(3) Slotted countersunk flat head screws (Quoted from GB/T 68—2016) 开槽沉头螺钉（摘自 GB/T 68—2016）

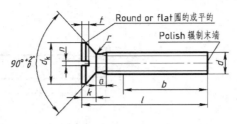

Mark examples

The slotted countersunk flat head screws (thread specification d = M5, l = 20, level of performance is 4.8, and no surface, grade A), is marked as:

Screw GB/T 68   M5×20

标记示例

螺纹规格 d = M5、公称长度 l = 20、性能等级为 4.8 级、不经表面处理的 A 级开槽沉头螺钉，其标记为：

螺钉 GB/T 68   M5×20

Table 2-5   Sizes of the slotted countersunk flat head screws   开槽沉头螺钉尺寸   （单位：mm）

Thread specification 螺纹规格 d	M1.6	M2	M2.5	M3	M4	M5	M6	M8	M10
P（Pitch 螺距）	0.35	0.4	0.45	0.5	0.7	0.8	1	1.25	1.5
b min	25	25	25	25	38	38	38	38	38
d_k 理论值 max	3.6	4.4	5.5	6.3	9.4	10.4	12.6	17.3	20
k max	1	1.2	1.5	1.65	2.7	2.7	3.3	4.65	5
n nom	0.4	0.5	0.6	0.8	1.2	1.2	1.6	2	2.5
r max	0.4	0.5	0.6	0.8	1	1.3	1.5	2	2.5
t max	0.5	0.6	0.75	0.85	1.3	1.4	1.6	2.3	2.6
Nominal length 公称长度 l	2.5—16	3—20	4—25	5—30	6—40	8—50	8—60	10—80	12—80
l（Series 系列）	2.5, 3, 4, 5, 6, 8, 10, 12, (14), 16, 20, 25, 30, 35, 40, 45, 50, (55), 60, (65), 70, (75), 80								

Notes 注：1. Socket screws M1.6—M3, nominal length $l \leqslant 30$, full threads; socket screws M4—M10, nominal length $l \leqslant 45$, full threads. M1.6~M3 的螺钉，公称长度 $l \leqslant 30$ 的，制出全螺纹；M4~M10 的螺钉，公称长度 $l \leqslant 45$ 的，制出全螺纹。

2. Better to use other values instead of the values in brackets. 尽可能不采用括号内的规格。

(4) Slotted set screws 开槽紧定螺钉

Slotted set screws with cone point
(Quoted from GB 71—1985)
开槽锥端紧定螺钉（摘自 GB 71—1985）

Slotted set screws with flat point
(Quoted from GB 73—1985)
开槽平端紧定螺钉（摘自 GB 73—1985）

Slotted set screws with long dog point
(Quoted from GB 75—1985)
开槽长圆柱端紧定螺钉（摘自 GB 75—1985）

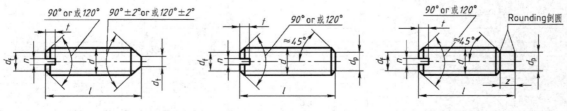

Mark examples

The slotted set screw with long dog point (d = M5, l = 20, level of performance is 14H, and oxidization as surface treatment) is marked as:

Screw GB 75   M5×20

标记示例

螺纹规格 d = M5、公称长度 l = 20、性能等级为 14H 级、表面氧化的开槽长圆柱端紧定螺钉，其标记为：

螺钉 GB 75   M5×20

Table 2-6 Sizes of slotted set screws 开槽紧定螺钉尺寸 （单位：mm）

Thread specification 螺纹规格 d		M1.6	M2	M2.5	M3	M4	M5	M6	M8	M10	M12
P（Pitch 螺距）		0.35	0.4	0.45	0.5	0.7	0.8	1	1.25	1.5	1.75
n nom		0.25	0.25	0.4	0.4	0.6	0.8	1	1.2	1.6	2
t max		0.74	0.84	0.95	1.05	1.42	1.63	2	2.5	3	3.6
d_t max		0.16	0.2	0.25	0.3	0.4	0.5	1.5	2	2.5	3
d_p max		0.8	1	1.5	2	2.5	3.5	4	5.5	7	8.5
z max		1.05	1.25	1.5	1.75	2.25	2.75	3.25	4.3	5.3	6.3
l	GB/T 71—1985	2—8	3—10	3—12	4—16	6—20	8—25	8—30	10—40	12—50	14—60
	GB/T 73—1985	2—8	2—10	2.5—12	3—16	4—20	5—25	6—30	8—40	10—50	12—60
	GB/T 75—1985	2.5—8	3—10	4—12	5—16	6—20	8—25	10—30	10—40	12—50	14—60
l（Series 系列）		2, 2.5, 3, 4, 5, 6, 8, 10, 12, (14), 16, 20, 25, 30, 35, 40, 45, 50, (55), 60									

Notes 注：1. Nominal length l. l 为公称长度。
2. Better to use other values instead of the values in brackets. 尽可能不采用括号内的规格。

4. Nut 螺母

（1）Hexagon nuts, style 1—Product grades A and B（Quoted from GB/T 6170—2015）1 型六角螺母—A 级和 B 级（摘自 GB/T 6170—2015）

Mark examples

Hexagon nuts, style 1 (D = M12, level of performance is 10, and oxidization as surface treatment, grades A) is marked as:

Nut GB/T 6170 M12

标记示例

螺纹规格 D = M12、性能等级为 10 级、表面氧化的 A 级 1 型六角螺母，其标记为：

螺母 GB/T 6170 M12

1) β=15°—30°.
2) Washer face should be indicated in the order. 垫圈面型应在订单中注明。
3) θ=90°—120°.

Table 2-7 Sizes of hexagon nuts, style 1 1 型六角螺母尺寸 （单位：mm）

Thread specification 螺纹规格 D		M1.6	M2	M2.5	M3	M4	M5	M6	M8	M10	M12
P		0.35	0.4	0.45	0.5	0.7	0.8	1	1.25	1.5	1.75
c	max	0.2	0.2	0.3	0.4	0.4	0.5	0.5	0.6	0.6	0.6
	min	0.1	0.1	0.1	0.15	0.15	0.15	0.15	0.15	0.15	0.15
d_a	max	1.84	2.3	2.9	3.45	4.6	5.75	6.75	8.75	10.8	13
	min	1.6	2.0	2.5	3.00	4.0	5.00	6.00	8.00	10.0	12
d_w min		2.4	3.1	4.1	4.6	5.9	6.9	8.9	11.6	14.6	16.6
e min		3.41	4.32	5.45	6.01	7.66	8.79	11.05	14.38	17.77	20.03
m	max	1.30	1.60	2.00	2.40	3.2	4.7	5.2	6.8	8.4	10.80
	min	1.05	1.35	1.75	2.15	2.9	4.4	4.9	6.44	8.04	10.37
m_w min		0.8	1.1	1.4	1.7	2.3	3.5	3.9	5.2	6.4	8.3
s	Nominal 公称 = max	3.20	4.00	5.00	5.50	7.00	8.00	10.00	13.00	16.00	18.00
	min	3.02	3.82	4.82	5.32	6.78	7.78	9.78	12.73	15.73	17.73

Thread specification 螺纹规格 D		M16	M20	M24	M30	M36	M42	M48	M56	M64
P		2	2.5	3	3.5	4	4.5	5	5.5	6
c	max	0.8	0.8	0.8	0.8	0.8	1.0	1.0	1.0	1.0
	min	0.2	0.2	0.2	0.2	0.2	0.3	0.3	0.3	0.3
d_a	max	17.3	21.6	25.9	32.4	38.9	45.4	51.8	60.5	69.1
	min	16.0	20.0	24.0	30.0	36.0	42.0	48.0	56.0	64.0
d_w min		22.5	27.7	33.3	42.8	51.1	60	69.5	78.7	88.2
e min		26.75	32.95	39.55	50.85	60.79	71.3	82.6	93.56	104.86
m	max	14.8	18.0	21.5	25.6	31.0	34	38.0	45.0	51.0
	min	14.1	16.9	20.2	24.3	29.4	32.4	36.4	43.4	49.1
m_w min		11.8	13.5	16.2	19.4	23.5	25.9	29.1	34.7	39.3
s	Nominal 公称 = max	24.00	30.00	36	46	55	65	75.0	85	95
	min	23.67	29.16	35	45	53.8	63.1	73.1	82.8	92.8

Notes 注：A class nuts are used for $D \leqslant$ M24 and $l \leqslant 10d$ or $\leqslant$ 150mm; B class nuts are used for $D >$ M24 and $l > 10d$ or $>$ 150mm.
A 级用于 $D \leqslant$ M24 和 $l \leqslant 10d$ 或 $\leqslant$ 150mm 的螺母；B 级用于 $D >$ M24 和 $l > 10d$ 或 $>$ 150mm 的螺母。

（2）Hexagon thin nuts（Quoted from GB/T 6172.1—2016）六角薄螺母（摘自 GB/T 6172.1—2016）

Mark examples

Hexagon thin nut（D = M12, level of performance is 04, and no surface treatment, product grade is A）is marked as:
Nut GB/T 6172.1 M20

标记示例

螺纹规格 D = M12、性能等级为 04 级、不经表面处理、产品等级为 A 级的六角薄螺母，其标记为：
螺母 GB/T 6172.1 M20

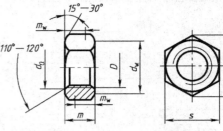

Table 2-8 Basic sizes of hexagon thin nuts 六角薄螺母尺寸（单位：mm）

Thread specification 螺纹规格	D	M1.6	M2	M2.5	M3	M4	M5	M6	M8	M10	M12	(M14)	M16	(M18)	M20	(M22)
	$D \times P$								M8×1.25	M10×1.5	M12×1.75	(M14×2)	M16×2	(M18×2.5)	M20×2.5	(M2×2.5)
d_a	min	1.6	2	2.5	3	4	5	6	8	10	12	14	16	18	20	22
d_w	min	2.4	3.1	4.1	4.6	5.9	6.9	8.9	11.6	14.6	16.6	19.6	22.5	24.9	27.7	31.4
e	min	3.41	4.32	5.45	6.01	7.66	8.79	11.05	14.38	17.77	20.03	23.36	26.75	29.56	32.95	37.29
m	max	1	1.2	1.6	1.8	2.2	2.7	3.2	4	5	6	7	8	9	10	11
s	max	3.2	4	5	5.5	7	8	10	13	16	18	21	24	27	30	32

Notes 注：1. Better to use other values instead of the values in brackets. 尽可能不采用括号内的规格。

2. P means Pitch. P 为螺距。

（3）Hexagon slotted and castle nuts, style 1—Product grades A and B（Quoted from GB 6178—1986）1 型六角开槽螺母—A 和 B 级（摘自 GB 6178—1986）

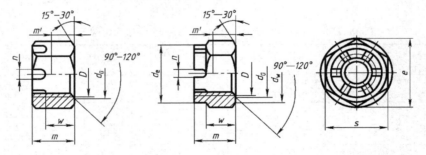

Mark examples

Hexagon slotted nuts, style 1（D = M5, level of performance is 8, and no surface treatment, product grade is grades as A）is marked as:
Nut GB 6178 M5

标记示例

螺纹规格 D = M5、性能等级为 8 级、不经表面处理、产品等级为 A 级的 1 型六角开槽螺母，其标记为：
螺母 GB 6178 M5

Table 2-9 Sizes of hexagon slotted nuts, style 1 1 型六角开槽螺母尺寸（单位：mm）

Thread specification 螺纹规格 D		M4	M5	M6	M8	M10	M12	(M14)	M16	M20	M24	M30	M36
d_e	max									28	34	42	50
e	min	7.66	8.79	11.05	14.38	17.77	20.03	23.35	26.75	32.95	39.55	50.85	60.79
m	max	5	6.7	7.7	9.8	12.4	15.8	17.8	20.8	24	29.5	34.6	40
n	min	1.2	1.4	2	2.5	2.8	3.5	3.5	4.5	4.5	5.5	7	7
s	max	7	8	10	13	16	18	21	24	30	36	46	55
w	max	3.2	4.7	5.2	6.8	8.4	10.8	12.8	14.8	18	21.5	25.6	31
Cotter pin 开口销		1×10	1.2×12	1.6×14	2×16	2.5×20	3.2×22	3.2×25	4×28	4×36	5×40	6.3×50	6.3×63

Notes 注：Better to use other values instead of using the value in brackets. 尽可能不采用括号内的规格。

5. Washer 垫圈

(1) Plain washers—Product grade A (Quoted from GB/T 97.1—2002) 平垫圈—A 级（摘自 GB/T 97.1—2002）

(2) Plain washers, chamfered—Product grade A (Quoted from GB/T 97.2—2002) 平垫圈 倒角型—A 级（摘自 GB/T 97.2—2002）

(3) Plain washers—Product grade C (Quoted from GB/T 95—2002) 平垫圈—C 级（摘自 GB/T 95—2002）

(4) Plain washers—Small series—Product grade A (Quoted from GB/T 848—2002) 小垫圈—A 级（摘自 GB/T 848—2002）

(5) Plain washers—Extra large series—Product grade C (Quoted from GB/T 5287—2002) 特大垫圈—C 级（摘自 GB/T 5287—2002）

(6) Plain washers- large series—Products grades A and C (Quoted from GB/T 96.1—2002 and GB/T 96.2—2002) 大垫圈—A 和 C 级（摘自 GB/T 96.1—2002 和 GB/T 96.2—2002）

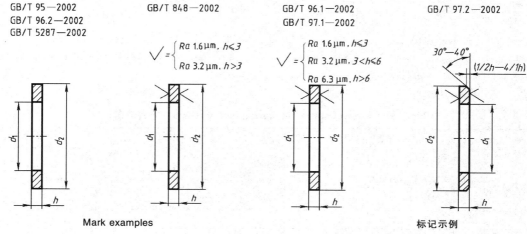

Mark examples 标记示例

Plain washer (standard series, nominal diameter 8mm, the hardness grade is 100HV, and no surface treatment, product grade C) is marked as:

Washer GB/T 95 8

Plain washers, chamfered (standard series, nominal diameter 8mm, the hardness of steel grade 200HV, and no surface treatment, product grade A) is marked as:

Washer GB/T 97.2 8

标准系列、公称规格 8mm、硬度等级为 100HV 级，不经表面处理、产品等级为 C 级的平垫圈的标记为：

垫圈 GB/T 95 8

标准系列、公称规格 8mm、由钢制造的硬度等级为 200HV 级，不经表面处理、产品等级为 A 级的倒角型平垫圈的标记为：

垫圈 GB/T 97.2 8

Table 2-10 Sizes of washers 垫圈尺寸 （单位：mm）

Thread specification 螺纹规格 d	GB/T 95—2002			GB/T 97.1—2002			GB/T 97.2—2002			GB/T 5287—2002			GB/T 96—2002			GB/T 848—2000		
	d_1	d_2	h	d_1	d_2	h	d_1	d_2	h	d_1	d_2	h	d_1	d_2	h	d_1	d_2	h
1.6	—	—	—	1.7	4	0.3	—	—	—	—	—	—	—	—	—	1.7	3.5	0.3
2	—	—	—	2.3	5	0.3	—	—	—	—	—	—	—	—	—	2.2	4.5	0.3
2.5	—	—	—	2.7	6	0.5	—	—	—	—	—	—	—	—	—	2.7	5	0.5
3	—	—	—	3.2	7	0.5	—	—	—	—	—	—	3.2	9	0.8	3.2	6	0.5
4	—	—	—	4.3	9	0.8	—	—	—	—	—	—	4.3	12	1	4.3	8	0.5
5	5.5	10	1	5.3	10	1	5.3	10	1	5.5	18	2	5.3	15	1.2	5.3	9	1
6	6.6	12	1.6	6.4	12	1.6	6.4	12	1.6	6.6	22	2	6.4	18	1.6	6.4	11	1.6
8	9	16	1.6	8.4	16	1.6	8.4	16	1.6	9	28	3	8.4	24	2	8.4	15	1.6
10	11	20	2	10.5	20	2	10.5	20	2	11	34	3	10.5	30	2.5	10.5	18	1.6
12	13.5	24	2.5	13	24	2.5	13	24	2.5	13.5	44	4	13	37	3	13	20	2
14	15.5	28	2.5	15	28	2.5	15	28	2.5	15.5	50	4	15	44	3	15	24	2.5
16	17.5	30	3	17	30	3	17	30	3	17.5	56	5	17	50	3	17	28	2.5
20	22	37	3	21	37	3	21	37	3	22	72	6	22	60	4	21	34	3
24	26	44	4	25	44	4	25	44	4	26	85	6	26	72	5	25	39	4
30	33	56	4	31	56	4	31	55	4	33	105	6	33	92	6	31	50	4
36	39	66	5	37	66	5	37	66	5	39	125	8	36	110	8	37	60	5

Notes 注：1. Grade A fits for finest assemble series, grade C fits for medium assemble series, grade C dose not require $Ra3.2\mu m$ and removing burr.

A 级适用于精装配系列，C 级适用于中等装配系列，C 级垫圈没有 $Ra3.2\mu m$ 和去毛刺的要求。

2. GB/T 848—2000 is used for column head screws, the others are used for hexagonal bolt, screw and nut.

GB/T 848—2000 主要用于带圆柱头螺钉，其他用于标准六角螺栓、螺钉和螺母。

（7）Single coil spring lock washers, Normal type (Quoted from GB 93—1987) 标准型弹簧垫圈（摘自 GB 93—1987）

（8）Single coil spring lock washers, Light type (Quoted from GB 859—1987) 轻型弹簧垫圈（摘自 GB 859—1987）

（9）Single coil spring lock washers, heavy type (Quoted from GB 7244—1987) 重型弹簧垫圈（摘自 GB 7244—1987）

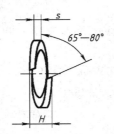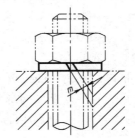

Mark examples 标记示例

Single coil spring lock washers, Normal type (16mm, material is 65Mn, and oxidization as surface treatment) is marked as:

Washer GB 93—87　16

规格 16mm、材料为 65Mn、表面氧化的标准型弹簧垫圈的标记为：

垫圈 GB 93—87　16

Table 2-11　Sizes of single coil spring lock washers　弹簧垫圈尺寸　（单位：mm）

Screw threads specfication 规格螺纹大径	d_1 min	GB 93—87				GB 859—87				GB 7244—87			
		s Nominal 公称	b Nominal 公称	H max	$m \leq$	s Nominal 公称	b Nominal 公称	H max	$m \leq$	s Nominal 公称	b Nominal 公称	H max	$m \leq$
2	2.1	0.5	0.5	1.25	0.25	—	—	—	—	—	—	—	—
2.5	2.6	0.65	0.65	1.63	0.33	—	—	—	—	—	—	—	—
3	3.1	0.8	0.8	0.4	0.6	1	1.5	0.3		—	—	—	—
4	4.1	1.1	1.1	2.75	0.55	0.8	1.2	2	0.4	—	—	—	—
5	5.1	1.3	1.3	3.25	0.65	1.1	1.5	2.75	0.55	—	—	—	—
6	6.1	1.6	1.6	4	0.8	1.3	2	3.25	0.65	1.8	2.6	4.5	0.9
8	8.1	2.1	2.1	5.25	1.05	1.6	2.5	4	0.8	2.4	3.2	6	1.2
10	10.2	2.6	2.6	6.5	1.3	2	3	5	1	3	3.8	7.5	1.5
12	12.2	3.1	3.1	7.75	1.55	2.5	3.5	6.25	1.25	3.5	4.3	8.75	1.75
(14)	14.2	3.6	3.6	9	1.8	3	4	7.5	1.5	4.1	4.8	10.25	2.05
16	16.2	4.1	4.1	10.25	2.05	3.2	4.5	8	1.6	4.8	5.3	12	2.4
(18)	18.2	4.5	4.5	11.25	2.25	3.6	5	9	1.8	5.3	5.8	13.25	2.65
20	20.2	5	5	12.5	2.5	4	5.5	10	2	6	6.4	15	3
(22)	22.5	5.5	5.5	13.75	2.75	4.5	6	11.25	2.25	6.6	7.2	16.5	3.3
24	24.5	6	6	15	3	5	7	12.25	2.5	7.1	7.5	17.75	3.55
(27)	27.5	6.8	6.8	17	3.4	5.5	8	13.75	2.75	8	8.5	20	4
30	30.5	7.5	7.5	18.75	3.75	6	9	15	3	9	9.3	22.5	4.5
(33)	33.5	8.5	8.5	21.25	4.25	—	—	—	—	9.9	10.2	24.75	4.95
36	36.5	9	9	22.5	4.5	—	—	—	—	10.8	11.1	27	5.4
(39)	39.5	10	10	25	5	—	—	—	—	—	—	—	—
42	42.5	10.5	10.5	26.25	5.25	—	—	—	—	—	—	—	—
(45)	45.5	11	11	27.5	5.5	—	—	—	—	—	—	—	—
48	48.5	12	12	30	6	—	—	—	—	—	—	—	—

Notes 注：1. Better to use other values instead of the values in brackets. 尽可能不采用括号内的规格。

2. $m > 0$. m 应大于零。

6. Key 键

(1) Square and rectangular keyways (Quoted from GB/T 1095—2003) 平键 键槽的剖面尺寸（摘自 GB/T 1095—2003）

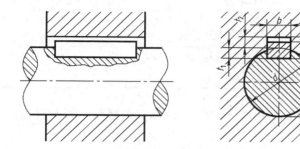
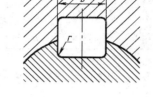

Table 2-12 Sizes of square and rectangular keyways 平键和键槽尺寸 （单位：mm）

Nominal diameter of shaft 轴的公称直径 d			From 自 6—8	>8 —10	>10 —12	>12 —17	>17 —22	>22 —30	>30 —38	>38 —44	>44 —50	>50 —58	>58 —65	>65 —75	>75 —85
Nominal size of keys 键的公称尺寸		b	2	3	4	5	6	8	10	12	14	16	18	20	22
		h	2	3	4	5	6	7	8	8	9	10	11	12	14
Keyway 键槽	Depth 深度	shaft 轴 t_1	1.2	1.8	2.5	3.0	3.5	4.0	5.0	5.0	5.5	6	7.0	7.5	9
		Hub 毂 t_2	1.0	1.4	1.8	2.3	2.8	3.3	3.3	3.3	3.8	4.3	4.4	4.9	5.4
	Radius 半径 r	max 最大	0.08			0.16				0.25				0.40	
		min 最小	0.16			0.25				0.40				0.60	

Notes 注：on engineering drawing, t_1 or $(d-t_1)$ denotes shaft slot depth, and $d+t_2$ denotes shaft hub depth.

在工作图中，轴槽深用 t_1 或 $(d-t_1)$ 标注，轴毂槽深用 $(d+t_2)$ 标注。

(2) Square and rectangular keys (Quoted from GB/T 1096—2003) 普通型 平键（摘自 GB/T 1096—2003）

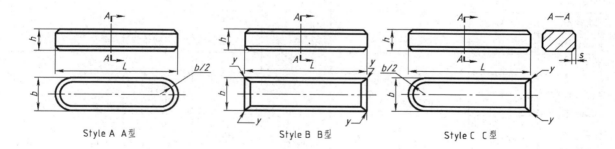

Style A A型   Style B B型   Style C C型

Mark examples

Square and rectangular key (style A), width $b=16$mm, height $h=10$mm, length $L=100$mm, is marked as:

GB/T 1096 Key 16×10×100

Square headed parallel key (style B), width $b=16$mm, height $h=10$mm, length $L=100$mm, is marked as:

GB/T 1096 Key B16×10×100

Square headed parallel key (style C), width $b=16$mm, height $h=10$mm, length $L=100$mm, is marked as:

GB/T 1096 Key C16×10×100

标记示例

宽度 $b=16$mm、高度 $h=10$mm、长度 $L=100$mm 普通A型平键的标记为：

GB/T 1096 键 16×10×100

宽度 $b=16$mm、高度 $h=10$mm、长度 $L=100$mm 普通B型平键的标记为：

GB/T 1096 键 B16×10×100

宽度 $b=16$mm、高度 $h=10$mm、长度 $L=100$mm 普通C型平键的标记为：

GB/T 1096 键 C16×10×100

Table 2-13 Sizes of square and rectangular keys 普通型平键尺寸　　　　　　　　　　（单位：mm）

b	2	3	4	5	6	8	10	12	14	16	18	20	22	25	28	32	36	40
h	2	3	4	5	6	7	8	8	9	10	11	12	14	14	16	18	20	22
s	0.16—0.25			0.25—0.40			0.40—0.60					0.60—1.0					1.0—1.2	
Length range 长度范围 L	6—20	6—36	8—45	10—56	14—70	18—90	22—110	28—140	36—160	45—180	50—200	56—220	63—250	70—280	80—320	90—360	100—400	100—400
L（Series 系列）	6，8，10，12，14，16，18，20，22，25，28，32，36，40，45，50，56，63，70，80，90，100，110，125，140，160，180，200，220，250，280，320，360，400																	

（3）Woodruff keys—Normal form（Quoted from GB/T 1099.1—2003）普通型　半圆键（摘自 GB/T 1099.1—2003）

Mark examples

Woodruff key—Normal form（width b = 6mm, height h = 10mm, Diameter D = 25mm）, is marked as:
GB/T 1099.1　Key 6×10×25

标记示例

宽度 b = 6mm、高度 h = 10mm、直径 D = 25mm 普通型半圆键标记为:
GB/T 1099.1　键 6×10×25

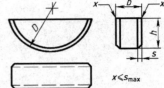

Table 2-14 Sizes of tolerance and woodruff keys—Normal form 普通型半圆键尺寸与公差

　　　（单位：mm）

Key size 键尺寸 b×h×D	Width 宽度 b		Height 高度 h		Diameter 直径 D		Chamfer or rounding 倒角或倒圆 s	
	Basic size 基本尺寸	Limiting deviation 极限偏差	Basic size 基本尺寸	Limiting deviation 极限偏差（h12）	Basic size 基本尺寸	Limiting deviation 极限偏差（h12）	Minor 最小	Major 最大
1×1.4×4	1		1.4		4	0 −0.120		
1.5×2.6×7	1.5		2.6	−0.10	7			
2×2.6×7	2		2.6		7	0 −0.150	0.16	0.25
2×3.7×10	2		3.7		10			
2.5×3.7×10	2.5		3.7	0 −0.12	10			
3×5×13	3		5		13	0 −0.180		
3×6.5×16	3		6.5		16			
4×6.5×16	4		6.5		16			
4×7.5×19	4	0 −0.025	7.5		19	0 −0.210		
5×6.5×16	5		6.5	0 −0.15	16	0 −0.180	0.25	0.40
5×7.5×19	5		7.5		19			
5×9×22	5		9		22	0 −0.210		
6×9×22	6		9		22			
6×10×25	6		10		25			
8×11×28	8		11		28		0.40	0.60
10×13×32	10		13	0 −0.18	32	0 −0.250		

7. Pin 销

（1）Parallel pins, of unhardend steel and austenitic stainless steel（Quoted from GB/T 119.1—2000）圆柱销　不淬硬钢和奥氏体不锈钢（摘自 GB/T 119.1—2000）

（2）Parallel pins, of hardend steel and martensitic stainless steel （Quoted from GB/T 119.2—2000）圆柱销　淬硬钢和马氏体不锈钢（摘自 GB/T 119.2—2000）

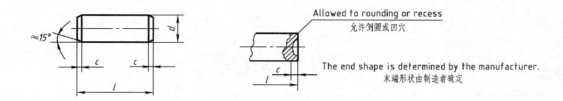

Mark examples

The parallel pin （nominal diameter $d=6$mm, tolerance m6, nominal length $l=30$mm, material is steel, without hardening and surface treatment） is marked as:

　　Pin GB/T 119.1　6m6×30

The parallel pin （nominal diameter, $d=6$mm, tolerance is m6, nominal length $l=30$mm, material is A1 austenitic stainless steel, simple surface treatment） is marked as:

　　Pin GB/T 119.1　6m6×30-A1

标记示例

公称直径 $d=6$mm、公差为 m6、公称长度 $l=30$mm、材料为钢、不经淬火、不经表面处理的圆柱销的标记：

　　销 GB/T 119.1　6m6×30

公称直径 $d=6$mm、公差为 m6、公称长度 $l=30$mm、材料为 A1 组奥氏体不锈钢、表面简单处理的圆柱销的标记：

　　销 GB/T 119.1　6m6×30-A1

Table 2-15　Sizes of parallel pins　圆柱销尺寸　　　　　（单位：mm）

d m6/h8①	0.6	0.8	1	1.2	1.5	2	2.5	3	4	5
$a \approx$	0.08	0.10	0.12	0.16	0.20	0.25	0.30	0.40	0.50	0.63
$c \approx$	0.12	0.16	0.20	0.25	0.30	0.35	0.40	0.50	0.63	0.80
l②	2—6	2—8	4—10	4—12	4—16	6—20	6—24	8—30	8—40	10—50
d m6/h8①	6	8	10	12	16	20	25	30	40	50
$a \approx$	0.80	1.0	1.2	1.6	2.0	2.5	3.0	4.0	5.0	6.3
$c \approx$	1.2	1.6	2.0	2.5	3.0	3.5	4.0	5.0	6.3	8.0
l②	12—60	14—80	18—95	22—140	26—180	35—200	50—200	60—200	80—200	95—200
l Series 系列	2, 3, 4, 6, 8, 10, 12, 14, 16, 18, 20, 22, 24, 26, 28, 30, 32, 35, 40, 45, 50, 55, 60, 65, 70, 75, 80, 85, 90, 95, 100, 120, 140, 160, 180, 200									

① Other tolerances shall be negotiated by both supplier and demender. 其他公差由供需双方协议。
② If nominal length is larger than 200mm, increment is up to 20mm. 公称长度大于 200mm，按 20mm 递增。

（3）Taper pins （Quoted from GB/T 117—2000）圆锥销（GB/T 117—2000）

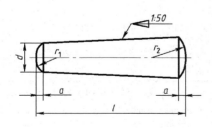

Style A (grinding): Cone surface roughness $Ra=0.8\mu m$.
A 型（磨削）：锥面表面粗糙度值 $Ra=0.8\mu m$。

Style B (cutting or cold emblem): cone surface roughness $Ra=3.2\mu m$.
B 型（切削或冷镦）：锥面表面粗糙度值 $Ra=3.2\mu m$。

$r_1 \approx d$
$r_2 \approx a/2 + d + (0.02)^2/8a$

Mark examples

The taper pin （nominal diameter $d=10$mm, nominal length $l=60$mm, material is 35 steel, heat treatment hardness 28—38 HRC, oxidization as surface treatment, style A） is marked as

　　Pin GB/T 117　10×60

标记示例

公称直径 $d=10$mm、公称长度 $l=60$mm、材料为 35 钢、热处理硬度为 28~38HRC、表面氧化处理的 A 型圆锥销的标记：

　　销 GB/T 117　10×60

Table 2-16 Sizes of the taper pins 圆锥销尺寸 （单位：mm）

d h10①	0.6	0.8	1	1.2	1.5	2	2.5	3	4	5
a ≈	0.08	0.1	0.12	0.16	0.2	0.25	0.3	0.4	0.5	0.63
l ②	4—8	5—12	6—16	6—20	8—24	10—35	10—35	12—45	14—55	18—60
d h10①	6	8	10	12	16	20	25	30	40	50
a ≈	0.8	1	1.2	1.6	2	2.5	3	4	5	6.3
l ②	22—290	22—120	26—160	32—180	40—200	45—200	50—200	55—200	60—200	65—200
l (Series 系列)	2, 3, 4, 5, 6, 8, 10, 12, 14, 16, 18, 20, 22, 24, 26, 28, 30, 32, 35, 40, 45, 50, 55, 60, 65, 70, 75, 80, 85, 90, 95, 100, 120, 140, 160, 180, 200									

① Other tolerances shall be negotiated by both supplier and demender. 其他公差由供需双方协议。

② If nominal length is larger than 200mm, increment is up to 20mm. 公称长度大于 200mm，按 20mm 递增。

（4）Split pins（Quoted from GB/T 91—2000）开口销（摘自 GB/T 91—2000）

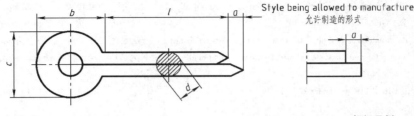

Mark examples　　　　　　　　　　　　　　　　　　　　　标记示例

The split pin (nominal diameter d=5mm, nominal length l=50mm, material is Q215 or Q235, without surface treatment) is marked as:　　Pin GB/T 91   5×50

公称直径 d=5mm、公称长度 l=50mm、材料为 Q215 或 Q235、不经表面处理的开口销的标记为：　　销 GB/T 91   5×50

Table 2-17 Sizes of cotter pins 开口销尺寸 （单位：mm）

Nominal specification 公称规格①		0.6	0.8	1	1.2	1.6	2	2.5	3.2	4	5	6.3	8	10	13	16	20
d	max	0.5	0.7	0.9	1	1.4	1.8	2.3	2.9	3.7	4.6	5.9	7.5	9.5	12.4	15.4	19.3
	min	0.4	0.6	0.8	0.9	1.3	1.7	2.1	2.7	3.5	4.4	5.7	7.3	9.3	12.1	15.1	19.0
a	max	1.6	1.6	1.6	2.50	2.50	2.50	2.50	3.2	4	4	4	4	6.3	6.3	6.3	6.3
	min	0.8	0.8	0.8	1.25	1.25	1.25	1.25	1.6	2	2	2	2	3.15	3.15	3.15	3.15
b ≈		2	2.4	3	3	3.2	4	5	6.4	8	10	12.6	16	20	26	32	40
c	max	1	1.4	1.8	2	2.8	3.6	4.6	5.8	7.4	9.2	11.8	15	19	24.8	30.8	38.5
	min	0.9	1.2	1.6	1.7	2.4	3.2	4	5.1	6.5	8.0	10.3	13.1	16.6	21.7	27.0	33.8
Applicable diameter 适用的直径②	Bolt 螺栓 >	—	2.5	3.5	4.5	5.5	7	9	11	14	20	27	39	56	80	120	170
	≤	2.5	3.5	4.5	5.5	7	9	11	14	20	27	39	56	80	120	170	—
	U bolt U形栓 >	—	2.5	3.5	4.5	5.5	7	9	11	12	17	23	29	44	69	110	160
	≤	2.5	3.5	4.5	5.5	7	9	11	14	17	23	29	44	69	110	160	—
Nominal length 商品规格范围公称长度 l		4—12	5—16	6—20	8—26	8—32	10—40	12—50	14—65	18—80	22—100	30—120	40—160	45—200	70—200	112—250	250—280
l (Series 系列)		4, 5, 6, 8, 10, 12, 14, 16, 18, 20, 22, 24, 26, 28, 30, 32, 36, 40, 45, 50, 55, 60, 65, 70, 75, 80, 85, 90, 95, 100, 120, 140, 160, 180, 200, 224, 250, 280														

① The nominal size is equal to the hole diameter of a cotter pin. The recommended tolerance on the pin hole diameter is as follows: Nominal size ≤ 1.2mm, H13; Nominal Size>1.2mm, H14. According to the agreement between both supplier and demender, the cotter pins with nominal size up to 3.6mm and 12mm are permitted to be exploited. 公称规格等于开口销的直径。对销孔直径推荐公差为：公称规格 ≤ 1.2mm, H13；公称规格 > 1.2mm, H14。根据供需双方协议，允许采公称规格为 3.6mm 和 12mm 的开口销。

② In the situations of bearing alternating stresses as in railway and U-shaped pin, cotter pins having larger grade of specifications than the following table are recommended for use. 用于铁道和在 U 形销中开口销承受交变横向力的场合，推荐使用的开口销规格应较本表规格规定的大一档。

8. Rolling bearing 滚动轴承

（1）Codes of bearings（Quoted from GB/T 272—1993）轴承类型代号（摘自 GB/T 272—1993）

Table 2-18　Types and codes of bearings　轴承的类型及代号

Code 代号	Types of bearings 轴承类型
0	Double row angular contact ball bearing 双列角接触球轴承
1	Self-aligning ball bearing 调心球轴承
2	Self-aligning roller bearing and thrust self-aligning roller bearing 调心滚子轴承和推力调心滚子轴承
3	Tapered roller bearing 圆锥滚子轴承
4	Double row deep groove ball bearing 双列深沟球轴承
5	Thrust ball bearing 推力球轴承
6	Deep groove ball bearing 深沟球轴承
7	Angular contact ball bearing 角接触球轴承
8	Thrust column roller bearing 推力圆柱滚子轴承
N	Column roller bearing, double or multiple rows represented by "NN" 圆柱滚子轴承，双列或多列用字母 NN 表示
U	External spherical surface ball bearing 外球面球轴承
QJ	Four-point contact ball bearing 四点接触球轴承

Notes 注：Different letters or numbers placed in prefixes or postfixes of codes represent specified structures of corresponding bearings. 在表中代号后或前加字母或数字表示该类轴承中的不同结构。

（2）Deep groove ball bearings（Quoted from GB/T 276—2013）深沟球轴承（摘自 GB/T 276—2013）

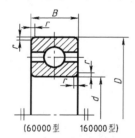

（60000型　160000型）

Mark examples　　　　　标记示例

Rolling bearing　6412　GB/T 276—2013　　　滚动轴承 6412　GB/T 276—2013

Table 2-19　Boundary dimensions of deep groove ball bearings　深沟球轴承外形尺寸

00 Series 系列					04 Series 系列						
Code of bearing 轴承代号	Boundary dimensions 外形尺寸/mm				Code of bearing 轴承代号		Boundary dimensions 外形尺寸/mm				
	d	D	B	r_{min}	6000 型	6000N 型	d	D	B	r_{smin}	r_{1smin}
16001	12	28	7	0.3	6403	6403N	17	62	17	1.1	0.5
16002	15	32	8	0.3	6404	6404N	20	72	19	1.1	0.5
16003	17	35	8	0.3	6405	6405N	25	80	21	1.5	0.5
16004	20	42	8	0.3	6406	6406N	30	90	23	1.5	0.5
16005	25	47	8	0.3	6407	6407N	35	100	25	1.5	0.5
16006	30	55	9	0.3	6408	6408N	40	110	27	2	0.5
16007	35	62	9	0.3	6409	6409N	45	120	29	2	0.5
16008	40	68	9	0.3	6410	6410N	50	130	31	2.1	0.5
16009	45	75	10	0.6	6411	6411N	55	140	33	2.1	0.5
16010	50	80	10	0.6	6412	6412N	60	150	35	2.1	0.5
16011	55	90	11	0.6	6413	6413N	65	160	37	2.1	0.5
16012	60	95	11	0.6	6414	6414N	70	180	42	3	0.5
16013	65	100	11	0.6	6415	6415N	75	190	45	3	0.5
16014	70	110	13	0.6	6416	6416N	80	200	48	3	0.5
16015	75	115	13	0.6	6417	6417N	85	210	52	4	0.5
16016	80	125	14	0.6	6418	6418N	90	225	54	4	0.5
16017	85	130	14	0.6	6419	6419N	95	240	55	4	0.5
16018	90	140	16	1	6420	6420N	100	250	58	4	0.5
16019	95	145	16	1	6422		110	280	65	4	—
16020	100	150	16	1							

(Contiunued 续)

17 Series 系列					37 Series 系列				
Code of bearing 轴承代号	Boundary dimensions 外形尺寸/mm				Code of bearing 轴承代号	Boundary dimensions 外形尺寸/mm			
	d	D	B	r_{smin}		d	D	B	r_{smin}
617/0.6	0.6	2	0.8	0.05	637/1.5	1.5	3	1.8	0.05
617/1	1	2.5	1	0.05	637/2	2	4	2	0.05
617/1.5	1.5	3	1	0.05	637/2.5	2.5	5	2.3	0.08
617/2	2	4	1.2	0.05	637/3	3	6	3	0.08
617/2.5	2.5	5	1.5	0.08	637/4	4	7	3	0.08
617/3	3	6	2	0.08	637/5	5	8	3	0.08
617/4	4	7	2	0.08	637/6	6	10	3.5	0.1
617/5	5	8	2	0.08	637/7	7	11	3.5	0.1
617/6	6	10	2.5	0.1	637/8	8	12	3.5	0.1
617/7	7	11	2.5	0.1	637/9	9	14	4.5	0.1
617/8	8	12	2.5	0.1	63700	10	15	4.5	0.1
617/9	9	14	3	0.1					
61700	10	15	3	0.1					

（3）Tapered roller bearings（Quoted from GB/T 297—2015）圆锥滚子轴承（摘自 GB/T 297—2015）

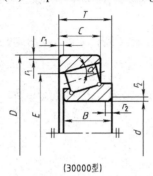

(30000型)

Mark examples　　　　　　　　　　　标记示例
Rolling bearing　30205 GB/T 297—2015　　滚动轴承　30205 GB/T 297—2015

Table 2-20　Boundary dimensions of tapered roller bearings　圆锥滚子轴承外形尺寸

02 Series 系列									
Code of bearing 轴承代号	Boundary dimensions 外形尺寸/mm								
	d	D	T	B	r_{smin}	C	r_{1smin}	a	E
30202	15	35	11.75	11	0.6	10	0.6	—	—
30203	17	40	13.25	12	1	11	1	12°57′10″	31.408
30204	20	47	15.25	14	1	12	1	12°57′10″	37.304
30205	25	52	16.25	15	1	13	1	14°02′10″	41.135
30206	30	62	17.25	16	1	14	1	14°02′10″	49.990
302/32	32	65	18.25	17	1	15	1	14°	52.500
30207	35	72	18.25	17	1.5	15	1.5	14°02′10″	58.844
30208	40	80	19.75	18	1.5	16	1.5	14°02′10″	65.730
30209	45	85	20.75	19	1.5	16	1.5	15°06′34″	70.440
30210	50	90	21.75	20	1.5	17	1.5	15°38′32″	75.78
30211	55	100	22.75	21	2	18	1.5	15°06′34″	84.197
30212	60	110	23.75	22	2	19	1.5	15°06′34″	91.876
30213	65	120	24.75	23	2	20	1.5	15°06′34″	101.934
30214	70	125	26.25	24	2	21	1.5	15°38′32″	105.748
30215	75	130	27.25	25	2	22	1.5	16°10′20″	110.408
30216	80	140	28.25	26	2.5	22	2	15°38′32″	119.169
30217	85	150	30.5	28	2.5	24	2	15°38′32″	126.685
30218	90	160	32.5	30	2.5	26	2	15°38′32″	134.901
30219	95	170	34.5	32	3	27	2.5	15°38′32″	143.385
30220	100	180	37	34	3	29	2.5	15°38′32″	151.310
30221	105	190	39	36	3	30	2.5	15°38′32″	159.795
30222	110	200	41	38	3	32	2.5	15°38′32″	168.548
30224	120	215	43.5	40	3	34	2.5	16°10′20″	151.310

（4）Thrust ball bearings（Quoted from GB/T 301—2015）推力球轴承（摘自 GB/T 301—2015）

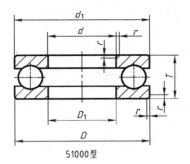

51000型

Mark examples
Rolling bearing 51210 GB/T 301—2015

标记示例
滚动轴承 51210 GB/T 301—2015

Table 2-21　Boundary dimensions of thrust ball bearings　推力球轴承外形尺寸

Code of bearing 轴承代号	11 Series 系列 Boundary dimensions 外形尺寸/mm						Code of bearing 轴承代号	12 Series 系列 Boundary dimensions 外形尺寸/mm					
	d	D	T	d_{1smin}	D_{1smin}	r_{smin}		d	D	T	d_{1smin}	D_{1smin}	r_{smin}
51100	10	24	9	11	24	0.6	51200	10	26	11	12	26	0.6
51101	12	26	9	13	26	0.6	51201	12	28	11	14	28	0.6
51102	15	28	9	16	28	0.6	51202	15	32	12	17	32	0.6
51103	17	32	9	18	30	0.6	51203	17	35	12	19	35	0.6
51104	20	35	10	21	35	0.6	51204	20	40	14	22	40	0.6
51105	25	40	11	26	42	0.6	51205	25	47	15	27	47	0.6
51106	30	47	11	32	47	0.6	51206	30	52	16	32	52	0.6
51107	35	52	12	37	52	1	51207	35	62	18	37	62	1
51108	40	62	13	42	60	1	51208	40	68	19	42	68	1
51109	45	65	14	47	65	1	51209	45	73	20	47	73	1
51110	50	70	14	52	70	1	51210	50	78	22	52	78	1
51111	55	78	16	57	78	1	51211	55	90	25	52	90	1
51112	60	85	17	62	85	1	51212	60	95	26	57	95	1
51113	65	90	18	67	90	1	51213	65	100	27	62	100	1
51114	70	95	18	72	95	1	51214	70	105	27	67	105	1
51115	75	100	19	77	100	1	51215	75	110	27	77	110	1
51116	80	105	19	82	105	1	51216	80	115	28	82	115	1
51117	85	115	19	87	110	1	51217	85	125	31	88	125	1
51118	90	120	22	92	120	1.1	51218	90	135	35	93	135	1.1
51120	100	135	25	102	135	1.1	51220	100	150	38	103	150	1.1
51122	110	145	25	112	145	1	51222	110	160	38	113	160	1.1
51124	120	155	25	122	155	1	51224	120	170	39	123	170	1.1
51126	130	170	30	132	170	1	51226	130	190	45	133	187	1.5
51128	140	180	31	142	178	1	51228	140	200	46	143	197	1.5
51130	150	190	31	152	188	1	51230	150	215	50	153	212	1.5
51132	160	200	31	162	198	1	51232	160	225	51	163	222	1.5
51134	170	215	34	172	213	1.1	51234	170	240	55	173	237	1.5
51136	180	225	34	183	222	1.1	51236	180	250	56	183	247	1.5
51138	190	240	37	193	237	1.1	51238	190	270	62	194	267	2
51140	200	250	37	203	247	1.1	51240	200	280	62	204	277	2
51144	220	270	37	223	267	1.1	51244	220	300	63	224	297	2
51148	240	300	45	243	297	1.5	51248	240	340	78	244	335	2.1
51152	260	320	45	263	317	1.5	51252	260	360	79	264	355	2.1
51156	280	350	53	283	347	1.5	51256	280	380	80	284	375	2.1
51160	300	380	62	304	376	2	51260	300	420	95	304	415	3
51164	320	400	63	324	396	2	51264	320	440	95	325	435	3
51168	340	420	64	344	416	2	51268	340	460	96	345	455	3
51172	360	440	65	364	436	2	51272	360	500	110	365	495	4
51176	380	460	65	384	456	2							
51180	400	480	65	404	476	2							
51184	420	500	65	424	495	2							
51188	440	540	80	444	535	2.1							
51192	460	560	80	464	555	2.1							
51196	480	580	80	484	575	2.1							
511/500	500	600	80	504	595	2.1							

(Contiunued 续)

13 Series 系列							14 Series 系列						
Codes of bearings 轴承代号	Boundary dimensions 外形尺寸/mm						Codes of bearings 轴承代号	Boundary dimensions 外形尺寸/mm					
	d	D	T	d_{1smin}	D_{1smin}	r_{smin}		d	D	T	d_{1smin}	D_{1smin}	r_{smin}
51304	20	47	18	22	47	1	51405	25	60	24	27	60	1
51305	25	52	18	27	52	1	51406	30	70	28	32	70	1
51306	30	60	21	32	60	1	51407	35	80	32	37	80	1.1
51307	35	68	24	37	68	1	51408	40	90	36	42	90	1.1
51308	40	78	26	42	78	1	51409	45	100	39	47	100	1.1
51310	50	95	31	52	95	1.1	51410	50	110	43	52	110	1.5
51311	55	105	35	57	105	1.1	51411	55	120	48	57	120	1.5
51312	60	110	35	62	110	1.1	51412	60	130	51	62	130	1.5
51313	65	115	36	67	115	1.1	51413	65	140	56	68	140	2
51314	70	125	40	72	125	1.1	51414	70	150	60	73	150	2
31315	75	135	44	77	135	1.5	51415	75	160	65	78	160	2
51316	80	140	44	82	140	1.5	51416	80	170	68	83	170	2.1
51317	85	150	49	88	150	1.5	51417	85	180	72	88	177	2.1
51318	90	155	50	93	155	1.5	51418	90	190	77	93	187	2.1
51320	100	170	55	103	170	1.5	51420	100	210	85	103	205	3
51322	110	190	63	113	187	2	51422	110	230	95	113	225	3
51324	120	210	70	123	205	2.1	51424	120	250	102	123	245	4
51326	130	225	75	134	220	2.1	51426	130	270	110	134	265	4
51328	140	240	80	144	235	2.1	51428	140	280	112	144	275	4
51330	150	250	80	154	245	2.1	51430	150	300	120	154	295	4
51332	160	270	87	164	265	3							
51334	170	280	87	174	275	3							
51336	180	300	95	184	295	3							
51338	190	320	105	195	315	4							
51340	200	340	110	205	335	4							
51344	220	360	112	225	355	4							
51348	240	380	112	145	375	4							

Appendix 3　Commonly Used Elements of Parts　附录3　常用的零件结构要素

1. Through hole and countersink hole 通孔与沉孔

（1）Clearance holes for bolts and screws (Qoted from GB/T 5277—1985) 螺栓和螺钉通孔（摘自 GB/T 5277—1985）

（2）Countersinks for countersink head screws (Quoted from GB/T 152.2—2014) 沉头用沉孔（摘自 GB/T 152.2—2014）

（3）Counterbores for hexagon socket head serews and slotted cheese head screws (Quoted from GB/T 152.3—1988) 圆柱头用沉孔（摘自 GB/T 152.3—1988）

（4）Counterbores for hexagon bolts and nuts (Quoted from GB/T 152.4—1988) 六角头螺栓和六角螺母用沉孔（摘自 GB/T 152.4—1988）

Table 3-1　Sizes of through holes and countersink holes for fasteners　紧固件通孔及沉孔尺寸

（单位：mm）

Screw threads specification 螺纹规格			M4	M5	M6	M8	M10	M12	M16	M20	M24	M30	M36
Through hole diameter 通孔直径 (GB/T 5277—1985)	d_h	Fine fit 精装配	4.3	5.3	6.4	8.4	10.5	13	17	21	25	31	37
		Medium fit 中等装配	4.5	5.5	6.6	9	11	13.5	17.5	22	26	33	39
		Coarse fit 粗装配	4.8	5.8	7	10	12	14.5	18.5	24	28	35	42
Countersinks for countersink head screws 沉头用沉孔 (GB/T 152.2—2014)	$t \approx$		2.55	2.68	3.13	4.28	4.65	—	—	—	—	—	—
	d_1		4.5	5.5	6.6	9.0	11.0	—	—	—	—	—	—
	d_2		9.6	10.65	12.85	17.55	20.3	—	—	—	—	—	—
Counterbores for hexagon socket head serews and slotted cheese head screws 圆柱头用沉孔 (GB/T 152.3—1988)	d_1		4.5	5.5	6.6	9.0	11.0	13.5	17.5	22.0	—	—	—
	d_2		8	10.0	11.0	15.0	18.0	20.0	26.0	33.0	40.0	48	57
	d_3		—	—	—	—	—	16	20	24	28	36	42
	t	①	4.6	5.7	6.8	9	11	13	17.5	21.5	25.5	32	38
		②	3.2	4	4.7	6	7	8	10.5	—	—	—	—
Counterbores for hexagon bolts and nuts 六角头螺栓和六角螺母用沉孔 (GB/T 152.4—1988)	d_2		10	11	13	18	22	26	33	40	48	61	71
	d_3		—	—	—	—	—	16	20	24	28	36	42
	d_1		4.5	5.5	6.6	9.0	11.0	13.5	17.5	22	26	33	39

2. Rounding and chamfer of parts (Quoted from GB/T 6403.4—2008) 零件倒角与倒圆（摘自 GB/T 6403.4—2008）

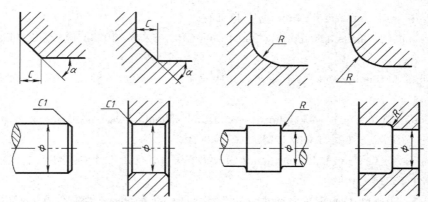

Table 3-2 The recommend parameters of the chamfer C, fillet rounding R corresponding to diameter ϕ
与直径 ϕ 相应的倒角 C、倒圆 R 的推荐值　　　　　　　　　　　　（单位：mm）

Diameter 直径 ϕ	>3	>3—6	>6—10	>10—18	>18—30	>30—50	>50—80	>80—120	>120—180	>120—250	>250—320	>320—400
C or 或 R	0.2	0.4	0.6	0.8	1.0	1.6	2.0	2.5	3.0	4.0	5.0	6.0

3. Relief grooves for grinding wheels (Quoted from GB/T 6403.5—2008) 砂轮越程槽（摘自 GB/T 6403.5—2008）

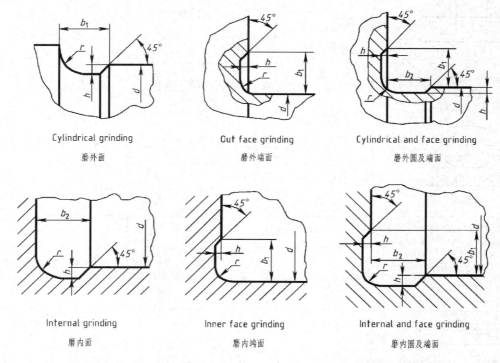

Table 3-3 Sizes of relief grooves for grinding wheels　砂轮越程槽尺寸　　　　　　（单位：mm）

b_1	0.6	1.0	1.6	2.0	3.0	4.0	5.0	8.0	10
b_2	2.0		3.0		4.0		5.0	8.0	10
h	0.1		0.2		0.3	0.4		0.6	0.8	1.2
r	0.2		0.5		0.8	1.0		1.6	2.0	3.0
d		—10			>10—50		>50—100		>100

Notes 注：1. The intersection of relief groove and line shall avoid causing sharp corners. 越程槽内与直线相交处，不允许产生尖角。

2. Relief groove depth h and radius r are required to satisfy $r \leqslant 3h$. 越程槽深度 h 与圆弧半径 r 要满足 $r \leqslant 3h$。

4. Run-outs, undercuts and chamfers for general purpose metric screw threads (Quoted from GB/T 3—1997) 普通螺纹收尾、肩距、退刀槽和倒角（摘自 GB/T 3—1997）

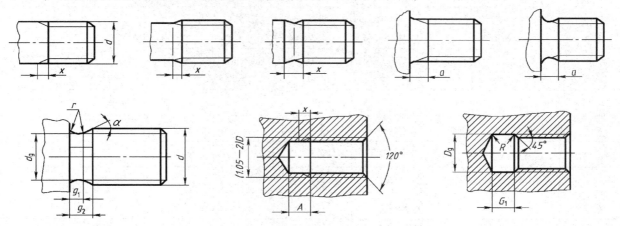

Table 3-4 Sizes of run-outs, undercuts and chamfers for general purpose metric screw threads
普通螺纹收尾、肩距、退刀槽和倒角尺寸 （单位：mm）

Pitch 螺距 P	Run-outs of external thread 外螺纹收尾和肩距					Undercuts of external thread 外螺纹退刀槽			
	Run-outs 收尾 x max		Shoulder distance 肩距 a max			g_2 max	g_1 max	d_g	r ≈
	Commonly 一般	Short 短的	Commonly 一般	Long 长的	Short 短的				
0.25	0.9	0.45	1.05	1.4	0.7	0.75	0.4	$d-0.4$	0.12
0.3	0.9	0.45	1.05	1.4	0.7	0.9	0.5	$d-0.5$	0.16
0.35	0.9	0.45	1.05	1.4	0.7	1.05	0.6	$d-0.6$	0.16
0.4	1	0.5	1.2	1.6	0.8	1.2	0.6	$d-0.7$	0.2
0.45	1.1	0.6	1.35	1.8	0.9	1.35	0.7	$d-0.7$	0.2
0.5	1.25	0.7	1.5	2	1	1.5	0.8	$d-0.8$	0.2
0.6	1.5	0.75	1.8	2.4	1.2	1.8	0.9	$d-1$	0.4
0.7	1.75	0.9	2.1	2.8	1.4	2.1	1.1	$d-1.1$	0.4
0.75	1.9	1	2.25	3	1.5	2.25	1.2	$d-1.2$	0.4
0.8	2	1	2.4	3.2	1.6	2.4	1.3	$d-1.3$	0.4
1	2.5	1.25	3	4	2	3	1.6	$d-1.6$	0.6
1.25	3.2	1.6	4	5	2.5	3.75	2	$d-2$	0.6
1.5	3.8	1.9	4.5	6	3	4.5	2.5	$d-2.3$	0.8
1.75	4.3	2.2	5.3	7	3.5	5.25	3	$d-2.6$	1
2	5	2.5	6	8	4	6	3.4	$d-3$	1
2.5	6.3	3.2	7.5	10	5	7.5	4.4	$d-3.6$	1.2
3	7.5	3.8	9	12	6	9	5.2	$d-4.4$	1.6
3.5	9	4.5	10.5	14	7	10.5	6.2	$d-5$	1.6
4	10	5	12	16	8	12	7	$d-5.7$	2
4.5	11	5.5	13.5	18	9	13.5	8	$d-6.4$	2.5
5	12.5	6.3	15	20	10	15	9	$d-7$	2.5
5.5	14	7	16.5	22	11	17.5	11	$d-7.7$	3.2
6	15	7.5	18	24	12	18	11	$d-8.3$	3.2
Reference 参考值	≈$2.5P$	≈$1.25P$	≈$3P$	=$4P$	$2P$	≈$3P$	—		—

(Continued 续)

Pitch 螺距 P	Run-outs of internal thread 内螺纹收尾和肩距				Undercuts of internal thread 内螺纹退刀槽			
	Run-outs 收尾 x max		Shoulder distance 肩距 a max		G_1		D_g	R ≈
	Commonly 一般	Short 短的	Commonly 一般	Long 长的	Commonly 一般	Short 短的		
0.25	1	0.5	1.5	2				
0.3	1.2	0.6	1.8	2.4				
0.35	1.4	0.7	2.2	2.8				
0.4	1.6	0.8	2.5	3.2				
0.45	1.8	0.9	2.8	3.6				
0.5	2	1	3	4	2	1		0.2
0.6	2.4	1.2	3.2	4.8	2.4	1.2		0.4
0.7	2.8	1.4	3.5	5.6	2.8	1.4	$D+0.3$	0.4
0.75	3	1.5	3.8	6	3	1.5		0.3
0.8	3.2	1.6	4	6.4	3.2	1.6		0.4
1	4	2	5	8	4	2		0.4
1.25	5	2.5	6	10	5	2.5		0.4
1.5	6	3	7	12	6	3		0.6
1.75	7	3.5	9	14	7	3.5		0.8
2	8	4	10	16	8	4		0.9
2.5	10	5	12	18	10	5		1
3	12	6	14	22	12	6	$D+0.5$	1.2
3.5	14	7	16	24	14	7		1.5
4	16	8	18	26	16	8		2
4.5	18	5.5	21	29	18	9		2.2
5	20	6.3	23	32	20	10		2.5
5.5	22	7	25	35	22	11		2.8
6	24	7.5	28	38	24	12		3
Reference 参考值	$=4P$	$=2P$	$≈6—5P$	$≈8—6.5P$	$=4P$	$=2P$	—	$≈0.5P$

Notes 注: "Normal" ending and shoulder-length distance are prioritizing to use; "short" ending and the "short" shoulder are only applied to thread pieces which structure is restricted; "long" shoulder distance can be applied to thread fasteners which qualified production grade B or C. 应优先选用"一般"长度的收尾和肩距;"短"收尾和"短"肩距仅用于结构受限制的螺纹件上;产品等级为 B 或 C 级的螺纹紧固件可采用"长"肩距。

Appendix 4　Surface Texture Parameter　附录4　表面结构参数

Table 4-1　Contour arithmetical mean deviation Ra　轮廓算术平均偏差 Ra 的数值（单位：μm）

1st Series 第1系列	2nd Series 第2系列	1st Series 第1系列	2nd Series 第2系列	1st Series 第1系列	2nd Series 第2系列	1st Series 第1系列	2nd Series 第2系列
	0.008						
	0.010						
0.012			0.125		1.25	12.5	
	0.016		0.16	1.60			16.0
	0.020	0.20			2.0		20
0.025			0.25	2.5		25	
	0.032		0.32	3.2			32
	0.040	0.40			4.0		40
0.050			0.50	5.0		50	
	0.063		0.63	6.3			63
0.100	0.080				8.0		80
		0.80	1.00		10.0	100	

Appendix 5　Limits and Fits　附录5　极限与配合

1. Standard tolerance grades（Quoted from GB/T 1800.2—2009）标准公差等级（摘自 GB/T 1800.2—2009）

Table 5-1　Standard tolerance values　标准公差数值

Nominal size 公称尺寸/mm		Standard tolerance grades 标准公差等级																
Over 大于	To 至	IT1	IT2	IT3	IT4	IT5	IT6	IT7	IT8	IT9	IT10	IT11	IT12	IT13	IT14	IT15	IT16	IT17	IT18
		μm											mm						
—	3	0.8	1.2	2	3	4	6	10	14	25	40	60	0.1	0.14	0.25	0.4	0.6	1	1.4
3	6	1	1.5	2.5	4	5	8	12	18	30	48	75	0.12	0.18	0.3	0.48	0.75	1.2	1.8
6	10	1	1.5	2.5	4	6	9	15	22	36	58	90	0.15	0.22	0.36	0.58	0.9	1.5	2.2
10	18	1.2	2	3	5	8	11	18	27	43	70	110	0.18	0.27	0.43	0.7	1.1	1.8	2.7
18	30	1.5	2.5	4	6	9	13	21	33	52	84	130	0.21	0.33	0.52	0.84	1.3	2.1	3.3
30	50	1.5	2.5	4	7	11	16	25	39	62	100	160	0.25	0.39	0.62	1	1.6	2.5	3.9
50	80	2	3	5	8	13	19	30	46	74	120	190	0.3	0.46	0.74	1.2	1.9	3	4.6
80	120	2.5	4	6	10	15	22	35	54	87	140	220	0.35	0.54	0.87	1.4	2.2	3.5	5.4
120	180	3.5	5	8	12	18	25	40	63	100	160	250	0.4	0.63	1	1.6	2.5	4	6.3
180	250	4.5	7	10	14	20	29	46	72	115	185	290	0.46	0.72	1.15	1.85	2.9	4.6	7.2
250	315	6	8	12	16	23	32	52	81	130	210	320	0.52	0.81	1.3	2.1	3.2	5.2	8.1
315	400	7	9	13	18	25	36	57	89	140	230	360	0.57	0.89	1.4	2.3	3.6	5.7	8.9
400	500	8	10	15	20	27	40	63	97	155	250	400	0.63	0.97	1.55	2.5	4	6.3	9.7
500	630	9	11	16	22	32	44	70	110	175	280	440	0.7	1.1	1.75	2.8	4.4	7	11
630	800	10	13	18	25	36	50	80	125	200	320	500	0.8	1.25	2	3.2	5	8	12.5
800	1000	11	15	21	28	40	56	90	140	230	360	560	0.9	1.4	2.3	3.6	5.6	9	14
1000	1250	13	18	24	33	47	66	105	165	260	420	660	1.05	1.65	2.6	4.2	6.6	10.5	16.5
1250	1600	15	21	29	39	55	78	125	195	310	500	780	1.25	1.95	3.1	5	7.8	12.5	19.5
1600	2000	18	25	35	46	65	92	150	230	370	600	920	1.5	2.3	3.7	6	9.2	15	23
2000	2500	22	30	41	55	78	110	175	280	440	700	1100	1.75	2.8	4.4	7	11	17.5	28
2500	3150	26	36	50	68	96	135	210	330	540	860	1350	2.1	3.3	5.4	8.6	13.51	21	33

Notes 注：1. When the nominal size>500mm and for grades from IT1 to IT5, the standard tolerances are for trial only. 公称尺寸大于500mm的 IT1~IT5 的标准公差数值为试行。

2. There are no standard tolerances from IT14 to IT18 when the nominal size<1mm. 公称尺寸小于1mm时，无 IT14~IT18。

2. Limit deviations for shafts（Quoted from GB/T 1800.2—2009）轴的极限偏差（摘自 GB/T 1800.2—2009）

Table 5-2 Limit deviations for shafts 轴的极限偏差数值　　　　　　　　　　（单位：μm）

Nominal size 公称尺寸/mm		Ordinary tolerance zone 常用公差带												
		a	b		c			d				e		
Over 大于	To 至	11	11	12	9	10	11	8	9	10	11	7	8	9
—	3	−270 −300	−140 −200	−140 −240	−60 −85	−60 −100	−60 −120	−20 −34	−20 −45	−20 −60	−20 −80	−14 −24	−14 −28	−14 −39
3	6	−270 −345	−140 −215	−140 −260	−70 −100	−70 −118	−70 −145	−30 −48	−30 −60	−30 −78	−30 −105	−20 −32	−20 −38	−20 −50
6	10	−280 −370	−150 −240	−150 −300	−80 −116	−80 −138	−80 −170	−40 −62	−40 −76	−40 −98	−40 −130	−25 −40	−25 −47	−25 −61
10	18	−290 −400	−150 −260	−150 −330	−95 −138	−95 −165	−95 −205	−50 −77	−50 −93	−50 −120	−50 −160	−32 −50	−32 −59	−32 −75
18	30	−300 −430	−160 −290	−160 −370	−110 −162	−110 −194	−110 −240	−65 −98	−65 −117	−65 −149	−65 −195	−40 −61	−40 −73	−40 −92
30	40	−310 −470	−170 −330	−170 −420	−120 −182	−120 −220	−120 −280	−80 −119	−80 −142	−80 −180	−80 −240	−50 −75	−50 −89	−50 −112
40	50	−320 −480	−180 −340	−180 −430	−130 −192	−130 −230	−130 −290							
50	65	−340 −530	−190 −380	−190 −490	−140 −214	−140 −260	−140 −330	−100 −146	−100 −174	−100 −220	−100 −290	−60 −90	−60 −106	−60 −134
65	80	−360 −550	−200 −390	−200 −500	−150 −224	−150 −270	−150 −340							
80	100	−380 −600	−220 −440	−220 −570	−170 −257	−170 −310	−170 −390	−120 −174	−120 −207	−120 −260	−120 −340	−72 −107	−72 −126	−72 −159
100	120	−410 −630	−240 −460	−240 −590	−180 −267	−180 −320	−180 −400							
120	140	−460 −710	−260 −510	−260 −660	−200 −300	−200 −360	−200 −450	−145 −208	−145 −245	−145 −305	−145 −395	−85 −125	−85 −148	−85 −185
140	160	−520 −770	−280 −530	−280 −680	−210 −310	−210 −370	−210 −460							
160	180	−580 −830	−310 −560	−310 −710	−230 −330	−230 −390	−230 −480							
180	200	−660 −950	−340 −630	−340 −800	−240 −355	−240 −425	−240 −530	−170 −242	−170 −285	−170 −355	−170 −460	−100 −146	−100 −172	−100 −215
200	225	−740 −1030	−380 −670	−380 −840	−260 −375	−260 −445	−260 −550							
225	250	−820 −1110	−420 −710	−420 −880	−280 −395	−280 −465	−280 −570							
250	280	−920 −1240	−480 −800	−480 −1000	−300 −430	−300 −510	−300 −620	−190 −271	−190 −320	−190 −400	−190 −510	−110 −162	−110 −191	−110 −240
280	315	−1050 −1370	−540 −860	−540 −1060	−330 −460	−330 −540	−330 −650							
315	355	−1200 −1560	−600 −960	−600 −1170	−360 −500	−360 −590	−360 −720	−210 −299	−210 −350	−210 −440	−210 −570	−125 −182	−125 −214	−125 −265
355	400	−1350 −1710	−680 −1040	−680 −1250	−400 −540	−400 −630	−400 −760							
400	450	−1500 −1900	−760 −1160	−760 −1390	−440 −595	−440 −690	−440 −840	−230 −327	−230 −385	−230 −480	−230 −630	−185 −198	−135 −232	−135 −250
450	500	−1650 −2050	−840 −1240	−840 −1470	−480 −635	−480 −730	−480 −880							

(Continued 续)

Nominal size 公称尺寸/mm		Ordinary tolerance zone 常用公差带														
		f					g			h							
Over 大于	To 至	5	6	7	8	9	5	6	7	5	6	7	8	9	10	11	12
—	3	−6 −10	−6 −12	−6 −16	−6 −20	−6 −31	−2 −6	−2 −8	−2 −12	0 −4	0 −6	0 −10	0 −14	0 −25	0 −40	0 −60	0 −100
3	6	−10 −15	−10 −18	−10 −22	−10 −28	−10 −40	−4 −9	−4 −12	−4 −16	0 −5	0 −8	0 −12	0 −18	0 −30	0 −48	0 −75	0 −120
6	10	−13 −19	−13 −22	−13 −28	−13 −35	−13 −49	−5 −11	−5 −14	−5 −20	0 −6	0 −9	0 −15	0 −22	0 −36	0 −58	0 −90	0 −150
10	18	−16 −24	−16 −27	−16 −34	−16 −43	−16 −59	−6 −14	−6 −17	−6 −24	0 −8	0 −11	0 −18	0 −27	0 −43	0 −70	0 −110	0 −180
18	30	−20 −29	−20 −33	−20 −41	−20 −53	−20 −72	−7 −16	−7 −20	−7 −28	0 −9	0 −13	0 −21	0 −33	0 −52	0 −84	0 −130	0 −210
30	50	−25 −36	−25 −41	−25 −50	−25 −64	−25 −87	−9 −20	−9 −25	−9 −34	0 −11	0 −16	0 −25	0 −39	0 −62	0 −100	0 −160	0 −300
50	80	−30 −43	−30 −49	−30 −60	−30 −76	−30 −104	−10 −23	−10 −29	−10 −40	0 −13	0 −19	0 −30	0 −46	0 −74	0 −120	0 −190	0 −300
80	120	−36 −51	−36 −58	−36 −71	−36 −90	−36 −123	−12 −27	−12 −34	−12 −47	0 −15	0 −22	0 −35	0 −54	0 −87	0 −140	0 −220	0 −350
120	180	−43 −61	−43 −68	−43 −83	−43 −106	−43 −143	−14 −32	−14 −39	−14 −54	0 −18	0 −25	0 −40	0 −63	0 −100	0 −160	0 −250	0 −400
180	250	−50 −70	−50 −79	−50 −96	−50 −122	−50 −165	−15 −35	−15 −44	−15 −61	0 −20	0 −29	0 −46	0 −72	0 −115	0 −185	0 −290	0 −460
250	315	−56 −79	−56 −88	−56 −108	−56 −137	−56 −186	−17 −40	−17 −49	−17 −69	0 −23	0 −32	0 −52	0 −81	0 −130	0 −210	0 −320	0 −520
315	400	−62 −87	−62 −98	−62 −119	−62 −151	−62 −202	−18 −43	−18 −54	−18 −75	0 −25	0 −36	0 −57	0 −89	0 −140	0 −230	0 −360	0 −570
400	500	−68 −95	−68 −108	−68 −131	−68 −165	−68 −223	−20 −47	−20 −60	−20 −83	0 −27	0 −40	0 −63	0 −97	0 −155	0 −250	0 −400	0 −630

(Continued 续)

Nominal size 公称尺寸/mm		Ordinary tolerance zone 常用公差带												
		js			k			m			n			p		
Over 大于	To 至	5	6	7	5	6	7	5	6	7	5	6	7	5	6	7
—	3	±2	±3	±5	+4 0	+6 +0	+10 +0	+6 +2	+8 +2	+12 +2	+8 +4	+10 +4	+14 +4	+10 +6	+12 +6	+16 +6
3	6	±2.5	±4	±6	+6 +1	+9 +1	+13 +1	+9 +4	+12 +4	+16 +4	+13 +8	+16 +8	+20 +8	+17 12	+20 +12	+24 +12
6	10	±3	±4.5	±7	+7 +1	+10 +1	+16 +1	+12 +6	+15 +6	+21 +6	+16 +10	+19 +10	+25 +10	+21 +15	+24 +15	+30 +15
10	18	±4	±5.5	±9	+9 +1	+12 +1	+19 +1	+15 +7	+18 +7	+25 +7	+20 +12	+23 +12	+30 +12	+26 +18	+29 +18	+36 +18
18	30	±4.5	±6.5	±10	+11 +2	+15 +2	+23 +2	+17 +8	+21 +8	+29 +8	+24 +15	+28 +15	+36 +15	+31 +22	+35 +22	+43 +22
30	50	±5.5	±8	±12	+13 +2	+18 +2	+27 +2	+20 +9	+25 +9	+34 +9	+28 +17	+33 +17	+42 +17	+37 +26	+42 +26	+51 +26
50	80	±6.5	±9.5	±15	+15 +2	+21 +2	+32 +2	+24 +11	+30 +11	+41 +11	+33 +20	+39 +20	+52 +20	+45 +32	+51 +32	+62 +32
80	120	±7.5	±11	±17	+18 +3	+25 +3	+38 +3	+28 +13	+35 +13	+48 +13	+38 +23	+45 +23	+58 +23	+52 +37	+59 +37	+72 +37
120	180	±9	±12.5	±20	+21 +3	+28 +3	+43 +3	+33 +15	+40 +15	+55 +15	+45 +27	+52 +27	+67 +27	+61 +43	+68 +43	+83 +43
180	250	±10	±14.5	±23	+24 +4	+33 +4	+50 +4	+37 +17	+46 +17	+63 +17	+51 +31	+60 +31	+77 +31	+70 50	+79 +50	+96 +50
250	315	±11.5	±16	±26	+27 +4	+36 +4	+56 +4	+43 +20	+52 +20	+72 +20	+57 +34	+66 +34	+86 +34	+79 +56	+88 +56	+108 +56
315	400	±12.5	±18	±28	+29 +4	+40 +4	+61 +4	+46 +21	+57 +21	+78 +21	+62 +37	+73 +37	+94 +37	+87 +62	+98 62	+119 +62
400	500	±13.5	±20	±31	+32 +5	+45 +5	+68 +5	+50 +23	+63 +23	+86 +23	+67 +40	+80 +40	+103 +40	+95 +68	+108 68	+148 +68

(Continued 续)

Nominal size 公称尺寸/mm		Ordinary tolerance zone 常用公差带													
		r			s			t			u		v	x	y	z
Over 大于	To 至	5	6	7	5	6	7	5	6	7	6	7	6	6	6	6
—	3	+14 +10	+16 +10	+20 +10	+18 +14	+20 +14	+24 +14	—	—	—	+24 +18	+28 +18	—	+26 +20	—	+32 +26
3	6	+20 +15	+23 +15	+27 +15	+24 +19	+27 +19	+31 +19	—	—	—	+31 +23	+35 +23	—	+36 +28	—	+43 +35
6	10	+25 +19	+28 +19	+34 +19	+29 +23	+32 +23	+38 +23	—	—	—	+37 +28	+43 +28	—	+43 +34	—	+51 +42
10	14	+31 +23	+34 +23	+41 +23	+36 +28	+39 +28	+46 +28	—	—	—	+44 +33	+51 +33	—	+51 +40	—	+61 +50
14	18	+31 +23	+34 +23	+41 +23	+36 +28	+39 +28	+46 +28	—	—	—	+44 +33	+51 +33	+50 +39	+56 +45	—	+71 +60
18	24	+37 +28	+41 +28	+49 +28	+44 +35	+48 +35	+56 +35	—	—	—	+54 +41	+62 +41	+60 +47	+67 +54	+76 +63	+86 +73
24	30	+37 +28	+41 +28	+49 +28	+44 +35	+48 +35	+56 +35	+50 +41	+54 +41	+62 +41	+61 +48	+69 +48	+68 +55	+77 +64	+88 +75	+101 +88
30	40	+45 +34	+50 +34	+59 +34	+54 +43	+59 +43	+68 +43	+59 +48	+64 +48	+73 +48	+76 +60	+85 +60	+84 +68	+96 +80	+110 +94	+128 +112
40	50	+45 +34	+50 +34	+59 +34	+54 +43	+59 +43	+68 +43	+65 +54	+70 +54	+79 +54	+86 +70	+95 +70	+97 +81	+113 +97	+130 +114	+152 +136
50	65	+54 +41	+60 +41	+71 +41	+66 +53	+72 +53	+83 +53	+79 +66	+85 +66	+96 +66	+106 +87	+117 +87	+121 +102	+141 +122	+163 +144	+191 +172
65	80	+56 +43	+62 +43	+73 +43	+72 +59	+78 +59	+89 +59	+88 +75	+94 +75	+105 +75	+121 +102	+132 +102	+139 +120	+165 +146	+193 +174	+229 +210
80	100	+66 +51	+73 +51	+86 +51	+86 +71	+93 +71	+106 +91	+106 +91	+113 +91	+126 +91	+146 +124	+159 +124	+168 +146	+200 +178	+236 +214	+280 +258
100	120	+69 +54	+76 +54	+89 +54	+94 +79	+101 +79	+114 79	+110 +104	+126 +104	+136 +104	+166 +144	+179 +144	+194 +172	+232 +210	+276 +254	+332 +310
120	140	+81 +63	+88 +63	+103 +63	+110 +92	+117 +92	+132 +92	+140 +122	+147 +122	+162 +122	+195 +170	+210 +170	+227 +202	+273 +248	+325 +300	+390 +365
140	160	+83 +65	+90 +65	+105 +65	+118 +100	+125 +100	+140 +100	+152 +134	+159 +134	+174 +134	+215 +190	+230 +190	+253 +228	+305 +280	+365 +340	+440 +415
160	180	+86 +68	+93 +68	+108 +68	+126 +108	+133 +108	+148 +108	+165 +146	+171 +146	+186 +146	+235 +210	+250 +210	+277 +252	+335 +310	+405 +380	+490 +465
180	200	+97 +77	+106 +77	+123 +77	+142 +122	+151 +122	+168 +122	+185 +166	+195 +166	+212 +166	+265 +236	+282 +236	+313 +284	+379 +350	+454 +425	+549 +520
200	225	+100 +80	+109 +80	+126 +80	+150 +130	+159 +130	+176 +130	+200 +180	+209 +180	+226 +180	+287 +258	+304 +258	+339 +310	+414 +385	+499 +470	+604 +575
225	250	+104 +84	+113 +84	+130 +84	+160 +140	+169 +140	+186 +140	+216 +196	+225 +196	+242 +196	+313 +284	+330 +284	+369 +340	+454 +425	+549 +520	+669 +640
250	280	+117 +94	+126 +94	+146 +94	+181 +158	+290 +158	+210 +158	+241 +218	+250 +218	+270 +218	+347 +315	+367 +315	+417 +385	+507 +475	+612 +580	+742 +710
280	315	+121 +98	130 +98	+150 +98	+193 +170	+202 +170	+222 +170	+263 +240	+272 +240	+292 +240	+382 +350	+402 +350	+457 +425	+557 +525	+682 +650	+822 +790
315	355	+133 +108	+144 +108	+165 +108	+215 +190	+226 +190	+247 +190	+293 +268	+304 +268	+325 +268	+426 +390	+447 +390	+511 +475	+626 +590	+766 +730	+936 +900
355	400	+139 +114	+150 +114	+171 +114	+233 +208	+244 +208	+265 +208	+319 +294	+330 +294	+351 +294	+471 +435	+492 +435	+566 +530	+696 +660	+856 +820	+1036 +1000
400	450	+153 +126	+166 +126	+189 +126	+259 +232	+272 +232	+295 +232	+357 +330	+370 +230	+351 +294	+530 +490	+553 +490	+635 +595	+780 +740	+960 +920	+1036 +1000
450	500	+159 +132	+172 +132	+195 +132	+279 +252	+292 +252	+315 +252	+387 +360	+400 +360	+423 +360	+580 +540	+603 +540	+700 +660	+860 +820	+1040 +1000	+1290 +1250

Notes 注：When the nominal size<1, a and b for all classes are unavailable. 公称尺寸<1mm 时，各级的 a 和 b 均不采用。

3. Limit tolerance of the hole（Quoted from GB/T 1800.2—2009）**孔的极限偏差**（摘自 GB/T 1800.2—2009）

Table 5-3 Limit deviations for holes 孔的极限偏差数值　　　　　　　　　　（单位：μm）

Nominal size 公称尺寸 mm		Ordinary tolerance zone 常用公差带													
		A*	B*		C	D				E		F			
Over 大于	To 至	11	11	12	11	8	9	10	11	8	9	6	7	8	9
—	3	+330 +270	+200 +140	+240 +140	+120 +60	+34 +20	+45 +20	+60 +20	+80 +20	+28 +14	+39 +14	+12 +6	+16 +6	+20 +6	+31 +6
3	6	+345 +270	+215 +140	+260 +140	+145 +70	+48 +30	+60 +30	+78 +30	+105 +30	+38 +20	+50 +20	+18 +10	+22 +10	+28 +10	+40 +10
6	10	+370 +280	+240 +150	+300 +150	+170 +80	+62 +40	+76 +40	+98 +40	+170 +40	+47 +25	+61 +25	+22 +13	+28 +13	+35 +13	+49 +13
10	14	+400 +290	+260 +150	+330 +150	+205 +95	+77 +50	+93 +50	+120 +50	+160 +50	+59 +32	+75 +32	+27 +16	+34 +16	+43 +16	+59 +16
14	18														
18	24	+430 +300	+290 +160	+370 +160	+240 +110	+98 +65	+117 +65	+149 +65	+195 +65	+73 +40	+92 +40	+33 +20	+41 +20	+53 +20	+72 +20
24	30														
30	40	+470 +310	+330 +170	+420 +170	+280 +170	+119 +80	+142 +80	+180 +80	+240 +80	+89 +50	+112 +50	+41 +25	+50 +25	+64 +25	+87 +25
40	50	+480 +320	+340 +180	+430 +180	+290 +180										
50	65	+530 +340	+389 +190	+490 +190	+330 +140	+146 +100	+170 +100	+220 +100	+290 +100	+106 +60	+134 +60	+49 +30	+60 +30	+76 +30	+104 +30
65	80	+550 +360	+330 +200	+500 +200	+340 +150										
80	100	+600 +380	+440 +220	+570 +220	+390 +170	+174 +120	+207 +120	+260 +120	+340 +120	+126 +72	+159 +72	+58 +36	+71 +36	+90 +36	+123 +36
100	120	+630 +410	+460 +240	+590 +240	+400 +180										
120	140	+710 +460	+510 +260	+660 +260	+450 +200	+208 +145	+245 +145	+305 +145	+395 +145	+148 +85	+135 +85	+68 +43	+83 +43	+106 +43	+143 +43
140	160	+770 +520	+530 +280	+680 +280	+460 +210										
160	180	+830 +580	+560 +310	+710 +310	+480 +230										
180	200	+950 +660	+630 +340	+800 +340	+530 +240	+242 +170	+285 +170	+355 +170	+460 +170	+172 +100	+215 +100	+79 +50	+96 +50	+122 +50	+165 +50
200	225	+1030 +740	+670 +380	+840 +380	+550 +260										
225	250	+1110 +820	+710 +420	+880 +420	+570 +280										
250	280	+1240 +920	+800 +480	+1000 +480	+620 +300	+271 +190	+320 +190	+400 +190	+510 +190	+191 +110	+240 +110	+88 +56	+108 +56	+137 +56	+186 +56
280	315	+1370 +1050	+860 +540	+1060 +540	+650 +330										
315	355	+1560 +1200	+960 +600	+1170 +600	+720 +360	+299 +210	+350 +210	+440 +210	+570 +210	+214 +125	+265 +125	+98 +62	+119 +62	+151 +62	+202 +62
355	400	+1710 +1350	+1040 +680	+1250 +680	+760 +400										

(Continued 续)

Nominal size 公称尺寸/mm		Ordinary tolerance zone 常用公差带																
		G		H							JS			K			M		
Over 大于	To 至	6	7	6	7	8	9	10	11	12	6	7	8	6	7	8	6	7	8
—	3	+8 +2	+12 +2	+6 0	+10 0	+14 0	+25 0	+40 0	+60 0	+100 0	±3	±5	±7	0 −6	0 −10	0 −11	−2 −8	−2 −12	−2 −16
3	6	+12 +4	−16 −4	+8 0	+12 0	+18 0	+30 0	+48 0	+75 0	+120 0	±4	±6	±9	+2 −6	+3 −9	+5 −1	−1 −9	0 −12	+2 −16
6	10	+14 +5	+20 +5	+9 0	+15 0	+22 0	+36 0	+58 0	+90 0	+150 0	±4.5	±7	±11	+2 −7	+5 −10	+6 −16	−3 −12	0 −15	+1 −21
10	18	+17 +6	+24 +6	+11 0	+18 0	+27 0	+43 0	+70 0	+110 0	+180 0	±5.5	±9	±13	+2 −9	+6 −12	+8 −19	−4 −15	0 −18	+2 −25
18	30	+20 +7	+28 +7	+13 0	+21 0	+33 0	+52 0	+84 0	+130 0	+210 0	±6.5	±10	±16	+2 −11	+6 −15	+10 −22	−4 −17	0 −21	+4 −29
30	50	+25 +9	+34 +9	+16 0	+25 0	+39 0	+62 0	+100 0	+160 0	+250 0	±8	±12	±19	+3 −13	+7 −18	+12 −27	−4 −20	0 −25	+5 −34
50	80	+29 +10	+40 +10	+19 0	+30 0	+46 0	+74 0	+120 0	+190 0	+300 0	±9.5	±15	±23	+4 −15	+9 −21	+14 −32	−5 −24	0 −30	+5 −41
80	120	+34 +12	+47 +12	+22 0	+35 0	+54 0	+87 0	+140 0	+220 0	+350 0	±11	±17	±27	+4 −18	+10 −25	+16 −33	−6 −28	0 −35	+6 −43
120	180	+39 +14	+54 +14	+25 0	+40 0	+63 0	+100 0	+160 0	+250 0	+400 0	±12.5	±20	±31	+4 −21	+12 −28	+20 −43	−8 −33	0 −40	+8 −35
180	250	+44 +15	+61 +15	+29 0	+46 0	+72 0	+115 0	+185 0	+290 0	+460 0	±14.5	±23	±36	+5 −24	+13 −33	+22 −50	−8 −37	0 −46	+9 −63
250	315	+49 +17	+69 +17	+32 0	+52 0	+81 0	+130 0	+210 0	+320 0	+520 0	±16	±26	±40	+5 −27	+16 −36	+25 −56	−9 −11	0 −52	+9 −72
315	400	+54 +18	+75 +18	+36 0	+57 0	+89 0	+140 0	+230 0	+360 0	+570 0	±18	±28	±44	+7 −29	+17 −40	+28 −61	−10 −46	0 −57	+11 −78
400	500	+54 +18	+75 +18	+36 0	+57 0	+89 0	+140 0	+230 0	+360 0	+630 0	±20	±31	±48	+7 −32	+17 −45	+28 −68	−10 −50	0 −63	+11 −86

(Continued 续)

Nominal size 公称尺寸/mm		Ordinary tolerance zone 常用公差带											
		N			P		R		S		T		U
Over 大于	To 至	6	7	8	6	7	6	7	6	7	6	7	7
—	3	-4 -10	-4 -14	-4 -18	-6 -12	-6 -16	-10 -16	-10 -20	-14 -20	-14 -24	—	—	-18 -28
3	6	-5 -13	-4 -16	-2 -20	-9 -17	-8 -20	-12 -20	-11 -23	-16 -24	-15 -27	—	—	-19 -31
6	10	-7 -16	-4 -19	-3 -25	-12 -21	-9 -24	-16 -25	-13 -28	-20 -29	-17 -32	—	—	-22 -37
10	18	-9 -20	-5 -23	-3 -30	-15 -26	-11 -29	-20 -31	-16 -34	-25 -36	-21 -39	—	—	-26 -44
18	24	-11 -24	-7 -28	-3 -36	-18 -31	-14 -35	-24 -37	-20 -41	-31 -44	-27 -48	—	—	-33 -54
24	30										-37 -50	-33 -54	-40 -61
30	40	-12 -28	-8 -33	-3 -42	-21 -37	-17 -42	-29 -45	-25 -50	-38 -54	-34 -59	-43 -59	-39 -64	-51 -76
40	50										-49 -65	-45 -70	-61 -76
50	65	-14 -33	-9 -39	-4 -50	-26 -45	-21 -51	-35 -54	-30 -60	-47 -66	-42 -72	-60 -79	-55 -85	-86 -106
65	80						-37 -56	-32 -62	-53 -72	-48 -78	-69 -88	-64 -94	-91 -121
80	100	-16 -38	-10 -45	-4 -58	-30 -52	-24 -59	-44 -66	-38 -73	-64 -86	-58 -93	-84 -106	-78 -113	-111 -146
100	120						-47 -69	-41 -76	-72 -94	-66 -101	-97 -119	-91 -126	-131 -166
120	140	-20 -45	-12 -52	-4 -67	-36 -61	-28 -68	-56 -81	-48 -88	-85 -110	-77 -117	-115 -140	-107 -147	-155 -195
140	160						-58 -83	-50 -90	-93 -118	-85 -125	-137 -152	-110 -159	-175 -215
160	180						-61 -86	-53 -93	-101 -126	-93 -133	-139 -164	-131 -171	-195 -235
180	200	-22 -51	-14 -60	-5 -77	-41 -70	-33 -79	-68 -97	-60 -106	-113 -142	-101 -155	-157 -186	-149 -195	-219 -265
200	225						-71 -100	-63 -109	-121 -150	-113 -159	-171 -200	-163 -209	-241 -287
225	250						-75 -104	-67 -113	-131 -160	-123 -169	-187 -216	-179 -225	-267 -313
250	280	-25 -57	-14 -66	-5 -86	-47 -79	-36 -88	-85 -117	-74 -126	-149 -181	-138 -190	-209 -241	-198 -250	-295 -347
280	315						-89 -121	-78 -130	-161 -193	-150 -202	-231 -263	-220 -272	-330 -382
315	355	-26 -62	-16 -73	-5 -94	-51 -87	-41 -98	-97 -133	-87 -144	-179 -215	-169 -226	-257 -293	-247 -304	-369 -426
355	400						-103 -139	-93 -150	-197 -233	-187 -244	-283 -319	-273 -330	-414 -471
400	450	-27 -67	-17 -80	-6 -103	-55 -95	-45 -108	-113 -153	-103 -166	-219 -259	-209 -272	-317 -357	-307 -370	-467 -530
450	500						-113 -159	-109 -172	-239 -279	-229 -292	-347 -387	-337 -400	-540 -637

Notes 注：When the nominal size<1，A and B for all classes are unavailable. 公称尺寸<1mm时，各级的 A 和 B 均不采用。

4. Hole-basis system of fits 基孔制配合

Table 5-4 Preferred and ordinary fits in hole-basis system 基孔制优先、常用配合

Basic hole 基准孔	Shaft 轴																			
	a	b	c	d	e	f	g	h	js	k	m	n	p	r	s	t	u	v	x	y	z
	Clearance fits 间隙配合								Transition fits 过渡配合				Interference fits 过盈配合							
H6						$\frac{H6}{f5}$	$\frac{H6}{g5}$	$\frac{H6}{h5}$	$\frac{H6}{js5}$	$\frac{H6}{k5}$	$\frac{H6}{m5}$	$\frac{H6}{n5}$	$\frac{H6}{p5}$	$\frac{H6}{r5}$	$\frac{H6}{s5}$	$\frac{H6}{t5}$				
H7						$\frac{H7}{f6}$	*$\frac{H7}{g6}$	*$\frac{H7}{h6}$	$\frac{H7}{js6}$	*$\frac{H7}{k6}$	$\frac{H7}{m6}$	*$\frac{H7}{n6}$	*$\frac{H7}{p6}$	$\frac{H7}{r6}$	*$\frac{H7}{s6}$	$\frac{H7}{t6}$	*$\frac{H7}{u6}$	$\frac{H7}{v6}$	$\frac{H7}{x6}$	$\frac{H7}{y6}$	$\frac{H7}{z6}$
H8					*$\frac{H8}{e7}$	*$\frac{H8}{f7}$	$\frac{H8}{g7}$	*$\frac{H8}{h7}$	$\frac{H8}{js7}$	$\frac{H8}{k7}$	$\frac{H8}{m7}$	$\frac{H8}{n7}$	$\frac{H8}{p7}$	$\frac{H8}{r7}$	$\frac{H8}{s7}$	$\frac{H8}{t7}$	$\frac{H8}{u7}$				
				$\frac{H8}{d8}$	$\frac{H8}{e8}$	$\frac{H8}{f8}$		$\frac{H8}{h8}$													
H9			$\frac{H9}{c9}$	*$\frac{H9}{d9}$	$\frac{H9}{e9}$	$\frac{H9}{f9}$		*$\frac{H9}{h9}$													
H10			$\frac{H10}{c10}$	$\frac{H10}{d10}$				$\frac{H10}{h10}$													
H11	$\frac{H11}{a11}$	$\frac{H11}{b11}$	*$\frac{H11}{c11}$	$\frac{H11}{d11}$				*$\frac{H11}{h11}$													
H12		$\frac{H12}{b12}$						$\frac{H12}{h12}$													

Notes 注：1. When the basic size < or = 3mm H6/n5 and H7/p6 are interference fits; When the basic size < or = 100mm, H8/r7 is transition one. 对于H6/n5、H7/p6，在公称尺寸小于或等于3mm 为间隙配合；对于H8/r7，当公称尺寸小于或等于100mm 时，为过渡配合。

2. Fits marked with "*" are priority ones. 标注"*"的配合为优先配合。

5. Shaft-basis system of fits 基轴制配合

Table 5-5 Preferred and ordinary fits in shaft-basis system 基轴制优先、常用配合

Basic shaft 基准轴	Hole 孔																			
	A	B	C	D	E	F	G	H	JS	K	M	N	P	R	S	T	U	V	X	Y	Z
	Clearance fits 间隙配合								Transition fits 过渡配合				Interference fits 过盈配合							
h5						$\frac{F6}{h5}$	$\frac{G6}{h5}$	$\frac{H6}{h5}$	$\frac{JS6}{h5}$	$\frac{K6}{h5}$	$\frac{M6}{h5}$	$\frac{N6}{h5}$	$\frac{P6}{h5}$	$\frac{R6}{h5}$	$\frac{S6}{h5}$	$\frac{T6}{h5}$					
h6						$\frac{F7}{h6}$	*$\frac{G7}{h6}$	*$\frac{H7}{h6}$	$\frac{JS7}{h6}$	*$\frac{K7}{h6}$	$\frac{M7}{h6}$	*$\frac{N7}{h6}$	*$\frac{P7}{h6}$	$\frac{R7}{h6}$	*$\frac{S7}{h6}$	$\frac{T7}{h6}$	*$\frac{U7}{h6}$				
h7					$\frac{E8}{h7}$	*$\frac{F8}{h7}$		*$\frac{H8}{h7}$	$\frac{JS8}{h7}$	$\frac{K8}{h7}$	$\frac{M8}{h7}$	$\frac{N8}{h7}$									
h8				$\frac{D8}{h8}$	$\frac{E8}{h8}$	$\frac{F8}{h8}$		$\frac{H8}{h8}$													
h9				*$\frac{D9}{h9}$	$\frac{E9}{h9}$	$\frac{F9}{h9}$		*$\frac{H9}{h9}$													
h10				$\frac{D10}{h10}$				$\frac{H10}{h10}$													
h11	$\frac{A11}{h11}$	$\frac{B11}{h11}$	*$\frac{C11}{h11}$	$\frac{D11}{h11}$				*$\frac{H11}{h11}$													
h12		$\frac{B12}{h12}$						$\frac{H12}{h12}$													

Notes 注：Fits marked with "*" are priority ones. 标注"*"的配合为优先配合。

Appendix 6 Commonly Used Materials 附录6 常用材料

1. Ferrous materials 黑色金属

Table 6-1 General ferrous materials 常用黑色金属

Name & standard 名称与标准	Grade 牌号	Performance and application 性能及应用举例	Remark 说明
Carbon structural steel 碳素结构钢 GB/T 700—2006	Q195	Rivets with small load, bolts, short shafts, central spindles, cams, cementite, weldment 负荷不大的铆钉、螺栓、短轴、心轴、凸轮、渗碳件、焊接件	Q is the first letter of pinyin "Qu", representing the yielding strength. the value represents the yielding strength (Unit: MPa). A, B, C and D represent grades of quality. F and Z represent boiling steel and killed steel, such as Q215FA Q为钢材屈服强度"屈"字汉语拼音首位字母,数值为屈服强度(单位:MPa),A、B、C、D表示质量等级,F和Z分别表示沸腾钢和镇静钢,如Q215FA
	Q215 A、B		
	Q235 Q275 A、B、C、D	Bolts, nuts, rods, pull rods, shafts, wedges, covers, weldment 螺栓、螺母、连杆、拉杆、轴、楔、盖、焊接件	
Quality carbon structural steel 优质碳素结构钢 GB/T 699—2015	10	Pull rods, drawing gears, washers, weldment 拉杆、卡头、垫圈、焊接件	The number represents the average mass fraction of carbon in the steel. For example, 45 represents that the average carbon content is 0.45% 牌号的数字表示平均含碳质量分数(w_C)。例如"45"表示平均含碳质量分数w_C为0.45%
	15 20	Cementite, fasteners, forging pieces, chemical engineering containers 渗碳件、紧固件、冲模锻件、化工储器	
	25 30 35	Crankshafts, rockers, pull rods, bolts, pins, keys 曲轴、摇杆、拉杆、螺栓、销、键	
	40 45 50	Gears, shafts, couplings, bushes, piston pins, chain wheels 齿轮、轴、联轴器、衬套、活塞销、链轮	
	55 60 65 70 75 80 85	Springs, vanes 弹簧、叶片	
	15Mn 20Mn 25Mn	Cemetite, weldment 渗碳件、焊接件	
	30Mn 40Mn 50Mn	Bolts, levers, dive flaps 螺栓、杠杆、制动器	The manganese content of 65Mn and 70Mn manganese steel is 0.70%~1.00% 65Mn、70Mn锰钢的锰含量为0.70%~1.00%
	60Mn 65Mn 70Mn	Springs, winding mechanisms 弹簧、发条	
Tool and mould steels 工模具钢 GB/T 1299—2014	T8	Due to its high toughness and hardness, it is used to manufacture tools that will have to withstand shock. For example, drills used on surfaces of medium hardness such as rocks, simple moulds and punch 有足够的韧性和较高的硬度,用于制造能承受振动的工具。如钻中等硬度岩石的钻头、简单模具、冲头等	This steel is represented with the average mass fraction of carbon by attaching 'carbon' or 'T'. For example T7—T13 represents that the average carbon content is 0.7%—1.3% 用"碳"或"T"后附以平均含碳量的千分数表示。例如T7~T13,平均含碳量为0.7%~1.3%

(Continued 续)

Name & standard 名称与标准	Grade 牌号	Performance and application 性能及应用举例	Remark 说明
High strength low alloy structional steels 低合金高强度结构钢 GB/T 1591—2008	Q345 Q390 Q420 (A、B、C、D、E) Q460 Q500 Q550 Q620 Q690 (C、D、E)	Bridges, shipbuilding, factory structures, oil tanks, pressure vessels, rolling stocks, hoisting apparatuses, mining machinery and other welded constructions to replace Q235 桥梁、造船、厂房结构、储油罐、压力容器、机车车辆、起重设备、矿山机械及其他代替 Q235 的焊接件	A, B, C, D and E represent the grade of quality The grade of symbols is composed of three parts, the phonetic alphabet, the value, and the quality grade of yield strength such as Q345D A、B、C、D 和 E 表示质量等级 牌号由钢的屈服强度拼音字母、屈服强度数值、质量等级符号三个部分组成，如 Q345D
Alloy structional steels 合金结构钢 GB/T 3077—2015	15Cr 20Cr	Carbon penetrated gears, cams, piston pins, clutches 渗碳齿轮、凸轮、活塞销、离合器	A certain amount of alloying element is added to ordinary steel to improve its mechanical properties as well as wear resistance and hardness penetration. It ensures the metal has high mechanical properties over large surfaces 钢中加入一定量的合金元素，提高了钢的力学性能和耐磨性，也提高了铁的淬透性，保证金属在较大的截面上获得高力学性能
	30Cr 40Cr	Important hardened tempered pieces: gears, axles, rockers, bolts 重要的调质件：齿轮、轮轴、摇杆、螺栓	
	45Cr 50Cr	High strength and wear resistance gears, axles, bolts 强度及耐磨性高的齿轮、轴、螺栓	
	15CrMo 20CrMo 40CrMo	Used for various wear resistance pieces, washers, bushes, pull rods, gears, axles 用于各种高耐磨性零件，如垫圈、套筒、连杆、齿轮、轴等	
	20CrMnTi	Important carbon penetrated pieces: gears 重要渗碳件：齿轮	
	30CrMnTi	Carbon penetrated gears with high strength used on the autos and tractors 汽车、拖拉机上强度特高的渗碳齿轮	
Heat-resisting steel bars 耐热钢棒 GB/T 1221—2007	1Cr18Ni9Ti	Used for various forging pieces, the spouts and collectors of the exhaust system in aeroengines 用于各种锻件，航空发动机排气系统的喷管及集合器等零件	Acid resistance. It can endure temperatures up to 600℃ and won't deform below 1000℃ 耐酸，在 600℃ 以下耐热，在 1000℃ 以下不起皮
Carbon steel castings for general engineering purpose 一般工程用铸造碳钢件 GB/T 11352—2009	ZG200-400 ZG230-450	Bases, boxes, brackets 机座、箱体、支架	"ZG" indicates cast steel. The number indicates the yield strength and tensile strength. (Unit: MPa) "ZG" 表示铸钢，数字表示屈服强度及抗拉强度值。（单位：MPa）
	ZG270-500 ZG310-570	Gears, flywheels, machine frames 齿轮、飞轮、机架	
	ZG340-640	Gears couplings for crane and transport 起重机、运输机中的齿轮、联轴器等重要的机件	

(Continued 续)

Name & standard 名称与标准	Grade 牌号	Performance and application 性能及应用举例	Remark 说明
Gray iron castings 灰铸铁件 GB/T 9439—2010	HT100 HT150	Used to manufacture end covers and pumps of steam turbines, bearing seats, valve shells, pipe and pipeline accessories, hand wheels, as well as the base, body, slide carriage, and workbench of general machines 用于制造端盖、汽轮机泵体、轴承座、阀壳、管子及管路附件、手轮，以及普通机床底座、机身、滑座、工作台等	"HT" indicates gray iron. The following number is the tensile strength. (Unit: MPa). For example, HT200 indicates gray iron with a tensile strength of 200 MPa "HT" 表示灰铸铁，其后数字表示抗拉强度值（单位为 MPa），如 HT200 表示抗拉强度为 200MPa 的灰铸铁
	HT200	Used for low-intensity castings, such as steam cylinders, gears, bases, machine bodies, flywheels, racks, bushes, as well as machine bodies cast with guide rails and hydraumatic tanks, hydraulic pumps, valve bodies of medium pressure 用于低强度铸件，如汽缸、齿轮、底架、机体、飞轮、齿条、衬筒，以及铸有导轨的机床床身、承受中等压力的液压筒、液压泵和阀体等	
	HT250	Used for parts requiring medium loads and wear resistance, such as valve shells, hydro-cylinders, steam cylinders, couplings, machine bodies, gears, shells of gearboxes, flywheels, bushes, cams, bearing seats, pistons, etc. 用于中等负荷和对耐磨性有一定要求的零件，如阀壳、液压缸、汽缸、联轴器、机体、齿轮、齿轮箱外壳、飞轮、衬套、凸轮、轴承座、活塞等	
	HT300 HT350	Used for the high-strength and wear-resistant castings, such as gears enduring large loads, bed ways, lathe chucks, shearing machine beds, forcing press beds, cams, high-pressure cylinders, hydraulic pumps and spool valve shells, punch die, etc. 用于高强度耐磨的铸件，如齿轮、床身导轨、车床卡盘、剪床床身、压力机的床身、凸轮、高压液压缸、液压泵和滑阀壳体、冲模模体等	

(Continued 续)

Name & standard 名称与标准	Grade 牌号	Performance and application 性能及应用举例	Remark 说明
Spheroidal graphite iron castings 球墨铸铁 GB/T 1348—2009	QT900-2 QT800-2 QT700-2	High strength, wear resistance, toughness. Used to manufacture spindles, cylinder sleeves, cylinder bodies, air compressors 具有较高强度、耐磨性和韧性，用于制造机床的主轴、缸套、缸体、空压机等	"QT" indicates ductile cast iron. The first group of numbers represents the tensile strength (Unit: MPa). The second group of numbers indicates the elongation (%). For example, QT500-15 represents ductile cast iron with a tensile strength of 500MPa and elongation of 15% "QT"表示球墨铸铁，其后第一组数字表示抗拉强度（单位：MPa），第二组数字表示断后伸长率（%）。如QT500-15即表示球墨铸铁的抗拉强度为500MPa，断后伸长率为15%
	QT600-3 QT550-5 QT500-7 QT450-10	Medium strength and toughness. Used to manufacture oil pump gears, cylinder gusset plates, valve bodies 具有中等强度和韧性，用于制造油泵齿轮、气缸隔板、阀门体等	
	QT400-18L QT400-18R QT400-18 QT400-15 QT350-22R QT350-22L QT350-22	High toughness, good hypothermia, corrosion resistance. Used to manufacture wheel hubs, shells, clutch shifters 韧性高，低温性能好，且有一定的耐蚀性，用于制造轮毂、壳体、离合器拨叉等	
Malleable iron castings 可锻铸铁 GB/T 9440—2010	KTH300-06 KTH350-10	Blackheart malleable cast iron, used to manufacture shock absorption devices such as vehicles, tractors, agricultural machinery casting 黑心可锻铸铁，用于承受冲击振动的零件：汽车、拖拉机、农机铸件	"KT" stands for malleable castiron, "H" for blackheart, "B" for whiteheart. The first group of digits indicates tensile strength (N/mm^2), the second group of digits indicates elongation (%). KTH300-06 is applicable to airtightness accessories "KT"表示可锻铸铁，"H"表示黑心，"B"表示白心，第一组数字表示抗拉强度值（MPa），第二组数字表示断后伸长率（%）。KTH300-06适用于气密性零件
	KTB350-04 KTB380-12 KTB400-05 KTB450-07	Whiteheart malleable cast iron, lower toughness, good abrasion performance and workability. Used to substitute devices made of mild steel, medium carbon steel, and lean alloy steel such as crank shaft, connecting rod, and accessories for machine tool, etc. 白心可锻铸铁，韧性较低，但强度高，耐磨性、加工性好。可代替低、中碳钢及低合金钢的重要零件，如曲轴、连杆、机床附件等	

2. Nonferrous metals and alloy materials 有色金属及合金材料

Table 6-2　General nonferrous metals and alloy materials　常用有色金属及合金材料

Standard 标准	Grade 牌号	Performance and application 性能及应用举例	Remark 说明
Designation and chemical composition of wrought copper and copper alloys 加工铜及铜合金牌号和化学成分 GB/T 5231—2012	H59 H62 H63 H65 H68 H70 H80 H85 H90 H96	Used for various loaded parts manufactured by elongation and bending, such as pins, washers, nuts, guide tubes, springs, rivets 适用于各种伸引和弯折制造的受力零件，如销钉、垫圈、螺母、导管、弹簧、铆钉等	H indicates brass. 62 indicates carbon content is 60.5%—63.5% H表示黄铜，62表示平均含铜量的百分数 w_{Cu}=60.5%~63.5%

321

(Continued 续)

Standard 标准	Grade 牌号	Performance and application 性能及应用举例	Remark 说明
Casting copper and copper alloy 铸造铜及铜合金 GB/T 1176—2013	Casting brassalloy 铸造铜合金 ZCuZn38	Heat sinks, washers, springs, nets, rivets and other parts 散热器、垫圈、弹簧、各种网、螺钉及其他零件	"Z" represents casting, the figures behind the chemical elements represent the mass fraction of the elements "Z" 表示铸造，后面的数字表示该元素的质量分数
	Silvel 锰黄铜 ZCuZn38Mn2Pb2	Used for bushes, sleeves and other antifriction parts 用于制造轴瓦、轴套及其他耐磨零件	
	Tin bronze 锡青铜 ZCuSn5Pb5Zn5	Used for parts enduring medium impact loads, liquid lubricating and antirust, such as bearings, bushes, worm wheels, nuts, steam and water spare parts below 1MPa 用于承受中等冲击负荷和液体润滑及耐腐蚀条件下工作的零件，如轴承、轴瓦、蜗轮、螺母，以及 1MPa 以下的蒸汽和水配件	
	Albronze 铝青铜 ZCuAl10Fe3	High strength, good wear resistance, corrosion resistance, pressure and castability. Used for parts working in steam and under water, parts bearing friction and corrosion, such as worm wheel sleeves 强度高、减磨性、耐蚀性、受压、铸造性均良好。用于在蒸汽和海水条件下工作的零件及受摩擦和腐蚀的零件，如蜗轮衬套等	
Casting aluminium alloy 铸造铝合金 GB/T 1173—2013	ZAlSi7Mg (ZL101) ZAlSi12 (ZL102) ZAlCu5Mg (ZL201) ZAlMg10 (ZL301) ZAlZn11Si7 (ZL401)	Medium wear resistance. used to manufacture thin walled parts to endure small loads 耐磨性中等，用于制造负荷不大的薄壁零件	ZL102 indicates silicon content is 10%—13%. The rest is an aluminum silicon alloy ZL102 表示含硅 10%～13%、其余为铝的铝硅合金
Wrought aluminium and aluminium alloy 变形铝及铝合金 GB/T 3190—2008	2Al0 (LY10) 2Al1 (LY11) 2Al2 (LY12)	Used to manufacture medium strength parts, which provide good weldability 适于制作中等强度的零件，焊接性好	2Al2 indicates the content of Si, Fe, Cu, Mn, Mg, Ni, Zn, Ti, and Al. The grade is shown in parentheses 2Al2 表示含硅、铁、铜、锰、镁 1.2%～1.8%、镍、锌、钛、铝的硬铝，括号内为牌号

3. Nonmetallic materials 非金属材料

Table 6-3　General nonmetallic materials　常用非金属材料

Name & standard 名称与标准	Type & code 类型或代号	Performance and application 性能及应用举例	Remark 说明
Industrial rubber sheet 工业用橡胶板 GB/T 5574—2008	Class A (Not resistant to oil) A 类 (不耐油)	Good wear resistance and elasticity. Used to manufacture washers, gland strips and backing plates, which require good wear resistance, impact and buffering 具有较好的耐磨性和弹性，适用于制作具有耐磨、耐冲击及缓冲性能好的垫圈、密封条、垫板	Medium hardness 中等硬度
	Class B (Medium resistant to oil) B 类 (中等耐油)		
	Class C (Oil-proof) C 类 (耐油)	It can work in $-30\sim100$℃ engine oil, and gasoline. It can be used to manufacture washers 可在 $-30\sim100$℃ 发动机油、汽油等介质中工作，可制作垫圈	High hardness. good dilatability 较高硬度，较好的耐溶剂膨胀性
Industrial felt 工业用毛毡 FZ/T 25001—2012	Fine felt 细毛毡	Used for airproofing, oil proofing, shock proofing, buffering washers 用作密封、防漏油、防振、缓冲衬垫等	T112-65 indicates specical white fine felt, with mass per unit volume 65g/cm^3 T112-65 表示白色细毛毡，单位体积质量为 65g/cm^3
	Semi-coarse felt 半粗毛毡		
	Coarse felt 粗毛毡		
	Roan 杂毛毡		
	Beast felt 兽毛毡		
	Synthetic felt 纯化纤毡		

(Continued 续)

Name & standard 名称与标准	Type & code 类型或代号	Performance and application 性能及应用举例	Remark 说明
Soft steel paper sheet 软钢纸板 QB/T 2200—1996	A	Application in making seal gaskets and other components used for the production of aircraft engines 供飞机发动机制作密封垫片及其他部件用	Paperboard thickness is 0.5—3.0mm 纸板厚度为0.5~3.0mm
	B	Application in making seal gaskets and other components used for productions of autos and tractors engines and other internal combustion engines 供汽车、拖拉机的发动机及其他内燃机制作密封垫片及其他部件用	
Polytetrafluorethylene colophony sheet 聚四氟乙烯板材 QB/T 3625—1999	SFB-1	Primarily use for electrical insulation 主要用于电器绝缘方面	Good chemical stability, high heat and cold resistance. Good self-lubrication 化学稳定性好，高耐热、耐寒性，自润滑性能好
	SFB-2	Primarily use for seal gaskets and lubrication materials in corrosion medium 主要用于腐蚀介质中的衬垫密封件及润滑材料	
	SFB-3	Primarily use for diaphragm and the mirrors in corrosion medium 主要用于腐蚀介质中的隔膜与视镜	
Poly (methyl methacrylate) cast sheets 浇铸型工业有机玻璃板材 GB/T 7134—2008	PMMA 平板	Used to manufacture parts with a certain transparency and strength, such as oil cups, nomenclature plates, pipes, insulators 制造一定透明度和强度的零件、油杯、标牌、管道、电气绝缘件等	Acid proof and alkali proof 耐酸、耐碱

Notes 注：FZ is the standard of Ministry of the Textile Industry; QB is the standard of Ministry of the Light Industry; HG is the standard of Ministry of the Light Industry. FZ 是纺织工业部标准；QB 是轻工业部标准；HG 是化学工业部标准。

Appendix 7 Definitions for General Heat Treatment and Surface Treatment
附录7 常用的热处理和表面处理名词解释

Name 名词	Code 代号	Remark 说明	Application 应用
Annealing 退火	511	Heat steel above the appropriate temperature (usually 710—715℃) hold at the temperature 30~50℃ for a period and cool slowly 将钢件加热到适当温度以上（一般是710~715℃），在30~50℃中保温一段时间，然后缓慢冷却	Used to eliminate the stress dislocations in castings, forgings and weldments and decrease the hardness thus improving machinability. It thins down the crystal grain, improves the microstructure and enhances toughness 用来消除铸、锻、焊零件的内应力，降低硬度，改善切削性能，细化金属晶粒，改善组织，增加韧性
Normalizing 正火	512	The process consists of heating above the critical temperature, holding at the temperature, then cooling to room temperature in the air. The cooling speed is slower than that during annealing 将钢件加热到临界温度以上，保温一段时间，然后在空气中冷却，冷却速度比退火为快	Used to thin down the microstructure, to increase strength and toughness, to decrease stress dislocations, and to improve machinability 用来处理低碳钢和中碳钢及渗碳零件，使其组织细化，增加强度与韧性，减少内应力，改善切削性能
Quenching 淬火	513	The process consists of heating above critical temperature, holding at the temperature and rapidly cooling in water, brine water or oil 将钢件加热到临界温度以上，保温一段时间，然后在水、盐水或油中急速冷却	Used to increase the hardness and strength limit, but quenching also causes internal stresses making the steel brittle. Therefore, tempering is required after quenching 用来提高钢的硬度和强度极限。但淬火会引起内应力，使钢变脆，所以淬火后必须回火
Tempering 回火	514	The process consists of heating quenched steel below the critical temperature, holding at the temperature, then cooling in air or oil 回火是将淬硬的钢件加热到临界点以下的温度，保温一段时间，然后在空气中或油中冷却	Used to eliminate brittleness and internal stresses caused by quenching to improve plasticity and impact toughness 用来消除淬火后的脆性和内应力，提高钢的塑性和冲击韧度
Harden & temper 调质	515	The process consists of tempering at a temperature of 450—650℃ after a quenching operation 淬火后在450~650℃进行高温回火，称为调质	Used to increase the toughness and strength, such as important gears, axles, and screws 用来使钢获得高的韧性和足够的强度，重要的齿轮、轴及丝杠等零件都需经调质处理
Surface hardening and hampering 表面淬火和回火	521	The process consists of rapidly heating the surface with a flame, then rapidly cooling down 用火焰将零件表面迅速加热至临界温度以上，急速冷却	It maintains steel with high hardness on the surface and certain toughness in the center to resist wear and impact. Usually used to treat the gears 使钢件表面获得高硬度，而心部保持一定的韧性，使零件既耐磨又能承受冲击。表面淬火常用来处理齿轮等
Chemical-heat treatment 化学热处理	530	Chemical heat treatment of metals and alloys is the process of changing the workpiece surface chemical composition, structure and properties of metal heat treatment 化学热处理是通过改变金属和合金工件表层的化学成分、组织和性能的金属热处理	The workpiece surface can obtain a high hardness, wear resistance and high strength, at the same time, the heart department to retain its good toughness, so that the workpiece being processed has the ability to impact load 在工件表层获得高硬度、耐磨损和高强度的同时，心部仍保持良好的韧性，使被处理工件具有抗冲击载荷的能力

(Continued 续)

Name 名词	Code 代号	Remark 说明	Application 应用
Carburize 渗碳	531	The process consists of heating to 900—950℃ in the carburizer, carburizing until the case depth is 0.5—2mm, then tempering after a quenching operation 在渗碳剂中将钢件加热到 900~950℃，使碳渗入钢的表面，深度为 0.5~2mm，再淬火、回火	It increases wear resistance, surface strength, tensile strength, and fatigue limit. It is applied to low and medium carbon steel pieces ($w_C<0.40\%$) 增加钢件的耐磨性、表面强度、抗拉强度及疲劳极限。适用于低碳、中碳（$w_C<0.40\%$）结构钢零件
Nitriding 渗氮	533	The process consists of heating in nitrogen until a nitride layer appears. The thickness of the nitride layer is 0.025—0.08mm and time required is 40—50h 将零件放入氮气内加热，使钢的表面获得含氮强化层的过程。氮化层厚度为 0.025~0.08mm，氮化时间需 40~50h	It increases the wear resistance, surface hardness, fatigue limit and corrosion resistance. Used in alloying steel, carbon steel and iron castings, such as spindles, screws and other parts working in damp conditions, soda water, or combustion gases 增加钢件的耐磨性、表面硬度、疲劳极限和抗蚀能力。适用于合金钢、碳钢、铸铁件，如机床主轴、丝杠以及在潮湿、碱水和燃烧气体介质的环境中工作的零件
Hardness 硬度	HB Brinell hardness 布氏硬度	Hardness is the ability for a material to resist other objects being pressed into it. Hardness is classified as Brinell, Rockwell, and Vickers, according to different measurements 材料抵抗硬的物体压入其表面的能力称"硬度"。根据测试方法不同，可分为布氏硬度、洛氏硬度和维氏硬度	Used to inspect annealed, normalized, hardened and tempered parts 用于退火、正火、调质的零件及铸件的硬度检验
	HRC Rockwell hardness 洛氏硬度		Used to check the hardness of quenched, tempered, surface carburized, and nitrided parts 用于经淬火、回火及表面渗碳、渗氮等处理的零件硬度检验
	HV Vickers hardness 维氏硬度		Used to check the hardness of the lamina hardened parts 用于薄层硬化零件的硬度检验
Aging 时效	Aging treatment 时效处理	The process consists of heating to 100—160℃ before a machining operation, holding at the temperature for 10—40 hours, then cooling down in air. Castings can apply natural aging (put in air for over one year) 精加工之前，加热到 100~160℃后，保持 10~40h，在空气中冷却。铸件也可用天然时效（放在露天一年以上）	Used to remove internal stresses of a work piece and to guarantee its shape. Used in measuring tools, precision screws, guide ways, lathe beds 使工件消除内应力和稳定形状，用于量具、精密丝杠、导轨、床身等
Color-hardening 着色硬化、发蓝、发黑		The process consists of heating in an alkali and oxidant liquor to have a protective layer of ferric oxide on the surface 将金属零件放在很浓的碱和氧化剂溶液中加热氧化，使表面形成一层氧化铁保护膜	Used to resist corrosion and to beautify. It is applied to standard pieces and other electric components 防腐蚀、美观。用于一般连接的标准件和其他电子类零件

Appendix 8 Commonly Used Terminologies and Abbreviations in Engineering Drawing

附录 8 工程制图中常用的专业术语及缩略语

(1) Terminologies 专业术语

A

凹坑	recessed surface, 157	

B

扳手	wrench, 223
板簧	leaf spring, 140
半径	radius (or radii), 17
保持架、隔离圈	cage, 143
泵	pump, 151
泵盖	pump cover, 207
泵体	pump body, 207
比例	scale, 6
标题栏	title block, 4
标准齿轮	standard gear, 127
标准公差	standard tolerance, 171
标准公差等级	standard tolerance grade, 171
标准螺纹	standard screw, 113
表面处理	surface treatment, 164
表面粗糙度	surface roughness, 163
表面结构	surface texture, 162
拨叉	transmission fork, 152

C

材料	material, 148
槽宽	width of space, 128
草图	sketch, 195
超级精细研磨	super fine grinding, 168
超声波焊	ultrasonic welding, 242
车内螺纹	internal thread lathing, 111
车外螺纹	external thread lathing, 111
尺寸	dimension, 10
尺寸公差	dimension tolerance, 170
尺寸公差（公差）	size tolerance, 170
尺寸界线	extension line, 10
尺寸偏差（偏差）	size deviation, 170
尺寸数字	dimension figure, 10
尺寸线	dimension line, 10
齿顶高	addendum, 128
齿顶圆	addendum circle, 127
齿高	tooth depth, 128
齿根高	dedendum, 128
齿根圆	dedendum circle, 127
齿厚	tooth thickness, 128
齿距	circular pitch, 128
齿宽	tooth width, 128
齿轮	gear, 126
齿数	number of teeth, 128
齿条	gear rack, 131
传动螺纹	transmission thread, 114
垂直度	perpendicularity, 179
粗车	rough turning, 168
粗刨	rough planning, 168
粗实线	continuous thick line, 8
粗铣	rough milling, 168
粗牙普通螺纹	coarse plain thread, 114

D

搭接接头	lap joint, 238
带轮	belt wheel, 151
单线螺纹	single-start thread, 112
弹簧	spring, 139
弹簧垫圈	spring lock washer, 119
弹簧节距	pitch, 140
弹簧内径	internal diameter, 140
弹簧丝展开长度	stretched length of wire, 140
弹簧外径	external diameter, 140
弹簧中径	effective diameter, 140
挡圈	ring, 218
导程	lead, 113
倒角	chamfer, 111
等离子弧焊	plasma arc welding, 242
第三角画法	third angle projection, 84
第一角画法	first angle projection, 84
点焊	spot welding, 242
电镀	plating, 183
电路图	circuit diagram, 272
电气图	electric diagram, 266
电器位置图	electric location plan, 274
电渣焊	electro-slag welding, 242
电子束焊	electron beam welding, 242
电阻对焊	butt resistance welding, 242
垫圈（片）	washer (or spacer), 119
丁字尺	T-square, 1

顶点	vertex, 33	间隙配合	clearance fit, 173
定位尺寸	location dimension, 17	渐开线齿轮	involute gear, 127
定形尺寸	size dimension, 17	键	key, 136
对接焊缝	butt weld, 238	键槽	keyway (slot), 136
对接接头	butt joint, 238	箭头	arrow (or arrowhead), 10
多线螺纹	multiple thread, 112	角焊缝	fillet weld, 238
		角接接头	corner joint, 238
		铰孔	reaming, 136
		接线图	wiring diagram, 275

F

阀	valve, 149	节流阀	throttle flap, 205
阀盖	valve cover, 222	金属	metal, 183
阀杆	valve handle, 223	精车	fine turning, 168
阀体	valve body, 222	精磨	fine grinding, 168
阀芯	valve core, 222	精刨	fine planning, 168
法兰盘；凸缘	flange, 151	精镗	fine boring, 168
方头螺栓	SQ screw, 203	精铣	fine milling, 168
分度（节）圆	pitch circle, 127		
腐蚀	corrosion, 165		

K

卡尺（钳）	caliper, 197
开槽沉头螺钉	slotted countersunk flat head screw, 119
开槽盘头螺钉	slotted pan head screw, 119
开槽圆柱头螺钉	slotted cheese head screw, 119
开口（尾）销	split (cotter) pin, 135
空气-乙炔焊	air-acetylene, 242

G

杠杆	pry bar, 152
公差带	tolerance zone, 171
公称尺寸	nominal size, 170
公称直径	nominal diameter, 112
攻内螺纹	tapping internal thread, 111
固定钳身	fixed clamp body, 218
管道布置图	piping layout, 263
管螺纹	pipe thread, 114
滚动体	rolling element, 143
滚动轴承	rolling bearing, 143
滚花	knurling, 108
过渡配合	transition fit, 173
过盈配合	interference fit, 173

L

肋（加强肋）板	rib, 108
棱柱	prism, 31
棱锥	pyramid, 33
裂纹（缝）	fissure, 155
六角开槽螺母	hexagon slotted nut, 119
六角螺母	hexagon nut, 119
六角头螺栓	hexagon head bolt, 119
露出端（旋螺母端）	nut end, 123
轮辐（辐条）	spoke, 108
轮毂	hub, 136
螺钉	screw, 119
螺杆，丝杠	screw rod, 218
螺距	pitch, 113
螺母	nut, 119
螺栓	bolt, 119
螺尾	vanish thread (or run-out), 111
螺纹	thread, 110
螺纹大径	major diameter, 112
螺纹规格	screw size, 120
螺纹紧固件	screw fasteners, 119
螺纹小径	minor diameter, 112
螺纹牙型	thread tooth profile, 112
螺纹中径	pitch (effective) diameter, 112
螺纹终止线	thread end line, 115

H

互换性	interchangeability, 169
滑动轴承	sliding bearing, 203
簧丝直径	wire diameter of spring, 140
灰铸铁	grey iron casting, 185
锪平（切鱼眼坑）	spotface, 194
活动钳身	movable clamp body, 218

J

机座	base parts, 154
激光焊	laser welding, 242
极限尺寸	limit size, 170
极限偏差	limiting deviation, 170
极限与配合	limits and fits, 169
几何公差	geometrical tolerance, 177
计算机辅助绘图	computer aided drawing, 279
技术要求	technical requirement, 148

螺旋阀	screw valve, 204	上极限尺寸	upper limit of size, 170
螺旋拉伸弹簧	helical extension spring, 140	上极限偏差	upper limit deviation, 170
螺旋扭转弹簧	helical torsion spring, 140	上填料	above stuffing, 202
螺旋压缩弹簧	helical compression spring, 140	设备布置图	equipment layout, 258
		深沟球轴承	deep groove ball bearing, 144
		施工流程图	construction flow diagram, 253

M

埋弧焊	submerged arc welding, 242	时效处理	ageing treatment, 185
冒（溢出）口	overflow, 155	手工电弧焊	manual arc welding, 242
米制锥螺纹	metric taper thread, 113	手轮	hand wheel, 205
密封垫（圈）	sealing gasket, 223	双头螺柱	double end stud, 119
模数	module, 128	双线螺纹	double-start thread 115
木螺钉螺纹	wooden nail thread, 113	缩孔	shrinkage hole, 155
		锁紧螺母	lock nut, 226
		锁紧套	lock sleeve, 226

N

内螺纹	internalthread, 110		
内衬圈	inner ring, 226		

T

内卡尺（钳）	internal (inside) caliper, 197	T形槽	tee groove (slot), 192
内圈	inner race, 143	T形接头	T-joint, 238
		碳弧焊	carbon arc welding, 242
		套螺母	sheathing nut, 218

P

抛光	polishing, 168	特殊螺纹	special thread, 113
配合	fitting, fit, 173	梯形螺纹	metric trapezoidal screw thread, 112
片簧	blade spring, 140		
平垫圈	plain washer, 119	填料	stuffing, 216
平口（台虎）钳	flat nose clamp, 218	填料垫	stuffing spacer, 223
普通（米制）螺纹	general (metric) thread, 114	填料函	stuffing box, 216
普通螺纹	metric thread, 112	填料压盖	packing gland, 205
普通型半圆键	woodruff key-normal form, 137	调节齿轮	adjusting gear, 226
普通型平键	square and rectangular key, 137	调整垫	spacer, 223
普通型楔键	taper key for general use, 137	通（穿透）孔	through hole (body size hole), 123

Q

		凸台	boss club, 157
起模斜度	draft (or draught), 155	推力球轴承	thrust ball bearing, 144
气瓶专用螺纹	thread for gas tanke, 113	退刀槽	undercut, 111
钳口板	clamp plank, 218		

V

切点	point of tangency, 16		
球阀	ball valve, 201	V形块	vee block, 193
曲线槽	curved slot, 193		

W

R

		外螺纹	externalthread, 110
		外径	outside diameter, 72
热处理	heat-treatment, 162	外卡尺（钳）	outside caliper, 197
人字齿轮	herringbone gear, 127	涡卷弹簧	turbination spring, 140
容差、加工余量	allowance, 164	蜗杆	worm, 127
软钎焊	soft soldering, 242	蜗轮	worm gear, 127

S

X

三角形螺纹	triangular thread, 112	细牙普通螺纹	fine plain thread, 114
砂轮越程槽	grinding undercut 161	下极限尺寸	lower limit of size, 170
砂型铸造	sand casting, 155	下极限偏差	lower limit deviation, 170
砂眼	blowhole, 196	下轴瓦	bottom bush, 203

线数	number of starts, 112	圆柱销	parallel (dowel) pin, 135
线型	line style, 7	圆锥滚子轴承	tapered roller bearing, 144
箱体	box, 154	圆锥体	cone, 37
销	pin, 135	圆锥销	taper pin, 135
斜齿轮	helical gear, 127		
斜度	slope (or incline), 15		**Z**
型腔（空穴）	cavity, 155	毡圈	felt ring, 216
旋塞	plug cock, 230	支承圈数	number of end coils, 140
		直齿轮	spur gear, 127
	Y	直径	diameter, 17
压盖	gland, 151	止动（锁紧）垫圈	lock washer, 119
压盖螺母	packing nut, 205	轴承	bearing, 143
压紧套	packing set, 202	轴承盖	bearing cover, 203
压力角	pressure angle, 129	轴承座	bearing seat, 203
牙底	root, 112	轴间角	axes angle, 80
牙顶	crest, 112	轴肩	shoulder, 192
研磨	grinding, 168	轴套	sleeve, 150
氧乙炔焊	oxygen-acetylene welding, 242	轴瓦固定套	bush fastness set, 203
印制电路板图	printed circuit board diagram, 276	铸（浇）口	pouring head, 155
		铸造（内）圆角	fillet, 156
硬钎焊	hard soldering, 242	铸造、铸件	casting, 155
油杯	oil cup, 203	状态行	status line, 281
游标卡尺	vernier (sliding) caliper, 197	锥齿轮	bevel gear, 127
右旋螺纹	right-hand thread, 113	锥度	taper, 15
原子氢焊	atomic-hydrogen welding, 242	锥形沉孔、埋头孔	countersunk, 194
圆头	round head, 111	自攻螺钉螺纹	self-drilling screw thread, 113
圆柱齿轮	cylindrical gear, 127	字体	lettering, 7
圆柱度	cylindricity, 179	钻孔	drilling, 111
圆柱头内六角螺钉	hexagon socket cap screw, 119	左旋螺纹	left-hand thread, 113

（2）Abbrevidtions 缩略语

AC	across comers	对角距
*AF	across fiats	对边距
AL	aluminum	铝
*ASSY	assembly	装配（图），部件
BHN	Brinell hardness number	布氏硬度值
*CBORE	counterbore	沉孔，埋头孔
*CHAM	chamfer	倒角
*CH HD	cheese head	开槽圆柱头
cir	circular	圆的
*CL or ₵	centerline	中心线
*CRS (C to C)	centers	中心距
*CSK	countersunk	锥孔，埋头孔
*CSK HD	countersunk head	沉（锥形）头
*CYL	cylinder or cylindrical	圆柱，圆柱体
° or deg	degree	度
*φ or DIA	diameter	直径
*DIM	dimension	尺寸
*DRG	drawing	制图，图样
EI or ei	来自法文 ecart inferieur	下偏差代号
ES or es	来自法文 ecart superieur	上偏差代号

* EQS or EQUI SP	equally spaced	均布
* EXT	external	外部的
* Flg.	figure	附图，插图
* HEX	hexagon	六角（边）形
* HEX HD	hexagon head	六角头
hp	horsepower	马力
hr.	hour	小时
I / D	inside diameter	内径
* INT	internal	内部的
* LH	left hand	左旋，左方向
* LG	long	长度
* LONG	longitude	经度（线）
* MATL	material	材料
* MAX	maximum	最大，最高值
* MIN	minimum	最小，最低值
* No.	number	号码，数字
O / D	outside diameter	外径
* PCD	pitch circle diameter	节圆直径
PATT No.	pattern number	型号，样（品）号
PCBD		印制板图
qty.	quantity	数量
* R or RAD	radius	半径
* RD HD	round head	半球头，圆顶
* RH	right hand	右旋，右方向
* SCR	screw (screwed)	螺钉，螺旋
* SH	sheet	片，张
* SPEC	specification	规格，说明
* Sϕ	spherical diameter	球直径
* SR	spherical radius	球半径
* SFACE	spotface	锪平，切鱼眼坑
* □ or ⊠ or SQ	square	平面（方），矩形
* STD	standard	标准，规格
t or * THK	thick	厚度
* THD	thread	螺纹
* TOL	tolerance	公差
* UCUT	undercut	退刀槽
* VOL	volume	体积
* WT	weight	重量

ISO = International Standardization Organization　国际标准化组织

　　注：Those abbreviations with "＊" are recognized in British Standards BS 8888. （带"＊"的为英国图制标准 BS 8888 中规定的缩略语。）

References 参考文献

[1] Simmons, Maguire. Manual of Engineering Drawing [M]. 2nd ed. London：Elsevier Newnes. Linacre House, Jordan Hill, 2004.

[2] Rhodes, Cook. Basic Engineering Drawing [M]. 2nd ed. Edinburgh：Addison Wesley Longman Limited, 1990.

[3] Parker & Fpickup. Engineering Drawing with Worked Examples [M]. London：Stanley Thornes Ltd., 1976.

[4] 齐玉来，韩群生. 机械制图（非机类）[M]. 天津：天津大学出版社，2004.

[5] 邹宜候，窦墨林. 机械制图（非机械类专业用）[M]. 北京：清华大学出版社，2001.

[6] 王成刚，张佑林，赵奇平. 工程图学简明教程 [M]. 武汉：武汉理工大学出版社，2002.

[7] 杨惠英，王玉坤. 机械制图（非机类）[M]. 北京：清华大学出版社，2002.

[8] 左晓明. 工程制图（近机及非机类）[M]. 北京：机械工业出版社，2004.

[9] 王槐德. 机械制图新旧标准代换教程（修订版）[M]. 北京：中国标准出版社，2004.

[10] 何铭新，钱可强. 机械制图 [M]. 5版. 北京：高等教育出版社，2004.

[11] 刘小年，郭克希. 机械制图（机械类、近机类）[M]. 北京：机械工业出版社，2005.